Acetic Acid Bacteria

Fundamentals and Food Applications

Books Published in *Food Biology* series

1. **Microorganisms and Fermentation of Traditional Foods**
   Ramesh C. Ray, Montet Didier (eds.), 2014
2. **Bread and Its Fortification: Nutrition and Health Benefits**
   Cristina M. Rosell, Joanna Bajerska, Aly F. El Sheikha (eds.), 2015
3. **Authenticity of Foods of Animal Origin**
   Ioannis Sotirios Arvanitoyannis (ed.), 2015
4. **Fermented Foods, Part I: Biochemistry and Biotechnology**
   Didier Montet, Ramesh C. Ray (eds.), 2015
5. **Foodborne Pathogens and Food Safety**
   Md. Latiful Bari, Dike O. Ukuku (eds.), 2015
6. **Fermented Meat Products: Health Aspects**
   Nevijo Zdolec (ed.), 2016
7. **Fermented Foods of Latin America: From Traditional Knowledge to Innovative Applications**
   Ana Lucia Barretto Penna, Luis A. Nero, Svetoslav D. Todorov (eds.), 2016
8. **Lactic Acid Fermentation of Fruits and Vegetables**
   Spiros Paramithiotis (ed.), 2016
9. **Microbial Enzyme Technology in Food Applications**
   Ramesh C. Ray, Cristina M. Rosell (eds.), 2016
10. **Acetic Acid Bacteria: Fundamentals and Food Applications**
    Ilkin Yucel Sengun (ed.), 2017
11. **Fermented Foods, Part II: Technological Interventions**
    Ramesh C. Ray, Montet Didier (eds.), 2017

Food Biology Series

Acetic Acid Bacteria

Fundamentals and Food Applications

Editor

Ilkin Yucel Sengun

Ege University
Engineering Faculty, Food Engineering Department
Izmir, Turkey

CRC Press
Taylor & Francis Group
Boca Raton London New York

CRC Press is an imprint of the
Taylor & Francis Group, an **informa** business
A SCIENCE PUBLISHERS BOOK

Cover credit
Cover illustration provided by Prof. Francois Barja. Designing assistance from Bengisu Şengün.

CRC Press
Taylor & Francis Group
6000 Broken Sound Parkway NW, Suite 300
Boca Raton, FL 33487-2742

Printed on acid-free paper
Version Date: 20161107

International Standard Book Number-13: 978-1-4987-6369-1 (Hardback)

Library of Congress Cataloging-in-Publication Data

Names: Sengun, Ilkin Yucel, editor.
Title: Acetic acid bacteria : fundamentals and food applications / editor, Ilkin Yucel Sengun, Egc [i.e. Ege] University, Engineering Faculty, Food Engineering, Department Izmir, Turkey.
Description: Boca Raton, FL : CRC Press, [2016] | Series: Food biology series | "A science publishers book." | Includes bibliographical references and index.
Identifiers: LCCN 2016050024| ISBN 9781498763691 (hardback : alk. paper) | ISBN 9781498763707 (e-book : alk. paper)
Subjects: LCSH: Acetobacter. | Food--Microbiology. | Bioreactors.
Classification: LCC QR82.P78 A34 2016 | DDC 579.3/3--dc23
LC record available at https://lccn.loc.gov/2016050024

Visit the Taylor & Francis Web site at
http://www.taylorandfrancis.com

and the CRC Press Web site at
http://www.crcpress.com

Preface to the Series

Food is the essential source of nutrients (such as carbohydrates, proteins, fats, vitamins, and minerals) for all living organisms to sustain life. A large part of daily human efforts is concentrated on food production, processing, packaging and marketing, product development, preservation, storage, and ensuring food safety and quality. It is obvious, therefore, our food supply chain can contain microorganisms that interact with the food, thereby interfering in the ecology of food substrates. The microbe-food interaction can be mostly beneficial (as in the case of many fermented foods such as cheese, butter, sausage, etc.) or in some cases, it is detrimental (spoilage of food, mycotoxin, etc.). The Food Biology series aims at bringing all these aspects of microbe-food interactions in form of topical volumes, covering food microbiology, food mycology, biochemistry, microbial ecology, food biotechnology and bio-processing, new food product developments with microbial interventions, food nutrification with nutraceuticals, food authenticity, food origin traceability, and food science and technology. Special emphasis is laid on new molecular techniques relevant to food biology research or to monitoring and assessing food safety and quality, multiple hurdle food preservation techniques, as well as new interventions in biotechnological applications in food processing and development.

The series is broadly broken up into food fermentation, food safety and hygiene, food authenticity and traceability, microbial interventions in food bio-processing and food additive development, sensory science, molecular diagnostic methods in detecting food-borne pathogens and food policy, etc. Leading international authorities with background in academia, research, industry and government have been drawn into the series either as authors or as editors. The series will be a useful reference resource base in food microbiology, biochemistry, biotechnology, food science and technology for researchers, teachers, students and food science and technology practitioners.

Ramesh C. Ray
Series Editor

Preface

Acetic acid bacteria are a group of microorganisms found widespread in Nature. They are characterized by their capability to oxidize sugars, alcohols and sugar alcohols into their corresponding organic acids. They can be involved in a variety of biological processes and are used for the production of fermented foods and beverages, because of their special and unique characteristic called 'oxidative fermentation'. The main well-known application of acetic acid bacteria is the production of acetic acid in the form of vinegar. Although acetic acid bacteria have an important role in producing special types of foods, it is not common to use them as a starter culture in food fermentations because of technological and economical reasons. Acetic acid bacteria were previously regarded as a small taxonomic group, which included only two genera, *Acetobacter* and *Gluconobacter*, but their classification has entirely changed in the last years due to the development of novel molecular detection and identification techniques. Currently, this important and diverse group of bacteria includes eighteen genera. However, there are only few sources dealing with their relevant reclassification.

This book presents a comprehensive and updated information on both fundamentals and food applications of acetic acid bacteria. It contains 13 chapters categorized under two parts. The first part gives detailed information on the general characteristics and current taxonomy of acetic acid bacteria. One chapter of particular interest describes the important findings that have emerged from genome studies of this diverse group of bacteria. The physiology and biochemistry, acetic acid resistance, exopolysaccharide production and thermotolerant properties of acetic acid bacteria are described separately in specific chapters. Two chapters are devoted to the latest identification and preservation techniques of acetic acid bacteria which are the most active fields of research today. Microbial collections of acetic acid bacteria and the available online databases are also given in this part. The second part of the book describes the importance of acetic acid bacteria in the food industry by giving information on the microbiological properties of fermented foods as well as their production procedures. A chapter is devoted to the microbiology of fermented foods, including dairy products, cereal-based products, fruit and vegetable products, meat products, locally produced traditional fermented foods and innovative functional foods. Several foods and beverages performed by acetic acid bacteria are discussed separately. Special attention is given to vinegar and cocoa fermentation, which are the most familiar and extensively used industrial applications of acetic acid bacteria. The chapter titled 'Vinegars' provides information on the types of vinegar, fermentation technologies, mass balance and yields, vinegar microbial community, vinegar spoilage and intended use of different vinegars. It is major concern in food science to provide safe foods and improve the health of consumers. Therefore, two chapters are devoted on detrimental and beneficial effects of acetic acid bacteria in terms of food safety and benefits to human health.

The chapters have been written by leading international authorities in the field with recent scientific data on microbiology, food science and technology and engineering. About thirty scientists from eleven countries have contributed to the preparation of this book. It is hoped that this book covers all the basic and applied aspects of acetic acid bacteria to satisfy the needs of readers, be the scientists, technologists, students or those working in this field. I want to thank the contributors for their kind support and sharing their valuable experiences, as well as to the production team at CRC Press for bringing out this book. I am also pleased to learn the reason why authors/editors always thank their families for the works completed successfully.

Ilkin Yucel Sengun
Editor

Contents

PART II: Importance of Acetic Acid Bacteria in Food Industry

List of Contributors

Albert Mas, Oenological Biotechnology Group. Department of Biochemistry and Biotechnology, Faculty of Oenology, University Rovira i Virgili, Marcel·li Domingo, 1. 43007, Tarragona, Spain, albert.mas@urv.cat

Ana Hospital Bravo, University of Geneva, Department of Botany and Plant Biology, Microbiology Unit, Quai Ernest-Ansermet 30 (Sciences III), 1211 Geneva 4, Geneva, Switzerland, Ana.Hospital@etu.unige.ch

Birce Mercanoglu Taban, Ankara University, Faculty of Agriculture, Department of Dairy Technology, Ankara, 06110, Turkey, btaban@ankara.edu.tr

Corinne Teyssier, University of Montpellier, TA B-95/16, 73, rue Jean-François Breton 34398 Montpellier cedex 5, France, corinne.teyssier@univ-montp1.fr

Cristina Andrés-Barrao, King Abdullah University of Science and Technology, Center for Desert Agriculture, Division of Biological and Environmental Sciences and Engineering, 4700 Thuwal 23955-6900, Saudi Arabia, cristina.andresbarrao@kaust.edu.sa

Didier Montet, Cirad, UMR 95 Qualisud TA B-95/16 73, rue Jean-François Breton 34398 Montpellier cedex 5, France, didier.montet@cirad.fr

François Barja, University of Geneva, Department of Botany and Plant Biology, Microbiology Unit, Quai Ernest-Ansermet 30 (Sciences III), 1211 Geneva 4, Geneva, Switzerland, François.Barja@unige.ch

Huong Thi Lan Vu, Vietnam National University-HCM City, University of Science, Faculty of Biology and Biotechnology, 227 Nguyen Van Cu Street, Ward 4, District 5, Hochiminh City, Vietnam, huong_vu14@yahoo.com.vn

Ilkin Yucel Sengun, Ege University, Faculty of Engineering, Department of Food Engineering, 35100, Bornova, Izmir, Turkey, ilkin.sengun@ege.edu.tr

Janja Trček, University of Maribor, Faculty of Natural Sciences and Mathematics, Department of Biology, Maribor, Slovenia, janja.trcek@um.si

Jure Škraban, University of Maribor, Faculty of Chemistry and Chemical Engineering, Maribor, Slovenia, jure.skraban@um.si

Luciana De Vero, Unimore Microbial Culture Collection (UMCC). Department of Life Sciences, University of Modena and Reggio Emilia, Via Amendola 2, 42122 Reggio Emilia, Italy, luciana.devero@unimore.it

Maria Gullo, Unimore Microbial Culture Collection (UMCC). Department of Life Sciences, University of Modena and Reggio Emilia, Via Amendola 2, 42122 Reggio Emilia, Italy, maria.gullo@unimore.it

Maria Jesús Torija, Oenological Biotechnology Group. Department of Biochemistry and Biotechnology, Faculty of Oenology, University Rovira i Virgili, Marcel·li Domingo, 1. 43007, Tarragona, Spain, mjesus.torija@urv.cat

Maria José Valera, Oenological Biotechnology Group. Department of Biochemistry and Biotechnology, Faculty of Oenology, University Rovira i Virgili, Marcel·li Domingo, 1. 43007, Tarragona, Spain, mariajose_valera_martinez@hotmail.com

Marie-Louise Chappuis, University of Geneva, Department of Botany and Plant Biology, Microbiology Unit, Quai Ernest-Ansermet 30 (Sciences III), 1211 Geneva 4, Geneva, Switzerland, Marie-Louise.Chappuis@unige.ch

Michael P. Doyle, University of Georgia, Center for Food Safety, Department of Food Science & Technology, 1109 Experiment Street, Griffin, GA 30223 USA, mdoyle@uga.edu

Natsaran Saichana, Mae Fah Luang University, School of Science, Chiang Rai – 57100, Thailand, natsaran.sai@mfu.ac.th

Paolo Giudici, Unimore Microbial Culture Collection (UMCC). Department of Life Sciences, University of Modena and Reggio Emilia, Via Amendola 2, 42122 Reggio Emilia, Italy, paolo.giudici@unimore.it

Pattaraporn Yukphan, National Science and Technology Development Agency (NSTDA), National Center for Genetic Engineering and Biotechnology (BIOTEC), BIOTEC Culture Collection (BCC), 113 Thailand Science Park, Pahonyothin Road, Khlong Nueng, Khlong Luang, Pathum Thani 12120, Thailand, pattaraporn@biotec.or.th

Ruben Ortega Pérez, University of Geneva, Department of Botany and Plant Biology, Microbiology Unit, Quai Ernest-Ansermet 30 (Sciences III), 1211 Geneva 4, Geneva, Switzerland, Ruben.OrtegaPerez@unige.ch

Sarah Braito, University of Geneva, Department of Botany and Plant Biology, Microbiology Unit, Quai Ernest-Ansermet 30 (Sciences III), 1211 Geneva 4, Geneva, Switzerland, Sarah.Braito@etu.unige.ch

Seniz Karabiyikli, Gaziosmanpaşa University, Faculty of Engineering and Natural Science, Department of Food Engineering, 60250, Tokat, Turkey, seniz.karabiyikli@gop.edu.tr

Seval Dağbağlı, Celal Bayar University, Faculty of Engineering, Department of Food Engineering, 45140, Muradiye Manisa, Turkey, seval.dagbagli@cbu.edu.tr

Somboon Tanasupawat, Chulalongkorn Univerisity, Faculty of Pharmaceutical Sciences, Department of Biochemistry and Microbiology, 254 Phayathai Road, Wangmai, Pathumwan, Bangkok 10330, Thailand, somboon.t@chula.ac.th

Taweesak Malimas, Organic Fertilizer Manufacturing Co., Ltd., 2 Moo 7, Nongpanchun, Bankha, Ratchaburi 70180, Thailand, malimas45@hotmail.com

Yasmine Hamdouche, Cirad, UMR 95 Qualisud TA B-95/16 73, rue Jean-François Breton 34398 Montpellier cedex 5, France, yasmine.hamdouche@cirad.fr

Yekta Göksungur, Ege University, Faculty of Engineering, Department of Food Engineering, 35100, Bornova, Izmir, Turkey, yekta.goksungur@ege.edu.tr

Yuki Muramatsu, National Institute of Technology and Evaluation, NITE Biological Resource Center (NBRC), 2-5-8 Kazusakamatari, Kisarazu, Chiba 292-0818, Japan, muramatsu-yuki@nite.go.jp

Yuzo Yamada, Shizuoka University, Faculty of Agriculture, Department of Applied Biological Chemistry, 836 Ohya, Suruga-ku, Shizuoka 422-8529, Japan, yamada333@kch.biglobe.ne.jp

PART I

DESCRIPTION OF ACETIC ACID BACTERIA

1

Systematics of Acetic Acid Bacteria

Taweesak Malimas[1], Huong Thi Lan Vu[2], Yuki Muramatsu[3], Pattaraporn Yukphan[4], Somboon Tanasupawat[5] and Yuzo Yamada[6,*]

[1] Organic Fertilizer Manufacturing Co., Ltd., 2 Moo 7, Nongpanchun, Bankha, Ratchaburi 70180, Thailand

[2] Vietnam National University-HCM City, University of Science, Faculty of Biology and Biotechnology, 227 Nguyen Van Cu Street, Ward 4, District 5, Hochiminh City, Vietnam

[3] National Institute of Technology and Evaluation, NITE Biological Resource Center (NBRC), 2-5-8 Kazusakamatari, Kisarazu, Chiba 292-0818, Japan

[4] National Science and Technology Development Agency (NSTDA), National Center for Genetic Engineering and Biotechnology (BIOTEC), BIOTEC Culture Collection (BCC), 113 Thailand Science Park, Pahonyothin Road, Khlong Nueng, Khlong Luang, Pathum Thani 12120, Thailand

[5] Chulalongkorn Univerisity, Faculty of Pharmaceutical Sciences, Department of Biochemistry and Microbiology, 254 Phayathai Road, Wangmai, Pathumwan, Bangkok 10330, Thailand

[6] Shizuoka University, Faculty of Agriculture, Department of Applied Biological Chemistry, 836 Ohya, Suruga-ku, Shizuoka 422-8529, Japan

Introduction

Acetic acid bacteria are Gram-negative, rod-shaped, and obligate aerobes and commonly known as vinegar-producing microorganisms (Asai 1968, Sievers and Swings 2005, Kersters et al. 2006, Komagata et al. 2014). The acetic acid bacteria in general have an ability to oxidize ethanol to acetic acid and glucose to gluconic acid and are classified in the family *Acetobacteraceae* Gillis and De Ley 1980 (Gillis and De Ley 1980, Sievers and Swings 2005, Kersters et al. 2006, Komagata et al. 2014). The members of the family are separated into two groups, i.e., the acetous and the acidophilic groups (Komagata et al. 2014). The acetic acid bacteria are included in the former group.

The genus *Acetobacter*, the type genus of the family *Acetobacteraceae* was introduced by Beijerinck (1898), with *Acetobacter aceti* (Pasteur 1864) Beijerinck 1898, the type species of the genus. Asai (1935) divided acetic acid bacteria into two genera. One was the genus *Acetobacter* Beijerinck 1898, and the other was the genus *Gluconobacter* Asai 1935. The former was comprised of the species that oxidized ethanol more intensely than glucose and an ability to oxidize acetic acid to carbon dioxide and water, and the latter was of the species that oxidized glucose more intensely than ethanol and no ability to oxidize acetic acid (Asai 1935). The genus *'Acetomonas'* Leifson 1954 was then proposed for the species that formed polar flagellation and were non acetate-oxidizing (Leifson 1954). On the other hand, the species of the genus *Acetobacter* formed peritrichous flagellation and oxidized

*Corresponding author: yamada333@kch.biglobe.ne.jp

"This chapter is a revised edition of Yuzo Yamada (2016) Systematics of Acetic Acid Bacteria, in K. Matsushita et al. (eds.) Acetic Acid Bacteria: Ecology and Physiology, ©Springer Japan 2016. Some of the texts are reused with some modification, with the kind permission of Springer."

acetic acid to carbon dioxide and water. The proposal of the two generic names caused confusion in classifying non acetate-oxidizing acetic acid bacteria (Asai and Shoda 1958, Shimwell 1958, Shimwell and Carr 1959, Carr and Shimwell 1960).

De Ley (1961) recognized the priority of *Gluconobacter,* but not of *'Acetomonas'* in the generic name. Since Asai (1935) did not designate the type species of the genus *Gluconobacter*, *Gluconobacter oxydans* (Henneberg 1897) De Ley 1961 was designated as the type species (De Ley 1961, De Ley and Frateur 1970).

Asai et al. (1964) reported two types of intermediate strains, in addition to the strains of the genera *Acetobacter* and *Gluconobacter*. One was composed of the strains that had peritrichous flagellation, and the other was of the strains that had polar flagellation in spite of "being acetate-oxidizing." The two types of the intermediate strains were designated by the ubiquinone system (Yamada et al. 1969a). The two genera were chemotaxonomically distinguished from each other by the presence of the major ubiquinone, i.e., UQ-9 in the genus *Acetobacter* and UQ-10 in the genus *Gluconobacter*. In the two types of the intermediate strains, however, the peritrichously flagellated intermediate strains that were acetate-oxidizing and once classified as *'Gluconobacter liquefaciens'* (= *Acetobacter liquefaciens*) (Asai 1935, Asai and Shoda 1958, Asai 1968, Gosselé et al. 1983a, Yamada 2016), had UQ-10, as found in the genus *Gluconobacter*, and the polarly flagellated intermediate strains that were "acetate-oxidizing" (Asai et al. 1964, Yamada et al. 1976a) and once classified as *'Acetobacter aurantius'* (Kondo and Ameyama 1958) had UQ-8, which was never found in any other strains of acetic acid bacteria (Yamada et al. 1969a). The polarly flagellated intermediate strains equipped with UQ-8 were later classified as *Frateuria aurantia* (ex Kondo and Ameyama 1958) Swings et al. 1980 (Swings et al. 1980). In addition to the peritrichously flagellated intermediate strains equipped with UQ-10, the strains of *Acetobacter xylinus* were UQ-10-equipped in the genus *Acetobacter* (Yamada et al. 1969a, b, 1976b, Yamada 1983).

In the Approved Lists of Bacterial Names 1980, the UQ-10-having peritrichously flagellated intermediate strains were classified as *Acetobacter aceti* subsp. *liquefaciens*, and the UQ-10-having *A. xylinus* strains were classified as *Acetobacter aceti* subsp. *xylinus* (Skerman et al. 1980). For the UQ-10-having *Acetobacter* strains mentioned above, the subgenus *Gluconacetobacter* Yamada and Kondo 1984 was proposed within the genus *Acetobacter* (Yamada and Kondo 1984). However, the subgenus was not accepted, together with the name of the genus *Acidomonas* Urakami et al. 1989 (Swings 1992, Sievers et al. 1994). The subgenus was later elevated to the generic level as the genus *Gluconacetobacter* Yamada et al. 1998, along with the recognition of the genus *Acidomonas* (Yamada et al. 1997).

In the genus *Gluconacetobacter*, there were two subclusters or the two subgroups, i.e., the *Gluconacetobacter liquefaciens* group and the *Gluconacetobacter xylinus* group (Franke et al. 1999, Yamada et al. 2000). The two groups were suggested to be distinguished from each other at the generic level based on morphological, physiological, ecological, and phylogenetical aspects (Yamada and Yukphan 2008). For the latter group, the genus *Komagataeibacter* Yamada et al. 2013 was introduced (Yamada et al. 2012a, b). Eighteen genera are recognized at present in the acetous group of the family *Acetobacteraceae*, i.e., *Acetobacter* Beijerinck 1898, *Gluconobacter* Asai 1935, *Acidomonas* Urakami et al. 1989 emend. Yamashita et al. 2004, *Gluconacetobacter* Yamada et al. 1998, *Asaia* Yamada et al. 2000, *Kozakia* Lisdiyanti et al. 2002, *Swaminathania* Loganathan and Nair 2004, *Saccharibacter* Jojima et al. 2004, *Neoasaia* Yukphan et al. 2006, *Granulibacter* Greenberg et al. 2006, *Tanticharoenia* Yukphan et al. 2008, *Ameyamaea* Yukphan et al. 2010, *Neokomagataea* Yukphan et al. 2011, *Komagataeibacter* Yamada et al. 2013, *Endobacter* Ramírez-Bahena et al. 2013, *Nguyenibacter* Vu et al. 2013, *Swingsia* Malimas et al. 2014, and *Bombella* Li et al. 2015 (Fig. 1).

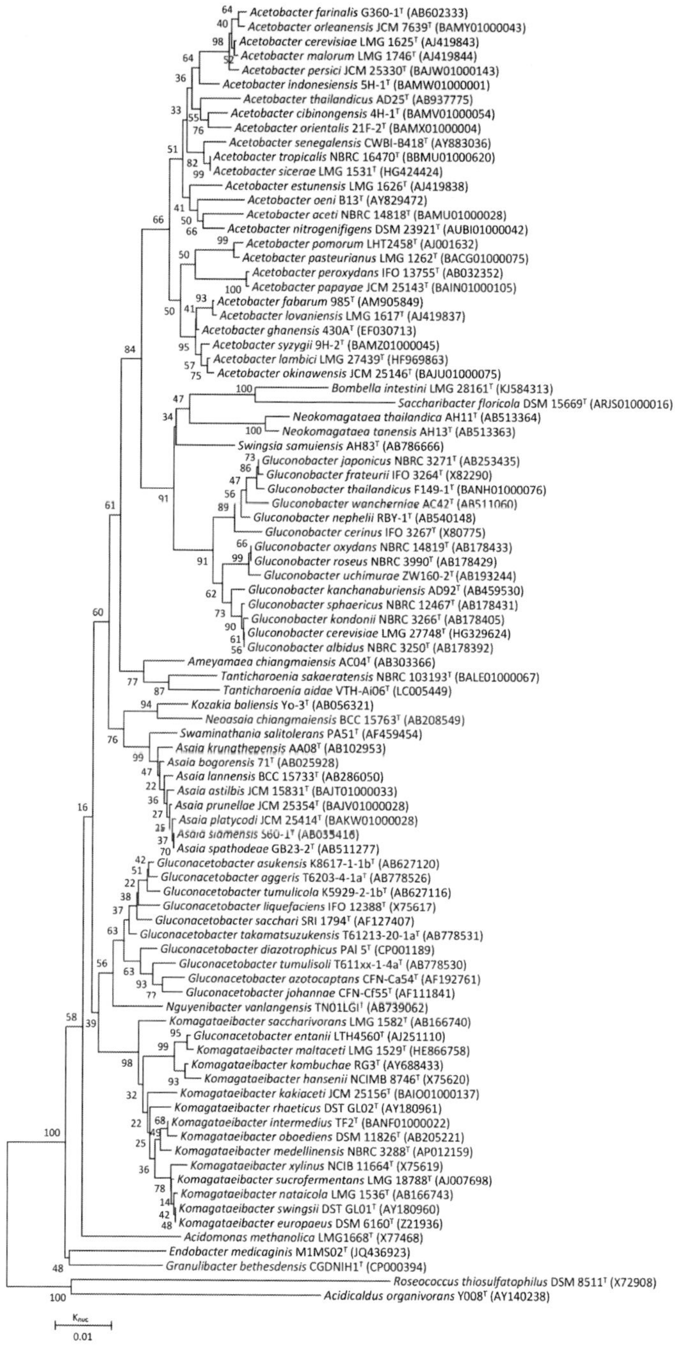

Figure 1: A neighbor-joining tree of acetic acid bacteria. The phylogenetic tree based on 16S rRNA gene sequences of 1213 bases was constructed by using MEGA6 (Tamura et al. 2013). The numerals at the nodes of respective branches indicate bootstrap values (per cent) derived from 1000 replications.

General Characteristics of Acetic Acid Bacteria

The acetic acid bacteria, obligate aerobes are quite unique. The strains assigned to the representative genera *Acetobacter* and *Gluconobacter* have a lack of the Embden/Meyerhof/

Parnas pathway, but the pentose phosphate cycle is functioning instead (Cheldelin 1961, De Ley 1961, Arai et al. 2016, Bringer and Bott 2016). In addition, the strains of the genus *Gluconobacter* lack the TCA cycle, differing from those of the genus *Acetobacter*, and the pentose phosphate cycle acts as a terminal oxidation system. In fact, glucose-6-phosphate dehydrogenase and 6-phospholuconate dehydrogenase reduce NAD except for NADP, and the two dehydrogenases play an important role in the respiratory chain phosphorylation (Cheldelin 1961).

The acetic acid bacteria additionally have the direct oxidation system for alcohols, sugars, and sugar alcohols and accumulate a large amount of the corresponding oxidation products, i.e., acetic acid from ethanol, gluconic acid, 2-ketogluconic acid, 5-ketogluconic acid, and 2,5-diketogluconic acid from glucose, fructose from mannitol, L-sorbose from sorbitol, and 5-ketofructose from fructose or sorbitol (Cheldelin 1961, De Ley 1961, Komagata et al. 2014). Such a partial or incomplete oxidation is traditionally called "oxidative fermentation" and carried out by the membrane-bound dehydrogenases that are linked to the energy-yielding or the non energy-yielding respiratory chain (Matsushita et al. 2004, Komagata et al. 2014, Adachi and Yakushi 2016, Matsushita and Matsutani 2016).

When acetic acid bacteria are grown on alcohols, sugars, or sugar alcohols, the biphasic growth is generally seen. In the 5-ketofructose fermentation of '*Gluconobacter suboxydans*' strain 1, e.g., the first phase of the growth seemed to be due to the oxidation of sorbitol to L-sorbose and then to 5-ketofructose catalyzed respectively by glycerol dehydrogenase [EC 1.1.99.22] and by L-sorbose 5-dehydrogenase [EC 1.1.99.12], which were linked to the energy-yielding respiratory chain (Sato et al. 1969a, Matsushita et al. 2004, Komagata et al. 2014, Adachi and Yakushi 2016, Matsushita and Matsutani 2016). In the second phase of the growth, the resulting 5-ketofructose was reduced to fructose catalyzed by 5-ketofructose reductase [EC 1.1.1.124], and then the resulting fructose flowed into the pentose phosphate cycle, which was coupled with the energy-yielding respiratory chain, probably via fructose-6-phosphate after phosphorylation (Fig. 2).

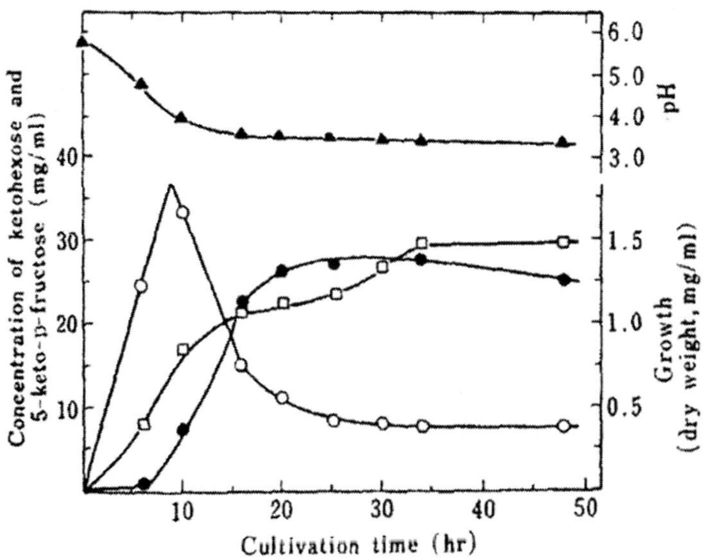

Figure 2: The biphasic growth of '*Gluconobacter suboxydans*' strain 1. The organism was cultured in 500 ml of a medium containing 4.0 per cent sorbitol (w/v) and defatted soybean extract (w/v) in a 5-liter conical flask. The cultivation was done at 30 °C on a rotary shaker at 160 rpm [The figure was cited from Sato et al. (1969b). ○, L-sorbose; ●, 5-ketofructose; □, growth; ▲, pH].

Isolation of Acetic Acid Bacteria

For the isolation of acetic acid bacteria, an enrichment culture approach is effectively used (Komagata et al. 2014). A medium for the enrichment culture procedure and for the isolation of acetic acid bacteria, designated as the pH 3.5 medium (Yamada et al. 1999), is composed, e.g., 1.0 per cent glucose (w/v), 0.5 per cent ethanol (99.8 per cent) (v/v), 0.3 per cent peptone (w/v), 0.2 per cent yeast extract (w/v), and 0.01 per cent cycloheximide (w/v) and adjusted to pH 3.5 with concentrated hydrochloric acid. In the isolation of acetic acid bacteria capable of fixing atmospheric nitrogen, the LGI medium that contains 10.0 per cent sucrose (w/v), 0.06 per cent KH_2PO_4 (w/v), 0.02 per cent K_2HPO_4 (w/v), 0.02 per cent $MgSO_4$ (w/v), 0.002 per cent $CaCl_2$ (w/v), 0.001 per cent $FeCl_3$ (w/v), and 0.0002 per cent Na_2MoO_4 (w/v) is used at pH 6.0 (Cavalcante and Döbereiner 1988). When microbial growth is seen in the LGI medium, the culture is transferred to the pH 3.5 medium mentioned above (Vu et al. 2013). To obtain and purify candidates of acetic acid bacteria, the culture in the pH 3.5 medium is streaked onto agar plates, which are composed of 2.0 per cent glucose (w/v), 0.5 per cent ethanol (99.8 per cent) (v/v), 0.3 per cent peptone (w/v), 0.3 per cent yeast extract (w/v), 0.7 per cent calcium carbonate (precipitated, e.g., by Japanese Pharmacopoeia) (w/v), and 1.5 per cent agar (w/v) (Yamada et al. 1999), and the resulting colonies that dissolve calcium carbonate on the agar plates are picked up and transferred onto agar slants containing the same in composition as the agar plates for temporary preservation. The strains isolated were examined again for growth on the pH 3.5 medium.

When the composition, especially of the carbon sources, of the medium in the enrichment culture procedure is changed, the selective isolation of acetic acid bacteria can be expected. In fact, the strains of *Asaia bogorensis* and *Asaia siamensis* were first isolated using sorbitol or dulcitol instead of glucose (Yamada et al. 2000, Katsura et al. 2001). Several kinds of media employed for the enrichment culture procedure result in the effective isolation of acetic acid bacteria (Lisdiyanti et al. 2003b, Suzuki et al. 2010).

To isolate acetic acid bacteria, sugary and alcoholic materials have widely been utilized as isolation sources. In some cases, the habitats of acetic acid bacteria are to be isolation sources (Sievers and Swings 2005, Kersters et al. 2006, Komagata et al. 2014, Li et al. 2015).

Most acetic acid bacteria can be maintained at 4 °C for one month on agar slants containing an appropriate medium. The long-term preservation of acetic acid bacteria can be achieved by lyophilization or by storage in liquid nitrogen, or by cryoconservation at -80 °C by the use of low-temperature refrigerators and appropriate cryoprotectants (Sievers and Swings 2005, Kersters et al. 2006, Komagata et al. 2014).

Identification of Acetic Acid Bacteria

To make a check whether the strains isolated are acetic acid bacteria or not, the growth test is applied at pH 3.5 (Yamada et al. 1999). The medium to be utilized has the same composition as the pH 3.5 medium mentioned above.

In the genera that are not monotypic, i.e., include more than several species and are therefore restricted only to *Acetobacter*, *Gluconobacter*, *Gluconacetobacter*, *Asaia*, and *Komagataeibacter*, which are supposed to be taxonomically and ecologically in common but not in rare existence, the generic-level, routine identification for certain strains of acetic acid bacteria can be achieved by the combination of only two conventional phenotypic tests comprised of acetate and lactate oxidation and production of acetic acid from ethanol (Yamada and Yukphan 2008).

In strains to be assigned to the genus *Acetobacter*, a deep blue color appears fast and clearly in the acetate and lactate oxidation tests, and acetic acid is produced in the acetic acid production test (Asai et al. 1964, Yamada and Yukphan 2008). In acetate and lactate oxidation, strains to be assigned to the genus *Gluconobacter* show a clear yellow color, and the color change to blue is not so vigorous in strains to be assigned to the genera *Gluconacetobacter* and *Komagataeibacter*, in contrast to the genus *Acetobacter*. The latter two genera, *Gluconacetobacter* and *Komagataeibacter* are additionally discriminated from each other by water-soluble brown pigment production and cell motility. Strains to be assigned to the former generally produce a water-soluble brown pigment, being motile, but strains to be assigned to the latter do not, being non motile. Strains to be assigned to the genus *Asaia* show no or little acetic acid production from ethanol, differing from the above-mentioned four genera, and the color change is very slow in acetate and lactate oxidation. The two conventional tests described above are very useful, especially when a large number of isolates are routinely identified or classified at the generic level.

The acetic acid bacteria can be routinely identified at the species level phylogenetically. The so-called partial 16S rRNA gene 800R-regions of the PCR products that are produced by the ordinal method are sequenced using the primer of 800R (5′-TACCAGGGTATCTAATCC-3′, positions 802-785; the numbering of the positions was based on the *Escherichia coli* numbering system, Brosius et al. 1981, accession number V00348), and a phylogenetic tree is constructed based on the sequence data obtained (Vu et al. 2013). On the other hand, the polyphasic taxonomic analysis is applicable to the identification (Cleenwerck and De Vos 2008). When a certain strain of acetic acid bacteria, e.g., a new isolate is precisely identified, DNA-DNA hybridization will be inevitable.

Genera and Species in Acetic Acid Bacteria

The acetic acid bacteria classified in the acetous group constitute the family *Acetobacteraceae*, the class *Alphaproteobacteria* Stackebrandt et al. 1988, along with the acidophilic group (Stackebrandt et al. 1988, Sievers and Swings 2005, Komagata et al. 2014). The type genus of the family is *Acetobacter* Beijerinck 1898. Eighteen genera are presently reported (Table 1).

1. *Acetobacter* Beijerinck 1898

A.ce.to.bac′ter. L. neut. n. *acetum*, vinegar; N. L. masc. n. *bacter*, rod; N. L. masc. n. *Acetobacter*, vinegar rod.

The genus *Acetobacter* is the oldest in the classification of acetic acid bacteria and the type genus of the family *Acetobacteraceae*. In the Approved Lists of Bacterial Names 1980, the three species, *Acetobacter aceti*, *Acetobacter pasteurianus*, and *Acetobacter peroxydans* were listed with their nine subspecies (Skerman et al. 1980). The genus is related phylogenetically to the genera *Gluconobacter*, *Neokomagataea*, *Swingsia*, *Saccharibacter*, and *Bombella*. In the genus *Acetobacter*, there are two phylogenetically different groups, i.e., the *Acetobacter aceti* group and the *Acetobacter pasteurianus* group.

Cells are Gram-negative, ellipsoidal to rod-shaped, measuring 0.4 to 1.0 by 1.2 to 3.0 μm, rarely longer cells. Cells occur singly or short chains and occasionally long chains. Peritrichously flagellated when motile, however, *Acetobacter nitrogenifigens* exceptionally has polar flagella (Dutta and Gachhui 2006). Colonies are generally circular, smooth, entire, convex, cream color to beige, opaque, and butyrous on glucose/ethanol/yeast extract/ peptone agar.

Strictly aerobic. Catalase positive, but negative in *Acetobacter peroxydans*. Oxidase negative. Acetic acid is produced from ethanol. Acetate and lactate are oxidized to carbon dioxide and water. Grows very weakly on mannitol agar while does not grow on

Table 1: Characteristics differentiating the genera of acetic acid bacteria

Characteristic	*Acetobacter*	*Gluconobacter*	*Acidomonas*	*Gluconacetobacter*	*Asaia*	*Kozakia*	*Swaminathania*	*Saccharibacter*	*Neoasaia*	*Granulibacter*	*Tanticharoenia*	*Ameyamaea*	*Neokomagataea*	*Komagataeibacter*	*Endobacter*	*Nguyenibacter*	*Swingsia*	*Bombella*
	1	2	3	4	5	6	7	8	9	10	11	12	13	14	15	16	17	18
Flagellation	per[a]	pol[a]	n[c]	per	per	n	per	n	n	n	n	pol	n	n	spol	per	n	n
Oxidation of																		
Acetate	+	-	+	+	w	w	w	-	-	w	-	+	-	+	-	+	-	-
Lactate	+	-	-	+	w	w	w	w	-	+	-	w	-	+	-	-	-	-
Growth on																		
30 per cent Glucose (w/v)	-	-[b]	+	-	+	-	+	+	+	nd	+	-	+	nd	nd	w	+	+
1 per cent Glucose (w/v)	+	+	+	+	+	+	nd	-	+	nd	+	+	+	+	+	+	+	nd
Glutamate agar	-	-	nd	+	+	-	+	-[e]	+	+	w	w	+	+	+	+	+	nd
Mannitol agar	vw	+	nd	+	+	+	+	-[e]	+	w	+	+	+	+	+	+	+	nd
Raffinose	-	-	-	-	+	w	nd	nd	+	nd	-	-	-	nd	nd	+	-	nd
Utilization of methanol	-	-	+	-	-	-	-	-	-	+	-	w	nd	nd	-	nd	nd	nd
Growth in the presence of																		
0.35 per cent acetic acid (w/v)	+	+	+	+	-	+	+	-	+	nd	+	+	-	+	nd	w	-	nd
1 per cent KNO_3 (w/v)	-	-	+	-	-	-	+	nd	-	nd	-	-	-	nd	nd	-	+	nd
Production of acetic acid from ethanol	+	+	+	+	-	+	+	w/-	+	vw	+	+	w	+	+	-	w	-

(*Contd.*)

Table 1: (*Contd.*)

Characteristic	*Acetobacter*	*Gluconobacter*	*Acidomonas*	*Gluconacetobacter*	*Asaia*	*Kozakia*	*Swaminathania*	*Saccharibacter*	*Neoasaia*	*Granulibacter*	*Tanticharoenia*	*Ameyamaea*	*Neokomagataea*	*Komagataeibacter*	*Endobacter*	*Nguyenibacter*	*Swingsia*	*Bombella*
	1	2	3	4	5	6	7	8	9	10	11	12	13	14	15	16	17	18
Water-soluble brown pigment production	-	-[b]	-	+	-	-	+	-	-	nd	+	-	-	-	nd	+	+	-
Production of dihydroxyacetone from glycerol	+	+	-	+	w	+	+	-	w	-	+	w	-	+	+	-	+	nd
Production of levan-like polysaccharide	-	-	-	-	-	+	nd	-	-	nd	-	-	-	-	nd	+	-	nd
Assimilation of ammoniac nitrogen on																		
Glucose	-	+	w	+	+	-	nd	-	-	+	-	vw	-	nd	+	-	-	-
Mannitol	-	+	w	+	+	-	nd	-	w	nd	-	vw	-	+	nd	w	+	-
Ethanol	w	-	w	-	-	-	nd	-	-	nd	-	vw	-	nd	nd	-	-	-
Production of																		
2-Ketogluconate	+	+	-	+	+	+	nd	+	+	nd	+	+	+	+	nd	+	+	+
5-Ketogluconate	+	+	-	+	+	+	nd	+	+	nd	+	+	+	+	nd	-	+	-
2,5-Diketogluconate	-	-[b]	-	+	-	-	nd	-	-	nd	+	-	+	-	nd	+	+	nd
Acid production from																		
Mannitol	-	+	w	-	+	-	-	+	w	-	-	-	-	-	-	-	+	+
Sorbitol	-	+	-	-	+(d)	-	+	-	+(d)	-	-	-	-	-	-	-	-	-

(*Contd.*)

Dulcitol	-	w	-	-	+(d)	-	v	-	w	-	-	-	-	nd	-	-	-	nd
Glycerol	-	+	+	-	+	+	+	-	+	w	+	w	-	nd	+	-	-	-
Raffinose	-	-	-	-	+	+	nd	-	+	nd	w	-	-	nd	nd	w	w	-
Ethanol	+	+	+	+	-	+	+	-	+	+	+	+	-	+	+	-	-	-
Major isoprenoid quinone	UQ-9	UQ-10	UQ-10	UQ-10	UQ-10	UQ-10	UQ-10	UQ-10	UQ-10	UQ-10	UQ-10	UQ-10	UQ-10	UQ-10	UQ-10	UQ-10	UQ-10	UQ-10
DNA G+C (mol per cent)	57.2	60.3	62	64.9	60.2	57.2	57.6-59.9[d]	52.3	63.1	59.1	65.6	66.0	56.8	62.5	60.3	66.8[f]	46.9	54.9

Abbreviations: pol, polar; per, peritrichous; spol, subpolar; n, none; +, positive; -, negative; w, weakly positive; vw, very weakly positive; d, delayed; v, variable; nd, not determined;

1, *Acetobacter aceti* NBRC 14818T; 2, *Gluconobacter oxydans* NBRC 14819T; 3, *Acidomonas methanolica* NRIC 0498T; 4, *Gluconacetobacter liquefaciens* NBRC 12388T; 5, *Asaia bogorensis* NBRC 16594T; 6, *Kozakia baliensis* NBRC 16664T; 7, *Swaminathania salitolerans* PA51T; 8, *Saccharibacter floricola* S-877T; 9, *Neoasaia chiangmaiensis* AC28T; 10, *Granulibacter bethesdensis* CGDNIH1T; 11, *Tanticharoenia sakaeratensis* AC37T; 12, *Ameyamaea chiangmaiensis* AC04T; 13, *Neokomagataea thailandica* AH11T; 14, *Komagataeibacter xylinus* JCM 7644T; 15, *Endobacter medicaginis* M1MS02T; 16, *Nguyenibacter vanlangensis* TN01LGIT; 17, *Swingsia samuiensis* AH83T; 18, *Bombella intestini* LMG 28161T.

[a]Some strains in the genus are non-motile. [b]Some strains in the genus are positive. [c]Some strains of the genus are polarly flagellated. [d]The DNA G+C content of the type strain was not recorded. [e]According to Jojima et al. (2004), growth was shown on 7 per cent glutamate but not on 1 per cent glutamate. [f]Data from N. Tanaka, NODAI, Japan.

glutamate agar. Dihydroxyacetone is not usually produced from glycerol, but produced by a few species. Gluconate is produced from glucose by all the species, 2-ketogluconate is produced by a considerable number of species, and 5-ketogluconate is by a few species. 2,5-Diketogluconate is not generally produced. Acid production depends on the kind of sugars, sugar alcohols, and alcohols as well as on the kinds of species and strains. In the type strain of *Acetobacter aceti*, acid is produced from L-arabinose, xylose, glucose, galactose, mannose, or ethanol. Ammoniac nitrogen is in general hardly utilized.

The optimal growth temperature is around 30 °C. Most of the species are able to grow at 37 °C, but not at 45 °C. Grows at pH 3.5. Most of the species are not able to grow on 30 per cent glucose (w/v). The major cellular fatty acid is $C_{18:1}\omega7c$. The major quinone is UQ-9. The DNA G+C content ranges from 53.5 to 60.7 mol per cent. For more details of characteristics, see Komagata et al. (2014).

The type species of the genus is *Acetobacter aceti* (Pasteur 1864) Beijerinck 1898. Twenty-six species are reported.

1.1. *Acetobacter aceti* (Pasteur 1864) Beijerinck 1898

For the characteristics of the species, refer to Gosselé et al. (1983b), Sievers and Swings (2005), Kersters et al. (2006), Komagata et al. (2014).

The type strain is ATCC 15973^T (= DSM 3508^T = JCM 7641^T = LMG 1261^T = LMG 1504^T = NBRC 14818^T = NCIMB 8621^T), isolated from beechwood shavings of a vinegar plant. The DNA G+C content of the type strain is 57.2 mol per cent.

1.2. *Acetobacter pasteurianus* (Hansen 1879) Beijerinck and Folpmers 1916

For the characteristics of the species, refer to Beijerinck and Folpmers (1916), Sievers and Swings (2005), Kersters et al. (2006), Komagata et al. (2014).

The type strain is LMG 1262^T (= ATCC 33445^T = DSM 3509^T = JCM 7640^T = LMD 22.1^T = LMG 1262^T), isolated from beer, Netherlands. The DNA G+C content of the type strain is 52.7 mol per cent.

1.3. *Acetobacter peroxydans* Visser't Hooft 1925

For the characteristics of the species, refer to Visser't Hooft (1925), Komagata et al. (2014).

The type strain is NBRC 13755^T (= ATCC 12874^T = JCM 25077^T = LMG 1635^T), isolated from ditch water, Delft, Netherlands. The DNA G+C content of the type strain is 60.3 mol per cent.

1.4. *Acetobacter pomorum* Sokollek, Hertel and Hammes 1998

For the characteristics of the species, refer to Sokollek et al. (1998).

The type strain is LTH 2458^T (= CIP 105762^T = DSM 11825^T = LMG 18848^T), isolated from a submerged cider vinegar fermentation at a factory in the southern part of Germany. The DNA G+C content of the type strain is 50.5 mol per cent.

1.5. *Acetobacter estunensis* (Carr 1958) Lisdiyanti, Kawasaki, Seki, Yamada, Uchimura and Komagata 2001

Basonym: *Acetobacter pasteurianus* subsp. *estunensis* (Carr 1958) De Ley and Frateur 1974.

For the characteristics of the species, refer to Lisdiyanti et al. (2000).

The type strain is NBRC 13751^T (= ATCC 23753^T = DSM 4493^T = JCM 21172^T = LMG 1626^T = NCIMB 8935^T), isolated from cider, Bristol. The DNA G+C content of the type strain is 59.7 mol per cent.

1.6. *Acetobacter lovaniensis* (Frateur 1950) Lisdiyanti, Kawasaki, Seki, Yamada, Uchimura and Komagata 2001

Basonym: *Acetobacter pasteurianus* subsp. *lovaniensis* (Frateur 1950) De Ley and Frateur 1974.

For the characteristics of the species, refer to Lisdiyanti et al. (2000).

The type strain is NBRC 13753^T (= ATCC 12875^T= DSM 4491^T = JCM 17121^T = LMG 1579^T = LMG 1617^T = NCIMB 8620^T), isolated from sewage on soil by J. Frateur in 1929. The DNA G+C content of the type strain is 58.6 mol per cent.

1.7. *Acetobacter orleanensis* (Henneberg 1906) Lisdiyanti, Kawasaki, Seki, Yamada, Uchimura and Komagata 2001

Basonym: *Acetobacter aceti* subsp. *orleanensis* (Henneberg 1906) De Ley and Frateur 1974.

For the characteristics of the species, refer to Lisdiyanti et al. (2000).

The type strain is NBRC 13752^T (= ATCC 12876^T = DSM 4492^T = JCM 7639^T = LMG 1583^T = NCIMB 8622^T), isolated from beer by J. Frateur in 1929. The DNA G+C content of the type strain is 58.6 mol per cent.

1.8. *Acetobacter indonesiensis* Lisdiyanti, Kawasaki, Seki, Yamada, Uchimura and Komagata 2001

For the characteristics of the species, refer to Lisdiyanti et al. (2000).

The type strain is 5H-1^T (= JCM 10948^T = LMG 19824^T = NBRC 16471^T = NRIC 0313^T), isolated from the fruit of zirzak (*Annona muricata*) in Indonesia. The DNA G+C content of the type strain is 53.7 mol per cent.

1.9. *Acetobacter tropicalis* Lisdiyanti, Kawasaki, Seki, Yamada, Uchimura and Komagata 2001

For the characteristics of the species, refer to Lisdiyanti et al. (2000).

The type strain is Ni-6b^T (= JCM 10947^T = LMG 19825^T = NBRC 16470^T = NRIC 0312^T), isolated from coconut (*Coccos nucifera*) in Indonesia. The DNA G+C content of the type strain is 55.9 mol per cent.

1.10. *Acetobacter cerevisiae* Cleenwerck, Vandemeulebroecke, Janssens and Swings 2002

For the characteristics of the species, refer to Cleenwerck et al. (2002).

The type strain is LMG 1625^T (= ATCC 23765^T = DSM 14362^T = JCM 17273^T = NCIMB 8894^T), isolated from beer (ale) in storage at Toronto, Canada. The DNA G+C content of the type strain is 57.6 mol per cent.

1.11. *Acetobacter malorum* Cleenwerck, Vandemeulebroecke, Janssens and Swings 2002

For the characteristics of the species, refer to Cleenwerck et al. (2002).

The type strain is LMG 1746^T (= DSM 14337^T = JCM 17274^T), isolated from a rotten apple in Ghent, Belgium. The DNA G+C content of the type strain is 57.2 mol per cent.

1.12. *Acetobacter cibinongensis* Lisdiyanti, Kawasaki, Seki, Yamada, Uchimura and Komagata 2002

For the characteristics of the species, refer to Lisdiyanti et al. (2001).

The type strain is 4H-1^T (= CIP 107380^T = DSM 15549^T = JCM 11196^T = NBRC 16605^T), isolated from mountain soursop (*Annona montana*) in Indonesia. The DNA G+C content of the type strain is 54.5 mol per cent.

1.13. *Acetobacter orientalis* Lisdiyanti, Kawasaki, Seki, Yamada, Uchimura and Komagata 2002

For the characteristics of the species, refer to Lisdiyanti et al. (2001).

The type strain is 21F-2^T (= CIP 107379^T = DSM 15550^T = JCM 11195^T = NBRC 16606^T = NRIC 0481^T), isolated from canna flower (*Canna hybrida*) in Indonesia. The DNA G+C content of the type strain is 52.3 mol per cent.

1.14. *Acetobacter syzygii* Lisdiyanti, Kawasaki, Seki, Yamada, Uchimura and Komagata 2002

For the characteristics of the species, refer to Lisdiyanti et al. (2001).

The type strain is 9H-2^T (= CIP 107378^T = DSM 15548^T = JCM 11197^T = NBRC 16604^T = NRIC 0483^T), isolated from fruit of Malay rose apple (*Syzygium malaccense*) in Indonesia. The DNA G+C content of the type strain is 55.3 mol per cent.

1.15. *Acetobacter nitrogenifigens* Dutta and Gachhui 2006

For the characteristics of the species, refer to Dutta and Gachhui (2006).

The type strain is RG1^T (= LMG 23498^T = MTCC 6912^T), isolated from Kombucha tea. The DNA G+C content of the type strain is 64.1 mol per cent.

1.16. *Acetobacter oeni* Silva, Cleenwerck, Rivas, Swings, Trujillo, Willems and Velázquez 2006

For the characteristics of the species, refer to Silva et al. (2006).

The type strain is B13^T (= CECT 5830^T = LMG 21952^T), isolated from spoiled red wine of the Dão region, Portugal. The DNA G+C content of the type strain is 58.1 mol per cent.

1.17. *Acetobacter ghanensis* Cleenwerck, Camu, Engelbeen, De Winter, Vandemeulebroecke, De Vos and De Vuyst 2007

For the characteristics of the species, refer to Cleenwerck et al. (2007).

The type strain is R-29337^T (= 430A^T = DSM 18895^T = LMG 23848^T), isolated from a traditional heap fermentation of Ghanaian cocoa beans. The DNA G+C content of the type strain is 57.3.mol per cent.

1.18. *Acetobacter senegalensis* Ndoye, Cleenwerck, Engelbeen, Dubois-Dauphin, Guiro, Van Trappen, Willems and Thonart 2007

For the characteristics of the species, refer to Ndoye et al. (2007).

The type strain is CWBI-B418^T (= DSM 18889^T = LMG 23690^T), isolated from mango fruit in Senegal (sub-Saharan Africa). The DNA G+C content of the type strain is 56.0 mol per cent.

1.19. *Acetobacter fabarum* Cleenwerck, González, Camu, Engelbeen, De Vos and De Vuyst 2008

For the characteristics of the species, refer to Cleenwerck et al. (2008).

The type strain is 985^T (= R-36330^T = DSM 19596^T = LMG 24244^T), isolated from Ghanaian cocoa heap fermentation. The DNA G+C content of the type strain is 57.6 mol per cent.

1.20. *Acetobacter farinalis* Tanasupawat, Kommanee, Yukphan, Muramatsu, Nakagawa and Yamada 2011

For the characteristics of the species, refer to Tanasupawat et al. (2011a).

The type strain is G360-1^T (= BCC 44845^T = NBRC 107750^T = PCU 319^T), isolated from fermented rice flour. The DNA G+C content of the type strain is 56.3 mol per cent.

1.21. *Acetobacter papayae* Iino, Suzuki, Kosako, Ohkuma, Komagata and Uchimura 2013

For the characteristics of the species, refer to Iino et al. (2012a).

The type strain is 1-25^T (= JCM 25143^T = LMG 26456^T = NRIC 0655^T), isolated from a papaya fruit, Okinawa, Japan. The DNA G+C content of the type strain is 60.5 mol per cent.

1.22. *Acetobacter okinawensis* Iino, Suzuki, Kosako, Ohkuma, Komagata and Uchimura 2013

For the characteristics of the species, refer to Iino et al. (2012a).

The type strain is 1-35^T (= JCM 25146^T = LMG 26457^T = NRIC 0658^T), isolated from a piece of the stem of sugarcane, Okinawa, Japan. The DNA G+C content of the type strain is 59.3 mol per cent.

1.23. *Acetobacter persici* corrig. Iino, Suzuki, Kosako, Ohkuma, Komagata and Uchimura 2013

For the characteristics of the species, refer to Iino et al. (2012a).

The type strain is T-120^T (= JCM 25330^T = LMG 26458^T), isolated from a peach fruit, Okinawa, Japan. The DNA G+C content of the type strain is 58.7 mol per cent.

1.24. *Acetobacter lambici* Spitaels, Li, Wieme, Balzarini, Cleenwerck, Van Landschoot, De Vuyst and Vandamme 2014

For the characteristics of the species, refer to Spitaels et al. (2014a).

The type strain is LMG 27439^T (= DSM 27328^T), isolated from fermenting lambic beer. The DNA G+C content of the type strain is 56.2 mol per cent

1.25. *Acetobacter sicerae* Li, Wieme, Spitaels, Balzarini, Nunes, Manaia, Van Landschoot, De Vuyst, Cleenwerck and Vandamme 2014

For the characteristics of the species, refer to Li et al. (2014).

The type strain is LMG 1531^T (= NCIMB 8941^T), isolated from traditionally produced kefir. The DNA G+C content of the type strain is 58.3 mol per cent.

1.26. *Acetobacter thailandicus* Pitiwittayakul, Yukphan, Chaipitakchonlatarn, Yamada and Theeragool 2016

For the characteristics of the species, refer to Pitiwittayakul et al. (2015).

The type strain is AD25^T (= BCC 15839^T = NBRC 103583^T), isolated from a flower of the blue trumpet vine. The DNA G+C content of the type strain is 51.4 mol per cent.

2. *Gluconobacter* Asai 1935

Glu.co.no.bac'ter. N. L. neut. n. *acidum gluconicum*, gluconic acid; N. L. masc. n. *bacter*, rod; N. L. masc. n. *Gluconobacter*, gluconate rod.

The genus *Gluconobacter* was proposed by Asai (1935), who selected a variety of fruits for isolation and found two taxonomic groups in the isolated strains on the oxidation of ethanol and glucose. One had intense ethanol oxidizability rather than glucose and oxidized acetic acid to carbon dioxide and water, and the other had intense glucose oxidizability rather than ethanol and did not oxidize acetic acid. For the latter group, the generic name *Gluconobacter* was given. In the Approved Lists of Bacterial Names 1980, the only species, *Gluconobacter oxydans* was listed with five subspecies (Skerman et al. 1980). However, the range of DNA G+C contents was 8.6 mol per cent from 54.2 to 62.8 mol per cent in the single species (Yamada et al. 1981b, 1984).

Cells are Gram-negative, ellipsoidal to rod-shaped, measuring 0.4 to 1.2 by 1.0 to 3.0 μm, and polarly flagellated when motile. Colonies are smooth, raised to convex, entire, and glistening on ethanol/glucose/yeast extract/calcium carbonate/agar. Some strains produce pink colonies.

Strictly aerobic. Catalase positive and oxidase negative. Acetic acid is produced from ethanol. Acetate and lactate are not oxidized. Grows on mannitol agar, but not on glutamate agar. Dihydroxyacetone is produced from glycerol. Gluconate, 2-ketogluconate, and 5-ketogluconate are produced from glucose, and a few strains produce 2,5-diketogluconate. A water-soluble brown pigment is produced in strains of a few species. Acid is produced from L-arabinose, xylose, glucose, galactose, mannose, fructose, melibiose, mannitol, sorbitol, glycerol, or ethanol. Grows on glucose, fructose, mannitol, sorbitol, and glycerol. Strains of several species require nicotinic acid for growth.

Optimum temperature for growth is between 25 °C and 30 °C. Many species grow at 35 °C, and a few species grow at 37 °C. Optimum pH for growth is around pH 5.5. Most of the species grow at pH 3.5. The major cellular fatty acid is $C_{18:1}\omega7c$. The major ubiquinone is UQ-10. The DNA G+C content ranges from 54.0 to 61.5 mol per cent. Strains of *Gluconobacter* are isolated from fruits, flowers and other sugar-rich materials. For more details of characteristics, see Komagata et al. (2014).

The type species of the genus is *Gluconobacter oxydans* (Henneberg 1897) De Ley 1961. Fourteen species are reported.

2.1. *Gluconobacter oxydans* (Henneberg 1897) De Ley 1961

For the characteristics of the species, refer to Gosselé et al. (1983a), Sievers and Swings (2005), Kersters et al. (2006), Komagata et al. (2014).

The type strain is ATCC 19357^T (= DSM 3503^T = DSM 7145^T = JCM 7642^T = LMG 1408^T = NBRC 14819^T = NCIMB 9013^T), isolated from beer by J. G. Carr. The DNA G+C content of the type strain is 60.3 mol per cent.

2.2. *Gluconobacter cerinus* (ex Asai 1935) Yamada and Akita 1984 emend. Katsura, Yamada, Uchimura and Komagata 2002

Synonym: *Gluconobacter asaii* Mason and Claus 1989.

For the characteristics of the species, refer to Yamada and Akita (1984), Yamada et al. (1984), Tanaka et al. (1999), Katsura et al. (2002).

The type strain is NBRC 3267^T (= ATCC 19441^T = DSM 9533^T = DSM 9534^T = LMG 1368^T = NRRL B-4241^T), isolated from cherry (*Prunus* sp.). The DNA G+C content of the type strain is 55.9 mol per cent.

2.3. *Gluconobacter frateurii* Mason and Claus 1989

For the characteristics of the species, refer to Mason and Claus (1989).

The type strain is Kondo 40^T (= NBRC 3264^T = ATCC 49207^T = DSM 7146^T = LMG 1365^T), isolated from strawberry (*Fragaria ananassa*). The DNA G+C content of the type strain is 55.1 mol per cent.

2.4. *Gluconobacter albidus* (ex Kondo and Ameyama 1958) Yukphan, Takahashi, Potacharoen, Tanasupawat, Nakagawa, Tanticharoen and Yamada 2005

For the characteristics of the species, refer to Yukphan et al. (2004a).

The type strain is NBRC 3250^T (= BCC 14434^T = JCM 20271^T), isolated from a flower of dahlia by Kondo and Ameyama (1958). The DNA G+C content of the type strain is 60.0 mol per cent.

2.5. *Gluconobacter thailandicus* Tanasupawat, Thawai, Yukphan, Moonmangmee, Itoh, Adachi and Yamada 2005

For the characteristics of the species, refer to Tanasupawat et al. (2004).

The type strain is F-149-1^T (= BCC 14116^T = JCM 12310^T = NBRC 100600^T = TISTR 1533^T), isolated from a flower of Indian cork tree (*Millingtonia hortensis*) in Bangkok, Thailand. The DNA G+C content of the type strain is 55.8 mol per cent.

2.6. *Gluconobacter kondonii* Malimas, Yukphan, Takahashi, Kaneyasu, Potacharoen, Tanasupawat, Nakagawa, Tanticharoen and Yamada 2007

For the characteristics of the species, refer to Malimas et al. (2007).

The type strain is Kondo 75^T (= BCC 14441^T = NBRC 3266^T), isolated from strawberry. The DNA G+C content of the type strain is 59.8 mol per cent.

2.7. *Gluconobacter roseus* (ex Asai 1935) Malimas, Yukphan, Takahashi, Muramatsu, Kaneyasu, Potacharoen, Tanasupawat, Nakagawa, Tanticharoen and Yamada 2008

For the characteristics of the species, refer to Malimas et al. (2008a).

The type strain is Asai G-2^T (= BCC 14456^T = JCM 20293^T = NBRC 3990^T), isolated from a fruit of kaki (persimmon, *Diasporas kaki*). The DNA G+C content of the type strain is 60.5 mol per cent.

2.8. *Gluconobacter sphaericus* (Ameyama 1975) Malimas, Yukphan, Takahashi, Muramatsu, Kaneyasu, Potacharoen, Tanasupawat, Nakagawa, Tanticharoen and Yamada 2008

Basonym: *Gluconobacter oxydans* subsp. *sphaericus* Ameyama 1975.

For the characteristics of the species, refer to Ameyama (1975), Malimas et al. (2008b).

The type strain is NBRC 12467^T (= BCC 14448^T = LMG 1414^T), isolated from fresh grapes by Ameyama (1975). The DNA G+C content of the type strain is 59.5 mol per cent.

2.9. *Gluconobacter kanchanaburiensis* Malimas, Yukphan, Lundaa, Muramatsu, Takahashi, Kaneyasu, Potacharoen, Tanasupawat, Nakagawa, Suzuki, Tanticharoen and Yamada 2009

For the characteristics of the species, refer to Malimas et al. (2009a).

The type strain is AD92^T (= BCC 15889^T = NBRC 103587^T), isolated from a spoiled fruit of jackfruit (*Artocarpus heterophyllus*). The DNA G+C content of the type strain is 59.5 mol per cent.

2.10. *Gluconobacter japonicus* Malimas, Yukphan, Takahashi, Muramatsu, Kaneyasu, Potacharoen, Tanasupawat, Nakagawa, Tanticharoen and Yamada 2009

For the characteristics of the species, refer to Malimas et al. (2009b).

The type strain is Kondo 7^T (= BCC 14458^T = NBRC 3271^T), isolated from a fruit of Chinese bayberry. The DNA G+C content of the type strain is 56.4 mol per cent.

2.11. *Gluconobacter wancherniae* Yukphan, Malimas, Lundaa, Muramatsu, Takahashi, Kaneyasu, Tanasupawat, Nakagawa, Suzuki, Tanticharoen and Yamada 2011

For the characteristics of the species, refer to Yukphan et al. (2010).

The type strain is AC42^T (= BCC 15775^T = NBRC 103581^T), isolated from unknown seed. The DNA G+C content of the type strain is 56.6 mol per cent.

2.12. *Gluconobacter uchimurae* Tanasupawat, Kommanee, Yukphan, Moonmangmee, Muramatsu, Nakagawa and Yamada 2011

For the characteristics of the species, refer to Tanasupawat et al. (2011b).

The type strain is ZW160-2^T (= BCC 14681^T = NBRC 100627^T), isolated from rakam fruit (*Zalacca wallichiana*). The DNA G+C content of the type strain is 60.5 mol per cent.

2.13. *Gluconobacter nephelii* Kommanee, Tanasupawat, Yukphan, Malimas, Muramatsu, Nakagawa and Yamada 2011

For the characteristics of the species, refer to Kommanee et al. (2011).

The type strain is RBY-1^T (= BCC 36733^T = NBRC 106061^T), isolated from rambutan (*Nephelium lappaceum*). The DNA G+C content of the type strain is 57.2 mol per cent.

2.14. *Gluconobacter cerevisiae* Spitaels, Wieme, Balzarini, Cleenwerck, Van Landschoot, De Vuyst and Vandamme 2014

For the characteristics of the species, refer to Spitaels et al. (2014b).

The type strain is LMG 27748^T (= DSM 27644^T), isolated from fermenting lambic beer. The DNA G+C content of the type strain is 58.0 mol per cent.

3. *Acidomonas* Urakami, Tamaoka, Suzuki and Komagata 1989 emend. Yamashita, Uchimura and Komagata 2004

A.ci.do.mo'nas. L. adj. *acidus*, sour or acid; L. fem. n. *monas*, unit or monad; *Acidomonas*, acidophilic monad.

The genus *Acidomonas* was introduced for the facultatively methylotrophic bacterium, *Acetobacter methanolicus* Uhlig et al. 1986 (Urakami et al. 1989). However, the generic name was not accepted for a long time (Swings 1992, Sievers et al. 1994). The phylogenetic relationship between the genus *Acidomonas* and other genera of acetic acid bacteria was remote from each other and enough to establish the new genus (Bulygina et al. 1992, Yamada et al. 1997, Yamashita et al. 2004).

Cells are Gram-negative, short rods, measuring 0.5 to 0.8 by 1.5 to 2.0 µm. Cells occur singly, in pairs or rarely in short chains and are either motile with a single polar flagellum or non-motile. Colonies are shiny, smooth, circular, convex, entire, and beige to pink, and 1–3 mm in diameter on glucose/peptone/yeast extract/malt extract (PYM) agar (pH 4.5) after five days incubation at 30 °C. Pellicles are produced in PYM broth.

Aerobic. Catalase positive and oxidase negative. Acetic acid is produced from ethanol. Acetate is oxidized, but lactate is not or only weakly oxidized. Dihydroxyacetone is not produced from glycerol. Gluconate is produced from glucose. 2-Ketogluconate, 5-ketogluconate, or 2,5-diketogluconate is not produced in culture media. Acid is produced from L-arabinose, xylose, ribose, glucose, galactose, mannose, glycerol, ethanol or methanol. Methanol, ethanol, glucose, mannose, glycerol, or succinic acid is utilized as a sole source of carbon. Pantothenic acid is essentially required for growth.

Grows on 30 per cent glucose (w/v) and 0.35 per cent acetic acid (v/v). Grows at pH 3.0. Grows at 30 °C but not at 45 °C. The major cellular fatty acids are $C_{18:1}\omega7c$, $C_{16:0}$, and $C_{18:1}2OH$. The major quinone is UQ-10. The DNA G+C content is from 62 to 63 mol per cent. Strains of *Acidomonas* were abundantly isolated from activated sludges, except for the type strain, but not from vegetables, fruit, decayed wood and leaves, manure, and paddy soil. For more details of characteristics, see Komagata et al. (2014).

3.1. *Acidomonas methanolica* (Uhlig et al. 1986) Urakami, Tamaoka, Suzuki and Komagata 1989. emend. Yamashita, Uchimura and Komagata 2004

Basonym: *Acetobacter methanolicus* Uhlig, Karbaum and Steudel 1986.

For the characteristics of the species, refer to Uhlig et al. (1986), Urakami et al. (1989), Yamashita et al. (2004).

The type strain is MB 58^T (= DSM 5432^T = JCM 6891^T = LMG 1668^T = NRIC 0498^T), isolated from a non-sterile fermentation process for the production of single cell protein (SCP) from methanol with *Candida* species. The cells of the type strain are non motile, and the DNA G+C content is 62 mol per cent.

4. *Gluconacetobacter* corrig. Yamada, Hoshino and Ishikawa 1998

Glu.con.a.ce.to.bac'ter. N. L. neut. n. *acetum gluconicum*, gluconic acid; L. neut. n. *acetum*, vinegar; N. L. masc. n. *bacter*, rod; N. L. masc. n. *Gluconacetobacter*, gluconate-vinegar rod.

The genus *Gluconacetobacter* was introduced by the elevation of the subgenus *Gluconacetobacter* corrig. (ex Asai 1935) Yamada and Kondo 1984 for the UQ-10-equipped *Acetobacter* species (Yamada et al. 1997). Phylogenetically, the genus *Gluconacetobacter* consisted of two groups. One was the *Gluconacetobacter liquefaciens* group, and the other was the *Gluconacetobacter xylinus* group. For the latter group, the genus *Komagataeibacter* Yamada et al. 2013 was proposed (Yamada et al. 2012a, b).

Cells are Gram-negative rods, measuring 0.6 to 0.9 by 1.2 to 2.0 μm, with peritrichous flagella when motile and occur singly or in pairs. Colonies are generally light brown to brown.

Aerobic. Catalase positive. Oxidase negative. Acid is produced from ethanol. Oxidizes acetate and lactate. Grows on glutamate agar and mannitol agar. A few species produce dihydroxyacetone from glycerol. 2-Ketogluconate is produced from glucose. Most of the species produce 2,5-diketogluconate, and a few species produce 5-ketogluconate. Most of the species produce a water-soluble brown pigment. Acid is produced from L-arabinose, xylose, glucose, mannose, or ethanol. Grows on glucose, fructose, sucrose, mannitol, or ethanol. Ammoniac nitrogen is used as a sole nitrogen source. Strains of the most species have the activity of nitrogen fixation.

Most of the species grow on 30 per cent glucose (w/v). Grows between 15 °C and 30 °C but not at 37 °C. The optimum growth temperature is around 30 °C. Grows at pH 3.0. The optimum growth pH is about 5.5. The major cellular fatty acid is $C_{18:1}\omega7c$. The major quinone is UQ-10. The DNA G+C content ranges from 58 to 65 mol per cent. For more details of characteristics, see Komagata et al. (2014).

The type species of the genus is *Gluconacetobacter liquefaciens* (Asai 1935) Yamada et al. 1998. Ten species are reported.

4.1. *Gluconacetobacter liquefaciens* (Asai 1935) Yamada, Hoshino and Ishikawa 1998

Basonym: *Acetobacter aceti* subsp. *liquefaciens* (Asai 1935) De Ley and Frateur 1974.

Synonyms: *Acetobacter liquefaciens* (Asai 1935) Gosselé, Swings, Kersters, Pauwels and De Ley 1983; '*Gluconobacter liquefaciens*' Asai 1935.

For the characteristics of the species, refer to Asai et al. (1964), Yamada and Kondo (1984), Yamada et al. (1997), Navarro and Komagata (1999), Sievers and Swings (2005), Kersters et al. (2006), Komagata et al. (2014).

The type strain is Asai G-1^T (= ATCC 14835^T = DSM 5603^T = JCM 17840^T = LMG 1381^T = LMG 1382^T = NBRC 12388^T), isolated from dried persimmon. The DNA G+C content of the type strain is 64.9 mol per cent.

4.2. *Gluconacetobacter diazotrophicus* (Gillis et al. 1989) Yamada, Hoshino and Ishikawa 1998

Basonym: *Acetobacter diazotrophicus* Gillis, Kersters, Hoste, Janssens, Kroppenstadt, Stephan, Teixeira, Döbereiner and De Ley 1989.

For the characteristics of the species, refer to Gillis et al. (1989).

The type strain is LMG 7603^T (= ATCC 49037^T = CCUG 37298^T = CIP 103539^T = DSM 5601^T), isolated from roots and stems of sugarcane in Alagoas, Brazil. The DNA G+C content of the type strain is 61 mol per cent.

4.3. *Gluconacetobacter sacchari* Franke, Fegan, Hayward, Leonard, Stackebrandt and Sly 1999

For the characteristics of the species, refer to Franke et al. (1999).

The type strain is SRI 1794^T (= CIP 106693^T = DSM 12717^T), isolated from the leaf sheath of sugar cane and from the pink sugar-cane mealy bug. The DNA G+C content of the type strain is 65 mol per cent.

4.4. *Gluconacetobacter johannae* Fuentes-Ramírez, Bustillos-Cristales, Tapia-Hernández, Jiménez-Salgado, Wang, Martínez-Romero and Caballero-Mellado 2001

For the characteristics of the species, refer to Fuentes-Ramírez et al. (2001).

The type strain is CFN-Cf55^T (= ATCC 700987^T = CIP 107160^T = DSM 13595^T), isolated from the rhizosphere of coffee plants. The DNA G+C content of the type strain is 57.96 mol per cent.

4.5. *Gluconacetobacter azotocaptans* Fuentes-Ramírez, Bustillos-Cristales, Tapia-Hernández, Jiménez-Salgado, Wang, Martínez-Romero and Caballero-Mellado 2001

For the characteristics of the species, refer to Fuentes-Ramírez et al. (2001).

The type strain is CFN-Ca54^T (= ATCC 700988^T = CIP 107161^T = DSM 13594^T), isolated from the rhizosphere of coffee plants. The DNA G+C content of the type strain is 64.01 mol per cent.

4.6. *Gluconacetobacter tumulicola* Tazato, Nishijima, Handa, Kigawa, Sano and Sugiyama 2012

For the characteristics of the species, refer to Tazato et al. (2012).

The type strain is K5929-2-1b^T (= JCM 17774^T = NCIMB 14760^T), isolated from a black viscous substance in a plaster hole at the center of the ceiling in the stone chamber of the Kitora Tumulus in Asuka village, Nara Prefecture, Japan. The DNA G+C content of the type strain is 64.7 mol per cent.

4.7. *Gluconacetobacter asukensis* Tazato, Nishijima, Handa, Kigawa, Sano and Sugiyama 2012

For the characteristics of the species, refer to Tazato et al. (2012).

The type strain is K8617-1-1b^T (= JCM 17772^T = NCIMB 14759^T), isolated from a brown viscous gel on the north-east area of the ceiling in the stone chamber of the Kitora Tumuli in Asuka village, Nara Prefecture, Japan. The DNA G+C content of the type strain is 65.4 mol per cent.

4.8. *Gluconacetobacter tumulisoli* Nishijima, Tazato, Handa, Tomita, Kigawa, Sano and Sugiyama 2013

For the characteristics of the species, refer to Nishijima et al. (2013).

The type strain is T611xx-1-4a^T (= JCM 19097^T = NCIMB 14861^T), isolated from clay soil taken from near the spider's web and an ant hole at a plugging stone directly under the plugging stone of the upper north side at the space adjacent to Takamatsuzuka Tumulus in Asuka village, Nara Prefecture, Japan. The DNA G+C content of the type strain is 66.5 mol per cent.

4.9. *Gluconacetobacter takamatsuzukensis* Nishijima, Tazato, Handa, Tomita, Kigawa, Sano and Sugiyama 2013

For the characteristics of the species, refer to Nishijima et al. (2013).

The type strain is T61213-20-1a^T (= JCM 19094^T = NCIMB 14859^T), isolated from soil taken from the left side wall of the west side in the stone chamber exterior during dismantling work of Takamatsuzuka Tumulus in Asuka village, Nara Prefecture, Japan. The DNA G+C content of the type strain is 66.6 mol per cent.

4.10. *Gluconacetobacter aggeris* Nishijima, Tazato, Handa, Tomita, Kigawa, Sano and Sugiyama 2013

For the characteristics of the species, refer to Nishijima et al. (2013).

The type strain is T6203-4-1a^T (= JCM 19092^T = NCIMB 14860^T), isolated from soil taken from 5 cm below the surface in a bamboo grove of the burial mound of Takamatsuzuka Tumulus in Asuka village, Nara Prefecture, Japan. The DNA G+C content of the type strain is 65.4 mol per cent.

5. *Asaia* Yamada, Katsura, Kawasaki, Widyastuti, Saono, Seki, Uchimura and Komagata 2000

A.sa'i.a. N. L. fem. n. *Asaia*, Asai, named after Professor Toshinobu Asai, a Japanese bacteriologist who contributed to the systematics of acetic acid bacteria.

The strains of the genus *Asaia* were first found and isolated from flowers collected in Indonesia. In the beginning, the distribution of the *Asaia* strains was supposed to be restricted only to the tropical zone, viz., in Thailand, the Philippines, and Indonesia (Yamada and Yukphan 2008). However, the *Asaia* strains were isolated in the temperate zone, viz., in Japan (Suzuki et al. 2010). The strains of the genus *Asaia* produced no or a very little amount of acetic acid from ethanol and did not grow in the presence of 0.35 per cent acetic acid (w/v).

Cells are Gram-negative, rod-shaped, measuring 0.4 to 1.0 by 1.0 to 2.5 μm, and motile with peritrichous flagella. Colonies are smooth, entire, raised, shiny, and light brown or pink to dark pinkish on glucose/peptone/yeast extract agar.

Aerobic. Catalase positive and oxidase negative. Produces no or a limited amount of acetic acid from ethanol. Oxidizes acetate and lactate to carbon dioxide and water. Grows on glutamate agar and mannitol agar. Dihydroxyacetone is generally produced. Produces 2-ketogluconate and 5-ketogluconate from glucose, but not 2,5-diketogluconate. Acid is produced from glucose, galactose, fructose, or other sugars and sugar alcohols. Grows on glucose, fructose, or mannitol. Ammoniac nitrogen is assimilated on glucose or mannitol.

Grows on 30 per cent glucose (w/v), but not in the presence of 0.35 per cent acetic acid (v/v). Growth generally occurs between 10 and 30 °C, but not at 37 °C. Grows at pH 3.0. The major cellular fatty acid is $C_{18:1}\omega 7c$. The major quinone is UQ 10. The DNA G+C content ranges from 58.6 to 61.0 mol per cent. For more details of characteristics, see Komagata et al. (2014).

The type species of the genus is *Asaia bogorensis* Yamada et al. 2000. Eight species are reported.

5.1. *Asaia bogorensis* Yamada, Katsura, Kawasaki, Widyastuti, Saono, Seki, Uchimura and Komagata 2000

For the characteristics of the species, refer to Yamada et al. (2000).

The type strain is 71^T (= JCM 19569^T = NRIC 0311^T), isolated from a flower of orchid tree (*Bauhinia purpurea*) in Bogor, Indonesia. The DNA G+C content of the type strain is 60.2 mol per cent.

5.2. *Asaia siamensis* Katsura, Kawasaki, Potacharoen, Saono, Seki, Yamada, Uchimura and Komagata 2001

For the characteristics of the species, refer to Katsura et al. (2001).

The type strain is S60-1^T (= JCM 10715^T = NBRC 16457^T = NRIC 0323^T), isolated from a flower of crown flower (*Calotropis gigantea*), in Bangkok, Thailand. The DNA G+C content of the type strain is 59.3 mol per cent.

5.3. *Asaia krungthepensis* Yukphan, Potacharoen, Tanasupawat, Tanticharoen and Yamada 2004

For the characteristics of the species, refer to Yukphan et al. (2004b).

The type strain is AA08^T (= BCC 12978^T = NBRC 100057^T = NRIC 0535^T = TISTR 1524^T), isolated from a heliconia flower (*Heliconia* sp.) in Bangkok, Thailand. The DNA G+C content of the type strain is 60.3 mol per cent.

5.4. *Asaia lannensis* corrig. Malimas, Yukphan, Takahashi, Kaneyasu, Potacharoen, Tanasupawat, Nakagawa, Tanticharoen and Yamada 2008

For the characteristics of the species, refer to Malimas et al. (2008c).

The type strain is AB92^T (= BCC 15733^T = NBRC 102526^T), isolated from a flower of spider lily (*Crynum asiaticum*) in Chiang Mai, Thailand. The DNA G+C content of the type strain is 60.8 mol per cent.

5.5. *Asaia spathodeae* Kommanee, Tanasupawat, Yukphan, Malimas, Muramatsu, Nakagawa and Yamada 2010

For the characteristics of the species, refer to Kommanee et al. (2010).

The type strain is GB23-2^T (= BCC 36458^T = NBRC 105894^T = PCU 307^T), isolated from a flower of the African tulip (*Spathodea campanulata*) in Thailand. The DNA G+C content of the type strain is 59.7 mol per cent.

5.6. *Asaia astilbis* corrig. Suzuki, Zhang, Iino, Kosako, Komagata and Uchimura 2010

For the characteristics of the species, refer to Suzuki et al. (2010).

The type strain is T-6133^T (= DSM 23030^T = JCM 15831^T), isolated from astilbe (*Astilbe thunbergii* var. *congesta*), Yamanashi Prefecture, Japan. The DNA G+C content of the type strain is 58.9 mol per cent.

5.7. *Asaia platycodi* Suzuki, Zhang, Iino, Kosako, Komagata and Uchimura 2010

For the characteristics of the species, refer to Suzuki et al. (2010).

The type strain is T-683^T (= JCM 25414^T = DSM 23029^T), isolated from balloon flower (*Platycodon grandiflorum*) in Akita Prefecture, Japan. The DNA G+C content of the type strain is 60.0 mol per cent.

5.8. *Asaia prunellae* Suzuki, Zhang, Iino, Kosako, Komagata and Uchimura 2010

For the characteristics of the species, refer to Suzuki et al. (2010).

The type strain is T-153^T (= DSM 23028^T = JCM 25354^T), isolated from self-heal (*Prunella vulgaris*) in Akita Prefecture, Japan. The DNA G+C content of the type strain is 58.9 mol per cent.

6. *Kozakia* Lisdiyanti, Kawasaki, Widyastuti, Saono, Seki, Yamada, Uchimura and Komagata 2002

Ko.za'ki.a. N. L. fem. n. *Kozakia*, Kozaki, named after Professor Michio Kozaki, a Japanese bacteriologist who contributed to the study of microorganisms in tropical regions, especially Southeast Asia.

The genus *Kozakia* was phylogenetically related to the genera *Asaia* and *Neoasaia*. However, the genus especially differed from the genus *Asaia* in oxidation of ethanol to acetic acid and in the production of a large amount of levan-like mucous substances from sucrose.

Cells are Gram-negative, rod-shaped, and non-motile, measuring 0.6 to 0.8 by 2.0 to 3.0 μm. Colonies are not pigmented.

Strictly aerobic. Catalase positive and oxidase negative. Acetic acid is produced from ethanol. Acetate and lactate are oxidized to carbon dioxide and water, but the activity is weak. Grows on mannitol agar but not on glutamate agar. Dihydroxyacetone is produced from glycerol. Gluconate, 2-ketogluconate, and 5-ketogluconate are produced from glucose, but 2,5-diketogluconate is not. A water-soluble brown pigment is not produced from glucose. Acid is produced from L-arabinose, xylose, glucose, galactose, mannose, melibiose, raffinose, *meso*-erythritol, glycerol, or ethanol. Methanol is not utilized. Ammoniac nitrogen is not assimilated on glucose, mannitol, or ethanol medium without vitamins. Levan-like mucous substance is produced from sucrose or fructose. γ-Pyrone is produced from fructose but not from glucose.

Growth is not inhibited by 0.35 per cent acetic acid (v/v) at pH 3.5. Does not grow on 30 per cent glucose (w/v). Grows at pH 3.0 and 30 °C. The major cellular fatty acid is $C_{18:1}\omega7c$. The major quinone is UQ-10. The DNA G+C content ranges from 56.8 to 57.2 mol per cent. For more details of characteristics, see Komagata et al. (2014).

6.1. *Kozakia baliensis* Lisdiyanti, Kawasaki, Widyastuti, Saono, Seki, Yamada, Uchimura and Komagata 2002

For the characteristics of the species, refer to Lisdiyanti et al. (2002).

The type strain is Yo-3^T (= DSM 14400^T = JCM 11301^T = NBRC 16664^T = NRIC 0488^T), isolated from palm brown sugar collected in Bali, Indonesia in 1996. The DNA G+C content of the type strain is 57.2 mol per cent.

7. *Swaminathania* Loganathan and Nair 2004

Swa.mi.na.tha'ni.a. N. L. fem. n. *Swaminathania*, Swaminathan, named after Swaminathan, an Indian biologist, the father of the Green Revolution in India.

The strains of the genus *Swaminathania*, which were isolated using a nitrogen-free semi-solid LGI medium at pH 5.5 from the rhizosphere, roots, and stems of salt-tolerant, mangrove-associated wild rice, was phylogenetically related especially to those of the genus *Asaia*. However, the genus was distinguished phenotypically from the genus *Asaia* by growth in the presence of 0.35 per cent acetic acid (v/v), 3 per cent NaCl (w/v), or 1 per cent KNO_3 (w/v).

Cells are Gram-negative, straight rods with round ends, measuring approximately 0.7 to 0.9 by 1.9 to 3.1 μm, and motile with peritrichous flagella. Colonies are initially yellowish and become dark orange later, smooth, and raised, with entire margin on LGI medium.

Aerobic. Catalase-positive and oxidase negative. Acetic acid is produced from ethanol under neutral and acidic conditions. Acetate and lactate are oxidized to carbon dioxide and water, but the activity was weak. Grows on mannitol agar and glutamate agar. Acid is produced from L-arabinose, glucose, galactose, mannose, sorbitol, glycerol, or ethanol. Methanol is not utilized. A water-soluble brown pigment is produced on glucose/calcium carbonate-containing agar. Strains are able to fix nitrogen. Solubilization of phosphate is shown. Grows intensely in the presence of 0.35 per cent acetic acid (v/v) at pH 3.5 and 3 per cent NaCl using 1 per cent KNO_3 (w/v) as a nitrogen source.

The major cellular fatty acid is $C_{18:1}\omega7c/\omega9t/\omega12t$. The major quinone is UQ-10. The DNA G+C content ranges from 57.6 to 59.9 mol per cent. For more details of characteristics, see Komagata et al. (2014).

7.1. *Swaminathania salitolerans* Loganathan and Nair 2004

For the characteristics of the species, refer to Loganathan and Nair (2004).

The type strain is PA51^T (= LMG 21291^T = MTCC 3852^T), isolated from mangrove-associated wild rice (*Porteresia coarctata*) in Pichavaram, Tamil Nadu, India. The DNA G+C content of the type strain is not reported.

8. *Saccharibacter* Jojima, Mihara, Suzuki, Yokozeki, Yamanaka and Fudo 2004

Sac.cha.ri.bac'ter. L. neut. n. *sacchrum* or *saccharon*, sugar; N. L. masc. n. *bacter*, rod; N. L. masc. n. *Saccharibacter*, a sugar rod or a rod that grows intensely in a sugar-rich environment.

The strains of the genus *Saccharibacter* that were isolated from the pollen of Japanese flower were quite remote phylogenetically from strains of any other genera of acetic acid bacteria. The strains of the genus were osmophilic, e.g., showing no growth on 1 per cent glutamate agar (w/v) but on 7 per cent glutamate agar (w/v). The phylogenetically related genera are *Bombella, Neokomagataea, Swingsia*, and *Gluconobacter*.

Cells are Gram-negative, straight rods, measuring 0.8 to 1.0 by 2.5 to 4.0 µm, and non motile. Colonies are circular, entire, and pale in color on yeast extract/glucose/peptone agar.

Strictly aerobic. Catalase positive and oxidase negative. Produces negligible or very little acetic acid from ethanol. Acetate is not oxidized to carbon dioxide and water, and lactate is weakly oxidized. Grows on mannitol agar and glutamate agar supplemented with 7 per cent substrates (w/v). Does not grow on common mannitol agar and glutamate agar with 1 per cent substrates (w/v). Dihydroxyacetone is not produced from glycerol. Gluconate, 2-ketogluconate, and 5-ketogluconate are produced from glucose. Acid is produced from L-arabinose, xylose, glucose, galactose, mannose, melibiose, sucrose, or mannitol. Methanol is not utilized. Ammoniac nitrogen is not assimilated on Hoyer–Frateur medium with glucose, mannitol, or ethanol. Cellulosic pellicles and water-soluble mucous substances are not produced. Not pigmented.

Grows in the glucose range between 2 to 40 per cent (w/v), with an optimum around 10 per cent (w/v). High glucose concentration, e.g., 10 per cent glucose (w/v), is preferable for growth. Osmophilic. No growth occurs in the presence of 0.35 per cent acetic acid (v/v) at pH 3.5. Temperature for growth ranges from 20 to 33 °C, and the optimum is around 25 to 30 °C. The growth pH ranges from pH 4.0 to pH 7.5, and the optimum pH is around pH 5.0 to pH 7.0. No growth is observed below pH 4.0. The major cellular fatty acids are $C_{16:0}$2OH (31.1-41.0 per cent) and $C_{18:1}\omega 7c$ (22.0-29.8 per cent). The major quinone is UQ-10. The DNA G+C content ranges from 52 to 53 mol per cent. For more details of characteristics, see Komagata et al. (2014).

8.1. *Saccharibacter floricola* Jojima, Mihara, Suzuki, Yokozeki, Yamanaka and Fudo 2004

For the characteristics of the species, refer to Jojima et al. (2004).

The type strain is S-877^T (= AJ 13480^T = DSM 15669^T = JCM 12116^T), isolated from pollen collected in Kanagawa Prefecture, Japan. The DNA G+C content of the type strain is 52.3 mol per cent.

9. *Neoasaia* Yukphan, Malimas, Potacharoen, Tanasupawat, Tanticharoen and Yamada 2006

Ne.o.a.sa'i.a. Gr. adj. *neos*, new; N. L. fem. n. *Asaia*, a bacterial name after Professor Asai, Japan; N. L. fem. n. *Neoasaia*, new *Asaia*.

The strain of the genus *Neoasaia* that was isolated from a flower of red ginger was closely related phylogenetically to those of the genera *Kozakia, Asaia*, and *Swaminathania*. However, the phenotypic characteristic showed no oxidation of acetate and lactate, differentiating from those of the three genera mentioned above.

Cells are Gram-negative, rod-shaped, measuring 0.8 to 1.0 by 1.0 to 2.0 µm, and non-motile. Colonies are smooth, raised, entire, shiny, and pink.

Aerobic. Acetic acid is produced from ethanol. Acetate and lactate are not oxidized. Grows on glutamate agar and mannitol agar. Dihydroxyacetone is weakly produced from glycerol. 2-Ketogluconate and 5-ketogluconate are produced from glucose. Acid is produced from arabinose (weakly), L-arabinose, xylose, L-rhamnose (weakly), fructose with delay, galactose, glucose, mannose, melibiose, sucrose, raffinose, mannitol (weakly), sorbitol with delay, dulcitol (weakly), *meso*-erythritol, glycerol, or ethanol.

Grows on arabinose (weakly), L-arabinose, xylose, fructose, L-sorbose, galactose, glucose, mannose (weakly), sucrose, raffinose, mannitol, sorbitol, dulcitol, *meso*-erythritol, or glycerol. Ammoniac nitrogen is hardly assimilated in the presence of glucose or mannitol as a carbon source. A water-soluble brown pigment is not produced on a glucose/peptone/yeast extract/calcium carbonate medium, and a levan-like polysaccharide is not produced on a sucrose medium. However, the production of fructan is reported in *Neoasaia chiangmaiensis* NBRC 101099^T (Jakob et al. 2013).

Grows on 30 per cent glucose (w/v) and in the presence of 0.35 per cent acetic acid (v/v), but not in the presence of 1.0 per cent KNO_3 (w/v). The major cellular fatty acid is $C_{18:1}\omega7c$. The major quinone is UQ-10. The DNA G+C content is 63.1 mol per cent. For more details of characteristics, see Komagata et al. (2014).

9.1. *Neoasaia chiangmaiensis* Yukphan, Malimas, Potacharoen, Tanasupawat, Tanticharoen and Yamada 2006

For the characteristics of the species, refer to Yukphan et al. (2005).

The type strain is AC28^T (= BCC 15763^T = NBRC 101099^T), isolated from a flower of red ginger (*Alpinia purpurata*) in Chiang Mai, Thailand in September 2002. The DNA G+C content of the type strain is 63.1 mol per cent.

10. *Granulibacter* Greenberg, Porcella, Stock, Wong, Conville, Murray, Holland and Zelazny 2006

Gra.nu.li.bac'ter. L. neut. n. *granulum*, grain; N. L. masc. n. *bacter*, rod; N. L. masc. n. *Granulibacter*, a rod that causes granules or granuloma formation.

The strain of the genus *Granulibacter* isolated first from three patients with chronic granulomatous disease was quite remote phylogenetically from other acetic acid bacteria. The strain grew at optimum temperatures of 35-37 °C and on methanol.

Gram-negative, coccobacillus to rod-shaped, and non motile. Colonies are convex, entire, smooth and non-diffusible yellow on a modified glucose/yeast extract/calcium carbonate.

Strictly aerobic. Catalase-positive and oxidase negative. Acetic acid is hardly produced from ethanol. Acetate and lactate are oxidized to carbon dioxide and water, but the activity of the former is weak. Grows on glutamate agar but weakly on mannitol agar. Dihydroxyacetone is not produced from glycerol. Acid is produced from glucose or ethanol and from glycerol weakly. Methanol can be used as a sole source of carbon. Ammoniac nitrogen is assimilated on glucose. High concentration of glucose, e.g., 5 per cent glucose (w/v), is preferable for growth.

Optimum temperature for the growth is 35 to 37 °C. Optimum pH for the growth is 5.0 to 6.5. Grows at pH 3.5. The major cellular fatty acids are $C_{18:1}\omega7c$ and $C_{16:0}$. The major quinone is UQ-10 (Yukphan et al. 2009). The DNA G+C content is 59.1 mol per cent. For more details of characteristics, see Komagata et al. (2014).

10.1. *Granulibacter bethesdensis* Greenberg, Porcella, Stock, Wong, Conville, Murray, Holland and Zelazny 2006

For the characteristics of the species, refer to Greenberg et al. (2006).

The type strain is CGDNIH1^T (= ATCC BAA-1260^T = DSM 17861^T), which was isolated from lymph node culture from a granulomatous disease patient in Bethesda, MD, USA, in 2003. The DNA G+C content of the type strain is 59.1 mol per cent.

11. *Tanticharoenia* Yukphan, Malimas, Muramatsu, Takahashi, Kaneyasu, Tanasupawat, Nakagawa, Suzuki, Potacharoen and Yamada 2008

Tan.ti.cha.ro.e'nia. N. L. fem. n. *Tanticharoenia*, named after Dr. Morakot Tanticharoen, Thailand, who contributed to studies of acetic acid bacteria.

The strains of the genus *Tanticharoenia* that were isolated from soil collected in Sakaerat, Nakhon Rachashima, Thailand constituted an independent cluster phylogenetically. The strains did not oxidize either acetate or lactate, but grew on 30 per cent glucose (w/v).

Cells are Gram-negative, rod-shaped, measuring 0.6 to 0.8 by 1.0 to 1.6 µm, and non motile. Colonies are creamy and smooth with entire margin when grown on glucose/ethanol/peptone/yeast extract/calcium carbonate agar.

Acetic acid is produced from ethanol. Acetate and lactate are not oxidized. Grows on glutamate agar weakly and on mannitol agar. Dihydroxyacetone is produced from glycerol. 2-Ketogluconate, 5-ketogluconate, and 2,5-diketogluconate are produced from glucose. A water-soluble brown pigment is intensely produced on glucose/peptone/yeast extract/calcium carbonate agar. Acid is produced from L-arabinose, xylose, fructose (weakly), galactose, glucose, mannose, melibiose, sucrose (weakly), raffinose (weakly), *meso*-erythritol, glycerol, or ethanol. Grows on L-arabinose, xylose, fructose, glucose, galactose, *meso*-erythritol, mannitol, sorbitol, glycerol, or sucrose. Ammoniac nitrogen is not assimilated in the presence of glucose, mannitol, or ethanol as a carbon source.

Grows in the presence of 0.35 per cent acetic acid (v/v), but not of 1 per cent KNO_3. Grows on 30 per cent glucose (w/v). The major cellular fatty acids is $C_{18:1}\omega7c$. The major quinone is UQ-10. The DNA G+C content ranges from 64.5 to 65.6 mol per cent. For more details of characteristics, see Komagata et al. (2014).

The type species of the genus is *Tanticharoenia sakaeratensis* Yukphan et al. 2008. Two species are reported.

11.1. *Tanticharoenia sakaeratensis* Yukphan, Malimas, Muramatsu, Takahashi, Kaneyasu, Tanasupawat, Nakagawa, Suzuki, Potacharoen and Yamada 2008

For the characteristics of the species, refer to Yukphan et al. (2008).

The type strain is AC37^T (= BCC 15772^T = NBRC 103193^T), isolated from soil collected at Sakaerat, Nakhon Ratchasima, Thailand. The DNA G+C content of the type strain is 65.6 mol per cent.

11.2. *Tanticharoenia aidae* Vu, Malimas, Chaipitakchonlatarn, Bui, Yukphan, Bui, Muramatsu, Sitdhipol, Tanasupawat, Duong, Nakagawa, Pham and Yamada 2016

For the characteristics of the species, refer to Vu et al. (2016).

The type strain is VTH-Ai06^T (= VTCC 910001^T = BCC 67839^T = NBRC 110637^T). The DNA G+C content of the type strain is 65.4 mol per cent.

12. *Ameyamaea* Yukphan, Malimas, Muramatsu, Takahashi, Kaneyasu, Potacharoen, Tanasupawat, Nakagawa, Hamana, Tahara, Suzuki, Tanticharoen and Yamada 2010

A.me.ya.ma'e.a. N. L. fem. n. *Ameyamaea*, Ameyama, named after Professor Minoru Ameyama, Japan, who contributed to studies of acetic acid bacteria, especially their biochemical and systematic studies.

The strains of the genus *Ameyamaea* that were isolated from flowers of red ginger collected in Chiang Mai, Thailand were closely related phylogenetically to strains of the genus *Tanticharoenia*. However, the strains showed oxidation of acetate and weak oxidation of lactate and no growth on 30 per cent glucose (w/v), differing from those of the genus *Tanticharoenia*.

Cells are Gram-negative, rods, measuring 0.6 to 0.8 by 1.0 to 1.8 μm, and motile with polar flagella. Colonies are creamy and smooth with entire margin on glucose/ethanol/peptone/yeast extract/calcium carbonate agar.

Acetic acid is produced from ethanol. Acetate is oxidized to carbon dioxide and water, but lactate is weakly oxidized. Grows on glutamate agar (weakly) and mannitol agar. Dihydroxyacetone is weakly produced from glycerol. 2-Ketogluconate and 5-ketogluconate are produced from glucose. A water-soluble brown pigment is not produced on glucose/peptone/yeast extract/calcium carbonate agar.

Acid is produced from arabinose (weakly), L-arabinose, xylose, L-rhamnose, glucose, mannose, galactose, *meso*-erythritol, glycerol (weakly), melibiose, or ethanol. Grows on glucose, mannose (very weakly), galactose, xylose, L-arabinose, L-rhamnose, fructose, L-sorbose, mannitol, sorbitol, dulcitol, *meso*-erythritol, glycerol, or melibiose (very weakly). Growth is weak on methanol. Ammoniac nitrogen is very weakly assimilated in the presence of glucose, mannitol, or ethanol as a carbon source.

Grows in the presence of 0.35 per cent acetic acid (v/v), but not on 30 per cent glucose (w/v). The major cellular fatty acids is $C_{18:1}\omega 7c$. The major quinone is UQ-10. The DNA G+C content is 66.0-66.1 mol per cent. For more details of characteristics, see Komagata et al. (2014).

12.1. *Ameyamaea chiangmaiensis* Yukphan, Malimas, Muramatsu, Takahashi, Kaneyasu, Potacharoen, Tanasupawat, Nakagawa, Hamana, Tahara, Suzuki, Tanticharoen and Yamada 2010

For the characteristics of the species, refer to Yukphan et al. (2009).

The type strain is AC04^T (= BCC 15744^T = NBRC 103196^T), isolated from a flower of red ginger (*Alpinia purpurea*) in Chiang Mai, Thailand. The DNA G+C content of the type strain is 66.0 mol per cent.

13. *Neokomagataea* Yukphan, Malimas, Muramatsu, Potacharoen, Tanasupawat, Nakagawa, Tanticharoen and Yamada 2011

Ne.o.ko.ma.ga.ta'ea. N. L. fem. n. *Neokomagataea*, new Komagata, named after Professor Kazuo Komagata, a Japanese microbiologist who contributed to bacterial systematics and phylogeny, especially of acetic acid bacteria.

The strains of the genus *Neokomagataea* that were isolated in Thailand from the flowers of lantana and candle bush were related phylogenetically to those of the genus *Gluconobacter*. The strains of the genus grew on 30 per cent glucose (w/v), but not in the presence of 0.35 per cent acetic acid (v/v), the latter of which differed from those of the genus *Gluconobacter*.

Cells are Gram-negative rods, measuring 0.6 to 0.8 by 1.0 to 1.6 μm, and non motile. Colonies are smooth, entire, and creamy on glucose/ethanol/peptone/yeast extract/calcium carbonate agar.

Acetic acid is weakly produced from ethanol. Acetate and lactate are not oxidized. Grows on glutamate agar and mannitol agar. Dihydroxyacetone is not produced from glycerol. 2-Ketogluconate, 5-ketogluconagte, and 2,5-diketogluconate are produced from glucose. A water-soluble brown pigment is not produced. Acid is produced from L-arabinose (weakly), xylose, glucose, galactose (weakly), fructose, or sucrose. Grows on glucose, L-rhamnose (weakly), or sucrose. Ammoniac nitrogen is not generally assimilated on glucose or ethanol as a source of carbon.

Grows between 1.0 per cent and 30 per cent glucose (w/v). Osmotolerant. Growth does not occur in the presence of 0.35 per cent acetic acid (v/v) or in the presence of 1.0 per cent or 2.0 per cent NaCl (w/v), or 1.0 per cent KNO_3 (w/v). The major cellular fatty acids are $C_{18:1}\omega7c$, $C_{16:0}$ and $C_{18:1}2OH$. The major quinone is UQ-10. The DNA G+C content ranges from 51.2 to 56.8 mol per cent. For more details of characteristics, see Komagata et al. (2014).

The type species of the genus is *Neokomagataea thailandica* Yukphan et al. 2011. Two species are reported.

13.1. *Neokomagataea thailandica* Yukphan, Malimas, Muramatsu, Potacharoen, Tanasupawat, Nakagawa, Tanticharoen and Yamada 2011

For the characteristics of the species, refer to Yukphan et al. (2011).

The type strain is AH11^T (= BCC 25710^T = NBRC 106555^T), isolated from a flower of lantana (*Lantana camera*) at Tan Island, Hat Khanom-Mu Ko Thale Thai National Park, Nakhon-Si-Thammarat, Thailand in 2007. The DNA G+C content of the type strain is 56.8 mol per cent.

13.2. *Neokomagataea tanensis* Yukphan, Malimas, Muramatsu, Potacharoen, Tanasupawat, Nakagawa, Tanticharoen and Yamada 2011

For the characteristics of the species, refer to Yukphan et al. (2011).

The type strain is AH13^T (= BCC 25711^T = NBRC 106556^T), isolated from a flower of candle bush (*Senna alata*) at Tan Island, Hat Khanom-Mu Ko Thale Thai National Park, Nakhon-Si-Thammarat, Thailand in 2007. The DNA G+C content of the type strain is 51.2 mol per cent.

14. *Komagataeibacter* Yamada, Yukphan, Vu, Muramatsu, Ochaikul, Tanasupawat and Nakagawa 2013

Ko.ma.ga.ta.e.i.bac'ter. N. L. fem. n. *Komagataea*, Komagata, the name of a Japanese microbiologist; N. L. masc. n. bacter, rod; N. L. masc. n. *Komagataeibacter*, Komagata rod, which is named after Professor Kazuo Komagata, Japan, who contributed to the bacterial systematics, especially of acetic acid bacteria.

The genus *Komagataeibacter* was introduced for the *Gluconacetobacter xylinus* group of the genus *Gluconacetobacter* based on 16S rRNA gene sequence analysis and morphological, physiological, and ecological characterizations (Yamada et al. 2012a, b). The eleven species of the genus *Gluconacetobacter* were transferred to the genus *Komagataeibacter* as new combinations. Recently, three new combinations were additionally reported. The phenotypic characteristics of the genus *Komagataeibacter* were generally no motility and no production of 2,5-diketogluconate from glucose and no water-soluble brown pigment production on glucose/peptone/yeast extract/calcium carbonate medium.

Cells are Gram-negative rods, measuring 0.5 to 0.8 by 1.0 to 3.0 μm, occurring singly, in pairs, or in chains. Non motile. Colonies are circular, smooth, or rough, raised to convex or umbonate, entire, glistening, and white-creamy to beige.

Aerobic. Catalase positive and oxidase negative. Acetic acid is produced from ethanol. Acetate and lactate are oxidized to carbon dioxide and water. Grows on glutamate agar and mannitol agar. Dihydroxyacetone is generally produced from glycerol. Gluconate, 2-ketogluconate, and/or 5-ketogluconate are produced from glucose, but 2,5-diketogluconate is not.

Acid is produced from L-arabinose, xylose, glucose, galactose, or ethanol. Grows on glucose, fructose, maltose, sucrose, or mannitol. Ammoniac nitrogen is generally assimilated on mannitol. Cellulosic materials are produced by some strains, for example, of *Komagataeibacter xylinus*, *Komagataeibacter nataicola*, and *Komagataeibacter medellinensis*. A water-soluble brown pigment is not produced on glucose/yeast extract/calcium carbonate medium. γ-Pyrone compounds are not produced.

Grows generally in the presence of 0.35 per cent acetic acid (v/v). Some species require acetic acid for growth. Grows at pH 3.0. The major cellular fatty acid is $C_{18:1}\omega 7c$. The major quinone is UQ-10. The DNA G+C content ranges from 58 to 64 mol per cent. For more details of characteristics, see Komagata et al. (2014).

The type species of the genus is *Komagataeibacter xylinus* (Brown 1886) Yamada et al. 2013. Fourteen species are reported.

14.1. *Komagataeibacter xylinus* (Brown 1886) Yamada, Yukphan, Vu, Muramatsu, Ochaikul, Tanasupawat and Nakagawa 2013

Basonym: *Acetobacter aceti* subsp. *xylinus* corrig. (Brown 1886) De Ley and Frateur 1974.

Synonyms: *Acetobacter xylinus* (Brown 1886) Yamada 1984; *Gluconacetobacter xylinus* (Brown 1886) Yamada, Hoshino and Ishikawa 1998; '*Bacterium xylinum*' Brown 1886.

For the characteristics of the species, refer to Gosselé et al. (1983b), Lisdiyanti et al. (2006), Navarro and Komagata (1999), Yamada (1983), Sievers and Swings (2005), Kersters et al. (2006), Yamada et al. (2012a, b), Komagata et al. (2014).

The type strain is NCIMB 11664^T (= DSM 6513^T = JCM 7644^T = LMG 1515^T = NBRC 15237^T = BCC 49175^T), isolated from a mountain ash berry by G. Bertrand. The DNA G+C content of the type strain is 62.5 mol per cent.

14.2. *Komagataeibacter hansenii* (Gosselé et al. 1983) Yamada, Yukphan, Vu, Muramatsu, Ochaikul, Tanasupawat and Nakagawa 2013

Basonym: *Acetobacter hansenii* Gosselé, Swings, Kersters, Pauwels and De Ley 1983.

Synonyms: *Gluconacetobacter hansenii* (Gosselé et al. 1983) Yamada, Hoshino and Ishikawa 1998.

For the characteristics of the species, refer to Gosselé et al. (1983b), Lisdiyanti et al. (2006), Dutta and Gachhui (2007), Cleenwerck et al. (2009), Komagata et al. (2014).

The type strain is NCIMB 8746^T (= DSM 5602^T = JCM 7643^T = LMG 1527^T = NBRC 14820^T = BCC 6318^T), isolated from a local vinegar in Jerusalem, Israel. The DNA G+C content of the type strain is 59.0 mol per cent.

14.3. *Komagataeibacter europaeus* (Sievers et al. 1992) Yamada, Yukphan, Vu, Muramatsu, Ochaikul, Tanasupawat and Nakagawa 2013

Basonym: *Acetobacter europaeus* Sievers, Sellmer and Teuber 1992.

Synonym: *Gluconacetobacter europaeus* (Sievers et al. 1992) Yamada, Hoshino and Ishikawa 1998.

For the characteristics of the species, refer to Sievers et al. (1992).

The type strain is DES11^T (= DSM 6160^T = JCM 16935^T = BCC 36446^T), isolated from a submerged culture vinegar generator at a factory in Esslingen in the southern part of Germany. The DNA G+C content of the type strain is not described. The range of DNA G+C content ranges from 56.2 to 57.3 mol per cent.

14.4. *Komagataeibacter oboediens* (Sokollek et al. 1998) Yamada, Yukphan, Vu, Muramatsu, Ochaikul, Tanasupawat and Nakagawa 2013

Basonym: *Acetobacter oboediens* Sokollek, Hertel and Hammes 1998.

Synonym: *Gluconacetobacter oboediens* (Sokollek et al. 1998) Yamada 2000.

For the characteristics of the species, refer to Sokollek et al. (1998), Yamada (2000).

The type strain is LTH 2460^T (= DSM 11826^T = JCM 16937^T = LMG 18849^T = BCC 36445^T), isolated from a submerged red wine vinegar fermentation at a factory in the southern part of Germany. The DNA G+C content of the type strain is 59.9 mol per cent.

14.5. *Komagataeibacter intermedius* (Boesch et al. 1998) Yamada, Yukphan, Vu, Muramatsu, Ochaikul, Tanasupawat and Nakagawa 2013

Basonym: *Acetobacter intermedius* Boesch, Trček, Sievers and Teuber 1998.

Synonym: *Gluconacetobacter intermedius* (Boesch et al. 1998) Yamada 2000.

For the characteristics of the species, refer to Boesch et al. (1998), Yamada (2000).

The type strain is TF2^T (= DSM 11804^T = JCM 16936^T = BCC 36447^T = LMG 18909^T), isolated from a commercially available tea fungus beverage (Kombucha) in Switzerland. The DNA G+C content of the type strain is 61.55 mol per cent.

14.6. *Gluconacetobacter entanii* Schüller, Hertel and Hammes 2000

For the characteristics of the species, refer to Schüller et al. (2000).

The type strain is LTH 4560^T (= BCRC 17196^T = DSM 13536^T = LMG 20950^T = LMG 21788^T), isolated from submerged high-acid industrial vinegar fermentations. The DNA G+C content of the type strain is 58 mol per cent.

The type strain is not available in any culture collections (Yamada et al. 2012b). This species is not listed as a new combination, according to Rule 27 of the Bacteriological Code (Tindall et al. 2006).

14.7. *Komagataeibacter swingsii* (Dellaglio et al. 2005) Yamada, Yukphan, Vu, Muramatsu, Ochaikul, Tanasupawat and Nakagawa 2013

Basonym: *Gluconacetobacter swingsii* Dellaglio, Cleenwerck, Felis, Engelbeen, Janssens and Marzotto 2005.

For the characteristics of the species, refer to Dellaglio et al. (2005).

The type strain is DST GL01^T (= DSM 16373^T = JCM 17123^T = LMG 22125^T = BCC 36451^T), isolated from apple juice in South Tyrol region, Italy. The DNA G+C content of the type strain is 61.7 mol per cent.

14.8. *Komagataeibacter rhaeticus* (Dellaglio et al. 2005) Yamada, Yukphan, Vu, Muramatsu, Ochaikul, Tanasupawat and Nakagawa 2013

Basonym: *Gluconacetobacter rhaeticus* Dellaglio, Cleenwerck, Felis, Engelbeen, Janssens and Marzotto 2005.

For the characteristics of the species, refer to Dellaglio et al. (2005).

The type strain is DST GL02^T (= DSM 16663^T = JCM 17122^T = LMG 22126^T = BCC 36452^T), isolated from apple juice in South Tyrol region, Italy. The DNA G+C content of the type strain is 63.4 mol per cent.

14.9. *Komagataeibacter saccharivorans* (Lisdiyanti et al. 2006) Yamada, Yukphan, Vu, Muramatsu, Ochaikul, Tanasupawat and Nakagawa 2013

Basonym: *Gluconacetobacter saccharivorans* Lisdiyanti, Navarro, Uchimura and Komagata 2006.

For the characteristics of the species, refer to Lisdiyanti et al. (2006).

The type strain is LMG 1582^T (= JCM 25121^T = NRIC 0614^T = BCC 36444^T), isolated from beet juice in Germany in 1927. The DNA G+C content of the type strain is 61 mol per cent.

14.10. *Komagataeibacter nataicola* (Lisdiyanti et al. 2006) Yamada, Yukphan, Vu, Muramatsu, Ochaikul, Tanasupawat and Nakagawa 2013

Basonym: *Gluconacetobacter nataicola* Lisdiyanti, Navarro, Uchimura and Komagata 2006.

For the characteristics of the species, refer to Lisdiyanti et al. (2006).

The type strain is LMG 1536^T (= JCM 25120^T = NRIC0616^T = BCC 36443^T), isolated from nata de coco in the Philippines. The DNA G+C content of the type strain is 62 mol per cent.

14.11. *Komagataeibacter kombuchae* (Dutta and Gachhui 2007) Yamada, Yukphan, Vu, Muramatsu, Ochaikul, Tanasupawat and Nakagawa 2013

Basonym: *Gluconacetobacter kombuchae* Dutta and Gachhui 2007.

For the characteristics of the species, refer to Dutta and Gachhui (2007), Cleenwerck et al. (2009), Komagata et al. (2014).

The type strain is RG-3^T (= LMG 23726^T = MTCC 6913^T), isolated from Kombucha tea. The DNA G+C content of the type strain is 55.8 mol per cent.

According to Cleenwerck et al. (2009), the species is a later heterotypic synonym of *Gluconacetobacter hansenii* (= *Komagataeibacter hansenii*).

14.12. *Komagataeibacter sucrofermentans* (Toyosaki et al. 1996) Yamada, Yukphan, Vu, Muramatsu, Ochaikul, Tanasupawat and Nakagawa 2013

Basonym: *Acetobacter xylinus* corrig. subsp. *sucrofermentans* Toyosaki, Kojima, Tsuchida, Hoshino, Yamada and Yoshinaga 1996.

Synonym: *Gluconacetobacter sucrofermentans* (Toyosaki et al. 1996) Cleenwerck, De Vos and De Vuyst 2010.

For the characteristics of the species, refer to Toyosaki et al. (1995), Cleenwerck et al. (2010).

The type strain is LMG 18788^T (= DSM 15973^T = JCM 9730^T = BCC 7227^T), isolated from a cherry. The DNA G+C content of the type strain is 62.7 mol per cent.

14.13. *Komagataeibacter kakiaceti* (Iino et al. 2012) Yamada 2014

Basonym: *Gluconacetobacter kakiaceti* Iino, Suzuki, Tanaka, Kosako, Ohkuma, Komagata and Uchimura 2012.

For the characteristics of the species, refer to Iino et al. (2012b), Yamada (2014).

The type strain is G5-1^T (= JCM 25156^T = NRIC 0798^T = LMG 26206^T), isolated from kaki vinegar collected in Kumamoto Prefecture, Japan in 2005. The DNA G+C content of the type strain is 63.6 mol per cent.

14.14. *Komagataeibacter medellinensis* (Castro et al. 2013) Yamada 2014

Basonym: *Gluconacetobacter medellinensis* Castro, Cleenwerck, Trček, Zuluaga, De Vos, Caro, Aguirre, Putaux and Gañán 2013.

For the characteristics of the species, refer to Castro et al. (2013), Yamada (2014).

The type strain is LMG 1693^T (= NBRC 3288^T = IFO 3288^T = Kondo 51^T), which was originally isolated from vinegar by K. Kondo, Japan as a cellulose-forming strain. The DNA G+C content of the type strain is 60.7 mol per cent.

The culture of strain IFO 3288 produced two types of colonies, i.e., the cellulose-forming and the cellulose-less (Yamada et al. 1969b, 1976b). The cellulose production of the strain was also found (Tanaka et al. 1998). The type strain designated as LMG 1693^T was therefore selected as the cellulose-less.

14.15. *Komagataeibacter maltaceti* (Slapšak et al. 2013) Yamada 2014

Basonym: *Gluconacetobacter maltaceti* Slapšak, Cleenwerck, De Vos and Trček 2013.

For the characteristics of the species, refer to Slapšak et al. (2013), Yamada (2014).

The type strain is LMG 1529^T (= NBRC 14815^T = NCIMB 8752^T), isolated from malt vinegar brewery acetifier by T. K. Walker in 1956. The DNA G+C content of the type strain is 62.5 mol per cent.

15. *Endobacter* Ramírez-Bahena, Teijedor, Martín, Velázques and Peix 2013

En.do.bac'ter. Gr. pref. *endo*, within; N. L. masc. n. *bacter*, rod; N. L. masc. n. *Endobacter*, a rod isolated from the inside of a root nodule of alfalfa.

The strain of the genus *Endobacter* was isolated from a surface sterilized nodule of alfalfa in Spain and quite remote phylogenetically from other acetic acid bacteria and constituted an independent cluster in a phylogenetic tree based on 16S rRNA gene sequences.

Cells are Gram-negative, coccoid to rod-shaped. Motile with subpolar flagella. Colonies are white and mucoid on the modified yeast extract mannitol agar.

Aerobic. Catalase-positive and oxidase negative. Acetate and lactate are not oxidized. Acetic acid is produced from ethanol. Grows on glutamate agar and mannitol agar. Dihydroxyacetone is produced from glycerol. Acid is produced from xylose, glucose, glycerol, or ethanol. Grows between 20 °C and 37 °C with an optimum temperature for growth of 28 °C. Ammoniac nitrogen is assimilated on glucose. Although optimal pH ranges from 5.0 to 7.0, the growth is observed at pH 3.5.

The major cellular fatty acids are $C_{18:1}\omega7c$ (39.94 per cent), $C_{19:0}$cyclo$\omega8c$ (12.15 per cent), and $C_{16:0}$ (13.40 per cent). The major quinone is UQ-10. The DNA G+C content is 60.3 mol per cent. For more details of characteristics, see Komagata et al. (2014).

15.1. *Endobacter medicaginis* Ramírez-Bahena, Teijedor, Martín, Valázques and Peix 2013

For the characteristics of the species, refer to Ramírez-Bahena et al. (2013).

The type strain is M1MS02^T (= CECT 8088^T = LMG 26838^T), isolated from a surface-sterilized nodule of alfalfa (*Medicago sativa*) in Spain. The DNA G+C content of the type strain is 60.3 mol per cent.

16. *Nguyenibacter* Vu, Yukphan, Chaipitakchonlatarn, Malimas, Muramatsu, Bui, Tanasupawat, Duong, Nakagawa, Pham and Yamada 2013

Ngu.ye.ni.bac'ter. N. L. masc. n. *Nguyenius*, Nguyen, the name of a famous Vietnamese microbiologist; N. L. masc. n. bacter, rod; N. L. masc. n. *Nguyenibacter*, a rod, which is named after Professor Dung Lan Nguyen, Vietnam, who contributed to the study of microorganisms, especially of strains isolated in Vietnam.

The two strains of the genus *Nguyenibacter* were isolated by the use of the nitrogen-free LGI medium. The strains were related phylogenetically to those of the genera *Gluconacetobacter* and *Acidomonas*.

Cells are Gram-negative rods, measuring 0.6 to 0.8 by 1.0 to 1.6 µm. Motile with peritrichous flagella. Colonies are smooth, entire, transparent and creamy to brownish.

Aerobic. Catalase positive and oxidase negative. Acetic acid is not produced from ethanol. Acetate is oxidized to carbon dioxide and water, but lactate is not oxidized. Grows on glutamate agar and mannitol agar. Dihydroxyacetone is not produced from glycerol. 2-Ketogluconate and 2,5-diketogluconate are produced from glucose. A water-soluble brown pigment is produced. Acid is produced from L-arabinose, xylose, glucose, galactose, fructose (weakly), maltose, melibiose, sucrose, or raffinose (weakly). Grows on glucose, galactose, L-arabinose (weakly), xylose (weakly), fructose (weakly), L-sorbose (weakly), maltose, melibiose (weakly), sucrose, raffinose, mannitol (weakly), sorbitol (weakly), or glycerol. Ammoniac nitrogen is utilized on mannitol, but not on glucose or ethanol. Growth occurs on N_2-free medium. γ-Pyrone compound is weakly produced. Levan-like polysaccharides are produced from sucrose.

Grows weakly on 30 per cent glucose (w/v) and weakly in the presence of 0.35 per cent acetic acid (v/v). Growth does not occur on 1.0 per cent KNO_3 (w/v). The major cellular fatty acid is $C_{18:1}\omega7c$. The major quinone is UQ-10. The DNA G+C content ranges from 68.1 to 69.4 mol per cent. For more details of characteristics, see Komagata et al. (2014).

16.1. *Nguyenibacter vanlangensis* Vu, Yukphan, Chaipitakchonlatarn, Malimas, Muramatsu, Bui, Tanasupawat, Duong, Nakagawa, Pham and Yamada 2013

For the characteristics of the species, refer to Vu et al. (2013).

The type strain is TN01LGI^T (= BCC 54774^T = NBRC 109046^T = VTCC-B-1198^T), isolated from the rhizosphere of Asian rice collected in Vietnam. The DNA G+C content of the type strain is 69.4 mol per cent.

17. *Swingsia* Malimas, Chaipitakchonlatarn, Vu, Yukphan, Muramatsu, Tanasupawat, Potacharoen, Nakagawa, Tanticharoen and Yamada 2014

Swing'si.a. N. L. fem. n. *Swingsia*, Swings, named after Professor Jean Swings, Belgium, who contributed to the systematics of bacteria, especially of acetic acid bacteria.

The two strains of the genus *Swingsia* were isolated from the flowers in Thailand and located at an intermediary position phylogenetically between the genera *Gluconobacter* and *Neokomagataea*. The strains grew on 30 per cent glucose (w/v) and at 37 °C, but not in the presence of 0.35 per cent acetic acid (v/v), and acetic acid was produced sometimes weakly from ethanol.

Cells are Gram-negative rods, measuring 0.6 to 0.8 by 1.0 to 1.8 μm. Non motile. Colonies are brownish and smooth with entire margin.

Aerobic. Catalase positive and oxidase negative. Acetic acid is produced weakly from ethanol. Acetate and lactate are not oxidized. Grows on glutamate agar and mannitol agar. Dihydroxyacetone is produced from glycerol. 2-Ketogluconate, 5-ketogluconate, and 2,5-diketogluconate are produced from glucose. A water-soluble pigment is produced. Acid is produced from L-arabinose (weakly), arabinose (weakly), xylose (weakly), L-rhamnose (weakly), glucose, mannose (weakly), galactose, fructose (weakly), arabitol (weakly), mannitol, maltose (weakly), lactose (weakly), melibiose, sucrose, or raffinose (weakly). Grows on glucose, fructose, sucrose, L-arabitol (weakly), arabitol (weakly), or mannitol. Ammoniac nitrogen is utilized on mannitol, but not on glucose or ethanol. Levan-like polysaccharides are not produced.

Grows on 30 per cent glucose (w/v), but not in the presence of 0.35 per cent acetic acid (v/v). Growth occurs in the presence of 1.0 per cent KNO_3 (w/v). The major cellular fatty acid is $C_{18:1}\omega7c$. The major quinone is UQ-10. The DNA G+C content ranges from 46.9 to 47.3 mol per cent.

17.1. *Swingsia samuiensis* Malimas, Chaipitakchonlatarn, Vu, Yukphan, Muramatsu, Tanasupawat, Potacharoen, Nakagawa, Tanticharoen and Yamada 2014

For the characteristics of the species, refer to Malimas et al. (2013).

The type strain is AH83^T (= BCC 25779^T = NBRC 107927^T), isolated from a flower of golden trumpet. The DNA G+C content of the type strain is 46.9 mol per cent.

18. *Bombella* Li, Praet, Borremans, Nunes, Manaia, Cleenwerck, Meeus, Smagghe, De Vuyst and Vandamme 2015

Bom.bel'la. N. L. fem. dim. n. *Bombella*, named after the bumblebee genus, *Bombus*.

A strain was isolated from the crop of a bumble bee (*Bombus lapidarius*) in Belgium. The strain, which was related phylogenetically to the genera *Saccharibacter*, *Swingsia*, *Neokomagataea*, and *Gluconobacter*, did not produce acetic acid from ethanol and not oxidize acetate and lactate (Li et al. 2015).

Cells are Gram-negative, straight rods, measuring 1 by 2-3 μm. Non motile. Colonies are round, smooth, brownish, and slightly raised. Catalase positive and oxidase negative. Ethanol is not oxidized to acetic acid. Acetate and lactate are not oxidized. 2-Ketogluconate is produced from glucose, but 5-ketogluconate is not. A water-soluble brown pigment is not produced. Acid is produced from L-arabinose (weakly), glucose, mannose (weakly), galactose, fructose, mannitol, or sucrose. Ammoniac nitrogen is not assimilated.

Grows on 30 per cent glucose (w/v) and at 37 °C. The predominant fatty acid is $C_{18:1}\omega7c$; other fatty acids in significant amounts are $C_{19:0}$ cyclo$\omega8c$, $C_{16:0}$, $C_{14:0}2OH$, $C_{14:0}$, $C_{16:0}2OH$, and $C_{16:0}3OH$. The major quinone is UQ-10. The DNA G+C content is 54.9 mol per cent.

18.1. *Bombella intestini* Li, Praet, Borremans, Nunes, Manaia, Cleenwerck, Meeus, Smagghe, De Vuyst and Vandamme 2015

For the characteristics of the species, refer to Li et al. (2015).

The type strain is LMG 28161^T (= DSM 28636^T = R-42587^T), isolated from the crop of a bumble bee (*Bombus lapidarius*). The DNA G+C content of the type strain is 54.9 mol per cent.

Genus and Species in Pseudacetic Acid Bacteria

Several strains were once isolated and named as '*Acetobacter aurantium*' by Kondo and Ameyama (1958). According to the description of the species, the strains were not able to oxidize acetate.

Asai et al. (1964) reinvestigated the strains for phenotypic characteristics and found that they had polar flagellation and oxidized acetate and lactate to carbon dioxide and water, and the strains were named the polarly flagellated intermediate strains. Additional three strains were then newly isolated and confirmed to have polar flagellation and the capability of oxidizing 'acetate' and lactate (Ameyama and Kondo 1967).

In the isoprenoid quinone analysis of the polarly flagellated intermediate strains, UQ-8 was detected as the major quinone, indicating that the quinone system obtained was quite different chemotaxonomically from either UQ-9 of *Acetobacter* strains or UQ-10 of *Gluconobacter* strains (Yamada et al. 1969a, 1976a). In the cellular fatty acid composition of the polarly flagellated intermediate strains, *iso*-$C_{15:0}$ acid was found as the major acid, indicating that the strains were quite different likewise from $C_{18:1}\omega7c$ acid of *Acetobacter* and *Gluconobacter* strains (Yamada et al. 1981a).

For such unique bacterial strains equipped with UQ-8 and *iso*-$C_{15:0}$ acid, the name of pseudacetic acid bacteria was given (Yamada 1979, Yamada et al. 1981a, b, Lisdiyanti et al. 2003a). The genus *Frateuria* was later introduced for the strains by Swings et al. (1980). The genus is accommodated in the class *Gammaproteobacteria* Stackebrandt et al. 1988.

1. ***Frateuria* Swings, Gillis, Kersters, De Vos, Gosselé and De Ley 1980 emend. Zhang, Liu and Liu 2011**

Fra.teu'ri.a. N. L. fem. n. *Frateuria*, Frateur, named after Professor Joseph Frateur, Belgium, especially in recognition of the study of acetic acid bacteria.

The genus *Frateuria* was monotypic for a long time. However, the second species was recently reported (Zhang et al. 2011).

Cells are Gram-negative, rod-shaped, measuring 0.4 to 0.8 by 0.8 to 2.0 μm, singly or in pairs, motile with a single polar or subpolar flagellum when motile. Shows luxuriant growth on glucose/yeast extract/calcium carbonate agar. Colonies are flat or circular, and medium turns to brown.

Aerobic. Catalase positive or negative. Oxidase negative or positive. Oxidizes lactate, but not acetate (Swings et al. 1980, Lisdiyanti et al. 2003a). Grows on glutamate agar and mannitol agar. Dihydroxyacetone is generally produced from glycerol. Produces gluconate, 2-ketogluconate and 2,5-diketogluconate from glucose but not 5-ketogluconate. Produces a water-soluble brown pigment. Acid is produced from arabinose, L-arabinose, ribose, xylose, L-rhamnose, galactose, glucose, mannose, fructose, glycerol, or ethanol. Does not grow on methanol. Assimilates ammoniac nitrogen on mannitol.

No growth is observed in the presence of 0.35 per cent acetic acid (v/v). Grows on 20 per cent glucose (w/v) at 34 °C and at pH 3.5. The major cellular fatty acid is *iso*-$C_{15:0}$. The major quinone is UQ-8. The DNA G+C content ranges from 62 to 68 mol per cent. *Frateuria* strains are isolated from flowers of lily, rose, ladybell, and coconut and fruits of raspberry, mango, rambai, and jackfruit. For more details of characteristics, see Komagata et al. (2014), Swings and Sievers (2005).

The type species is *Frateuria aurantia* (ex Kondo and Ameyama 1958) Swings et al. 1980. Two species are reported.

1.1. *Frateuria aurantia* (ex Kondo and Ameyama 1958) Swings, Gillis, Kersters, De Vos, Gosselé and De Ley 1980

Synonym: '*Acetobacter aurantius*' corrig. Kondo and Ameyama 1958.

For the characteristics of the species, refer to Swings et al. (1980), Yamada et al. (1969a, 1976a, 1981a, b), Lisdiyanti et al. (2003a, b).

The type strain is G-6^T (= Kondo 67^T = NBRC 3245^T = ATCC 33424^T = DSM 6220^T = LMG 1558^T), isolated from a flower of lily. The DNA G+C content of the type strain is 65.0 mol per cent.

1.2. *Frateuria terrea* Zhang, Liu and Liu 2011

For the characteristics of the species, refer to Zhang et al. (2011).

The type strain is VA24^T (= CGMCC 1.7053^T = NBRC 104236^T), isolated from forest soil of the Changbai Mountains, Heilongjiang province, China. The DNA G+C content of the type stran is 67.4 mol per cent.

Conclusion

More than one hundred years have already passed, since the genus *Acetobacter* Beijerinck 1898 was first introduced with the only species, *Acetobacter aceti* (Pasteur 1864) Beijerinck 1898 for vinegar-producing acetic acid bacteria. Up to 1960, acetic acid bacteria were believed to constitute a quite small taxonomic group, i.e., the only one genus. However, their circumstances have entirely changed. The range of DNA G+C contents is in fact almost 20 mol per cent from 46.9 to 66.8 mol per cent in the acetous group (Table 1).

The acetic acid bacteria, including the vinegar-producing and their relatives have been found in large numbers by expanding their living environments to be looked for, viz., activated sludge, rhizosphere soils, soils, pollen, patients, mosquitoes, stone chambers of tumuluses, nodules of plants, and insect guts, in addition to sugary and alcoholic materials.

Up to the present, the acetic acid bacteria have numbered eighteen genera and eighty-eight species. These numbers will be greatly increased in the future, and many of new taxa, i.e., new genera and new species will be reported.

References

Adachi, O. and Yakushi, T. (2016). Membrane-bound dehydrogenases of acetic acid bacteria. *In*: M. Matsushita, H. Toyama, N. Tonouchi and A. Okamoto-Kainuma (eds.), Acetic Acid Bacteria: Ecology and Physiology. Springer, Japan, Tokyo, pp. 273-297.

Ameyama, M. and Kondo, K. (1967). Carbohydrate metabolism by acetic acid bacteria. Part VI. Characteristics of the intermediate type strains. Agricultural and Biological Chemistry 31: 724-737.

Ameyama, M. (1975). *Gluconobacter oxydans* subsp. *sphaericus*, new subspecies isolated from grapes. International Journal of Systematic Bacteriology 25: 365-370.

Arai, H., Sakurai, K. and Ishii, M. (2016). Metabolic features of *Acetobacter aceti*. *In*: M. Matsushita, H. Toyama, N. Tonouchi and A. Okamoto-Kainuma (eds.), Acetic Acid Bacteria: Ecology and Physiology. Springer Japan, Tokyo, pp. 255-271.

Asai, T. (1935). Taxonomic study of acetic acid bacteria and allied oxidative bacteria in fruits and a new classification of oxidative bacteria. Nippon Nogeikagaku Kaishi 11: 674-708 (in Japanese).

Asai, T. and Shoda, K. (1958). The taxonomy of *Acetobacter* and allied oxidative bacteria. Journal of General and Applied Microbiology 4: 289-311.

Asai, T., Iizuka, H. and Komagata, K. (1964). The flagellation and taxonomy of genera *Gluconobacter* and *Acetobacter* with reference to the existence of intermediate strains. Journal of General and Applied Microbiology 10: 95-126.

Asai, T. (1968). Acetic Acid Bacteria: Classification and Biochemical Activities. University of Tokyo Press, Tokyo.

Beijerinck, M.W. (1898). Über die Arten der Essigbakterien. Zentralblatt für Bakteriologie, Parasitenkunde, Infektionskrankheiten und Hygiene Abteilung II 4: 209-216.

Beijerinck, M.W. and Folpmers, T. (1916). Formation of pyruvic acid from malic acid by microbes. Verslag van de Gewone Vergaderingen der Wis- en Natuurkundige Afdeeling der Koninklijke Akademie van Wetenschappen te Amsterdam 18: 1198-1200.

Boesch, C., Trček, J., Sievers, M. and Teuber, M. (1998). *Acetobacter intermedius* sp. nov. Systematic and Applied Microbiology 21: 220-229.

Bringer, S. and Bott, M. (2016). Central carbon metabolism and respiration in *Gluconobacter oxydans*. *In*: M. Matsushita, H. Toyama, N. Tonouchi and A. Okamoto-Kainuma (eds.), Acetic Acid Bacteria: Ecology and Physiology. Springer Japan, Tokyo, pp. 235-253.

Brosius, J., Dull, T.J., Sleeter, D.D. and Noller, H.F. (1981). Gene organization and primary structure of a ribosomal RNA operon from *Escherichia coli*. Journal of Molecular Biology 148: 107-127.

Bulygina, E.S., Gulikova, O.M., Dikanskaya, E.M., Netrusov, A.I., Tourova, T.P. and Chumakov, K.M. (1992). Taxonomic studies of the genera *Acidomonas*, *Acetobacter* and *Gluconobacter* by 5S ribosomal RNA sequencing. Journal of General Microbiology 138: 2283-2286.

Carr, J.G. and Shimwell, J.L. (1960). Pigment-producing strains of *Acetobacter aceti*. Nature 186: 331-332.

Castro, C., Cleenwerck, I., Trček, J., Zuluaga, R., De Vos, P., Caro, G., Aguirre, R., Putaux, J.-L. and Gañán, P. (2013). *Gluconacetobacter medellinensis* sp. nov., cellulose- and non-cellulose-producing acetic acid bacteria isolated from vinegar. International Journal of Systematic and Evolutionary Microbiology 63: 1119-1125.

Cavalcante, V.A. and Döbereiner, J. (1988). A new acid-tolerant nitrogen-fixing bacterium associated with sugar cane. Plant and Soil 108: 23-31.

Cheldelin, V.H. (1961). Metabolic Pathways in Microorganisms. John Wiley & Sons, Inc., New York.

Cleenwerck, I., Vandemeulebroecke, K., Janssens, D. and Swings, J. (2002). Re-examination of the genus *Acetobacter*, with descriptions of *Acetobacter cerevisiae* sp. nov. and *Acetobacter malorum* sp. nov. International Journal of Systematic and Evolutionary Microbiology 52: 1551-1558.

Cleenwerck, I., Camu, N., Engelbeen, K., De Winter, T., Vandemeulebroecke, K., De Vos, P. and De Vuyst, L. (2007). *Acetobacter ghanensis* sp. nov., a novel acetic acid bacterium isolated from traditional heap fermentations of Ghanaian cocoa beans. International Journal of Systematic and Evolutionary Microbiology 57: 1647-1652.

Cleenwerck, I. and De Vos, P. (2008). Polyphasic taxonomy of acetic acid bacteria: An overview of the currently applied methodology. International Journal of Food Microbiology 125: 2-14.

Cleenwerck, I., González, Á., Camu, N., Engelbeen, K., De Vos, P. and De Vuyst, L. (2008). *Acetobacter fabarum* sp. nov., an acetic acid bacterium from a Ghanaian cocoa bean heap fermentation. International Journal of Systematic and Evolutionary Microbiology 58: 2180-2185.

Cleenwerck, I., De Wachter, M., González, Á., De Vuyst, L. and De Vos, P. (2009). Differentiation of species of the family *Acetobacteraceae* by AFLP DNA fingerprinting: *Gluconacetobacter kombuchae* is a later heterotypic synonym of *Gluconacetobacter hansenii*. International Journal of Systematic and Evolutionary Microbiology 59: 1771-1786

Cleenwerck, I., De Vos, P. and De Vuyst, L. (2010). Phylogeny and differentiation of species of the genus *Gluconacetobacter* and related taxa based on multilocus sequence analyses of housekeeping genes and reclassification of *Acetobacter xylinus* subsp. *sucrofermentans* as *Gluconacetobacter sucrofermentans* (Toyosaki et al. 1996) sp. nov., comb. nov. International Journal of Systematic and Evolutionary Microbiology 60: 2277-2283.

Dellaglio, F., Cleenwerck, I., Felis, G.E., Engelbeen, K., Janssens, D. and Marzotto, M. (2005). Description of *Gluconacetobacter swingsii* sp. nov. and *Gluconacetobacter rhaeticus* sp. nov., isolated from Italian apple fruit. International Journal of Systematic and Evolutionary Microbiology 55: 2365-2370.

De Ley, J. (1961). Comparative carbohydrate metabolism and a proposal for a phylogenetic relationship of the acetic acid bacteria. Journal of General Microbiology 24: 31-50.

De Ley, J. and Frateur, J. (1970). The status of the generic name *Gluconobacter*. International Journal of Systematic Bacteriology 20: 83-95.

Dutta, D. and Gachhui, R. (2006). Novel nitrogen-fixing *Acetobacter nitrogenifigens* sp. nov., isolated from Kombucha tea. Internationl Journal of Systematic and Evolutionary Microbiology 56: 1899-1903.

Dutta, D. and Gachhui, R. (2007). Nitrogen-fixing and cellulose-producing *Gluconacetobacter kombuchae* sp. nov., isolated from Kombucha tea. Internationl Journal of Systematic and Evolutionary Microbiology 57: 353-357.

Franke, I.H., Fegan, M., Hayward, C., Leonard, G., Stackebrandt, E. and Sly, L.I. (1999). Description of *Gluconacetobacter sacchari* sp. nov., a new species of acetic acid bacterium isolated from the leaf sheath of sugar cane and from the pink sugar cane mealy bug. International Journal of Systematic Bacteriology 49: 1681-1693.

Fuentes-Ramírez, L.E., Bustillos-Cristales, R., Tapía-Hernández, A., Jiménez-Salgado, T., Wang, E.T., Martínez-Romero, E. and Caballero-Mellado, J. (2001). Novel nitrogen-fixing acetic acid bacteria, *Gluconacetobacter johannae* sp. nov. and *Gluconacetobacter azotocaptans* sp. nov., associated with coffee plants. Internationl Journal of Systematic and Evolutionary Microbiology 51: 1305-1314.

Gillis, M. and De Ley, J. (1980). Intra- and intergeneric similarities of the ribosomal ribonucleic acid cistrons of *Acetobacter* and *Gluconobacter*. International Journal of Systematic Bacteriology 30: 7-27

Gillis, M., Kersters, K., Hoste, B., Janssens, D., Kroppenstedt, R.M., Stephan, M.P., Teixeira, K.R.S., Döbereiner, J. and De Ley, J. (1989). *Acetobacter diazotrophicus* sp. nov., a nitrogen-fixing acetic acid bacterium associated with sugarcane. International Journal of Systematic Bacteriology 39: 361-364.

Gosselé, F., Swings, J., Kersters, K. and De Ley, J. (1983a). Numerical analysis of phenotypic features and protein gel electropherograms of *Gluconobacter* Asai 1935 emend. mut. char. Asai, Iizuka, and Komagata 1964. International Journal of Systematic Bacteriology 33: 65-81.

Gosselé, F., Swings, J., Kersters, K., Pauwels, P. and De Ley, J. (1983b). Numerical analysis of phenotypic features and protein gel electrophoregrams of a wide variety of *Acetobacter* strains. Proposal for the improvement of the taxonomy of the genus *Acetobacter* Beijerinck 1898 215. Systematic and Applied Microbiology 4: 338-368.

Greenberg, D.E., Porcella, S.F., Stock, F., Wong, A., Conville, P.S., Murray, P.R., Holland, S.M. and Zelazny, A.M. (2006). *Granulibacter bethesdensis* gen. nov., sp. nov., a distinctive pathogenic acetic acid bacterium in the family Acetobacteraceae. Internationl Journal of Systematic and Evolutionary Microbiology 56: 2609-2616.

Iino, T., Suzuki, R., Kosako, Y., Ohkuma, M., Komagata, K. and Uchimura, T. (2012a). *Acetobacter okinawensis* sp. nov., *Acetobacter papayae* sp. nov., and *Acetobacter persicus* sp. nov.: Novel acetic acid bacteria isolated from stems of sugarcane, fruits and a flower in Japan. Journal of General and Applied Microbiology 58: 235-243.

Iino, T., Suzuki, R., Tanaka, N., Kosako, Y., Ohkuma, M., Komagata, K. and Uchimura, T. (2012b). *Gluconacetobacter kakiaceti* sp. nov., an acetic acid bacterium isolated from a traditional Japanese fruit vinegar. Internationl Journal of Systematic and Evolutionary Microbiology 62: 1465-1469.

Jakob, F., Phaff, A., Novoa-Carballal, R., Rübsam, H., Becker, T. and Vogel, R.F. (2013). Structural analysis of fructans produced by acetic acid bacteria reveals a relation to hydrocolloidal function. Carbohydrate Polymers 92: 1234-1242.

Jojima, Y., Mihara, Y., Suzuki, S., Yokozeki, K., Yamanaka, S. and Fudou, R (2004). *Saccharibacter floricola* gen. nov., sp. nov., a novel osmophilic acetic acid bacterium isolated from pollen. International Journal of Systematic and Evolutionary Microbiology 54: 2263-2267.

Katsura, K., Kawasaki, H., Potacharoen, W., Saono, S., Seki, T., Yamada, Y., Uchimura, T. and Komagata, K. (2001). *Asaia siamensis* sp. nov., an acetic acid bacterium in the α-Proteobacteria. International Journal of Systematic and Evolutionary Microbiology 51: 559-563.

Katsura, K., Yamada, Y., Uchimura, T. and Komagata, K. (2002). *Gluconobacter asaii* Mason and Claus 1989 is a junior subjective synonym of *Gluconobacter cerinus* Yamada and Akita 1984. Internationl Journal of Systematic and Evolutionary Microbiology 52: 1635-1640.

Kersters, K., Lisdiyanti, P., Komagata, K. and Swings, J. (2006). The family Acetobacteraceae: The genera *Acetobacter, Acidomonas, Asaia, Gluconacetobacter, Gluconobacter* and *Kozakia*. *In*: M. Dworkin, S. Falkow, E. Rosenberg, K.-H. Schleifer and E. Stackebrandt (eds.), The Prokaryotes, 3rd edn, vol. 5. Springer, New York, pp. 163-200.

Komagata, K., Iino, T. and Yamada, Y. (2014). 1. The family Acetobacteraceae. *In*: E. Rosenberg, E.F. De Long, S. Lory, E. Stackebrandt and F. Thompson (eds.), The Prokaryotes. Alphaproteobacteria and Betaproteobacteria. Springer-Verlag, Berlin Heidelberg, pp. 3-78.

Kommanee, J., Tanasupawat, S., Yukphan, P., Malimas, T., Muramatsu, Y., Nakagawa, Y. and Yamada, Y. (2010). *Asaia spathodeae* sp. nov., an acetic acid bacterium in the α-Proteobacteria. Journal of General and Applied Microbiology 56: 81-87.

Kommanee, J., Tanasupawat, S., Yukphan, P., Malimas, T., Muramatsu, Y., Nakagawa, Y. and Yamada, Y. (2011). *Gluconobacter nephelii* sp. nov., an acetic acid bacterium in the class Alphaproteobacteria. Internationl Journal of Systematic and Evolutionary Microbiology 61: 2117-2122.

Kondo, K. and Ameyama, M. (1958). Carbohydrate metabolism by *Acetobacter* species. Part 1. Oxidative activity for various carbohydrates. Bulletin of Agricultural Chemical Society of Japan 22: 369-372.

Leifson, E. (1954). The flagellation and taxonomy of species of *Acetobacter*. Antonie van Leeuwenhoek 20: 102-110.

Li, L., Wieme, A., Spitaels, F., Balzarini, T., Nunes, O.C., Manaia, C.M., Van Landschoot, A., De Vuyst, L., Cleenwerck, I. and Vandamme, P. (2014). *Acetobacter sicerae* sp. nov., isolated from cider and kefir, and identification of species of the genus *Acetobacter* by *dna*K, *gro*EL and *rpo*B sequence analysis. International Journal of Systematic and Evolutionary Microbiology 64: 2407-2415.

Li, L., Praet, J., Borremans, W., Nunes, O.C., Manaia, C.M., Cleenwerck, I., Meeus, I., Smagghe, G., De Vuyst, L. and Vandamme, P. (2015). *Bombella intestini* gen. nov., sp. nov., an acetic acid bacterium isolated from bumble bee crop. International Journal of Systematic and Evolutionary Microbiology 65: 267-273.

Lisdiyanti, P., Kawasaki, H., Seki, T., Yamada, Y., Uchimura, T. and Komagata, K. (2000). Systematic study of the genus *Acetobacter* with descriptions of *Acetobacter indonesiensis* sp. nov., *Acetobacter tropicalis* sp. nov., *Acetobacter orleanensis* (Henneberg 1906) comb. nov., *Acetobacter lovaniensis* (Frateur 1950) comb. nov. and *Acetobacter estunensis* (Carr 1958) comb. nov. Journal of General and Applied Microbiology 46: 147-165.

Lisdiyanti, P., Kawasaki, H., Seki, T., Yamada, Y., Uchimura, T. and Komagata, K. (2001). Identification of *Acetobacter* strains isolated from Indonesian sources, and proposals of *Acetobacter syzygii* sp. nov., *Acetobacter cibinongensis* sp. nov. and *Acetobacter orientalis* sp. nov. Journal of General and Applied Microbiology 47: 119-131.

Lisdiyanti, P., Kawasaki, H., Widyastuti, Y., Saono, S., Seki, T., Yamada, Y., Uchimura, T. and Komagata, K. (2002). *Kozakia baliensis* gen. nov., sp. nov., a novel acetic acid bacterium in the α-Proteobacteria. International Journal of Systematic and Evolutionary Microbiology 52: 813-818.

Lisdiyanti, P., Yamada, Y., Uchimura, T. and Komagata, K. (2003a). Identification of *Frateuria aurantia* strains isolated from Indonesian sources. Microbiology and Culture Collections 19: 81-90.

Lisdiyanti, P., Katsura, K., Potacharoen, W., Navarro, R.R., Yamada, Y., Uchimura, T. and Komagata, K. (2003b). Diversity of acetic acid bacteria in Indonesia, Thailand and the Philippines. Microbiology and Culture Collections 19: 91-99.

Lisdiyanti, P., Navarro, R.R., Uchimura T. and Komagata, K. (2006). Reclassification of *Gluconacetobacter hansenii* strains and proposals of *Gluconacetobacter saccharivorans* sp. nov. and *Gluconacetobacter nataicola* sp. nov. International Journal of Systematic and Evolutionary Microbiology 56: 2101-2111.

Loganathan, P. and Nair, S. (2004). *Swaminathania salitolerans* gen. nov., sp. nov., a salt-tolerant, nitrogen-fixing and phosphate-solubilizing bacterium from wild rice (*Porteresia coarctata* Tateoka). International Journal of Systematic and Evolutionary Microbiology 54: 1185-1190.

Malimas, T., Yukphan, P., Takahashi, M., Kaneyasu, M., Potacharoen, W., Tanasupawat, S., Nakagawa, Y., Tanticharoen, M. and Yamada, Y. (2007). *Gluconobacter kondonii* sp. nov., an acetic acid bacterium in the α-Proteobacteria. Journal of General and Applied Microbiology 53: 301-307.

Malimas, T., Yukphan, P., Takahashi, M., Muramatsu, Y., Kaneyasu, M., Potacharoen, W., Tanasupawat, S., Nakagawa, Y., Tanticharoen, M. and Yamada, Y. (2008a). *Gluconobacter roseus* (ex Asai 1935) sp. nov., nom. rev., a pink-colored acetic acid bacterium in the Alphaproteobacteria. Journal of General and Applied Microbiology 54: 119-125.

Malimas, T., Yukphan, P., Takahashi, M., Muramatsu, Y., Kaneyasu, M., Potacharoen, W., Tanasupawat, S., Nakagawa, Y., Tanticharoen, M. and Yamada, Y. (2008b). *Gluconobacter sphaericus* (Ameyama 1975) comb. nov., a brown pigment-producing acetic acid bacterium in the Alphaproteobacteria. Journal of General and Applied Microbiology 54: 211-220.

Malimas, T., Yukphan, P., Takahashi, M., Kaneyasu, M., Potacharoen, W., Tanasupawat, S., Nakagawa, Y., Tanticharoen, M. and Yamada, Y. (2008c). *Asaia lannaensis* sp. nov., a new acetic acid bacterium in the Alphaproteobacteria. Bioscience, Biotechnology and Biochemistry 72: 666-671.

Malimas, T., Yukphan, P., Lundaa, T., Muramatsu, Y., Takahashi, M., Kaneyasu, M., Potacharoen, W., Tanasupawat, S., Nakagawa, Y., Suzuki, K., Tanticharoen, M. and Yamada, Y. (2009a). *Gluconobacter kanchanaburiensis* sp. nov., a brown pigment-producing acetic acid bacterium for Thai isolates in the Alphaproteobacteria. Journal of General and Applied Microbiology 55: 247-254.

Malimas, T., Yukphan, P., Takahashi, M., Muramatsu, Y., Kaneyasu, M., Potacharoen, W., Tanasupawat, S., Nakagawa, Y., Tanticharoen, M. and Yamada, Y. (2009b). *Gluconobacter japonicus* sp. nov., an acetic acid bacterium in the Alphaproteobacteria. International Journal of Systematic and Evolutionary Microbiology 59: 466-471.

Malimas, T., Chaipitakchonlatarn, W., Vu, H.T.L., Yukphan, P., Muramatsu, Y., Tanasupawat, S., Potacharoen, W., Nakagawa, Y., Tanticharoen, M. and Yamada, Y. (2013). *Swingsia samuiensis* gen. nov., sp. nov., an osmotolerant acetic acid bacterium in the α-Proteobacteria. Journal of General and Applied Microbiology 59: 375-384.

Mason, L.M. and Claus, G.W. (1989). Phenotypic characteristics correlated with deoxyribonucleic acid sequence similarities for three species of *Gluconobacter*: *G. oxydans* (Henneberg 1897) De Ley 1961, *G. frateurii* sp. nov. and *G. asaii* sp. nov. International Journal of Systematic Bacteriology 39: 174-184.

Matsushita, K., Toyama, H. and Adachi, O. (2004). Respiratory chain in acetic acid bacteria: Membran-bound periplasmic sugar and alcohol respirations. *In*: D. Zannoni (ed.), Respiration of Archaea and Bacteria. Springer, Dordrecht, pp. 81-99.

Matsushita, K. and Matsutani, K. (2016). Distribution, evolution and physiology of oxidative fermentation. *In*: M. Matsushita, H. Toyama, N. Tonouchi and A. Okamoto-Kainuma (eds.), Acetic Acid Bacteria: Ecology and Physiology. Springer Japan, Tokyo, pp. 159-178.

Navarro, R.R. and Komagata, K. (1999). Differentiation of *Gluconacetobacter liquefaciens* and *Gluconacetobacter xylinus* on the basis of DNA base composition, DNA relatedness and oxidation products from glucose. Journal of General and Applied Microbiology 45: 7-15.

Ndoye, B., Cleenwerck, I., Engelbeen, K., Dubois-Dauphin, R., Guiro, A.T., van Trappen, S., Willems, A. and Thonart, P. (2007). *Acetobacter senegalensis* sp. nov., a thermotolerant acetic acid bacterium isolated in Senegal (sub-Saharan Africa) from mango fruit (*Mangifera indica* L.). International Journal of Systematic and Evolutionary Microbiology 57: 1576-1581.

Nishijima, M., Tazato, N., Handa, Y., Tomita, J., Kigawa, R., Sano, C. and Sugiyama, J. (2013). *Gluconacetobacter tumulisoli* sp. nov., *Gluconacetobacter takamatsuzukensis* sp. nov. and *Gluconacetobacter aggeris* sp. nov., isolated from Takamatsuzuka Tumulus samples before and during the dismantling work in 2007. International Journal of Systematic and Evolutionary Microbiology 63: 3981-3988.

Pitiwittayakul, N., Yukphan, P., Chaipitakchonlatarn, W., Yamada, Y. and Theeragool, G. (2015). *Acetobacter thailandicus* sp. nov., for a strain isolated in Thailand. Annals of Microbiology 65: 1855-1863.

Ramírez-Bahena, M.H., Tejedor, C., Martín, I., Velázquez, E. and Peix, A. (2013). *Endobacter medicaginis* gen. nov., sp. nov., isolated from alfalfa nodules in acidic soil. International Journal of Systematic and Evolutionary Microbiology 63: 1760-1765.

Sato, K., Yamada, Y., Aida, K. and Uemura, T. (1969a). Enzymatic studies on the oxidation of sugar and sugar alcohol. Part VIII. Particle-bound L-sorbose dehydrogenase from *Gluconobacter suboxydans*. Journal of Biochemistry 66: 521–527.

Sato, K., Yamada, Y., Aida, K. and Uemura, T. (1969b). Enzymatic studies on the oxidation of sugar and sugar alcohol. Part VII. On the catabolism of D-sorbitol by way of 5-keto-D-fructose in *Gluconobacter suboxydans*. Agricultural and Biological Chemistry 33: 1612-1618.

Schüller, G., Hertel, C. and Hammes, W.P. (2000). *Gluconacetobacter entanii* sp. nov., isolated from submerged high-acid industrial vinegar fermentations. International Journal of Systematic and Evolutionary Microbiology 50: 2013-2020.

Shimwell, J.L. (1958). Flagellation and taxonomy of *Acetobacter* and *Acetomonas*. Antonie van Leeuwenhoek 24: 187-192.

Shimwell, J.L. and Carr, J.G. (1959). The genus *Acetomonas*. Antonie van Leeuwenhoek 25: 353-368.

Sievers, M., Sellmer, S. and Teuber, M. (1992). *Acetobacter europaeus* sp. nov., a main component of industrial vinegar fermenters in central Europe. Systematic and Applied Microbiology 15: 386-392.

Sievers, M., Ludwig, W. and Teuber, M. (1994). Revival of the species *Acetobacter methanolicus* (ex Uhlig et al. 1986) nom. rev. Systematic and Applied Microbiology 17: 352-354.

Sievers, M. and J. Swings (2005). Family II. Acetobacteraceae Gillis and De Ley 1980 *In*: G.M. Garrity, D.J. Brenner, N.R. Krieg, J.T. Staley (eds.), Bergey's Manual of Systematic Bacteriology, 2nd edn, vol. 2. The Proteobacteria. Part C. The Alpha-, Beta-, Delta- and Epsilonproteobacteria. Springer, New York, pp. 41-95.

Silva, L.R., Cleenwerck, I., Rivas, P., Swings, J., Trujillo, M.E., Willems, A. and Velázquez, E. (2006). *Acetobacter oenis* sp. nov., isolated from spoiled red wine. International Journal of Systematic and Evolutionary Microbiology 56: 21-24.

Skerman, V.B.D., McGowan, V. and Sneath, P.H.A. (1980). Approved Lists of Bacterial Names. International Journal of Systematic Bacteriology 30: 225-420.

Slapšak, N., Cleenwerck, I., De Vos, P. and Trček, J. (2013). *Gluconacetobacter maltaceti* sp. nov., a novel vinegar producing acetic acid bacterium. Systematic and Applied Microbiology 36: 17-21.

Sokollek, S.J., Hertel, C. and Hammes, W.P. (1998). Description of *Acetobacter oboediens* sp. nov. and *Acetobacter pomorum* sp. nov., two new species isolated from industrial vinegar fermentations. International Journal of Systematic Bacteriology 48: 935-940.

Spitaels, F., Li, L., Wieme, A., Balzarini, T., Cleenwerck, I., Van Landschoot, A., De Vuyst, L. and Vandamme, P. (2014a). *Acetobacter lambici* sp. nov., isolated from fermenting lambic beer. International Journal of Systematic and Evolutionary Microbiology 64: 1083-1089.

Spitaels, F., Wieme, A., Balzarini, T., Cleenwerck, I., Van Landschoot, A., De Vuyst, L. and Vandamme, P. (2014b). *Gluconobacter cerevisiae* sp. nov., isolated from brewery environment. International Journal of Systematic and Evolutionary Microbiology 64: 1134-1141.

Stackebrandt, E., Murray, R.G.E. and Trüper, H.G. (1988). Proteobacteria classis nov., a name for the phylogenetic taxon that includes the 'purple bacteria and their relatives.' International Journal of Systematic and Bacteriology 38: 321-325.

Suzuki, R., Zhang, Y., Iino, T., Kosako, Y., Komagata, K. and Uchimura, T. (2010). *Asaia astilbes* sp. nov., *Asaia platycodi* sp. nov. and *Asaia prunellae* sp. nov., novel acetic acid bacteria isolated from flowers in Japan. Journal of General and Applied Microbiology 56: 339-346.

Swings, J., Gillis, M., Kersters, K., De Vos, P., Gosselé, F. and De Ley, J. (1980). *Frateuria*, a new genus for "*Acetobacter aurantius*." International Journal of Systematic Bacteriology 30: 547-556.

Swings J. (1992). The genera *Acetobacter* and *Gluconobacter In*: A. Balows, H.G. Trüper, M. Dworkin, W. Harder and K.-H. Schleifer (eds.), The Prokaryotes, 2nd edn, vol. 3. Springer-Verlag, New York, pp. 2268-2286.

Swings, J. and Sievers, M. (2005). Genus II. *Frateuria* Swings, Gillis, Kersters, De Vos, Gosselé and De Ley 1980 *In*: G.M. Garrity, D.J. Brenner, N.R. Krieg, J.T. Staley (eds.) Bergey's Manual of Systematic Bacteriology, 2nd edn, vol 2. The Proteobacteria Part B The Gammaproteobacteria. Springer, New York, pp 91-93.

Tamura, K., Stecher, G., Peterson, D., Filipski, A. and Kumar, S. (2013). MEGA6: Molecular evolutionary genetics analysis version 6.0. Molecular Biology and Evolution 30: 2725-2729.

Tanaka, M., Yoshida, M., Murakami, S., Aoki, K. and Shinke, R. (1998). The characterization of phenotypic features of the cellulose-forming acetic acid bacteria. Science Reports of Faculty of Agriculture Kobe University 23: 65-74.

Tanaka, M., Murakami, S., Shinke, R. and Aoki, K. (1999). Reclassification of the strains with low G+C contents of DNA belonging to the genus *Gluconobacter* Asai 1935 (Acetobacteraceae). Bioscience, Biotechnology and Biochemistry 63: 989-992.

Tanasupawat, S., Thawai, C., Yukphan, P., Moonmangmee, D., Itoh, T., Adachi, O. and Yamada, Y. (2004). *Gluconobacter thailandicus* sp. nov., an acetic acid bacterium in the α-Proteobacteria. Journal of General and Applied Microbiology 50: 159-167.

Tanasupawat, S., Kommanee, J., Yukphan, P., Muramatsu, Y., Nakagawa, Y. and Yamada, Y. (2011a). *Acetobacter farinalis* sp. nov., an acetic acid bacterium in the α-Proteobacteria. Journal of General and Applied Microbiology 57: 159-167.

Tanasupawat, S., Kommanee, J., Yukphan, P., Moonmangmee, D., Muramatsu, Y., Nakagawa, Y. and Yamada, Y. (2011b). *Gluconobacter uchimurae* sp. nov., an acetic acid bacterium in the α-Proteobacteria. Journal of General and Applied Microbiology 57: 293-301.

Tazato, N., Nishijima, M., Handa, Y., Kigawa, R., Sano, C. and Sugiyama, J. (2012). *Gluconacetobacter tumulicola* sp. nov. and *Gluconacetobacter asukensis* sp. nov., isolated from the stone chamber interior of the Kitora Tumulus. International Journal of Systematic and Evolutionary Microbiology 62: 2032-2038.

Tindall, B.J., Kämpfer, P., Euzéby, J.P. and Oren, A. (2006). Valid publication of names of prokaryotes according to the rules of nomenclature: Past history and current practice. International Journal of Systematic and Evolutionary Microbiology 56: 2715-2720.

Toyosaki, H., Kojima, Y., Tsuchida, T., Hoshino, K., Yamada, Y. and Yoshinaga, F. (1995). The characterization of an acetic acid bacterium useful for producing bacterial cellulose in agitation cultures: The proposal of *Acetobacter xylinum* subsp. *sucrofermentans* subsp. nov. Journal of General and Applied Microbiology 41: 307-314.

Uhlig, H., Karbaum, K. and Steudel, A. (1986). *Acetobacter methanolicus* sp. nov., an acidophilic facultatively methylotrophic bacterium. International Journal of Systematic Bacteriology 36: 317-322.

Urakami, T., Tamaoka, J., Suzuki, K. and Komagata, K. (1989). *Acidomonas* gen. nov., incorporating *Acetobacter methanolicus* as *Acidomonas methanolica* comb. nov. International Journal of Systematic Bacteriology 39: 50-55.

Visser't Hooft, F. (1925). Biochemische onderzoekingen over het geslacht *Acetobacter*. Dissertation, Technical University Meinema, Delft, pp 1-129.

Vu, H.T.L., Yukphan, P., Chaipitakchonlatarn, W., Malimas, T., Muramatsu, Y., Bui, U.T.T., Tanasupawat, S., Duong, K.C., Nakagawa, Y., Pham, H.T. and Yamada, Y. (2013). *Nguyenibacter vanlangensis* gen. nov., sp. nov., an unusual acetic acid bacterium in the α-Proteobacteria. Journal of General and Applied Microbiology 59: 153-166.

Vu, H.T.L., Malimas, T., Chaipitakchonlatarn, W., Bui, V.T.T., Yukphan, P., Bui, U.T.T., Muramatsu, Y., Sitdhipol, J., Tanasupawat, S., Duong, K.C., Nakagawa, Y., Pham, H.T. and Yamada, Y. (2016). *Tanticharoenia aidae* sp. nov., for acetic acid bacteria isolated in Vietnam. Annals of Microbiology 66: 417-423.

Yamada, Y., Aida, K. and Uemura, T. (1969a). Enzymatic studies on the oxidation of sugar and sugar alcohol. V. Ubiquinone of acetic acid bacteria and its relation to classification of genera

Gluconobacter and *Acetobacter*, especially of the so-called intermediate strains. Journal of General and Applied Microbiology 15: 181-196.

Yamada, Y., Nakazawa, E., Nozaki, A. and Kondo, K. (1969b). Characterization of *Acetobacter xylinum* by ubiquinone system. Agricultural and Biological Chemistry 33: 1659-1661.

Yamada, Y., Okada, Y. and Kondo, K. (1976a). Isolation and characterization of "polarly flagellated intermediate strains" in acetic acid bacteria. Journal of General and Applied Microbiology 22: 237-245.

Yamada, Y., Nakazawa, E., Nozaki, A. and Kondo, K. (1976b). Characterization of *Acetobacter xylinum* by ubiquinone system. Journal of General and Applied Microbiology 22: 285-292.

Yamada, Y. (1979). Classification of microorganisms based on the molecular species of the respiratory quinones. Hakko to Kogyo 37: 940-954 (in Japanese).

Yamada, Y., Nunoda, M., Ishikawa, T. and Tahara, Y. (1981a). The cellular fatty acid composition in acetic acid bacteria. Journal of General and Applied Microbiology 27: 405-417.

Yamada, Y., Ishikawa, T., Yamashita, M., Tahara, Y., Yamasato, K. and Kaneko, T. (1981b). Deoxyribonucleic acid base composition and deoxyribonucleic acid homology in the polarly flagellated intermediate strains. Journal of General and Applied Microbiology 27: 465-475.

Yamada, Y. (1983). *Acetobacter xylinus* sp. nov., nom. rev., for the cellulose-forming and cellulose-less, acetate-oxidizing acetic acid bacteria with the Q-10 system. Journal of General and Applied Microbiology 29: 417-420.

Yamada, Y. and Akita, M. (1984). An electrophoretic comparison of enzymes in strains of *Gluconobacter* species. Journal of General and Applied Microbiology 30: 115-126.

Yamada, Y. and Kondo, K. (1984). *Gluconoacetobacter*, a new subgenus comprising the acetate-oxidizing acetic acid bacteria with ubiquinone-10 in the genus *Acetobacter*. Journal of General and Applied Microbiology 30: 297-303.

Yamada, Y., Itakura, N., Yamashita, M. and Tahara, Y. (1984). Deoxyribonucleic acid homologies in strains of *Gluconobacter* species. Journal of Fermentation Technology 62: 595-600.

Yamada, Y., Hoshino, K. and Ishikawa, T. (1997). The phylogeny of acetic acid bacteria based on the partial sequences of 16S ribosomal RNA: The elevation of the subgenus *Gluconoacetobacter* to the generic level. Bioscience, Biotechnology and Biochemistry 61: 1244-1251.

Yamada, Y., Hosono, R., Lisdiyanti, P., Widyastuti, Y., Saono, S., Uchimura, T. and Komagata, K. (1999). Identification of acetic acid bacteria isolated from Indonesian sources, especially of isolates classified in the genus *Gluconobacter*. Journal of General and Applied Microbiology 45: 23-28.

Yamada, Y. (2000). Transfer of *Acetobacter oboediens* Sokollek et al. 1998 and *Acetobacter intermedius* Boesch et al. 1998 to the genus *Gluconacetobacter* as *Gluconacetobacter oboediens* comb. nov. and *Gluconacetobacter intermedius* comb. nov. International Journal of Systematic and Evolutionary Microbiology 50: 2225-2227.

Yamada, Y., Katsura, K., Kawasaki, H., Widyastuti, Y., Saono, S., Seki, T., Uchimura, T. and Komagata, K. (2000). *Asaia bogorensis* gen. nov., sp. nov., an unusual acetic acid bacterium in the α-Proteobacteria. International Journal of Systematic and Evolutionary Microbiology 50: 823-829.

Yamada, Y. and Yukphan, P. (2008). Genera and species in acetic acid bacteria. International Journal of Food Microbiology 125: 15-24.

Yamada, Y., Yukphan, P., Vu, H.T.L., Muramatsu, Y., Ochaikul, D. and Nakagawa, Y. (2012a). Subdivision of the genus *Gluconacetobacter* Yamada, Hoshino and Ishikawa 1998: The proposal of *Komagatabacter* gen. nov., for strains accommodated to the *Gluconacetobacter xylinus* group in the α-Proteobacteria. Annals of Microbiology 62: 849-859.

Yamada, Y., Yukphan, P., Vu, H.T.L, Muramatsu, Y., Ochaikul, D., Tanasupawat, S. and Nakagawa, Y. (2012b). Description of *Komagataeibacter* gen. nov., with proposals of new combinations (Acetobacteraceae). Journal of General and Applied Microbiology 58: 397-404.

Yamada, Y. (2014). Transfer of *Gluconacetobacter kakiaceti*, *Gluconacetobacter medellinensis* and *Gluconacetobacter maltaceti* to the genus *Komagataeibacter* as *Komagataeibacter kakiaceti* comb. nov., *Komagataeibacter medellinensis* comb. nov. and *Komagataeibacter maltaceti* comb. nov. International Journal of Systematic and Evolutionary Microbiology 64: 1670-1672.

Yamada, Y. (2016). Systematics of acetic acid bacteria. *In*: M. Matsushita, H. Toyama, N. Tonouchi and A. Okamoto-Kainuma (eds.), Acetic Acid Bacteria: Ecology and Physiology. Springer Japan, Tokyo, pp. 1-50.

Yamashita, S., Uchimura, T. and Komagata, K. (2004). Emendation of the genus *Acidomonas* Urakami, Tamaoka, Suzuki and Komagata 1989. International Journal of Systematic and Evolutionary Microbiology 54: 865-870.

Yukphan, P., Takahashi, M., Potacharoen, W., Tanasupawat, S., Nakagawa, Y., Tanticharoen, M. and Yamada, Y. (2004a). *Gluconobacter albidus* (ex Kondo and Ameyama 1958) sp. nov., nom. rev., an acetic acid bacterium in the α-Proteobacteria. Journal of General and Applied Microbiology 50: 235-242.

Yukphan, P., Potacharoen, W., Tanasupawat, S., Tanticharoen, M. and Yamada, Y. (2004b). *Asaia krungthepensis* sp. nov., an acetic acid bacterium in the α-Proteobacteria. International Journal of Systematic and Evolutionary Microbiology 54: 313-316.

Yukphan, P., Malimas, T., Potacharoen, W., Tanasupawat, S., Tanticharoen, M. and Yamada, Y. (2005). *Neoasaia chiangmaiensis* gen. nov., sp. nov., a novel osmotolerant acetic acid bacterium in the α-Proteobacteria. Journal of General and Applied Microbiology 51: 301-311.

Yukphan, P., Malimas, T., Muramatsu, Y., Takahashi, M., Kaneyasu, M., Tanasupawat, S., Nakagawa, Y., Suzuki, K., Potacharoen, W. and Yamada, Y. (2008). *Tanticharoenia sakaeratensis* gen. nov., sp. nov., a new osmotolerant acetic acid bacterium in the α-Proteobacteria. Bioscience, Biotechnology and Biochemistry 72: 672-676.

Yukphan, P., Malimas, T., Muramatsu, Y., Takahashi, M., Kaneyasu, M., Potacharoen, W., Tanasupawat, S., Nakagawa, Y., Hamana, K., Tahara, Y., Suzuki, K., Tanticharoen, M. and Yamada, Y. (2009). *Ameyamaea chiangmaiensis* gen. nov., sp. nov., an acetic acid bacterium in the α-Proteobacteria. Bioscience, Biotechnology and Biochemistry 73: 2156-2162.

Yukphan, P., Malimas T., Lundaa, T., Muramatsu, Y., Takahashi, M., Kaneyasu, M., Tanasupawat, S., Nakagawa, Y., Suzuki, K., Tanticharoen, M. and Yamada, Y. (2010). *Gluconobacter wancherniae* sp. nov., an acetic acid bacterium from Thai isolates in the α-Proteobacteria. Journal of General and Applied Microbiology 56: 67-73.

Yukphan, P., Malimas, T., Muramatsu, Y., Potacharoen, W., Tanasupawat, S., Nakagawa, Y., Tanticharoen, M. and Yamada, Y. (2011). *Neokomagataea* gen. nov., with descriptions of *Neokomagataea thailandica* sp. nov. and *Neokomagataea tannensis* sp. nov., osmotolerant acetic acid bacteria of the α-Proteobacteria. Bioscience, Biotechnology and Biochemistry 75: 419-426.

Zhang, J.-Y., Liu, X.-Y. and Liu, S.-J. (2011). *Frateuria terrea* sp. nov., isolated from forest soil and emended description of the genus *Frateuria*. International Journal of Systematic and Evolutionary Microbiology 61: 443-447.

Addendum in General Characteristics of Acetic Acid Bacteria

The flow rate of the pentose phosphate cycle in the resting cells of '*Gluconobacter suboxydans*' strain 1 cultured for 15 h was calculated as 13-17 microatoms/hr/mg dry cells of oxygen consumed by the manometric method (Sato et al. 1969b).

2

Comparative Genomics of *Acetobacter* and other Acetic Acid Bacteria

Jure Škraban[1] and Janja Trček[1,2,*]

[1] University of Maribor, Faculty of Natural Sciences and Mathematics, Department of Biology, Maribor, Slovenia

[2] University of Maribor, Faculty of Chemistry and Chemical Engineering, Maribor, Slovenia

Introduction

Acetic acid bacteria (AAB) are a group of *Alphaproteobacteria* belonging to the family *Acetobacteraceae* (Kersters et al. 2006). They are frequent colonizers of tropical flowers and fruits, and food fermenters (Sievers and Swings 2005, Cleenwerck et al. 2009). The food industry is interested in them because of their ability to produce large amounts of acetic acid from ethanol during vinegar production, kombucha tea, or cocoa bean fermentation (Trček and Barja 2015, De Vuyst and Weckx 2016). They are also used for the production of specific chemicals, such as ascorbic acid and bacterial cellulose (Adachi et al. 2003, Bremus et al. 2006, Römling et al. 2015). Besides acting as fermenting agents, some members of AAB form symbiotic relationships with insects (Crotti et al. 2010, Shin et al. 2011, Chouaia et al. 2014), or cause opportunistic infections in humans (Greenberg et al. 2007, Kawai et al. 2015). Economic importance and the influence of AAB on our lives warrant a better understanding of their physiological/phenotypical properties. New sequencing technologies have made the whole genome sequencing readily available to a wider scientific community. As a result, quite a few complete AAB genomes have been analyzed in the last decade, providing an important source of information regarding their metabolic and phenotypic potential. The review aims to summarize the important findings that emerged from genome studies in this important and diverse group of bacteria.

General Characteristics of Complete Genome Sequences of Acetic Acid Bacteria

The group of AAB has been revised many times over the past years. At the moment (May 2016), it includes 19 genera (Trček and Barja 2015) if counting also the proposed genus *Commensalibacter* which still needs to be taxonomically validated (Roh et al. 2008). Among them, *Acetobacter* and *Komagataeibacter* are the most studied because of their importance in vinegar production (Matsushita et al. 2005a, Gullo et al. 2006, Trček et al. 2007, Hidalgo et al. 2010). Altogether, 18 complete genomes belonging to seven species of AAB have been published to date (March 2016) (Table 1), among which eleven completely sequenced genomes belong to the species *A. pasteurianus* (including all the seven substrains and

*Corresponding author: janja.trcek@um.si

Table 1: General characteristics of the complete genomes of acetic acid bacteria

	Chromosome size (bp)	*Chromosome GC content (per cent)*	*Number of plasmids*	*Number of transposases*	*tRNA[1] genes*	*rRNA[2] operons*	*CDS[3] (total)*	*Source*
Acetobacter pausterianus 386B	2,818,679	52.91	7	50	57	5	2,875	Illeghems et al. 2013
Acetobacter pausterianus IFO 3283-01	2,907,495	50.70	6	285	57	5	3,050	Azuma et al. 2009
Acetobacter pasteurianus CICC 20001	2,865,612	52.94	10	5	118	34 (genes)	3,623	Wang et al. 2015
Acetobacter pasteurianus CGMCC 1.41	2,928,931	52.95	7	4	111	30 (genes)	3,250	Wang et al. 2015
Gluconobacter oxydans H24	3,602,424	56.25	1	80	59	5	3,732	Ge et al. 2013
Gluconobacter oxydans 612H	2,702,173	60.80	5	98	55	4	2,664	Prust et al. 2005
Gluconacetobacter diazotrophicus Pal5	3,944,163	66.19	2	190	55	4	3,938	Bertalan et al. 2009
Komagataeibacter xylinus E25	3,447,725	62.64	5	81	57	5	3,156	Kubiak et al. 2014
Komagataeibacter medellinensis NBRC 3288	3,136,818	60.92	7	130	60	5	3,185	Ogino et al. 2011
Granulibacter bethesdensis CGDNIH1	2,708,355	59.07	0	16	52	3	2,437	Greenberg et al. 2007
Asaia bogorensis NBRC 16594	3,198,265	59.80	0	14	58	5	2,758	Kawai et al. 2015

[1]transfer RNA; [2]ribosomal RNA; [3] protein coding sequence

one mutant strain of *A. pasteurianus* IFO 3283 sequenced by Azuma et al. 2009). The remaining seven genomes belong to *Komagataeibacter xylinus* E25, *K. medellinensis* NBRC 3288, *Gluconacetobacter diazotrophicus* Pal5, *Gluconobacter oxydans* 612H and H24, and two opportunistic AAB, *Granulibacter bethesdensis* CGDNIH1-4 and *Asaia bogorensis* NBRC 16594 (Table 1).

Acetobacter species are important in biotechnological processes because they can oxidize ethanol to acetic acid. They secrete large amounts of acetic acid into the medium, which they can tolerate up to 6 per cent (v/v) (Trcek et al. 2006). The genus includes important vinegar producers, such as *A. pasteurianus*, *A. aceti* and *A. pomorum* (Sokollek et al. 1998, Gullo et al. 2006, Hidalgo et al. 2010). In addition to producing vinegar, some *A. pasteurianus* strains also participate in fermentation of cocoa bean pulp by oxidizing ethanol to acetic acid, and finally acetic acid, together with lactic acid, produced by lactic acid bacteria (LAB), to carbon dioxide and water (De Vuyst and Weckx 2016). *Acetobacter* species use an alternative tricarboxylic acid (TCA) cycle, which enables the removal of the cytosolic acetate, via acetyl-CoA oxidation, as carbon dioxide (Mullins et al. 2008). In addition to their usefulness in the food industry, some members, such as *A. cerevisiae*, on the other hand, act as spoiling agents during fermentation of beer or wine (Bartowsky and Henschke 2008, Wieme et al. 2014).

From the genus *Acetobacter*, only *A. pasteurianus* 386B (Illeghems et al. 2013), *A. pasteurianus* IFO 3283 (substrains -01, -03, -07, -12, -22, -26, -32, and the adapted strain IFO 3283-01-42C) (Azuma et al. 2009), *A. pasteurianus* CICC20001 and CGMCC 1.41 (Wang et al. 2015) genomes have been completely sequenced. *A. pasteurianus* 386B genome is composed of a circular chromosome of 2,818,679 bp and seven plasmids of different sizes (3,851, 6,548, 9,914, 11,105, 15,601, 18,169, 194,780 bp). The chromosome contains 2,595 coding sequences (CDS) and the plasmids 280 CDS (4, 5, 11, 9, 11, 20, and 220, respectively). The chromosome also has five ribosomal RNA operons and 57 tRNA genes (Table 1) (Illeghems et al. 2013). *A. pasteurianus* IFO 3283 maintenance by serial subculturing every third month over the course of 21 years (between 1954 and 1974) resulted in the formation of a multi-phenotype cell complex. The cell complex was subjected to genome sequencing, which showed that it was composed from seven different *A. pasteurianus* IFO 3283 substrains (mentioned above) based on phenotypic differences (smooth and rough colonies) and 13 mutation loci. The differences in chromosomes arose from three single nucleotide mutations, four transposon insertions and four different hyper-mutable tandem repeats. In addition to the chromosome, the largest plasmid also had one transposon insertion and variability in one hyper-mutable tandem repeat (Azuma et al. 2009). The basic genome structure of *A. pasteurianus* IFO 3283 substrains is the same, except for the 13 variable loci. The IFO 3283-01 genome is composed of a chromosome of 2,907,495 bp and six plasmids (1,815, 3,035, 3,204, 49,961, 182,940, and 191,799 bp). Like *A. pasteurianus* 386B, *A. pasteurianus* IFO 3283-01 also contains five rRNA operons and 57 tRNA genes. Out of the total of 3,050 CDS, 2,628 are located on the chromosome and 422 on plasmids (Table 1). Seventy-five per cent of the CDS code for proteins and the rest are hypothetical genes (Azuma et al. 2009). Comparative analysis of *A. pasteurianus* 386B and *A. pasteurianus* IFO 3283-01 chromosome sequences shows a highly conserved order of orthologous genes (Illeghems et al. 2013). Additionally, comparisons of the two complete genomes (*A. pasteurianus* IFO 3283-01 and 386B) and draft genomes of other strains (*A. pasteurianus* NBRC 101655, *A. pasteurianus* subsp. *pasteurianus* LMG 1262 and *A. pasteurianus* 3P3) found 2,019 common orthologous CDS, representing approximately 68 per cent of the predicted proteins from *A. pasteurianus* 386B (Illeghems et al. 2013). Recently, complete genomes of *A. pasteurianus* CICC 20001 and CGMCC 1.41 strains were published (Wang et al. 2015). Both are widely used in vinegar production in China and have different fermentation characteristics in respect of acetic acid production and tolerance. The genome of *A. pasteurianus* CICC 20001 is composed of a chromosome of 2,865,612 bp and 10 plasmids (one of which is a very large plasmid of 474,484 bp). The chromosome codes for 34 rRNA and 118 tRNA genes. From the total of 3,623 CDS, 3,006 are present on the chromosome and 617 on plasmids (Table 1). Out of the 3,623 CDS, 2,450 (67.6 per cent) have been assigned a specific function and the rest are hypothetical genes (Wang et al. 2015). The genome of *A. pasteurianus* CGMCC 1.41 contains a slightly larger chromosome of 2,928,931 bp and seven plasmids. The chromosome has 30 rRNA and 111 tRNA genes. From the total of 3,250 CDS, 3,012 lie on the chromosome and 237 on plasmids (Table 1) (Wang et al. 2015). High homology between the chromosomes was confirmed for all the completely sequenced *A. pasteurianus* strains. Orthologues, which represent more than 100 essential genes present in all the strains, are frequently grouped into clusters separated by several large regions longer than 300 kb (Wang et al. 2015). Essential genes basically form the core genome of *A. pasteurianus* strains and are critical for survival as they code for proteins primarily involved in maintaining cell structure, DNA replication, translation, mediating transport across membrane and controlling central metabolism (Wang et al. 2015). In addition to similarities, the chromosomes of *Acetobacter* strains also contain some deletions, amplifications, insertions, inversions and translocations. Compared to the *A. pasteurianus* IFO 3283 substrains and *A. pasteurianus* 386B, some of the fragments are missing in *A. pasteurianus* CICC 20001 and CGMCC 1.41. The chromosome of *A. pasteurianus* CGMCC 1.4 is also slightly bigger than the chromosomes of other sequenced strains (Wang et al. 2015). While the chromosomes display general homology, the plasmids of different *A. pasteurianus* strains on the other hand have different genotypes. For instance, a comparison of the two largest plasmids of *A. pasteurianus* IFO 3283, pAPA01-011 (178 CDS) and pAPA01-020 (174 CDS) with the largest plasmid of *A. pasteurianus* 386B, APA386B_1P (220 CDS), reveals that they share only 44 and 16 CDS, respectively. Furthermore, APA386B_1P contains 165 unique genes not present on the two largest plasmids of *A. pasteurianus* IFO 3283 substrains (Illeghems et al. 2013). Similarly, *A. pasteurianus* CICC 20001 has a large plasmid of 474,484 bp, which shows almost no homology to plasmids from other *A. pasteurianus* strains (Wang et al. 2015). It contains clustered, regularly interspaced, short palindromic repeat (CRISPR) elements (discussed later), which means that it could play a role in preserving genomic integrity of *A. pasteurianus* CICC 20001 (Wang et al. 2015). In fact, in addition to chromosomes, the plasmids of *A. pasteurianus* IFO 3283 substrains also display great genetic stability after storage for several decades. The relative positions of the genes on the plasmids of the substrains are identical except for one transposase (APA01_40012) present on the large plasmid of *A. pasteurianus* IFO 3283-01 (Azuma et al. 2009). Displayed variability between the genomes of different *A. pasteurianus* strains is mainly the result of a high number of mobile elements (transposons). *A. pasteurianus* IFO 3283 contains more than 280 transposons (more than 150 in the chromosome), which amounts to approximately nine per cent of the total genes in the genome (Azuma et al. 2009). Seventy-five of them belong to the IS*1380* type, an insertion sequence abundant in many other *A. pasteurianus* strains and implicated in genetic instability of AAB (Takemura et al. 1991, Azuma et al. 2009). The very high number of transposons and simultaneous high number of plasmids could be connected to hyper-mutability of *A. pasteurianus* IFO 3283 (Azuma et al. 2009). The transposons truncate 32 genes with an assigned function, which may contribute to specific metabolic features of different strains (Azuma et al. 2009). For example, nitrate reductase catalytic subunit (APA01_42100) within the nitrate assimilation gene cluster is truncated by a transposon, which also causes a frameshift, resulting in deactivation of two other genes within the cluster, nitrate reductase electron-transfer subunit (APA01_42120) and a nitrate transporter (APA01_42150). This suggests that *A. pasteurianus* IFO 3283 is unable to reduce nitrate via nitrite to ammonium, and so is incapable of growing on nitrate as the sole source of nitrogen (Azuma et al. 2009). *A. pasteurianus* 386B on the other hand contains only 50 transposases (26 in the chromosome) (Illeghems et al. 2013). Transposons of the IS*1380* type, which are abundant in *A. pasteurianus* IFO 3283, are absent in *A. pasteurianus* 386B (Illeghems et al. 2013). This, with the fact that they both have a similar number of plasmids, implies a higher genetic stability of the 386B strain. The lower number of transposons also means less truncated genes. The genes not truncated in *A. pasteurianus* 386B which include D-galactonate transporter (APA386B_203), oxalyl-CoA decarboxylase (APA386B_504), transporter (APA386B_1075), acetyl-CoA: propinoate CoA-transferase (APA386B_2066) and an oxidoreductase (APA386B_1P21) could contribute to specific properties of the strain (Illeghems et al. 2013). *A. pasteurianus* CICC 20001 and *A. pasteurianus* CGMCC 1.41 contain even less (five and four) transposase genes, respectively, implying high genetic stability of these two strains (Wang et al. 2015). In addition to transposons, integrated bacteriophages are also a source of genomic variations among *A. pasteurianus* strains. The chromosomal synteny between *A. pasteurianus* 386B and *A. pasteurianus* IFO 3283 is interrupted by the presence of transposases and a prophage incorporated in the genome of *A. pasteurianus* 386B. The prophage genomic segment has approximately 28.8 kbp and 61 genes (APA386B_370 – APA386B_430). Several prophage genes found in *A. pasteurianus* 386B have homologues in *A. pasteurianus* IFO 3283, *Gluconacetobacter diazotrophicus* Pal5 and *Gluconobacter oxydans* 621H (Illeghems et al. 2013). The prophage region in *A. pasteurianus* 386B contains an integrase (APA386B_430) with a homologue also found in *G. oxydans* 621H (AAW62050.1) and *Granulibacter bethesdensis* CGDNIH2 (AHJ69369.1), a phage terminase (APA386B_403) and a lambda family phage portal protein (APA386B_401). It, however, lacks virulence-associated genes, genes coding for a head maturation protease, a coat, or a tail measure protein, which indicates that the prophage is defective (Canchaya et al. 2003, Illegehems et al. 2013).

Long terminal repeat (LTR) retrotransposons are another indicator of genetic instability. *A. pasteurianus* CGMCC 1.41 and *A. pasteurianus* IFO 3283 substrains contain one LTR of 1269 and 833 bp, respectively. The LTR of *A. pasteurianus* CGMCC 1.41 is located on the chromosome and the LTR of the IFO 3283 substrains is located on a plasmid. LTRs, on the other hand, are absent from the genomes of *A. pasteurianus* 386B and CICC 20001 (Wang et al. 2015). As in other organisms, instability and stability factors direct the balance of the genome in AAB, thereby ensuring the survival of organisms and their offspring (Saier 2008). Genomic instability plays two roles in the survival of organisms. Factors such as mobile elements and recombination events can cause genome instability and generate phenotype variations. On the other hand, restriction-modification (RM) systems, which are present in all the completely sequenced genomes of *A. pasteurianus* strains (Wang et al. 2015), and clustered, regularly interspaced, short palindromic repeat (CRISPR) elements, use genome instability to fight phages and mobile elements (Lin et al. 2014, Caliando and Voight 2015). One CRISPR element is present at an almost identical position on the largest plasmid of all *A. pasteurianus* 3283 substrains. It is composed of 1431 bp with 23 spacers and a direct repeat (DR) consensus sequence GTGTTCCCCGCACACGCGGGGATGAACCG. *A. pasteurianus* CGMCC 1.41 and *A. pasteurianus* CICC 20001 contain one and two putative CRISPR elements on the chromosome, respectively. In addition, *A. pasteurianus* CICC 20001 contains one putative and two confirmed CRISPR elements on the large plasmid, one of the size 1527 bp with 25 spacers and a DR consensus CCAGACCGCCGCATAGGCGGTTTAGAAA, and the other composed of 334 bp with five spacers and a DR consensus CGGTTAAACCCCGCAGACGCGGGGAAGACT. *A. pasteurianus* 386B has no CRISPR elements in its genome. The large plasmid could, therefore, contribute to the observed genetic stability of *A. pasteurianus* CICC 20001. The balance of stability (RM and CRISPR) and instability (mobile elements, LTR) factors at any given moment probably determines the overall genome stability of *A. pasteurianus* strains (Wang et al. 2015).

The first completely sequenced AAB was *Gluconobacter oxydans* 621H (Prust et al. 2005). Genomic information proved indispensable in reconstructing metabolic properties of this acetic acid bacterium, which has an extraordinary potential to oxidize a variety of carbohydrates, alcohols and other organic compounds. It is also well adapted to environments with high osmotic concentrations (Deppenmeier et al. 2002, Matsutani et al. 2011). The size of the genome of *G. oxydans* 621H is 2,922,384 bp and consists of a circular chromosome (2,702,173 bp) and five plasmids (163,186, 26,568, 14,547, 13,223, and 2,687 bp). The chromosome contains four rRNA operons and 55 tRNA genes. Almost 90 per cent of the genome encodes proteins or stable RNA. The total number of predicted CDS is 2,664 (2,432 in the chromosome and 232 in the plasmids) (Table 1), 1,877 (70.5 per cent) of which were assigned to specific functions (Prust et al. 2005). The genome also contains 82 insertion sequences (IS) (10 copies belong to IS*12528* and eight copies to IS*1032* type (Mahillon and Chandler 1998)) and 98 transposase genes. Some of the copies are partially deleted and possibly defective. Most of them, however, seem to be functional and probably contribute to the marked genetic and physiological instability as seen not only in *Gluconobacter*, but also in other AAB (Takemura et al. 1991, Kondo and Horinouchi 1997a, b, Azuma et al. 2009). Additionally, two putative regions (GOX2318-2357 and GOX1211-1226), possibly representing prophages, were recognized (Prust et al. 2005). The genome of *G. oxydans* 621H also contains five plasmids, which are poorly characterized. As in *A. pasteurianus* strains, the five plasmids show no homology to known plasmids of other *G. oxydans* strains, for instance pAG5 from *G. oxydans* IFO 3171 (Tonouchi et al. 2003) and pGO128 from *G. oxydans* DSM 3504 (Sievers M., direct submission to GenBank, Gen ID: NC_003374). However, the smallest plasmid pGOX5 shows substantial similarities to the plasmid pJK2-1 from *Komagataeibacter europaeus* DSM 13109 (Trček et al. 2000, Prust et al. 2005). Most open reading frames (70 per cent) of the plasmid encode hypothetical proteins of unknown function. Coding sequences, with predicted functions, code for proteins of plasmid replication (DNA helicase II (*umuD*)), a restriction modification system, a conjugation system, C_4-dicarboxylate transporter, a heavy metal resistance system and two alcohol dehydrogenases of unknown substrate specificity (GOX2318-2357 and GOX1211-1226) (Prust et al. 2005).

G. oxydans 621H utilizes a wide range of sugars because it possesses a wide range of membrane-bound dehydrogenases that supply electrons for the respiratory chain composed of a membrane-bound transhydrogenase (GOX0310-0312), a non-proton-translocating NADH: ubiquinone oxidoreductase (GOX1675) and two quinol oxidases, bo_3 (GOX1911-1914) and *bd* (GOX0278-0279). The respiratory chain however lacks a proton translocating NADH: ubiquinone oxidoreductase (complex I) and, like most other AAB, also lacks a cytochrome *c* oxidase (complex IV) (Matsushita et al. 1992, Prust et al. 2005, Sievers and Swings 2005), which limits the capacity to create the electrochemical membrane potential and lowers the energy-transducing efficiency, resulting in a very slow growth observed in chemostat cultures (Olijve and Kok 1979). At least 75 genes in the genome of *G. oxydans* 621H were identified as potential oxidoreductases. Three have previously been characterized as membrane-bound quinoprotein dehydrogenases (quinohemoprotein alcohol dehydrogenase complex, a major polyol dehydrogenase and a glucose dehydrogenase). Four additional putative dehydrogenases were found – one predicted to be located in the periplasmic space and the other three in the cytoplasmic membrane (Prust 2004), one of the latter presumably a quinoprotein *myo*-inositol dehydrogenase (Hölscher et al. 2007). Many of the putative dehydrogenases are still poorly characterized and of unknown substrate specificity, although 15 are predicted to be membrane bound and use NAD^+ as a cofactor (Prust et al. 2005). The large oxidative potential of this organism is, therefore, still not fully explored.

Recently, an industrial strain of *G. oxydans* H24 was completely sequenced and its genome published (Ge et al. 2013). The strain is widely used in the production of vitamin C because of its ability to produce high amounts of L-sorbose, a trait enhanced from a wild-type through decades of selection (Ge et al. 2013). The genome of this strain consists of a circular chromosome of 3,602,424 bp and one plasmid of 213,808 bp. As in the strain 621H, almost 90 per cent of the genome encodes proteins or stable RNA. There are a total of 3,732 putative CDS (3,469 on the chromosome and 263 on the plasmid), 59 tRNA genes and five full rRNA operons (Table 1) (Ge et al. 2013). *G. oxydans* H24 is able to produce very high levels of L-sorbose. The genome carries information for membrane-bound pyrroloquinoline quinon (PQQ)-dependent D-sorbitol dehydrogenase (PQQ-SLDH) (*sldhAB*) (AFW01111 and AFW01112), flavin adenine dinucleotide (FAD)-dependent D-sorbitol dehydrogenase (FAD-SLDH) (*sldhSLC*) (AFW02569) and NADP-dependent D-sorbitol dehydrogenase (NADP-SLDH) (*sldH*) (AFW00649). Sequence comparisons to the homologues found in *G. oxydans* 621H show 79.4 per cent, 63.6 per cent and 29.2 per cent similarity, respectively (Ge et al. 2013).

Genus *Komagataeibacter* (previously *Gluconacetobacter*) contains species with high acetic acid production ability and acetic acid tolerance. Such species are *K. oboediens*, *K. europaeus* and *K. intermedius*, but their genomes are yet to be sequenced completely. For the moment, only two complete genomes of *Komagataeibacter xylinus* have been published. One of them is a well-known bacterial nanocellulose (BNC) producer and regarded as a model organism for biosynthesis of bacterial cellulose (Kubiak et al. 2014). BNC is used in the cosmetics industry for BNC-based cosmetic masks and wound dressings (Czaja et al. 2006). The food industry uses BNC in the production of nata de coco, a low-calorie dessert very popular in the East (Gama et al. 2012), and there is even intensive ongoing research in the field of medicine in the hope of developing BNC-based blood vessels and bone replacements (Gama et al. 2012, Kowalska-Ludwicka et al. 2013, Tang et al. 2015). *K. xylinus* E25 is currently the only cellulose-producing strain to be completely sequenced (Kubiak et al. 2014). The complete genome consists of a 3,447,725 bp circular chromosome and five plasmids of varying sizes (336,138, 87,176, 26,296, 5,531 and 2,216 bp). The total number of CDS is 3,228, of which 3,156 are protein-coding genes, 57 tRNA and five rRNA operons. Putative function was assigned to 2,378 (72.5 per cent) protein coding genes (Table 1) (Kubiak et al. 2014). *K. xylinus* E25 carries a large plasmid of 336,138 bp, which was not found in other *Komagataeibacter* representatives. However, such mega plasmids are frequently present in other *Alphaproteobacteria*, where they usually carry genes that are essential for survival in unfavorable conditions (e.g. exopolysaccharide production and heavy metal defence systems) (Finan et al. 2001, Andres et al. 2013, Mazur et al. 2013, Weidner et al. 2013). In the case of *K. xylinus* E25, genes responsible for arsenite metabolism were identified (H845_3442 - 3444) and several transport systems, but no genes were found for exopolysaccharide production (Kubiak et al. 2014). The most interesting aspect of *K. xylinus* E25 is the synthesis of BNC. The genome contains two bacterial cellulose synthase (*bcs*) operons (Saxena et al. 1990), although only one is functional (*bcsI*) (Kubiak et al. 2014). The functional operon consists of seven genes coding for endo-1,4-beta-glucanase (CMCax homologue) (H845_448), a putative beta-glucosidase endoglucanase (H845_449), a putative cpc homologue, four cellulose synthase (CS) subunits (BcsA, BcsB, BcsC and BcsD) (H845_450-3), and a beta-glucosidase (H845_454) (Kubiak et al. 2014). The exact role of the *bcs* operon in cellulose biosynthesis remains unclear despite great efforts undertaken to characterize its endoglucanases (Koo et al. 1998, Yasutake et al. 2006). Comparison of the four cellulose synthase subunits from *K. xylinus* E25 to the other known *bcs* sequences from *Komagataeibacter* representatives showed the highest similarity between *bcsI* operons of *K. xylinus* E25 and *K. xylinus* JCM 7664 (Kubiak et al. 2014). The most conservative regions of the *bcsI* operon lie within the *bcsA* gene (250-1450 nucleotide), which codes for the catalytic subunit (CS) and the whole *bcsD* gene whose function remains unclear (Kubiak et al. 2014). Another strain (*K. medellinensis* NBRC 3288) has recently been completely sequenced (Ogino et al. 2011). Unlike *K. xylinus* E25, this one is unable to synthesize cellulose. Its genome consists of a chromosome of 3,136,818 bp and seven plasmids (255,866, 76,071, 28,572, 4,776, 4,615, 4,255 and 2,218 bp). The chromosome has five rRNA operons and 60 tRNA genes. Putative function was assigned to 2,358 (73.9 per cent) from the total of 3,185 CDS (Table 1) (Ogino et al. 2011). Even though it is unable to produce cellulose, the genome contains two operons with 11 genes related to cellulose synthesis. The genes for endoglucanase (GLX_25040 and GLX25050) and cellulose synthase catalytic subunits (GLX_25060, GLX_25070, GLX_25080, GLX_25090 and GLX_25100) were found (Ogino et al. 2011). GLX_25070 and GLX_25080 show great similarities to the catalytic subunit *bcsB1* in *K. xylinus* JCM 7664 (BAA77586.1) (Umeda et al. 1998). *bcsB* is vital for the catalytic activity of cellulose synthase (Wong et al. 1990, Omadjela et al. 2013). A nonsense mutation (TGA stop codon at position 514 of the GLX_25070) in *K. xylinus* NBRC 3288, however, caused the splitting of *bcsB* into GLX_25070 and GLX_25080. Because *bcsB* gene is indispensable in cellulose synthesis (Wong et al. 1990), the single mutation could affect the cellulose production of this strain (Ogino et al. 2011).

Gluconacetobacter diazotrophicus Pal5 is an endophytic nitrogen-fixing AAB that lives in close association with sugarcane roots, stems and leaves. The intercellular space of the plant tissues provides a niche where it can grow to high concentrations (Cavalcante and Dobereiner 1988, James et al. 1994). It cannot grow on nitrate as a sole nitrogen source, but it can fix N_2 at a pH as low as 2.5, and in the presence of ammonium and high sugar concentration (Cavalcante and Dobereiner 1988). Other biotechnologically desirable features include promotion of plant growth by producing growth-promoting substances (*nif* mutants) (Sevilla et al. 2001), inhibition of the sugarcane pathogen *Xanthomonas albilineans* by excretion of a lysozyme-like bacteriocin (Blanco et al. 2005) and antifungal activity against *Fusarium* sp. and *Helmintosporium carbonum* (Mehnaz and Lazarovits 2006). It is the only endophytic AAB that has been sequenced completely (Bertalan et al. 2009). Its genome is particularly interesting because it can reveal novel information about the metabolic routes, organization and regulation of genes involved in nitrogen fixation and in the establishment of close associations with plant tissues (Bertalan et al. 2009).

The complete genome of *Ga. diazotrophicus* Pal5 is composed of one circular chromosome (3,944,163 bp) and two plasmids (38,818 and 16,610 bp). The chromosome contains four rRNA operons and 55 tRNA genes (Table 1) (Bertalan et al. 2009). The genome encodes 3938 putative CDS, of which 2,861 have an unknown function and 1,077 encode hypothetical proteins (Table 1). The larger of the two plasmids has 53 CDS, 37 of which encode hypothetical proteins and 11 code for the Type IV secretion system (T4SS). The rest of the CDS code for proteins which are involved in plasmid maintenance and replication. The small plasmid has 21 CDS and about half of them code for hypothetical proteins (Bertalan et al. 2009). Genomes of endophytic bacteria generally contain very few mobile elements (integrases and transposases), ranging approximately 0.5-1.5 per cent of the total CDS (Bertalan et al. 2009). The low number of mobile elements is assumed to be an adaptation to a more stable environment within a host (Krause et al. 2006). Surprisingly, the complete genome of *Ga. diazotrophicus* contains 245 transposable elements (55 integrases and 190 transposases), which represent 6.23 per cent of the total CDS. The high percentage of mobile elements could be a sign of only recent evolutionary adaptation to endophytic lifestyle (Parkhill et al. 2003). Alternatively, because *Ga. diazotrophicus* is found, albeit in low frequencies in the rhizosphere, it could acquire transposable elements from other bacteria occupying the same niche (Bertalan et al. 2009). Little less than a fifth (17 per cent) of the total chromosomal CDS are predicted to be highly expressed and 318 of them were identified (Bertalan et al. 2009). Not surprisingly, they belong to transcription and translation factors, ribosomal proteins, chaperones and different proteases (Bertalan et al. 2009). Among highly expressed CDS are also 50 transporters or transport related proteins, 27 of which are putative ABC transporters and six are TonB-dependent receptors. Proteins involved in ammonium metabolism are also highly expressed, which is consistent with the fact that *Ga. diazotrophicus* uses ammonium as the preferred nitrogen source when available (Bertalan et al. 2009). The genome comparison of *Ga. diazotrophicus* Pal5 to closely related *Gluconobacter oxydans* 621H, *Granulibacter bethesdensis* CGDNIH1 and *Acidiphilium cryptum* JF-5 identified 894 CDS as core and 851 CDS as accessory genes. Accessory regions cover around 24 per cent of the complete *G. dizotrophicus* Pal5 genome and are located in 28 distinct genome islands (Bertalan et al. 2009). The accessory regions have a higher share of conserved hypothetical proteins, hypothetical proteins, phage and IS elements, and pseudogenes compared to core or total genome, which indicate that they were acquired by horizontal gene transfer (Bertalan et al. 2009).

Comparison of Genomes among Species Isolated from Different Vinegars

Some acetic acid bacteria (AAB) can thrive in extreme acidic environments containing high concentrations of acetic acid (up to 18 per cent) (Trček et al. 2007) while most other microorganisms cease to grow under much lower concentration (0.5 per cent) (Conner and Kotrola 1995). The degree of acetic acid tolerance varies among AAB. Species traditionally used in the production of vinegars tolerate higher concentrations of acetic acid than other AAB (*A. pomorum*, *A. pasteurianus*, *K. europaeus*, *K. intermedius* and *K. oboediens*) (Sievers et al. 1992, Boesch et al. 1998, Sokollek et al. 1998, Schuller et al. 2000, Trček et al. 2000). Acetic acid increases the concentration of acetate anions and decreases the pH in the cytoplasm, thereby disrupting the transmembrane proton gradient essential for ATP synthesis, leading to the shutdown of cell metabolism (Diez-Gonzalez and Russell 1997, Trček et al. 2015). Some research regarding acetic acid resistance of AAB has been done.

To elucidate molecular mechanisms of acetic acid tolerance in *A. pasteurianus*, Wang et al. (2015) undertook a complete genome sequencing of *A. pasteurianus* strains CICC 20001 and CGMCC 1.41 and compared their genomes to some other complete and draft genomes of high acetic acid tolerant AAB (*A. pasteurianus* 386B, *A. pasteurianus* IFO 3283-32, *K. europaeus* 5P3 and *K. oboediens* 174Bp2). *A. pasteurianus* strains CICC 20001 and CGMCC 1.41 have different fermentation characteristics. They respond differently to initial levels of ethanol and acetic acid and produce different maximum levels of acetic acid during fermentation (6.35 and 7.15 per cent, respectively) (Wang et al. 2015). Genome comparisons between *A. pasteurianus* CICC 20001, CGMCC 1.41, 386B and the seven IFO 3283 substrains (Azuma et al. 2009) showed high homology between the chromosomes and great differences in plasmids (Wang et al. 2015). Regardless of the differences in plasmid genotypes, the overall genome similarities mean that they probably have similar mechanisms protecting them from deleterious effects of acetic acid. Membrane-bound PQQ-dependent alcohol dehydrogenase (PQQ-ADH) and aldehyde dehydrogenase (PQQ-ALDH) are the enzymes responsible for extracellular acetic acid production and resistance in AAB. The acetic acid is produced by sequential oxidation of alcohol to aldehyde and aldehyde to extracellular acetic acid, respectively (Yakushi and Matsushita 2010). Both enzymes are connected to the respiratory chain in AAB and are involved in energy metabolism. Enzymatic activities of PQQ-ADH were shown to be correlated to the ability of different strains of AAB to produce high concentrations of acetic acid (Trcek et al. 2006). In addition, the purified form of PQQ-ADH from a highly acetic acid-resistant strain of *K. europaeus* showed a relatively high stability in acid environment (Trcek et al. 2006). High levels of PQQ-ADH, therefore, contribute not only to greater production of acetic acid but also to increased tolerance of the extreme acid environment. The loss of the genes coding for the membrane-bound PQQ-ADH in *A. aceti* (Okumura et al. 1985) and *A. pasteurianus* (Takemura et al. 1993) resulted in diminished growth in the presence of higher concentrations of acetic acid compared to their wild-type counterparts. Disruption of the gene coding for PQQ-ADH in a strain of *A. pasteurianus* similarly resulted in the loss of tolerance for acetic acid (Chinnawirotpisan et al. 2003). AAB species used in vinegar production (*K. europaeus, K. oboediens* and *A. pasteurianus*) are capable of producing very high concentrations of acetic acid (18, 8 and 6 per cent, respectively) (Sievers et al. 1992, Trcek et al. 2006, Trček et al. 2015). Comparison of the total (cytosolic and membrane-bound) number of genes encoding alcohol dehydrogenases in efficient acetic acid producers (*A. pasteurianus* CICC 20001, *A. pasteurianus* CGMCC 1.41, *A. pasteurianus* 386B, *A. pasteurianus* IFO 3283-32, *K. europaeus* 5P3 and *K. oboediens* 174Bp2) has shown that *A. pasteurianus* strains possess fewer number of genes coding for alcohol dehydrogenases (10, 8, 13 and 8, respectively), than *K. europaeus* 5P3 and *K. oboediens* 174Bp2 (20 and 14, respectively) (Wang et al. 2015). The genomes of the same strains contain 1, 1, 2, 2, 7 and 6 genes coding membrane-bound alcohol dehydrogenases, respectively. Highly acetic acid-resistant *K. europaeus* 5P3 and *K. oboediens* 174Bp2, which can tolerate 18 and 8 per cent of acetic acid, respectively, therefore, contain more than three times the genes coding for the membrane-bound alcohol dehydrogenases than more sensitive *A. pasteurianus* strains (Sievers et al. 1992, Wang et al. 2015). Trcek et al. (2006) have linked the activity of membrane-bound PQQ-ADH to acetic acid production and resistance. Most of the putative dehydrogenases present in the genomes of AAB are, however, still poorly characterized. Whether there is any link between the number of membrane-bound alcohol dehydrogenases and acetic acid resistance remains to be resolved.

Part of acetic acid resistance in AAB is efficient assimilation of acetic acid through the enzymes of the tricarboxylic acid (TCA) cycle, such as acetate kinase, acetyl-CoA synthetase, citrate synthase, aconitate hydratase and phosphate acetyltransferase. Some AAB contain *aarC*, an additional gene involved in acetic acid assimilation (Fukaya et al. 1993), identified as succinyl-coenzyme A (CoA): acetate CoA transferase (AarC), which replaces succinyl-CoA synthetase in a modified TCA cycle (Mullins et al. 2008). Succinyl-CoA and acetate are thereby converted to acetyl-CoA and succinate, which increase the assimilation of acetate (Mullins et al. 2008). *aarC* is present in AAB, which produce and tolerate high acetic acid concentrations, such as *K. europaeus* 5P3, *K. oboediens* 174Bp2, *A. pasteurianus* 386B, *A. pasteurianus* IFO 3283-32, *A. pasteurianus* CICC 20001 and *A. pasteurianus* CGMCC 1.41 (Wang et al. 2015), but is absent in *Gluconobacter oxydans*, or opportunists *Granulibacter bethesdensis* and *Asaia bogorensis* (Table 2).

Genomes of AAB contain many genes coding for the proteins previously implicated in acid stress response in *Escherichia coli* (Lin et al. 1996, Foster 2004, Nakano and Fukaya 2008, Kanjee and Houry 2013, Wang et al. 2015). In the presence of various stress conditions such as high temperature, ethanol or acetic acid, AAB induce the expression of several stress proteins, such as GroES and GroEL, which play an important role in the proper folding of proteins and confer temperature, ethanol and acetic acid resistance (Okamoto-Kainuma et al. 2002, Andrés-Barrao et al. 2012). In addition, molecular chaperones, DnaK, DnaJ and GrpE, act cooperatively to stabilize protein-folding intermediates, help in the protein assembly and disassembly, secretion and degradation and play a key role under stress (Gething and Sambrook 1992). Overexpression of the *dnaKJ* operon in *A. aceti* leads to an increased resistance against high temperature and ethanol, but not acetic acid (Okamoto-Kainuma et al. 2002). An additional protein GrpE, which acts as a co-chaperone to DnaK, is needed

Table 2: Protein BLAST pairwise comparison (percentages of identity) of selected *Acetobacter pasteurianus* 386B homologues, implicated in acetic acid resistance, to other acetic acid bacteria

Locus	*MFS transporter*	*polE*	*aarC*	*ureC*	*ureB*	*ureA*	*ureF*	*ureG*	*ureD*
Acetobacter pasteurianus 386B	*WP_020944044.1*	*APA386B_1398*	*APA386B_2589*	*WP_003624697.1*	*WP_003624699.1*	*CCT59278.1*	*CCT59274.1*	*WP_003624691.1*	*CCT59279.1*
Acetobacter pasteurianus IFO 3283-32	99 per cent WP_003630289.1	99 per cent BAI15930.1	100 per cent WP_003622979.1	100 per cent WP_003624697.1	100 per cent WP_003624699.1	100 per cent WP_003624701.1	100 per cent WP_003624693.1	100 per cent WP_003624691.1	97 per cent WP_014457471.1
Acetobacter aceti NBRC 14818	82 per cent WP_010667228.1	77 per cent WP_052013474.1	97 per cent WP_042786640.1	89 per cent WP_010666820.1	70 per cent WP_010666821.1	81 per cent WP_010666822.1	57 per cent WP_010666818.1	86 per cent WP_010666817.1	46 per cent WP_010666823.1
Acetobacter pomorum DM001	absent	90 per cent EGE48485.1	99 per cent WP_006115151.1	98 per cent WP_006115794.1	96 per cent WP_006115793.1	98 per cent WP_006115792.1	84 per cent EGE48373.1	99 per cent WP_006115797.1	84 per cent WP_006115791.1
Komagataeibacter europaeus 5P3	83 per cent WP_019090497.1	absent	68 per cent WP_019091418.1	absent	absent	absent	absent	absent	absent
Komagataeibacter oboediens 174Bp2	84 per cent WP_010515909.1	absent	68 per cent WP_010514013.1	absent	absent	absent	absent	absent	50 per cent WP_010513378.1
Komagataeibacter intermedius TF2	85 per cent WP_039736174.1	absent	68 per cent WP_039735384.1	93 per cent WP_039736086.1	75 per cent WP_039736085.1	89 per cent WP_039736084.1	72 per cent WP_039736112.1	90 per cent WP_039736088.1	63 per cent WP_039736083.1
Komagataeibacter xylinus E25	84 per cent AHI24651.1	absent	68 per cent WP_025438701.1	absent	absent	absent	absent	absent	absent
Komagataeibacter medellinensis NBRC 3288	absent	absent	68 per cent WP_014104663.1	absent	absent	absent	absent	absent	absent
Gluconacetobacter diazotrophicus PAl 5	absent	absent	66 per cent WP_012225388.1	absent	absent	absent	absent	absent	absent

(Contd.)

Gluconobacter oxydans H24	absent	46 per cent WP_015072892.1	absent	absent	absent	absent	absent	absent	absent
Gluconobacter oxydans 621H	absent	45 per cent WP_011252611.1	absent	absent	absent	absent	absent	absent	absent
Granulibacter bethesdensis CGDNIH1	absent	45 per cent WP_011630814.1	absent	75 per cent WP_043453157.1	63 per cent WP_011632863.1	85 per cent WP_011632862.1	43 per cent WP_011632866.1	77 per cent WP_011632867.1	34 per cent Q0BQ43.1
Asaia bogorensis NBRC 16594	absent	45 per cent BAT18330.1	absent	absent	absent	absent	absent	absent	absent

(80-100), (60-80) and (30-60) percentage of identity.

to increase the resistance of *A. pasteurianus* against acetic acid stress during fermentation (Ishikawa et al. 2010a). ClpB is another molecular chaperone, which acts cooperatively with DnaK, DnaJ and GrpE in solubilizing and refolding of aggregated proteins during heat shock, but whose exact role in acetic acid resistance is currently unknown (Ishikawa et al. 2010b). In *A. pasteurianus* NBRC 3283, a regulator gene *rpoH* controls the expression of the *groEL, dnaKJ, grpE* and *clpB* genes and its disruption leads to increased acetic acid sensitivity (Okamoto-Kainuma et al. 2011). Genes coding for the enzymes conferring acetic acid resistance often cluster in the chromosomes of *A. pasteurianus* strains. Such clusters of functionally-related genes include combinations of ADH and ALDH, enzymes of the TCA cycle (phosphate acetyltransferase together with acetate kinase or the combination of citrate synthase and acetyl-CoA synthetase), and combinations of chaperones DnaK, GrpE and DnaJ (Wang et al. 2015).

In addition to more efficient assimilation of acetic acid and adaptation of cell proteins to a higher acid environment, AAB also effectively decrease toxic effects of acetic acid by pumping acetic acid from the cytosol to the periplasmic space, outside the cell. ATP independent and dependent transport systems involved in the export of acetic acid have been found in AAB. The membrane-bound PQQ-ADH and PQQ-ALDH, which are linked to the respiratory chain, transfer electrons via ubiquinone to O_2 when it is oxidized by ubiquinone oxidase. In the process, four protons from ubiquinone are released to the periplasmic space, resulting in a higher concentration of protons outside of the cell. *A. pasteurianus* IFO 3283 contains a putative transporter, which couples the proton motive force in the direction of the cytosol to an ATP-independent efflux of acetate (Matsushita et al. 2005b). Multidrug resistance transporters (MDR) of the major facilitator superfamily (MFS) found in *Saccharomyces cerevisiae* (Azr1, Aqr1, Tpo2 and Tpo3) have been shown to increase acid tolerance of yeast by pumping acetate out of the cells (Tenreiro et al. 2000, 2002, Fernandes et al. 2005). Interestingly, the genome of *A. pasteurianus* IFO 3283 encodes an orthologue (APA01_15230) of Azr1 (Trček et al. 2015). Similar proteins are also present in the genomes of *A. pasteurianus* 386B (WP_020944044.1), *A. aceti* NBRC 14818 (WP_010667228.1), *K. europaeus* 5P3 (WP_019090497.1), *K. oboediens* 174Bp2 (WP_010515909.1), *K. intermedius* TF2 (WP_039736174.1) and *K. xylinus* E25 (AHI24651.1) (Table 2), all of which can tolerate high acetic acid concentrations and are used in vinegar production (Sievers et al. 1992, Boesch et al. 1998). Genomes of some of the acid-sensitive AAB, on the other hand, lack this orthologue (Table 2). In future, it would be interesting to investigate and determine the exact role of these so far poorly characterized MFS transporters, to see to what extent they confer acetic acid tolerance to AAB.

In addition to ATP-independent MFS transporters, ATP-binding cassette (ABC) transporters could be involved in acetic acid resistance by actively pumping acetic acid outside of the cell. ABC transporters, which exist in prokaryotes and eukaryotes, are transmembrane proteins that use energy from the hydrolysis of ATP to transport various substrates, such as drugs, amino acids, ions, sugars, lipids and weak acids across membranes, thereby maintaining stable intracellular environment (Rees et al. 2009). They were shown to be involved in the transport of weak acids in *Saccharomyces cerevisiae* (Piper et al. 1998). Expression studies of *A. aceti* membrane proteins in the presence of varying concentrations of acetic acid showed an induction of a 60 kDa membrane protein, named AatA and classified as a type B ABC transporter. Overexpression of this protein on a multicopy plasmid increased the strain's ability to grow at higher rates and tolerate higher concentrations of acetic acid (Nakano et al. 2006). In addition to acetic acid, it seems to confer also formic and lactic acid resistance (Nakano and Fukaya 2008). Genome comparisons of *K. europaeus* 5P3 and *K. oboediens* 174Bp2, and *A. pasteurianus* 386B, IFO 3283-32, CICC 20001 and CGMCC 1.41 strains show that highly resistant *Komagataeibacter* strains contain more genes coding for putative ABC transporters than the more susceptible *Acetobacter* species. *K. europaeus* 5P3 and *K. oboediens* 174Bp2 have 86 and 93, and the genomes of *A. pasteurianus* strains 386B, IFO 3283-32, CICC 20001 and CGMCC 1.41 contain 43, 21, 50 and 56 genes, respectively. The correlation seems to suggest that the strains with a higher number of genes for putative ABC transporters cope better under high acid stress conditions (Wang et al. 2015). The lack of experimental data, however, prevents us from drawing more definite conclusions about the proposed correlation, as their exact function and involvement in acetic acid transport still has to be elucidated.

Bacteria frequently produce extracellular hetero- or homo-polysaccharides, which either remain associated with the cell wall and form capsular polysaccharides (CPS), or are secreted into the medium as free extracellular polysaccharides (EPS) (Deppenmeier and Ehrenreich 2009). Many of AAB produce strain specific hetero-polysaccharides (HePS), which often contain monomers of glucose, rhamnose, galactose, mannose and glucuronic acid (Kornmann et al. 2003, Ali et al. 2011, Serrato et al. 2013). During the growth of *Acetobacter tropicalis* SKU1100, some cells start to produce capsular polysaccharides (CPS) on the cell surface and form a pellicle of tightly associated cells. The cells with CPS start to dominate in the late exponential and stationary growth phase, but are absent in the early phases of acetate oxidation, indicating that the polysaccharides covering the cells offer protection by reducing diffusion of acetic acid into the cells (Kanchanarach et al. 2010). The genes responsible for pellicle formation in AAB are organized in *polABCDE* operon (Deeraksa et al. 2005, 2006), which shows a significant homology to the *rfbBACD* genes involved in the dTDP-rhamnose synthesis pathway in other Gram-negative bacteria (Marolda and Valvano 1995, Mitchison et al. 1997, Tsukioka et al. 1997, Boels et al. 2004). The dTDP-rhamnose synthesis pathway usually consists of four sequential steps; initially, glucose 1-phosphate is converted to dTDP-glucose by glucose-1-phosphate thymidyl-transferase (RfbA), then to dTDP-4-keto-6-deoxy-D-glucose by dTDP-glucose 4, 6-dehydratase (RfbB), then to dTDP 6 deoxy-L-mannose by dTDP-4-dehydrorhamnose-3,5 epimerase (RfbC), and finally to dTDP-L-rhamnose by dTDP-4-dehydrorhamnose reductase (RfbD). *Acetobacter* species have an additional *polE* gene downstream (Table 2). Nonfunctional *polE* gene in *A. tropicalis* SKU1100 causes a shift from cell surface CPS to secretory EPS, maintaining the general sugar composition of secreted polysaccharides. Disruption of *polB*, on the other hand, results in defective CPS and EPS synthesis. Strains with disrupted either *polB* or *polE* gene cannot form the pellicle at the top of the medium and also exhibit a higher acetic acid sensitivity (Deeraksa et al. 2005). A sequence of seven cytosine nucleotides in the *polE* gene was identified as the hot spot for frameshift mutations, which is the result of a slippage during replication (Deeraksa et al. 2005), a phenomenon described in other bacteria (Lovett 2004). An addition or deletion of a single cytosine nucleotide results in a frameshift, leading to interchange of phenotype between rough (CPS) and smooth (EPS) strains of *Acetobacter tropicalis* SKU1100 (Deeraksa et al. 2005).

AAB use several strategies to cope with acidic environment. In the presence of acetic acid, they upregulate the enzymes for the acetic acid assimilation, eliminate acetic acid from cytoplasm by pumping it out of the cell, modify cell morphology and membrane composition, and regulate the expression of molecular chaperons (Matsushita et al. 2005b, Nakano et al. 2006, Nakano and Fukaya et al. 2008). AAB share some mechanisms of acetic acid resistance with other well-studied bacteria. *Escherichia coli* and AAB have similar metabolic responses, transportation systems and stress proteins, which help to assimilate, eliminate, or adapt to high levels of acetic acid in the cytosol (Rollan et al. 2003, Foster 2004, Trček et al. 2015). The whole genome sequencing and subsequent comparative analysis is an effective method for analyzing gene functions and deciphering metabolic pathways, leading to better understanding of the overall metabolic blueprint of acid tolerance mechanisms. Still, the overall acid resistance mechanisms of AAB remain unclear.

Specific Genome Characteristics of Potentially Opportunistic Human Pathogens among AAB (*Asaia bogorensis* and *Granulibacter bethesdensis*)

Asaia bogorensis is a member of AAB which primarily colonises tropical flowers. Its special feature is that it oxidizes acetate to carbon dioxide and water, but at the same time lacks enzymes to convert ethanol to acetic acid (Yamada et al. 2000). It is able to oxidize a variety of sugars and alcohols with enzymes such as quinoprotein glycerol dehydrogenase, but at the same time lacks quinoprotein alcohol dehydrogenase (PQQ-ADH). The respiratory chain contains cytochrome bo_3 ubiquinol oxidase as the sole terminal oxidase (Ano et al. 2008). The bacterium was first reported as a human opportunistic pathogen in a patient with a peritoneal dialysis showing a persistent peritonitis (Snyder et al. 2004). Later it was often isolated from the blood of patients on intravenous drugs. The clinical isolates were frequently resistant to many antibiotics used against Gram-negative bacteria, except gentamycin and doxycycline (Tuuminen et al. 2006, 2007). It is, therefore, important to understand the genetic basis of pathogenicity of this organism. Comparison of 42 different prokaryotic universal genes (Kawai et al. 2015) and 70 AAB core genes (Chouaia et al. 2014) shows that it is phylogenetically related to *Gluconobacter*. The complete genome sequence of *As. bogorensis* NBRC 16594 consists of a single chromosome of 3,198,265 bp but lacks plasmids. The chromosome contains 2,758 CDS, 58 tRNA genes and five rRNA operons (Table 1).

As. bogorensis is not the only AAB associated with human infections. The other emerging pathogen is *Granulibacter bethesdensis*. It was isolated from patients with chronic granulomatous disease (CGD). CGD is the consequence of an impaired formation of superoxide within the phagocytes as a result of a genetic mutation. Patients are frequently inflicted with serious life-threatening bacterial and fungal infections caused by *Staphylococcus aureus*, *Serratia marcescens*, *Burkholderia cepacia* complex, *Nocardia* sp. and *Aspergillus* sp. (Segal et al. 2000). *Gr. bethesdensis* was found to cause fever and lymphadenitis (Greenberg et al. 2006a, b). Tests conducted on animal models showed that CGD mice were susceptible to *Gr. bethesdensis*, with a similar pathology as CGD patients. In contrast, *Gluconobacter oxydans* caused very little pathology in CGD mouse model (Greenberg et al. 2006a). The whole genome sequencing of *Gr. bethesdensis* CGDNIH1 and comparison to other CGD pathogens helped to reveal virulence factors associated to CGD (Greenberg et al. 2007). *Gr. bethesdensis* CGDNIH1 is composed of a chromosome of 2,708,355 bp and lacks plasmids. Chromosome contains a total of 2,437 protein coding sequences, 1,832 (75.17 per cent) of them with predicted function, 52 tRNAs and three rRNA operons. A comparison of *Gr. bethesdensis* CGDNIH1 and *G. oxydans* 621H identified 152 ORFs specific to *Gr. bethesdensis* (Greenberg et al. 2007). Unlike *G. oxydans*, it contains several *trb* (*trbF* (GbCGDNIH1_1062), *trbG* (GbCGDNIH1_1063), *trbI* (GbCGDNIH1_1064) and *trbL* (GbCGDNIH1_1061)) and *tra* (*traD* (GbCGDNIH1_1512) and *traI* (GbCGDNIH1_1514)) genes encoding the type IV DNA conjugation system that delivers DNA or proteins into bacterial cells (Burns 2003). In addition to the conjugation system, *Gr. bethesdensis* contains one putative phage region of 39 CDS, with 12 CDS showing similarity to *Haemophilus influenzae* (Greenberg et al. 2007). Majority of transposases (11 out of 16) are located in a 25 kb region with a higher than average GC content compared to the rest of the genome, which implies that it was acquired recently (Greenberg et al. 2007) and could be a hot spot for recombinations (Lawrence and Ochman 1997).

An interesting feature of *Gr. bethesdensis* is the presence of genes coding for a putative methanol dehydrogenase complex (GbCGDNIH1_0122, GbCGDNIH1_0651, GbCGDNIH1_0344 and GbCGDNIH1_1922), which is absent from *G. oxydans* (Greenberg et al. 2007). Moreover, *Gr. bethesdensis* is capable of growing on methanol as the sole carbon source (Greenberg et al. 2006a, b). Gram-negative bacteria have surface lipopolysacharides (LPS) and capsules which are known virulence factors (Miller et al. 2005). *Gr. bethesdensis* and *G. oxydans* also contain genes encoding LPS as well as cell surface capsular polysaccharides. However, *Gr. bethesdensis* also contains a gene for a glucosyltransferase (GbCGDNIH1_0741) absent in *G. oxydans* (Greenberg et al. 2007). *Gr. bethesdensis* also has some other pathogenic features absent in *G. oxydans*. A variety of genes for an efficient uptake of iron were found, including a heme oxygenase gene (GbCGDNIH1_0284), which serves to degrade heme to iron and biliverdin, and an *iucABCD* (iron uptake chelate) gene cluster for aerobactin synthesis (GbCGDNIH1_2375-78), which acts as a siderophore (Greenberg et al. 2007). Genes coding for adhesin involved in diffuse adherence (AIDA) were identified as virulence factors in certain strains of *Escherichia coli* and shown to be involved in binding to human and non-human cells (Sherlock et al. 2004). Many AIDA adhesins are present in *Gr. bethesdensis* (GbCGDNIH1_1601, GbCGDNIH1_0627, GbCGDNIH1_0615, GbCGDNIH1_1413 and GbCGDNIH1_1465), which are absent in *G. oxydans*. Homologues of GbCGDNIH1_0627 and GbCGDNIH1_0615 are also present in other CGD pathogens, like *Burkholderia cepacia* and the former in virtually pathognomonic *Chromobacterium violaceum* (Greenberg et al. 2007). *Gr. bethesdensis* also contains several hemagglutinin related genes (GbCGDNIH1_1408, GbCGDNIH1_1474 and GbCGDNIH1_1135), which show similarities to AIDA and filamentous hemagglutinin of *Bordetella pertussis* (Relman et al. 1989, Greenberg et al. 2007). A hemolysin activator protein homologue (GbCGDNIH1_1664) of ShlB in the CGD pathogen *Serratia marcescens* (Walker et al. 2004) and a hemolysin homologue (GbCGDNIH1_1663) to the filamentous hemagglutinin FhaB in *Bordetella pertussis* (Relman et al. 1989) are also found in *Gr. bethesdensis* (Greenberg et al. 2007). *Gr. bethesdensis* also has the genes coding for the type I protein secretion system. The system is composed of an ATP-binding subunit (GbCGDNIH1_1527), an ATPase-encoding component (GbCGDNIH1_1528) and an adaptor-encoding protein (GbCGDNIH1_1526). Homologues of GbCGDNIH1_1527 are present in other CGD pathogens (*Burkholderia cepacia, Chromobacterium violaceum, Nocardia farcinica* and *Staphylococcus aureus*). The type I secretion system might be involved in the secretion of hemolysins and as such, it could be an important factor in pathogenesis of *Gr. bethesdensis* (Greenberg et al. 2007). Other genes that could contribute to pathogenesis of *Gr. bethesdensis* in CGD patients are genes involved in oxidative and acid stress protection. In addition to superoxide dismutase (GbCGDNIH1_0954) and catalase encoding genes (GbCGDNIH1_1969 and GbCGDNIH1_1598), *Gr. bethesdensis* contains a unique peroxidase gene (GbCGDNIH1_1677) which is absent in *G. oxydans*. The importance of these genes in pathogenicity of CGD patients was demonstrated by rapid upregulation in healthy individuals, but not in CGD patients (Greenberg et al. 2015). Efficient neutralisation of reactive oxygen species produced by the host may be a way *Gr. bethesdensis* increases its chances of survival (Greenberg et al. 2007, 2015). Another protective mechanism could be conversion of urea to ammonia and carbon dioxide. Several chromosomal loci have been identified in *Gr. bethesdensis*, which could be involved in pH-neutralization. The relevant loci are urease locus *ureABCDEFG* (GbCGDNIH1_2161-67), amide-urea transport locus (GbCGDNIH1_1831-36), allophanate hydrolase (GbCGDNIH1_1744) and urea carboxylase locus (GbCGDNIH1_1745). The first two loci are absent in closely related *G. oxydans*, which is not a CGD pathogen. Interestingly, other CGD pathogens (*B. cepacia, N. farcinica* and *S. aureus*) contain all the three loci. This could mean that CGD pathogens, including *Gr. bethesdensis*, may be more resistant against acid within phagocytic vacuoles in macrophages and neutrophils (Greenberg et al. 2007).

The general chromosomal features of *As. bogorensis* are similar to other AAB (Table 1). *As. bogorensis* and *Gr. bethesdensis*, however, share a few interesting common features, which distinguish them from other AAB. They both lack plasmids and have only few transposase genes in their genomes (Table 1). Phylogenetic trees based on core genes also show that they are only distantly related. This suggests that their pathogenicity and common traits evolved independently during evolutionary adaptation (Kawai et al. 2015). Characteristic genes that could play a role in pathogenicity of AAB were identified by comparing the genomes of the two AAB known to form close association with eukaryotic organisms and free-living AAB. The genes found in both *As. bogorensis* and *Gr. bethesdensis* include a large adhesin protein, haemolysin activator protein, several peroxidase genes and genes involved in antibiotic-resistance (beta-lactamase and penicillin amidase). Both bacteria also contain a type II secretion system (Kawai et al. 2015). Large adhesin gene family and type II secretion system are common among animal and plant pathogens (Rojas et al. 2002, Douzi et al. 2012). Comparative transcriptome analysis under co-culture with mammalian cells showed increased expression of 101 and diminished expression of 78 genes out of the total 2,833 genes in *As. bogorensis* (Kawai et al. 2015). *As. bogorensis* genome contains two complete operons of *cyoABCD* encoding bo_3 ubiquinol oxidase (*cyoABCD-1* and *cyoABCD-2*). Kawai et al. analyzed 705 bacterial species belonging to 705 different genera and they detected *cyoABCD* in 98 species. Besides *As. bogorensis*, eight other analysed species contained two or more complete *cyo* operons. It is interesting to note that many of them are able to form close associations with eukaryotic organisms (*Bordetella bronchiseptica* RB50, *Burkholderia pseudomallei* K96243, *Stenotrophomonas maltophilia* K279a and *Ralstonia solanacearum* GMI100). The *cyoABCD-1* operon of *As. bogorensis* forms a monophyletic clade with other AAB, whereas the *cyoABCD-2* shows similarities to other various *Proteobacteria* (Kawai et al. 2015). Transcription studies in *As. bogorensis* showed higher transcription levels of the *cyoABCD-2* operon compared to *cyoABCD-1*, under co-culture with mammalian cells and the opposite under non-co-culture conditions. This suggests that the alternating *cyoABCD-1* and *cyoABCD-2* expression might be advantageous when grown on animal tissue (Kawai et al. 2015). *As. bogorensis* and some other AAB (*A. pasteurianus, K. medellinensis* and *Gr. bethesdensis*) contain one *nuo* operon which codes for a type I NADH dehydrogenase (complex I), *Ga. diazotrophicus* contains two and *G. oxydans* lacks this operon completely. Phylogenetic analysis of the *nuo* operons (*nuoHIJKLN*) based on amino acid sequences indicates that *As. bogorensis* and *Ga. diazotrophicus* form a clade separated from *A. pasteurianus, K. medellinensis, Gr. bethesdensis,* and from the second operon of *Ga. diazotrophicus*. This, together with the phylogenetic analysis of prokaryotic core genes, indicates that *As. bogorensis* and *Ga. diazotrophicus* acquired an extra *nuo* operon and *As. bogorensis* and *G. oxydans* lost the one characteristic for AAB. In addition, *As. bogorensis* also has *ndh* (encoding type II NADH-dehydrogenase), *sdh* (encoding membrane-associated FAD-dependent succinate dehydrogenase) and *atp* (encoding F1F0-type ATP synthase) operons involved in the electron transport. The expressions of these operons also change under co-culture conditions with mammalian cells, resulting in downregulation (*sdh* and *atp*) or upregulation (*ndh*) under the same conditions (Kawai et al. 2015).

Like other AAB (Prust et al. 2005, Ogino et al. 2011), *As. bogorensis* genome contains several genes encoding membrane-associated dehydrogenases. The six genes code for sorbitol dehydrogenase (SLDH), gluconate 2 dehydrogenase (GADH), glucose dehydrogenase (GDH), glycerol dehydrogenase (GLDH), inositol dehydrogenase (IDH) and D-lactate dehydrogenase (LDH) (Illeghems et al. 2013, Chouaia et al. 2014). At the same time, *As. bogorensis* lacks the genes coding for membrane-associated alcohol dehydrogenase (ADH) and aldehyde dehydrogenase (ALDH). Because of this, *As. bogorensis* cannot oxidise ethanol to acetic acid (Yamada et al. 2000, Ano et al. 2008). *As. bogorensis* also contains a unique pyruvate oxidase operon (Asbog_02184-Asbog_02187). The genes in the operon code for a thiamine- and FAD-binding dehydrogenase, a putative protein with seven transmembrane helices, a gluconate 2-dehydrogenase subunit 3, which is similar to FAD-binding small subunits of primary dehydrogenases (*sldh* and *gadh*), and a glucose-methanol-choline family oxidoreductase similar to FAD-binding large subunits of primary dehydrogenases (*sldh* and *gadh*). One possibility is that it is located on the periplasmic side of the cytoplasmic membrane, but it could also localise on the cytoplasmic side of the membrane. The exact topology of the pyruvate oxidase is, therefore, still questionable (Kawai et al. 2015). Under co-culture with mammalian cells, the operon for the pyruvate oxidase is highly expressed compared to the other primary dehydrogenases, whose expression levels remain the same under different conditions. The expression of pyruvate oxidase may be related to nutrition, concentration of oxygen, medium, temperature, as well as interactions with mammalian cells. The functional implications of the altered expression levels are unknown (Kawai et al. 2015). The majority of genes that had the highest expression levels under co-culture conditions are encoding hypothetical very small proteins (<150 amino acids) with an unknown function, most of them unique to *As. bogorensis*. Small proteins were previously implicated in the stress response and adaptation to stressful environments (Setlow et al. 2007, Hemm et al. 2010). The other highly expressed genes under co-culture conditions included *rpoH* (heat shock factor Sigma 32), five genes for a heavy-metal resistance domain protein, thioredoxin peroxidase, stress response protein CsbD, cold shock transcriptional regulator and a starvation-inducible DNA-binding protein (Kawai et al. 2015). In *As. bogorensis*, the expression of heat shock protein genes (*groEL*, *groES* and *dnaJK*) diminished under co-culture conditions. Bacterial heat shock proteins (HSP) are immunogenic and therefore, stimulate immune responses from their hosts through increased secretion of pro-inflammatory cytokines (La Verda et al. 1999, Maguire et al. 2002). The lower expression of HSP under co-culture conditions could be a way to avoid the immune response of the host (Kawai et al. 2015).

Comparison of Genomes Between Species Isolated as Insects' Symbionts and Other Free-Living Acetic Acid Bacteria

AAB bacteria live in nutrient rich environments, especially food matrices and plant tissue. In addition to free-living representatives, some were found to exist in symbiotic relationships within insect gut (Crotti et al. 2010). *Asaia platycodi* was found to colonise the gut and salivary glands and reproductive systems of both sexes of the mosquito *Anopheles stephensi*. Colonisation of these compartments enables *Asaia* to spread horizontally and vertically (Damiani et al. 2008, Crotti et al. 2009, Gonella et al. 2012). Benefits to the host include larvae development (Chouaia et al. 2012) through modulation of gene expression affecting cuticle development (Mitraka et al. 2013). In *Drosophila*, AAB affect innate immunity, which in turn inhibits the growth of pathogens (Ryu et al. 2008). In the same host, *Acetobacter pomorum* affects the insulin signalling pathway, which regulates development, body size, energy metabolism and intestinal activity of the host (Shin et al. 2011).

To reveal the traits associated with a symbiotic lifestyle in insect gut, Chouaia et al. (2014) sequenced the genomes of insect symbionts *As. platycodi* and *Saccharibacter* sp. which colonise *Anopheles stephensi* and *Apis mellifera*, respectively and compared them to 14 previously published AAB genomes. Both symbionts were found to contain traits compatible with a symbiotic lifestyle. Secretion systems Sec-SRP and Tat were present in both genomes. In addition, genome of *As. platycodi* contains type IV secretion system and

ABC transport system. Both could be involved in communication between the bacterium and the host. Surface structures, which could have a role in the colonisation of the host epithelium include genes for flagellar components (MotA, MotB, FlaA, FlaB, FlgC, FlgD, FlgE2, FlgH and FtsI), fimbriae (Sf-Chap and Sf-UshP) and glycan synthesis. Although these traits could help the association between the bacterium in the host, they are not enough to establish symbiosis, as they were found in symbionts and free-living AAB (Chouaia et al. 2014).

Production of metabolites, which act as insect pheromones (acetoin) (Tolasch et al. 2003) and modulators of innate immunity (2-3 butandiol) (Bari et al. 2011) could also influence insect physiology and facilitate the establishment of symbiotic relationship. The genes for acetoin and 2-3 butandiol synthesis, which are present in genomes of symbiotic and free-living AAB could provide AAB with a propensity for symbiotic lifestyle in insect host (Chouaia et al. 2014).

AAB are aerobic bacteria found in oxygen-rich environments (Kersters et al. 2006). Symbiotic AAB, on the other hand, inhabit niches in insect gut with low levels of oxygen. All AAB, including symbiotic *As. platycodi* and *Saccharibacter* sp., possess respiratory chains with high oxygen affinity terminal oxidases cytochrome bo_3 (*CyoA-D*) and *bd* ubiquinol oxidase (*CydAB*). This gives them potential to thrive in oxygen-depleted animal intestine (Chouaia et al. 2014). The phylogenetic trees based only on cytochrome bo_3 subunits and AAB core genome match each other. In both trees, the two symbionts branch from the genus *Gluconobacter*. Cytochrome bo_3 oxidase thus seems to follow the evolutionary history of AAB from the common ancestor. The phylogenetic tree based on *bd* ubiquinol oxidase, on the other hand, mismatches the topology of the tree inferred from the core genome. In this case, *Saccharibacter* sp. branches with *Acetobacter* instead of *Gluconobacter*, implying that the horizontal gene transfer of the *bd* oxidase might have occurred during the evolution of AAB (Chouaia et al. 2014). Phylogenetic comparisons also indicate that terminal oxidases with high oxygen affinity were present in the common ancestor of AAB, conferring capacity for thriving in microoxic environments, such as insect gut (Sudakaran et al. 2012).

Conclusion

AAB are a group of incredibly diverse organisms with distinctive metabolism capable of producing large amounts of organic acids from a wide variety of substrates. Comparing the available genomes among AAB and other organisms helps to broaden the knowledge of their metabolic potential by discovering putative factors involved in acetic acid resistance (alternative TCA cycle, transport systems, factors involved in increased stability of proteins under acidic conditions) and identify genes responsible for the synthesis of different exopolysaccharides. High metabolic versatility of some AAB coupled with unstable phenotype has always indicated a relative instability of the genetic material—a phenomenon now easier to explain thanks to the presence of various genomic instability factors like high number of plasmids and mobile genetic elements. On the other hand, strains of *A. pasteurianus*, which proved stable through decades also contain CRISPR and restriction modification systems, known to increase genomic stability. Although AAB are generally regarded as nonpathogenic, some have been described as opportunistic pathogens. This could be due to acquired virulence factors, such as adhesion molecules, hemolysins and iron intake system. An exciting new group of insect gut symbionts has recently been discovered which expand our notion of versatility among AAB even further. Genomic comparisons have shown the presence of factors that could be involved in forming tighter association with other organisms. The available genomic sequences enable us to investigate the genetic potential of AAB in more detail than it was ever possible before. The growing pool of sequenced strains will bring new insights into common traits and unique properties of the ever-expanding group of AAB.

References

Adachi, O., Moonmangmee, D., Toyama, H., Yamada, M., Shinagawa, E. and Matsushita, K. (2003). New developments in oxidative fermentation. *Applied Microbiology and Biotechnology* **60:** 643-653.

Ali, I.A.I., Akakabe, Y., Moonmangmee, S., Deeraksa, M., Matsutani, M., Yakushi, T., Yamada, M. and Matsushita, K. (2011). Structural characterization of pellicle polysaccharides of *Acetobacter tropicalis* SKU1100 wild type and mutant strains. *Carbohydrate Polymers* **86:** 1000-1006.

Andrés-Barrao, C., Saad, M.M., Chappuis, M.L., Boffa, M., Perret, X., Perez, R.O. and Barja, F. (2012). Proteome analysis of *Acetobacter pasteurianus* during acetic acid fermentation. *Journal of Proteomics* **75:** 1701-1717.

Andres, J., Arsène-Ploetzel, F., Barbe, V., Brochier-Armanet, C., Cleiss-Arnold, J., Coppée, J.Y., Dillies, M.A., Geist, L., Joublin, A., Koechler, S., Lassalle, F., Marchal, M., Médigue, C., Muller, D., Nesme, X., Plewniak, F., Proux, C., Ramírez-Bahena, M.H., Schenowitz, C., Sismeiro, O., Vallenet, D., Santini, J.M. and Bertin, P.N. (2013). Life in an arsenic-containing gold mine: Genome and physiology of the autotrophic arsenite-oxidizing bacterium *Rhizobium* sp. NT-26. *Genome Biology and Evolution* **5:** 934-953.

Ano, Y., Toyama, H., Adachi, O. and Matsushita, K. (2008). Energy metabolism of a unique acetic acid bacterium, *Asaia bogorensis*, that lacks ethanol oxidation activity. *Bioscience, Biotechnology, and Biochemistry* **72:** 989-997.

Azuma, Y., Hosoyama, A., Matsutani, M., Furuya, N., Horikawa, H., Harada, T., Hirakawa, H., Kuhara, S., Matsushita, K., Fujita, N. and Shirai, M. (2009). Whole-genome analyses reveal genetic instability of *Acetobacter pasteurianus*. *Nucleic Acids Research* **37:** 5768-5783.

Bari, W., Song, Y.J. and Yoon, S.S. (2011). Suppressed induction of proinflammatory cytokines by a unique metabolite produced by *Vibrio cholerae* O1 El Tor biotype in cultured host cells. *Infection and Immunity* **79:** 3149-3158.

Bartowsky, E.J. and Henschke, P.A. (2008). Acetic acid bacteria spoilage of bottled red wine—A review. *International Journal of Food Microbiology* **125:** 60-70.

Bertalan, M., Albano, R., De Pádua, V., Rouws, L., Rojas, C., Hemerly, A., Teixeira, K., Schwab, S., Araujo, J., Oliveira, A., França, L., Magalhães, V., Alquéres, S., Cardoso, A., Almeida, W., Loureiro, M.M., Nogueira, E., Cidade, D., Oliveira, D., Simão, T., Macedo, J., Valadão, A., Dreschsel, M., Freitas, F., Vidal, M., Guedes, H., Rodrigues, E., Meneses, C., Brioso, P., Pozzer, L., Figueiredo, D., Montano, H., Junior, J., De Souza Filho, G., Flores, V.M.Q., Ferreira, B., Branco, A., Gonzalez, P., Guillobel, H., Lemos, M., Seibel, L., Macedo, J., Alves-Ferreira, M., Sachetto-Martins, G., Coelho, A., Santos, E., Amaral, G., Neves, A., Pacheco, A.B., Carvalho, D., Lery, L., Bisch, P., Rössle, S.C., Ürményi, T., Pereira, A.R., Silva, R., Rondinelli, E., Von Krüger, W., Martins, O., Baldani, J.I. and Ferreira, P.C.G. (2009). Complete genome sequence of the sugarcane nitrogen-fixing endophyte *Gluconacetobacter diazotrophicus* Pal5. *BMC Genomics* **10:** 450.

Blanco, Y., Blanch, M., Pinon, B., Legaz, M.E. and Vicente, C. (2005). Antagonism of *Gluconacetobacter diazotrophicus* (a sugarcane endosymbiont) against *Xanthomonas albilineans* (pathogen) studied in alginate-immobilized sugarcane stalk tissues. *Journal of Bioscience and Bioengineering* **99:** 366-371.

Boels, I.C., Beerthuyzen, M.M., Kosters, M.H.W., Van Kaauwen, M.P.W., Kleerebezem, M. and De Vos, W.M. (2004). Identification and functional characterization of the *Lactococcus lactis rfb* operon, required for dTDP-rhamnose biosynthesis. *Journal of Bacteriology* **186:** 1239-1248.

Boesch, C., Trček, J., Sievers, M. and Teuber, M. (1998). *Acetobacter intermedius*, sp. nov. *Systematic and Applied Microbiology* **21:** 220-229.

Bremus, C., Herrmann, U., Bringer-Meyer, S. and Sahm, H. (2006). The use of microorganisms in L-ascorbic acid production. *Journal of Biotechnology* **124:** 196-205.

Burns, D.L. (2003). Type IV transporters of pathogenic bacteria. *Current Opinion in Microbiology* **6:** 29-34.

Caliando, B.J. and Voigt, C.A. (2015). Targeted DNA degradation using a CRISPR device stably carried in the host genome. *Nature Communications* **6:** 6989.

Canchaya, C., Proux, C., Fournous, G., Bruttin, A. and Brussow, H. (2003). Prophage genomics. *Microbiology and Molecular Biology Reviews* **67:** 238-276.

Cavalcante, V.A. and Dobereiner, J. (1988). A new acid-tolerant nitrogen-fixing bacterium associated with sugarcane. *Plant and Soil* **108:** 23-31.

Chinnawirotpisan, P., Theeragool, G., Limtong, S., Toyama, H., Adachi, O.O. and Matsushita, K. (2003). Quinoprotein alcohol dehydrogenase is involved in catabolic acetate production, while NAD-dependent alcohol dehydrogenase in ethanol assimilation in *Acetobacter pasteurianus* SKU1108. *Journal of Bioscience and Bioengineering* **96:** 564-571.

Chouaia, B., Rossi, P., Epis, S., Mosca, M., Ricci, I., Damiani, C., Ulissi, U., Crotti, E., Daffonchio, D., Bandi, C. and Favia, G. (2012). Delayed larval development in *Anopheles* mosquitoes deprived of *Asaia* bacterial symbionts. *BMC Microbiology* **12** (Suppl 1): S2.

Chouaia, B., Gaiarsa, S., Crotti, E., Comandatore, F., Degli Esposti, M., Ricci, I., Alma, A., Favia, G., Bandi, C. and Daffonchio, D. (2014). Acetic acid bacteria genomes reveal functional traits for adaptation to life in insect guts. *Genome Biology and Evolution* **6:** 912-920.

Cleenwerck, I., De Wachter, M., Gonzalez, A., De Vuyst, L. and De Vos, P. (2009). Differentiation of species of the family Acetobacteraceae by AFLP DNA fingerprinting: *Gluconacetobacter kombuchae* is a later heterotypic synonym of *Gluconacetobacter hansenii*. *International Journal of Systematic and Evolutionary Microbiology* **59:** 1771-1786.

Conner, D.E. and Kotrola, J.S. (1995). Growth and survival of *Escherichia coli* O157:H7 under acidic conditions. *Applied and Environmental Microbiology* **61:** 382-385.

Crotti, E., Damiani, C., Pajoro, M., Gonella, E., Rizzi, A., Ricci, I., Negri, I., Scuppa, P., Rossi, P., Ballarini, P., Raddadi, N., Marzorati, M., Sacchi, L., Clementi, E., Genchi, M., Mandrioli, M., Bandi, C., Favia, G., Alma, A. and Daffonchio, D. (2009). *Asaia*, a versatile acetic acid bacterial symbiont, capable of cross-colonizing insects of phylogenetically distant genera and orders. *Environmental Microbiology* **11:** 3252-3264.

Crotti, E., Rizzi, A., Chouaia, B., Ricci, I., Favia, G., Alma, A., Sacchi, L., Bourtzis, K., Mandrioli, M., Cherif, A., Bandi, C. and Daffonchio, D. (2010). Acetic acid bacteria, newly emerging symbionts of insects. *Applied and Environmental Microbiology* **76:** 6963-6970.

Czaja, W., Krystynowicz, A., Bielecki, S. and Brown, R.M. (2006). Microbial cellulose—The natural power to heal wounds. *Biomaterials* **27:** 145-151.

Damiani, C., Ricci, I., Crotti, E., Rossi, P., Rizzi, A., Scuppa, P., Esposito, F., Bandi, C., Daffonchio, D. and Favia, G. (2008). Paternal transmission of symbiotic bacteria in malaria vectors. *Current Biology* **18:** 1087-1088.

Deeraksa, A., Moonmangmee, S., Toyama, H., Yamada, M., Adachi, O. and Matsushita, K. (2005). Characterization and spontaneous mutation of a novel gene, *polE*, involved in pellicle formation in *Acetobacter tropicalis* SKU 1100. *Microbiology* **151:** 4111-4120.

Deeraksa, A., Moonmangmee, S., Toyama, H., Adachi, O. and Matsushita, K. (2006). Conversion of capsular polysaccharide, involved in pellicle formation, to extracellular polysaccharide by *galE* deletion in *Acetobacter tropicalis*. *Bioscience, Biotechnology, and Biochemistry* **70:** 2536-2539.

Deppenmeier, U., Hoffmeister, M. and Prust, C. (2002). Biochemistry and biotechnological applications of *Gluconobacter* strains. *Applied Microbiology and Biotechnology* **60:** 233-242.

Deppenmeier, U. and Ehrenreich, A. (2009). Physiology of acetic acid bacteria in light of the genome sequence of *Gluconobacter oxydans*. *Journal of Molecular Microbiology and Biotechnology* **16:** 69-80.

De Vuyst, L. and Weckx, S. (2016). The cocoa bean fermentation process: From ecosystem analysis to starter culture development. *Journal of Applied Microbiology* (in press).

Diez-Gonzalez, F. and Russell, J.B. (1997). The ability of *Escherichia coli* O157:H7 to decrease its intracellular pH and resist the toxicity of acetic acid. *Microbiology* **143:** 1175-1180.

Douzi, B., Filloux, A. and Voulhoux, R. (2012). On the path to uncover the bacterial type II secretion system. Philosophical transactions of the Royal Society of London Series B. *Biological Sciences* **367:** 1059-1072.

Fernandes, A.R., Mira, N.P., Vargas, R.C., Canelhas, I. and Sa-Correia, I. (2005). *Saccharomyces cerevisiae* adaptation to weak acids involves the transcription factor Haa1p and Haa1p-regulated genes. *Biochemical and Biophysical Research Communications* **337:** 95-103.

Finan, T.M., Weidner, S., Wong, K., Buhrmester, J., Chain, P., Vorhölter, F.J., Hernandez-Lucas, I., Becker, A., Cowie, A., Gouzy, J., Golding, B. and Pühler, A. (2001). The complete sequence of the 1,683-kb pSymB megaplasmid from the N_2-fixing endosymbiont *Sinorhizobium meliloti*. *Proceedings of the National Academy of Sciences of the USA* **98:** 9889-9894.

Foster, J.W. (2004). *Escherichia coli* acid resistance: Tales of an amateur acidophile. *Nature Reviews Microbiology* **2:** 898-907.

Fukaya, M., Takemura, H., Tayama, K., Okumura, H., Kawamura, Y., Horinouchi, S. and Beppu, T. (1993). The *aarC* gene responsible for acetic acid assimilation confers acetic acid resistance on *Acetobacter aceti*. *Journal of Fermentation and Bioengineering* **76:** 270-275.

Gama, M., Gatenholm, P. and Klemm, D. (2012). *Bacterial NanoCellulose: A Sophisticated Multifunctional Material*. CRC Press, Boca Raton, Florida.

Ge, X., Zhao, Y., Hou, W., Zhang, W., Chen, W., Wang, J., Zhao, N., Lin, J., Wang, W., Chen, M., Wang, Q., Jiao, Y., Yuan, Z. and Xiong, X. (2013). Complete genome sequence of the industrial strain *Gluconobacter oxydans* H24. *Genome Announcements* **1(1):** e00003-13.

Gething, M.J. and Sambrook, J. (1992). Protein folding in the cell. *Nature* **355:** 33-45.

Gonella, E., Crotti, E., Rizzi, A., Mandrioli, M., Favia, G., Daffonchio, D. and Alma, A. (2012). Horizontal transmission of the symbiotic bacterium *Asaia* sp. in the leafhopper *Scaphoideus titanus* Ball (Hemiptera: Cicadellidae). *BMC Microbiology* **12** (Suppl. 1): S4.

Greenberg, D.E., Ding, L., Zelazny, A.M., Stock, F., Wong, A., Anderson, V.L., Miller, G., Kleiner, D.E., Tenorio, A.R., Brinster, L., Dorward, D.W., Murray, P.R. and Hollandet, S.M. (2006a). A novel bacterium associated with lymphadenitis in a patient with chronic granulomatous disease. *PLoS Pathogens* **2:** 260-267.

Greenberg, D.E., Porcella, S.F., Stock, F., Wong, A., Conville, P.S. and Murray, P.R. (2006b). *Granulibacter bethesdensis* gen. nov., sp nov., a distinctive pathogenic acetic acid bacterium in the family Acetobacteraceae. *International Journal of Systematic and Evolutionary Microbiology* **56:** 2609-2616.

Greenberg, D.E., Porcella, S.F., Zelazny A.M., Virtaneva, K., Sturdevant, D.E., Kupko III, J.J., Barbian, K.D., Babar, A., Dorward, D.W. and Holland, S.M. (2007). Genome sequence analysis of the emerging human pathogenic acetic acid bacterium *Granulibacter bethesdensis*. *Journal of Bacteriology* **189.** 8727-8736.

Greenberg, D.E., Sturdevant, D.E., Marshall-Batty, K.R., Chu, J., Pettinato, A.M., Virtaneva, K., Lane, J., Geller, B.L., Porcella, S.F., Gallin, J.I., Holland, S.M. and Zarember, K.A. (2015). Simultaneous host-pathogen transcriptome analysis during *Granulibacter bethesdensis* infection of neutrophils from healthy subjects and patients with chronic granulomatous disease. *Infection and Immunity* **83:** 4277-4292.

Gullo, M., Caggia, C., De Vero, L. and Giudici, P. (2006). Characterization of acetic acid bacteria in 'traditional balsamic vinegar'. *International Journal of Food Microbiology* **106:** 209-212.

Hemm, M.R., Paul, B.J., Miranda-Rios, J., Zhang, A., Soltanzad, N. and Storz, G. (2010). Small stress response proteins in *Escherichia coli*: Proteins missed by classical proteomic studies. *Journal of Bacteriology* **192:** 46-58.

Hidalgo, C., Vegas, C., Mateo, E., Tesfaye, W., Cerezo, A. B., Callejón, R. M., Poblet, M., Guillamón, J. M., Mas, A. and Torija, M. J. (2010). Effect of barrel design and the inoculation of *Acetobacter pasteurianus* in wine vinegar production. *International Journal of Food Microbiology* **141:** 56-62.

Hölscher, T., Weinert-Sepalage, D. and Görisch, H. (2007). Identification of membrane-bound quinoprotein inositol dehydrogenase in *Gluconobacter oxydans* ATCC 621H. *Microbiology* **153:** 499-506.

Illeghems, K., De Vuyst, L. and Weckx, S. (2013). Complete genome sequence and comparative analysis of *Acetobacter pasteurianus* 386B, a strain well-adapted to the cocoa bean fermentation ecosystem. *BMC Genomics* **14:** 526.

Ishikawa, M., Okamoto-Kainuma, A., Jochi, T., Suzuki, I., Matsui, K., Kaga, T. and Koizumi, Y. (2010a). Cloning and characterization of *grpE* in *Acetobacter pasteurianus* NBRC 3283. *Journal of Bioscience and Bioengineering* **109:** 25-31.

Ishikawa, M., Okamoto-Kainuma, A., Matsui, K., Takigishi, A., Kaga, T. and Koizumi, Y. (2010b). Cloning and characterization of *clpB* in *Acetobacter pasteurianus* NBRC 3283. *Journal of Bioscience and Bioengineering* **110:** 69-71.

James, E.K., Reis, V.M., Olivares, F.L., Baldani, J.I. and Dobereiner, J. (1994). Infection of sugar cane by the nitrogen-fixing bacterium *Acetobacter diazotrophicus*. *Journal of Experimental Botany* **45:** 757-766.

Kanchanarach, W., Theeragool, G., Inoue, T., Yakushi, T., Adachi, O. and Matsushita, K. (2010). Acetic acid fermentation of *Acetobacter pasteurianus*: Relationship between acetic acid resistance and pellicle polysaccharide formation. *Bioscience, Biotechnology and Biochemistry* **74:** 1591-1597.

Kanjee, U. and Houry, W.A. (2013). Mechanisms of acid resistance in *Escherichia coli*. *Annual Review of Microbiology* **67:** 65-81.

Kawai, M., Higashiura, N., Hayasaki, K., Okamoto, N., Takami, A., Hirakawa, H., Matsushita, K. and Azuma, Y. (2015). Complete genome and gene expression analyses of *Asaia bogorensis* reveal unique responses to culture with mammalian cells as a potential opportunistic human pathogen. *DNA Research* **22:** 357-366.

Kersters, K., Lisdiyanti, P., Komagata, K. and Swings, J. (2006). The Family Acetobacteraceae: The Genera *Acetobacter*, *Acidomonas*, *Asaia*, *Gluconacetobacter*, *Gluconobacter*, and *Kozakia*. *In:* M. Dworkin, S. Falkow, E. Rosenberg, K.H. Schleifer and E. Stackebrandt (Eds.). *The Prokaryotes*, 3rd edn, Vol. 5, Proteobacteria: Alpha and Beta Subclasses. Springer, New York, pp. 163-200.

Kondo, K. and Horinouchi, S. (1997a). Characterization of an insertion sequence, IS*12528*, from *Gluconobacter suboxydans*. *Applied and Environmental Microbiology* **63:** 1139-1142.

Kondo, K. and Horinouchi, S. (1997b). A new insertion sequence IS*1452* from *Acetobacter pasteurianus*. *Microbiology* **143:** 539-546.

Koo, H.M., Song, S.H., Pyun, Y.R. and Kim, Y.S. (1998). Evidence that a beta-1,4-endoglucanase secreted by *Acetobacter xylinum* plays an essential role for the formation of cellulose fiber. *Bioscience, Biotechnology and Biochemistry* **62:** 2257-2259.

Kornmann, H., Duboc, P., Marison, I. and Von Stockar, U. (2003). Influence of nutritional factors on the nature, yield, and composition of exopolysaccharides produced by *Gluconacetobacter xylinus* I-2281. *Applied and Environmental Microbiology* **69:** 6091-6098.

Kowalska-Ludwicka, K., Cala, J., Grobelski, B., Sygut, D., Jesionek-Kupnicka, D., Kolodziejczyk, M., Bielecki, S. and Pasieka, Z. (2013). Modified bacterial cellulose tubes for regeneration of damaged peripheral nerves. *Archives of Medical Science* **9:** 527-534.

Krause, A., Ramakumar, A., Bartels, D., Battistoni, F., Bekel, T., Boch, J., Böhm, M., Friedrich, F., Hurek, T., Krause, L., Linke, B., McHardy, A.C., Sarkar, A., Schneiker, S., Syed, A.A., Thauer, R., Vorhölter, F.J., Weidner, S., Pühler, A., Reinhold-Hurek, B., Kaiser, O. and Goesmann, A. (2006). Complete genome of the mutualistic, N_2-fixing grass endophyte *Azoarcus* sp. strain BH72. *Nature Biotechnology* **24:** 1385-1391.

Kubiak, K., Kurzawa, M., Jędrzejczak-Krzepkowska, M., Ludwicka, K., Krawczyk, M., Migdalski, A., Kacprzak, M.M., Loska, D., Krystynowicz, A. and Bielecki, S. (2014). Complete genome sequence of *Gluconacetobacter xylinus* E25 strain—valuable and effective producer of bacterial nanocellulose. *Journal of Biotechnology* **176:** 18-19.

La Verda, D., Kalayoglu, M.V. and Byrne, G.I. (1999). Chlamydial heat shock proteins and disease pathology: New paradigms for old problems? *Infectious Diseases in Obstetrics and Gynecology* **7:** 64-71.

Lawrence, J.G. and Ochman, H. (1997). Amelioration of bacterial genomes: Rates of change and exchange. *Journal of Molecular Evolution* **44:** 383-397.

Lin, J., Smith, M.P., Chapin, K.C., Baik, H.S., Bennett, G.N. and Foster, J.W. (1996). Mechanisms of acid resistance in enterohemorrhagic *Escherichia coli*. *Applied and Environmental Microbiology* **62:** 3094-3100.

Lin, S., Staahl, B., Alla, R.K. and Doudna, J.A. (2014). Enhanced homology-directed human genome engineering by controlled timing of CRISPR/Cas9 delivery. eLIFE 3: e04766.

Lovett, S.T. (2004). Encoded errors: Mutations and rearrangements mediated by misalignment at repetitive DNA sequences. *Molecular Microbiology* **52:** 1243-1253.

Maguire, M., Coates, A.R. and Henderson, B. (2002). Chaperonin 60 unfolds its secrets of cellular communication. *Cell Stress and Chaperones* **7:** 317-329.

Mahillon, J. and Chandler, M. (1998). Insertion sequences. *Microbiology and Molecular Biology Reviews* **62:** 725-774.

Marolda, C.L. and Valvano, M.A. (1995). Genetic analysis of the dTDP-rhamnose biosynthesis region of the *Escherichia coli* VW187 (O7/K1) *rfb* gene cluster: Identification of functional homologs of *rfbB* and *rfbA* in the *rff* cluster and correct location of the *rffE* gene. *Journal of Bacteriology* **177:** 5539-5546.

Matsushita, K., Takahashi, K., Takahashi, M., Ameyama, M. and Adachi, O. (1992). Methanol and ethanol oxidase respiratory chains of the methylotrophic acetic acid bacterium, *Acetobacter methanolicus*. *The Journal of Biochemistry* **111:** 739-747.

Matsushita, K., Inoue, T., Theeragol, G., Trček, J., Toyama, H. and Adachi, O. (2005a). Acetic acid production in acetic acid bacteria leading to their death and survival. *In:* M. Yamada (Ed.). *Survival and Death in Bacteria*. Research Signpost, Trivandrum-695 023, Kerala, pp. 169-181.

Matsushita, K., Inoue, T., Adachi, O. and Toyama, H. (2005b). *Acetobacter aceti* possesses a proton motive force-dependent efflux system for acetic acid. *Journal of Bacteriology* **187:** 4346-4352.

Matsutani, M., Hirakawa, H., Yakushi, T. and Matsushita, K. (2011). Genome-wide phylogenetic analysis of *Gluconobacter*, *Acetobacter*, and *Gluconacetobacter*. *FEMS Microbiology Letters* **315:** 122-128.

Mazur, A., Stasiak, G., Wielbo, J., Koper, P., Kubik-Komar, A. and Skorupska, A. (2013). Phenotype profiling of *Rhizobium leguminosarum* bv. *trifolii* clover nodule isolates reveal their both versatile and specialized metabolic capabilities. *Archives of Microbiology* **195:** 255-267.

Mehnaz, S. and Lazarovits, G. (2006). Inoculation effects of *Pseudomonas putida*, *Gluconacetobacter azotocaptans*, and *Azospirillum lipoferum* on corn plant growth under greenhouse conditions. *Microbial Ecology* **51:** 326-335.

Miller, S.I., Ernst, R.K. and Bader, M.W. (2005). LPS, TLR4 and infectious disease diversity. *Nature Reviews Microbiology* **3:** 36-46.

Mitchison, M., Bulach, D.M., Vinh, T., Rajakumar, K., Faine, S. and Adler, B. (1997). Identification and characterization of the dTDP-rhamnose biosynthesis and transfer genes of the lipopolysaccharide-related *rfb* locus in *Leptospira interrogans* serovar Copenhageni. *Journal of Bacteriology* **179:** 1262-1267.

Mitraka, E., Stathopoulos, S., Siden-Kiamos, I., Christophides, G.K. and Louis, C. (2013). *Asaia* accelerates larval development of *Anopheles gambiae*. *Pathogens and Global Health* **107:** 305-311.

Mullins, E.A., Francois, J.A. and Kappock, T.J. (2008). A specialized citric acid cycle requiring succinyl-coenzyme A (CoA): acetate CoA-transferase (AarC) confers acetic acid resistance on the acidophile *Acetobacter aceti*. *Journal of Bacteriology* **190:** 4933-4940.

Nakano, S., Fukaya, M. and Horinouchi, S. (2006). Putative ABC transporter responsible for acetic acid resistance in *Acetobacter aceti*. *Applied and Environmental Microbiology* **72:** 497-505.

Nakano, S. and Fukaya, M. (2008). Analysis of proteins responsive to acetic acid in *Acetobacter*: Molecular mechanisms conferring acetic acid resistance in acetic acid bacteria. *International Journal of Food Microbiology* **125:** 54-59.

Ogino, H., Azuma, Y., Hosoyama, A., Nakazawa, H., Matsutani, M., Hasegawa, A., Otsuyama, K., Matsushita, K., Fujita, N. and Shirai, M. (2011). Complete genome sequence of NBRC 3288, a unique cellulose-nonproducing strain of *Gluconacetobacter xylinus* isolated from vinegar. *Journal of Bacteriology* **193:** 6997-6998.

Okamoto-Kainuma, A., Yan, W., Kadono, S., Tayama, K., Koizumi, Y. and Yanagida, F. (2002). Cloning and characterization of *groESL* operon in *Acetobacter aceti*. *Journal of Bioscience and Bioengineering* **94:** 140-147.

Okamoto-Kainuma, A., Ishikawa, M., Nakamura, H., Fukazawa, S., Tanaka, N., Yamagami, K. and Koizumi, Y. (2011). Characterization of *rpoH* in *Acetobacter pasteurianus* NBRC3283. *Journal of Bioscience and Bioengineering* **111:** 429-432.

Okumura, H., Uozumi, T. and Beppu, T. (1985). Biochemical characteristics of spontaneous mutants of *Acetobacter aceti* deficient in ethanol oxidation. *Agricultural and Biological Chemistry* **49:** 2485-2487.

Olijve, W. and Kok, J.J. (1979). An analysis of the growth of *Gluconobacter oxydans* in chemostat cultures. *Archives of Microbiology* **121:** 291-297.

Omadjela, O., Narahari, A., Strumillo, J., Melida, H., Mazur, O., Bulone, V. and Zimmer, J. (2013). BcsA and BcsB form the catalytically active core of bacterial cellulose synthase sufficient for in vitro cellulose synthesis. *Proceedings of the National Academy of Sciences of the United States of America* **110:** 17856-17861.

Parkhill, J., Sebaihia, M., Preston, A., Murphy, L.D., Thomson, N., Harris, D.E., Holden, M.T.G., Churcher, C.M., Bentley, S.D., Mungall, K.L., Cerdeño-Tárraga, A.M., Temple, L., James, K.,

Harris, B., Quail, M.A., Achtman, M., Atkin, R., Baker, S., Basham, D., Bason, N., Cherevach, I., Chillingworth, T., Collins, M., Cronin, A., Davis, P., Doggett, J., Feltwell, T., Goble, A., Hamlin, N., Hauser, H., Holroyd, S., Jagels, K., Leather, S., Moule, S., Norberczak, H., O'Neil, S., Ormond, D., Price, C., Rabbinowitsch, E., Rutter, S., Sanders, M., Saunders, D., Seeger, K., Sharp, S., Simmonds, M., Skelton, J., Squares, R., Squares, S., Stevens, K., Unwin, L., Whitehead, S., Barrell, B.G. and Maskell, D.J. (2003). Comparative analysis of the genome sequences of *Bordetella pertussis, Bordetella parapertussis* and *Bordetella bronchiseptica. Nature Genetics* **35:** 32-40.

Piper, P., Mahé, Y., Thompson, S., Pandjaitan, R., Holyoak, C., Egner, R., Mühlbauer, M., Coote, P. and Kuchler, K. (1998). The Pdr12 ABC transporter is required for the development of weak organic acid resistance in yeast. *The EMBO Journal* **17:** 4257-4265.

Prust, C. (2004). *Entschlusselung des genoms von Gluconobacter oxydans 621H – Einem bakterium von industriellem interesse.* Thesis, Georg-August University, Goettingen.

Prust, C., Hoffmeister, M., Liesegang, H., Wiezer, A., Fricke, W.F., Ehrenreich, A., Gottschalk, G. and Deppenmeier, U. (2005). Complete genome sequence of the acetic acid bacterium *Gluconobacter oxydans. Nature Biotechnology* **23:** 195-200.

Rees, D.C., Johnson E. and Lewinson, O. (2009). ABC transporters: The power to change. Nature Reviews. *Molecular Cell Biology* **10:** 218-227.

Relman, D.A., Domenighini, M., Tuomanen, E., Rappuoli, R. and Falkow, S. (1989). Filamentous Hemagglutinin of *Bordetella pertussis*: Nucleotide sequence and crucial role in adherence. *Proceedings of the National Academy of Sciences of the United States of America* **86:** 2637-2641.

Roh, W.S., Nam, Y.D., Chang, H.W., Kim, K.H., Kim, M.S., Ryu, J.H., Kim, S.H., Lee, W.J. and Bae. J.W. (2008). Phylogenetic characterization of two novel commensal bacteria involved with innate immune homeostasis in *Drosophila melanogaster. Applied and Environmental Microbiology* **74:** 6171-6177.

Rojas, C.M., Ham, J.H., Deng, W.L., Doyle, J.J. and Collmer, A. (2002). HecA, a member of a class of adhesins produced by diverse pathogenic bacteria, contributes to the attachment, aggregation, epidermal cell killing and virulence phenotypes of *Erwinia chrysanthemi* EC16 on *Nicotiana clevelandii* seedlings. *Proceedings of the National Academy of Sciences of the United States of America* **99:** 13142-13147.

Rollan, G., Lorca, G.L. and De Valdez, G.F. (2003). Arginine catabolism and acid tolerance response in *Lactobacillus reuteri* isolated from sourdough. *Food Microbiology* **20:** 313-319.

Römling, U. and Galperin, M.Y. (2015). Bacterial cellulose biosynthesis: Diversity of operons, subunits, products, and functions. *Trends in Microbiology* **23 (9):** 545-557.

Ryu, J.H., Kim, S.H., Lee, H.Y., Bai, J.Y., Nam, Y.D., Bae, J.W., Lee, D.G., Shin, S.C., Ha, E.M. and Lee, W.J. (2008). Innate immune homeostasis by the homeobox gene *caudal* and commensal-gut mutualism in *Drosophila. Science* **319:** 777-782.

Saier, M.H. (2008). The bacterial chromosome. *Critical Reviews in Biochemistry and Molecular Biology* **43:** 89-134.

Saxena, I.M., Lin, F.C. and Brown, R.M. (1990). Cloning and sequencing of the cellulose synthase catalytic subunit gene of *Acetobacter xylinum. Plant Molecular Biology* **15:** 673-683.

Schuller, G., Hertel, C. and Hammes, W.P. (2000). *Gluconacetobacter entanii* sp. nov., isolated from submerged high-acid industrial vinegar fermentations. *International Journal of Systematic and Evolutionary Microbiology* **50:** 2013-2020.

Segal, B.H., Leto, T.L., Gallin, J.I., Malech, H.L. and Holland, S.M. (2000). Genetic, biochemical, and clinical features of chronic granulomatous disease. *Medicine* **79:** 170-200.

Serrato, R.V., Meneses, C.H., Vidal, M.S., Santana-Filho, A.P., Iacomini, M., Sassaki, G.L. and Baldani, J.I. (2013). Structural studies of an exopolysaccharide produced by *Gluconacetobacter diazotrophicus* Pal5. *Carbohydrate Polymers* **98:** 1153-1159.

Setlow, P. (2007). I will survive: DNA protection in bacterial spores. *Trends in Microbiology* **15:** 172-180.

Sevilla, M., Burris, R.H., Gunapala, N. and Kennedy, C. (2001). Comparison of benefit to sugarcane plant growth and $^{15}N_2$ incorporation following inoculation of sterile plants with *Acetobacter diazotrophicus* wild-type and Nif^- mutant strains. *Molecular Plant-Microbe Interactions* **14:** 358-366.

Sherlock, O., Schembri, M.A., Reisner, A. and Klemm, P. (2004). Novel roles for the AIDA adhesin from diarrheagenic *Escherichia coli*: Cell aggregation and biofilm formation. *Journal of Bacteriology* **186:** 8058-8065.

Shin, S.C., Kim, S.H., You, H., Kim, B., Kim, A.C., Lee, K.A., Yoon, J.H., Ryu, J.H. and Lee, W.J. (2011). Drosophila microbiome modulates host developmental and metabolic homeostasis via insulin signaling. *Science* **334:** 670-674.

Sievers, M., Sellmer, S. and Teuber, M. (1992). *Acetobacter europaeus* sp. nov., a main component of industrial vinegar fermenters in Central Europe. *Systematic and Applied Microbiology* **15:** 386-392.

Sievers, M. and Swings, J. (2005). Family II. Acetobacteriaceae. *In:* GM Garrity (Ed.). *Bergey´s Manual of Systematic Bacteriology*, vol. 2. Springer, New York, pp. 41-95.

Snyder, R.W., Ruhe, J., Kobrin, S., Wasserstein, A., Doline, C., Nachamkin, I. and Lipschutz, J.H. (2004). *Asaia bogorensis* peritonitis identified by 16S ribosomal RNA sequence analysis in a patient receiving peritoneal dialysis. *American Journal of Kidney Diseases: The Official Journal of the National Kidney Foundation* **44:** e15-17.

Sokollek, S.J., Hertel, C. and Hammes, W.P. (1998). Description of *Acetobacter oboediens* sp. nov. and *Acetobacter pomorum* sp. nov., two new species isolated from industrial vinegar fermentations. *International Journal of Systematic Bacteriology* **48:** 935-940.

Sudakaran, S., Salem, H., Kost, C. and Kaltenpoth, M. (2012). Geographical and ecological stability of the symbiotic mid-gut microbiota in European firebugs, *Pyrrhocoris apterus* (Hemiptera, Pyrrhocoridae). *Molecular Ecology* **21:** 6134-6151.

Takemura, H., Horinouchi, S. and Beppu, T. (1991). Novel insertion sequence IS*1380* from *Acetobacter pasteurianus* is involved in loss of ethanol-oxidizing ability. *Journal of Bacteriology* **173:** 7070-7076.

Takemura, H., Kondo, K., Horinouchi, S. and Beppu, T. (1993). Induction by ethanol of alcohol dehydrogenase activity in *Acetobacter pasteurianus*. *Journal of Bacteriology* **175:** 6857-6866.

Tang, J., Bao, L., Li, X., Chen, L. and Hong, F.F. (2015). Potential of PVA-doped bacterial nano-cellulose tubular composites for artificial blood vessels. *Journal of Materials Chemistry* **3:** 8537-8547.

Tenreiro, S., Nunes, P.A., Viegas, C.A., Neves, M.S., Teixeira, M.C., Cabral, M.G. and Sa-Correia, I. (2002). AQR1 gene (ORF YNL065w) encodes a plasma membrane transporter of the major facilitator superfamily that confers resistance to short-chain monocarboxylic acids and quinidine in *Saccharomyces cerevisiae*. *Biochemical and Biophysical Research Communications* **292**: 741-748.

Tenreiro, S., Rosa, P.C., Viegas, C.A. and Sa-Correia, I. (2000). Expression of the AZR1 gene (ORF YGR224w), encoding a plasma membrane transporter of the major facilitator superfamily, is required for adaptation to acetic acid and resistance to azoles in *Saccharomyces cerevisiae*. *Yeast* **16:** 1469-1481.

Tolasch, T., Solter, S., Toth, M., Ruther, J. and Francke, W. (2003). (R)-acetoin-female sex pheromone of the summer chafer *Amphimallon solstitiale* (L.). *Journal of Chemical Ecology* **29:** 1045-1050.

Tonouchi, N., Sugiyama, M. and Yokozeki, K. (2003). Construction of a vector plasmid for use in *Gluconobacter oxydans*. *Bioscience, Biotechnology and Biochemistry* **67:** 211-213.

Trček, J., Raspor, P. and Teuber, M. (2000). Molecular identification of *Acetobacter* isolates from submerged vinegar production, sequence analysis of plasmid pJK2-1 and application in the development of a cloning vector. *Applied Microbiology and Biotechnology* **53:** 289-295.

Trcek, J., Toyama, H., Czuba, J., Misiewicz, A. and Matsushita, K. (2006). Correlation between acetic acid resistance and characteristics of PQQ-dependent ADH in acetic acid bacteria. *Applied Microbiology and Biotechnology* **70:** 366-373.

Trček, J., Jernejc, K. and Matsushita, K. (2007). The highly tolerant acetic acid bacterium *Gluconacetobacter europaeus* adapts to the presence of acetic acid by changes in lipid composition, morphological properties and PQQ-dependent ADH expression. *Extremophiles* **11:** 627-635.

Trček, J. and Barja, F. (2015). Updates on quick identification of acetic acid bacteria with a focus on the 16S-23S rRNA gene internal transcribed spacer and the analysis of cell proteins by MALDI-TOF mass spectrometry. *International Journal of Food Microbiology* **196:** 137-144.

Trček, J., Mira, N.P. and Jarboe, L.R. (2015). Adaptation and tolerance of bacteria against acetic acid. *Applied Microbiology and Biotechnology* **99:** 6215-6229.

Tsukioka, Y., Yamashita, Y., Oho, T., Nakano, Y. and Koga, T. (1997). Biological function of the dTDP-rhamnose synthesis pathway in *Streptococcus mutans*. *Journal of Bacteriology* **179:** 1126-1134.

Tuuminen, T., Heinasmaki, T. and Kerttula, T. (2006). First report of bacteremia by *Asaia bogorensis*, in a patient with a history of intravenous-drug abuse. *Journal of Clinical Microbiology* **44:** 3048-3050.

Tuuminen, T., Roggenkamp, A. and Vuopio-Varkila, J. (2007). Comparison of two bacteremic *Asaia bogorensis* isolates from Europe. *European Journal of Clinical Microbiology and Infectious Diseases* **26:** 523-524.

Umeda, Y., Hirano, A., Hon-Nami, K. and Inoue, Y. (1998). Conversion of CO_2 into cellulose by gene manipulation of microalgae: Cloning of cellulose synthase genes from *Acetobacter xylinum*. *Studies in Surface Science and Catalysis* **114:** 653-656.

Walker, G., Hertle, R. and Braun, V. (2004). Activation of *Serratia marcescens* hemolysin through a conformational change. *Infection and Immunity* **72:** 611-614.

Wang, B., Shao, Y.C., Chen, T., Chen, W.P. and Chen, F.S. (2015). Global insights into acetic acid resistance mechanisms and genetic stability of *Acetobacter pasteurianus* strains by comparative genomics. *Scientific Reports* **5:** 18330.

Weidner, S., Baumgarth, B., Göttfert, M., Jaenicke, S., Pühler, A., Schneiker-Bekel, S., Serrania, J., Szczepanowski, R. and Becker, A. (2013). Genome sequence of *Sinorhizobium meliloti* Rm41. *Genome Announcements* **1:** e00013-12.

Wieme, A.D., Spitaels, F., Aerts, M., De Bruyne, K., Van Landschoot, A. and Vandamme, P. (2014). Identification of beer-spoilage bacteria using matrix-assisted laser desorption/ionization time-of-flight mass spectrometry. *International Journal of Food Microbiology* **185:** 41-50.

Wong, H.C., Fear, A.L., Calhoon, R.D., Eichinger, G.H., Mayer, R., Amikam, D., Benziman, M., Gelfand, D.H., Meade, J.H., Emerick, A.W., Bruner, R., Ben-Bassat, A. and Tal, R. (1990). Genetic organization of the cellulose synthase operon in *Acetobacter xylinum*. *Proceedings of the National Academy of Sciences of the United States of America* **87:** 8130-8134.

Yakushi, T. and Matsushita, K. (2010). Alcohol dehydrogenase of acetic acid bacteria: Structure, mode of action, and applications in biotechnology. *Applied Microbiology and Biotechnology* **86:** 1257-1265.

Yamada, Y., Katsura, K., Kawasaki, H., Widyastuti, Y., Saono, S., Seki, T., Uchimura, T. and Komagata, K. (2000). *Asaia bogorensis* gen. nov., sp. nov., an unusual acetic acid bacterium in the α-Proteobacteria. *International Journal of Systematic and Evolutionary Microbiology* **50:** 823-829.

Yasutake, Y., Kawano, S., Tajima, K., Yao, M., Satoh, Y., Munekata, M. and Tanaka, I. (2006). Structural characterization of the *Acetobacter xylinum* endo-beta-1,4-glucanase CMCax required for cellulose biosynthesis. *Proteins* **64:** 1069-1077.

3

Physiology and Biochemistry of Acetic Acid Bacteria

Birce Mercanoglu Taban[1] **and Natsaran Saichana**[2,*]

[1] Ankara University, Faculty of Agriculture, Department of Dairy Technology, Ankara – 06110, Turkey

[2] Mae Fah Luang University, School of Science, Chiang Rai – 57100, Thailand

Introduction

Eighteen genera in the family *Acetobacteraceae* are classified into the group of acetic acid bacteria (AAB). Generally, AAB can produce acetic acid from ethanol and widely inhabit in sugary or acidic substances in Nature and sometimes they are found as spoilers in alcoholic beverages (Gupta et al. 2001, Moonmangmee et al. 2002, Bartowsky and Henschke 2008, Raspor and Goranovič 2008, Sengun and Karabiyikli 2011, Mamlouk and Gullo 2013). Besides the preference to grow under sugary or acidic or alcoholic substances, AAB can be found elsewhere. Enrichment, by using selective media containing sugars, ethanol or sugar alcohols as the sole carbon sources, is needed for isolation of AAB from their natural habitats (Carr 1968, Cirigliano 1982, González et al. 2006). AAB grow and produce valuable products from incomplete oxidative fermentation system, called 'oxidative fermentation', in which O_2 is constantly required. Oxygen availability is a key factor that affects the fermentation process and thus the productivity (Qi et al. 2013, Schlepütz et al. 2013, Gullo et al. 2014). It has been known that a temperature of 30°C is optimal for the growth and production of some valuable products by most AAB and changing of the temperature can severely affect the production.

In this chapter, we will describe the physiology and biochemistry of AAB relating to the growth and production of several valuable products.

Phenotypic Characteristics and Sources of Acetic Acid Bacteria

AAB are a group of Gram-negative or Gram-variable bacteria, obligately aerobic, ellipsoidal to rod-shaped, non-spore forming and either peritrichously or polarly flagellated microorganisms that belong to the family of *Acetobacteraceae* as part of the order of *Rhodospirillales* of class *Alphaproteobacteria*. They form enlarged and irregular cells that occur singly, in pairs or chains. Their sizes vary between 0.4 and 1 μm wide and 0.8 and 4.5 μm long (Cleenwerck and De Vos 2008, Sengun and Karabiyikli 2011). Most AAB possess the C18:1ω7 straight-chain unsaturated acid as the most important fatty acid in the composition of their cell envelopes (Kersters et al. 2006). They in general have special

*Corresponding author: natsaran.sai@mfu.ac.th

and rapid oxidative fermentation abilities of oxidizing ethanol into acetic acid besides oxidizing a wide range of sugars and sugar alcohols to corresponding organic acids in the presence of oxygen, although some of them are very weak producers. This special and unique metabolism imparts them their name (Katsura et al. 2001, Deppenmeier et al. 2002, Kersters et al. 2006, Guillamón and Mas 2009, Sengun and Karabiyikli 2011, Saichana et al. 2015, Wang et al. 2015). Ethanol oxidation is pronounced in *Acetobacter* and *Komagataeibacter* strains, while sugar oxidation is more efficient in *Gluconobacter* and *Gluconacetobacter* strains (Deppenmeier et al. 2002). *Neokomagataea* and *Tanticharoenia* strains cannot oxidize acetate or lactate. Although the first ones cannot grow even on 0.35 per cent acetic acid, the latter ones can grow on 30 per cent D-glucose (Yukphan et al. 2008, 2011). *Ameyamaea* strains present fast acetate oxidation rate, which is almost same as the acetate oxidation rate of *Acetobacter aceti* (Yukphan et al. 2011).

Previously, AAB were classified into two main genera—*Acetobacter* and *Gluconobacter*—but the taxonomy of them has undergone various changes in the last 35 years due to the development of novel analysis technologies and so, nowadays, they are members of the following acidophilic genera: *Acetobacter*, *Acidomonas*, *Ameyamaea*, *Asaia*, *Bombella*, *Endobacter*, *Gluconacetobacter*, *Gluconobacter*, *Granulibacter*, *Komagataeibacter*, *Kozakia*, *Neoasaia*, *Neokomagataea*, *Nguyenibacter*, *Saccharibacter*, *Swaminathania*, *Swingsia* and *Tanticharoenia* (Malimas et al. 2013, Thi Lan Vu et al. 2013, Li et al. 2015, Saichana et al. 2015, Sengun 2015, Wang et al. 2015) of which most can grow at pH between 3 and 4.5 with a strict oxidative metabolism, although their optimum growth pH varies between 5 and 6.5 (Du Toit and Pretorius 2002, Sengun and Karabiyikli 2011). Acetic acid tolerance of AAB varies from species to species. Some *Acetobacter* and *Komagataeibacter* species, such as *Acetobacter pomorum*, *Komagataeibacter intermedius*, *K. europaeus* and *K. oboediens* can survive at higher concentrations of acetic acid than other AAB species (Wang et al. 2015). Moreover, their resistance to high acetic acid concentrations and low pH can be supported by prolonged and gradual exposure to low pH, which is essential for them to produce large amounts of acids (Kösebalaban and Özilgen 1992).

Although some species, such as *Acetobacter tropicalis* and *A. pasteurianus*, are known as thermotolerant, their optimum growth temperature ranges between 25°C and 30°C, yet some strains can grow very slowly at 10°C (Joyeux et al. 1984, Ndoye et al. 2006, Gullo and Giudici 2008). They are catalase positive and oxidase negative bacteria (due to the lack of a cytochrome *c* oxidase), except *Acidomonas methanolica*, which is oxidase positive during growth on methanol (Matsushita et al. 1994, Sievers and Swings 2005). AAB can produce several valuable products, such as 2-keto-D-gluconate and 5-keto-D-gluconate, the precursor for Vitamin C production (Gupta et al. 2001, Deppenmeier et al. 2002), high-quality cellulose, produced by *K. xylinus* (Czaja et al. 2006, 2007) and L-sorbose, produced by *Gluconobacter oxydans* and *G. frateurii* (Adachi et al. 2003, Hattori et al. 2012), yellow and brown pigments (Malimas et al. 2009), surfactants (Desai and Banat 1997) and various exopolysaccharides (EPS) (Jakob et al. 2012, Hermann et al. 2015).

AAB are found widespread in sugary or acidic substances, such as fruits and flowers in the environment and are involved in the production of vinegar, kefir, Kombucha tea (a slightly acidulous and effervescent beverage obtained from black tea fermentation by symbiotic culture of AAB and yeasts), nata de coco (an edible form of bacterial cellulose produced from AAB grown in coconut water), pulque (a beverage from agave) and cocoa-based products, or the spoilage of many foods and alcoholic beverages [vinegary flavor and oily or moldy surface in beers and wines, acetification of cider (apple wine) and cider sickness] and fruit juices (off-flavor, pack swelling and haze) (Dutta and Gachhui 2006, Bartowsky and Henschke 2008, Mamlouk and Gullo 2013). Therefore, they are also known as 'living oxidative catalysts' (Deppenmeier and Ehrenreich 2009). Most *Gluconobacter* strains occur in sugar-enriched environments, whereas *Acetobacter* strains occur in alcohol-enriched environments (Raspor and Goranovič 2008). Although AAB are non-pathogenic to humans or animals (Gupta et al. 2001), *Granulibacter bethesdensis*, *A. cibinongensis* and some *Asaia* and *Gluconobacter* species are reported as emerging opportunistic human pathogens (Greenberg et al. 2006, Gouby et al. 2007, Alauzet et al. 2010). The first documented case of bacteremia caused by *Asaia bogorensis* was reported in a patient with peritoneal dialysis showing persistent peritonitis (Snyder et al. 2004) and later in a young patient with an intravenous-drug abuse (Tuuminen et al. 2006). Besides, some of them are plant-associated, as *Gluconacetobacter*, *Swaminathania* and *Acetobacter* include some N_2-fixing species such as, *Gluconacetobacter johannae*, *Ga. azotocaptans*, *Ga. diazotrophicus*, *Swaminathania salitolerans*, *A. peroxydans* and *A. nitrogenifigens* (Fuentes-Ramirez et al. 2001, Loganathan and Nair 2004, Dutta and Gachhui 2006, Pedraza 2008).

Respiratory Chains of AAB

Aerobic bacteria generate energy from the respiratory chains that use molecular oxygen as the terminal electron acceptor and the process ends with the formation of H_2O along with generation of proton gradient. Terminal oxidases in aerobic bacteria are classified into two major groups—heme–copper oxidases and *bd*-type oxidases. The first group has a binuclear O_2-reducing site containing one heme (heme *a*, heme *o*, or heme *b*) and a copper (Cu_B), which are divided into cytochrome *c* oxidase (COX) and ubiquinol oxidase (UOX). COX receives electrons from cytochrome *c*, whereas UOX receives electrons from ubiquinol (Matsutani et al. 2014). Respiratory machineries of AAB consist of a membrane-bound proton pumping transhydrogenase, a non-proton translocating NADH: ubiquinone oxidoreductase (type II) for regeneration of $NADP^+$ and NAD^+, respectively, and two quinol oxidases of bo_3 and *bd*-type oxidases (Prust et al. 2005). Genes for cytochrome bc_1 complex, *qcrABC*, were also found in the genome of AAB with *cycA* gene for soluble cytochrome *c*. However, the function of this protein complex is unclear as the cytochrome *c* oxidase (complex IV) is missing in the genome of *Gluconobacter* and the reduced cytochrome *c* generated from the cytochrome bc_1 complex cannot be reoxidized (Fig. 1). But later studies suggested that in AAB, COX was originally present with the cytochrome bc_1 complex for electron transport via cytochrome *c* similar to other α-Proteobacteria. During evolution, parts of COX genes, except for the heme *o*/*a* synthase genes, were replaced by UOX from some β/γ-Proteobacteria via horizontal gene transfer (Matsutani et al. 2014). The respiratory chain of AAB is uniquely truncated since it has been revealed to directly link to the aerobic microbial fermentation, so-called oxidative fermentation (Matsushita et al. 1994). The simple respiratory chain linked to oxidative fermentation of AAB consists of quinoproteins or flavoproteins or other periplasmic dehydrogenases and terminal ubiquinol oxidases (Matsushita et al. 1994). Electron abstracted from the oxidation reaction are transferred to ubiquinone (UQ) turning it to ubiquinol which is then oxidized by the ubiquinol oxidases, resulting in the generation of proton gradients. In only the genus *Acetobacter*, the main respiratory quinone is UQ-9, although others have UQ-10 (Roh et al. 2008, Malimas et al. 2013, Thi Lan Yu et al. 2013, Li et al. 2015, Saichana et al. 2015). The truncated respiratory chain in AAB with UOX generates a reduced proton-motive force with a higher electron transfer ability (Matsutani et al. 2014). This respiratory system contributes to incomplete oxidation in AAB, which results in rapid oxidation of many sugar alcohols or sugars present in the natural habitats that lead to high accumulation of acid products in their environment (Matsutani et al. 2014). The overall process gives benefits to AAB to grow dominantly in Nature where there are many competitive microorganisms.

The cytochrome bo_3 ubiquinol oxidase of *G. oxydans* NBRC 3172 (formerly *G. suboxydans* IFO 12528) is characterized and its genes were reported as *cyoBACD* (Matsushita et al.

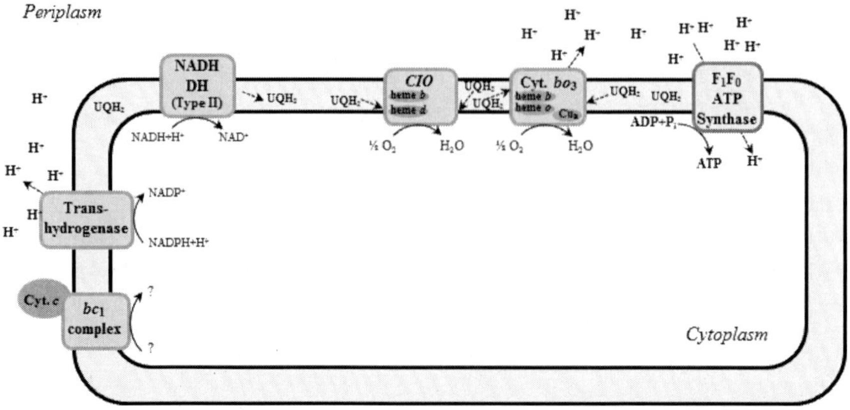

Figure 1: Respiratory chains on the cytoplasmic membrane of AAB. Electrons abstracted from oxidative fermentation process are transferred via ubiquinol (UQH_2) to O_2 as terminal electron acceptor by ubiquinol oxidases (cytochrome bo_3 oxidase or *CIO*). Proton gradients generated from the process are then used for generation of ATP through F_1F_0-type ATP synthase. ? indicates unknown component (adapted from Prust et al. 2005, Matsutani et al. 2014).

1987, Prust et al. 2005, Richhardt et al. 2013). This ubiquinol oxidase contains heme *b*, heme *o* and a copper ion (Cu_B) as prosthetic groups, and it belongs to the heme-copper oxidase superfamily. Cytochrome bo_3 ubiquinol oxidase catalyzes a reaction, which contributes to the generation of proton motive force as ubiquinol is oxidized at the periplasmic side of the membrane and protons are released into the periplasm and the electrons are then transferred to the binuclear heme o_3-Cu center to form a water molecule (Richhardt et al. 2013). The electrochemical proton gradient is then used for generation of ATP via an F_1F_o-type ATP synthase (Prust et al. 2005). *A. lovaniensis* NBRC 3284 (previously known as *A. pasteurianus* IFO 3284 or *A. aceti* IFO 3284) has a cytochrome ba_3 ubiquinol oxidase (formerly termed cytochrome a_1 oxidase) during growth in shaking culture or cytochrome bo_3 ubiquinol oxidase in static culture (Matsushita et al. 1990, Matsushita et al. 1992). The cytochrome ba_3 and bo_3 ubiquinol oxidase have the same proteins, but carry different hemes in the binuclear center as heme *a* or heme *o*, respectively.

Cytochrome *bd* quinol oxidase in *G. oxydans* NBRC 3172 is reported to encode from *cioAB* genes and the enzyme is classified as a cyanide-insensitive oxidase (*CIO*) (Miura et al. 2013). Cytochrome *bd* is a heterodimeric quinol oxidase, which catalyzes the two-electron oxidation of either ubiquinol (in Gram-negative bacteria) or menaquinol (in Gram-positive bacteria) and the four-electron reduction of O_2 to water. This quinol oxidase is completely unrelated to the proton-pumping cytochrome *c* oxidase and cytochrome bo_3 quinol oxidase. *CIO* is found to have heme b_{558}, b_{595} and *d*, but the protein backbone lacks the C-terminal half of the long hydrophillic Q-loop, which connects transmembrane helices 6 and 7 in subunit I of cytochrome *bd* of Gram-negative bacteria (Miura et al. 2013). *CIO* from *G. oxydans* NBRC 3172 was compared with cytochrome *bd* oxidase of *Escherichia coli* and it was found that, besides a much lower sensitivity to cyanide, the *CIO* has a higher sensitivity to azide with an exceptionally high turnover number and a K_M value for oxygen compared to *E. coli* cytochrome *bd* oxidase and the *G. oxydans* cytochrome bo_3 oxidase (Miura et al. 2013).

The lack of *cyoBACD* genes for cytochrome bo_3 oxidase severely affects the growth rate and the cell yield while the lack of cytochrome *bd* oxidase has no influence on respiration activity and growth under either oxygen excess or oxygen limitation conditions (Richhardt et al. 2013). The high quinol oxidase activity of *G. oxydans CIO* at low pH allows efficient oxidation of substrates by primary dehydrogenases through reoxidation of ubiquinol in the cytoplasmic membrane. This property has physiological meaning as AAB cells inhabited in fruitful environment (high sugar or alcohol) need to make their environment acidic to suppress the growth of other microorganisms. Thus *CIO*, as an effective oxidase linked to the several dehydrogenases, allows efficient oxidation when the pH decreased to acidic range. In *G. oxydans*, cytochrome bo_3 may serve as a major terminal oxidase at the early growth phase, of which the culture pH is neutral, and when pH is decreased as a result of the production of large amount of oxidized products, *CIO* keeps the oxidative fermentation under acidic conditions by complementing the function of cytochrome bo_3 oxidase (Miura et al. 2013).

AAB have respiratory systems linked to oxidative fermentation that allows faster production of oxidized products to compete with other microorganisms in the enriched environment and these respiratory systems result in a lower biomass production. Reoxidations of NADPH and NADH are catalyzed by a membrane-bound transhydrogenase, which pumps out protons to periplasm and a non-proton pumping NADH: ubiquinone oxidoreductase (type II). It was found in *G. oxydans* DSM 3504 that the membrane-bound dehydrogenases are competing with the NADH: ubiquinone oxidoreductase in channeling electrons in the electron transport chain and the additional NADH: ubiquinone oxidoreductase is responsible for the increased growth yields. The less active NADH: ubiquinone oxidoreductase is more active as membrane-bound polyol dehydrogenase (Kostner et al. 2015).

Oxidative Fermentations

Carbon Sources Oxidation (Incomplete and Complete Fermentations)—Membrane-bound Enzymes

Aerobic microorganisms normally oxidize the carbon sources completely to CO_2 and H_2O. However, this oxidation process remains incomplete (partial) due to the presence of inhibitory substances, excess of carbon substrates and abnormal physiological and growth conditions. On the other hand, incomplete oxidation occurs even under normal physiological and growth conditions of AAB (Deppenmeier et al. 2002, Deppenmeier and Ehrenreich 2009). Therefore, AAB perform special and unique oxidation reactions via the process of one-step incomplete oxidation ('oxidative fermentation') and channel the released electrons to molecular oxygen (Mamlouk and Gullo 2013).

Food and biotechnology industries make use of the benefits of oxidative fermentation of AAB for a long time since they provide an important economic increment and are therefore of great interest to them (Raspor and Goranovič 2008, De Vero 2010). However, information on the reaction mechanism of oxidative fermentation is not clear since there had been confusion over the enzyme localization (membrane-bound or cytosolic) for oxidative fermentation. This confusion impeded the optimization of biotechnological processes until the characterization of various membrane-bound dehydrogenases (Adachi et al. 2003, Deppenmeier and Ehrenreich 2009).

Membrane-bound enzymes oxidize various alcohols, sugar alcohols, sugars and organic acids to supply the primary energy, although cytosolic enzymes assist in assimilation of the oxidation products in the next phase, which is related to biomass formation (Saichana et al. 2015). In other words, membrane-bound enzymes are responsible for product formation, but cytosolic enzymes facilitate product assimilation. Oxidative fermentation is catalyzed by membrane-bound dehydrogenases located on the outer surface of the cytoplasmic membrane of AAB. All enzyme activities are connected to the terminal ubiquinol oxidase by ubiquinone in the respiratory chain of AAB (Matsushita et al. 1994, Adachi et al. 2007).

There are 32 membrane-bound dehydrogenases in the genome of *G. oxydans* 621H with 11 known and 21 unknown substrate specificities (Prust et al. 2005, Richhardt et al. 2013). Studies on membrane-bound enzymes found in the periplasmic space or on the outer surface of the cytoplasmic membrane of AAB showed that either pyrroloquinoline quinone (PQQ) or flavin adenine dinucleotide (FAD) is the functioning coenzyme in oxidative fermentation, whereas the cytosolic nicotinamide adenine dinucleotide phosphate [NAD(P)]-dependent dehydrogenases have no role on oxidative fermentation (Matsushita et al. 1994, Adachi et al. 2003, Baldrian 2006, Hölscher and Görisch 2006). In addition, the coenzyme of a membrane-bound aldehyde dehydrogenase (ALDH) was suspected to be a molypdopterin. However, this has not been proved and needs further investigation (Richhardt et al. 2013). NAD(P)-dependent enzymes play roles both in the maintenance of cells in the stationary growth phase and in the synthesis of biosynthetic precursors (Matsushita et al. 1994, Deppenmeier et al. 2002). In other words, PQQ-dependent dehydrogenases (quinoproteins and quinoprotein-cytochrome *c* complexes) and FAD-dependent dehydrogenases (flavoprotein-cytochrome *c* complexes) are the membrane-bound enzymes involved in oxidative fermentation of AAB and some of them work with the same substrate, but produce different oxidation products. For example, in the production of vinegar, L-sorbose and D-gluconate (GA) are catalyzed by quinoprotein dehydrogenases, but the production of 2-keto-D-gluconate (2KGA), 5-keto-D-fructose (5KF) and 2,5-diketo-D-gluconate (25KGA) are catalyzed by flavoprotein dehydrogenases (Adachi et al. 2003).

Accordingly, oxidative fermentation is considered a process of incomplete oxidation where ethanol, sugars and sugar alcohols are oxidized by membrane-bound dehydrogenases and the oxidized products, such as various organic acids, aldehydes and ketones, are then released and accumulated as final products in the surrounding media (Matsushita et al. 2003, Guillamón and Mas 2009, Saichana et al. 2015). The electrons abstracted from the substrate are released with the cofactor of the specific enzyme as mediator and then transferred to the terminal oxidases through respiratory ubiquinone to create a proton motive force for generation of ATP (Matsushita et al. 1994, 2002). Oxygen availability is of prime importance and profoundly affects the fermentation rate and thus the productivity (Gullo et al. 2014, Saichana et al. 2015). The most important and classic examples of oxidative fermentation include production of vinegar and L-sorbose since they are incomplete oxidation reactions and similar amounts of oxidized substrates are accumulated in the surrounding media (Adachi et al. 2003, Sengun and Karabiyikli 2011).

In addition, the metabolism of AAB comprises a tricarboxylic acid cycle (TCA) function which incites them to completely oxidize (overoxidation) acetic acid to carbon dioxide and water. In other words, strains of *Acetobacter*, *Acidomonas*, *Ameyamaea* *Gluconacetobacter*, *Komagataeibacter* and *Nguyenibacter* strongly oxidize acetate further to CO_2 and H_2O through the TCA cycle and the glyoxylic acid shunt, just after complete oxidation of ethanol. But, oxidation is weak in *Asaia*, *Kozakia*, *Granulibacter* and *Swaminathania* (Sievers and Swings 2005, Kersters et al. 2006, Malimas et al. 2013). However, the composition of the media affects this overoxidizing ability and thus this metabolic pathway is repressed due to the inhibition of entry into the TCA cycle by the presence of ethanol (O'Toole and Kun 2003, Raspor and Goranovič 2008). On the other hand, strains of *Gluconobacter*, *Neoasaia*, *Neokomagataea*, *Saccharibacter*, *Swingsia* and *Tanticharoenia* do not overoxidize acetate (Malimas et al. 2013) since they do not have the glyoxylic acid shunt and their TCA cycle is incomplete (Deppenmeier and Ehrenreich 2009, Mamlouk and Gullo 2013). In addition, strong overoxidation of lactate occurs in *Acetobacter*, *Gluconacetobacter*, *Granulibacter* and *Komagataeibacter* while it is weak in *Ameyamaea*, *Asaia*, *Kozakia*, *Swaminathania* and *Saccharibacter*; the remaining genera do not show overoxidation for lactate (Malimas et al. 2013).

Ethanol Oxidation

AAB oxidize ethanol and yield acetic acid as the product in oxidative fermentation. It has been known that *Acetobacter* spp. and *Komagataeibacter* spp. have a strong ability to produce acetic acid and both genera also show high resistance to ethanol and acetic acid, which are suitable for industrial vinegar production. The oxidation of ethanol is catalyzed by two membrane-bound enzymes located on the outer surface of the cytoplasmic membrane. Ethanol is firstly oxidized to acetaldehyde by a membrane-bound pyrroloquinoline quinon (PQQ)-dependent alcohol dehydrogenase (ADH) and then the acetaldehyde is further oxidized to acetic acid by a membrane-bound aldehyde dehydrogenase (ALDH) (Fig. 2a). Acetaldehyde and acetic acid are accumulated in the culture medium, resulting in a decrease of pH during the course of oxidation. PQQ-ADH is a primary dehydrogenase in AAB and was found to be constitutively expressed even under non-ethanol conditions. PQQ-ADH consists of three subunits, a large dehydrogenase subunit (subunit I) containing PQQ in the active site and one heme *c* as an electron mediator, a cytochrome *c* subunit (subunit II) consisting three hemes *c* involving in ubiquinone reduction and ubiquinol oxidation and a small subunit (subunit III), which helps in the assembly of the functional enzyme (Adachi et al. 2007, Matsushita et al. 2008, Masud et al. 2010). On the other hand, ADH from *K. europaeus* was reported to have only subunits I and II (Trcek et al. 2006). PQQ-ADH of *Gluconobacter* strain also contains a high affinity quinone binding site (Q_H) with a bound ubiquinone-10 (UQ-10) besides the low affinity quinone reduction (Q_N) and quinol oxidation (Q_L) sites. The UQ-10 in the Q_H site is involved in electron transfer between heme *c* moieties and the quinone and quinol in the low affinity sites (Fig. 2b) (Matsushita et al. 2008). PQQ-ADH is reportedly stable in a broad pH range as more than 90 per cent of the activity was detected after incubation at the pH of 2.3-8.0 on ice for 30 minutes (Kanchanarach et al. 2010). As mentioned earlier, most membrane-bound enzymes in AAB were reported with either PQQ or FAD as the prosthetic group, though the nature of the prosthetic group of ALDH is still unclear. It is controversial that ALDH of AAB is a membrane-bound hetero-dimeric or trimeric enzyme as ALDH from *A. aceti* and *K. europaeus* contain three subunits, whereas ALDH reported from *G. suboxydans*, *A. rancens* and *A. polyoxogenes* have only two subunits (Gómez-Manzo et al. 2010). It was reported that a PQQ-deficient mutant of *Acetobacter* sp. BPR2001 lost activities of all known PQQ-dependent enzymes, but ALDH activity remained intact, suggesting that ALDH was not PQQ-dependent (Takemura et al. 1994). Moreover, verification of the prosthetic groups of ALDH in AAB is still needed as Thurner et al. (1997) reported that ALDH of *K. europaeus* contained heme *b*, [2Fe-2S] cluster and molybdopterin as the prosthetic groups, while Gómez-Manzo et al. (2010) later demonstrated that ALDH from *Gluconacetobacter diazotrophicus* possessed PQQ, heme *b* and heme *c* as the prosthetic groups. ALDH was reported to be very stable at acidic pH and the enzyme remained more than 50 per cent of its activity after heating at 60°C for 30 minutes (Adachi et al. 2007) while PQQ-ADH is less heat stable. The vinegar production at higher temperature is limited by the PQQ-ADH stability (Kanchanarach et al. 2010).

Sugars Oxidation

AAB catalyze sugars (mostly glucose, but also fructose, arabinose, ribose, sorbose, xylose, galactose and mannose) by cytoplasmic pentose phosphate pathway (hexose monophosphate pathway) (Drysdale and Fleet 1988, Sievers and Swings 2005). No glycolysis occurs since they do not have the enzyme phosphofructokinase (Sievers and Swings 2005). Besides using the pentose phosphate pathway, *Acetobacter* species can use Embden-Meyerhof-Parnas (EMP) and Entner-Doudoroff (ED) pathways in further

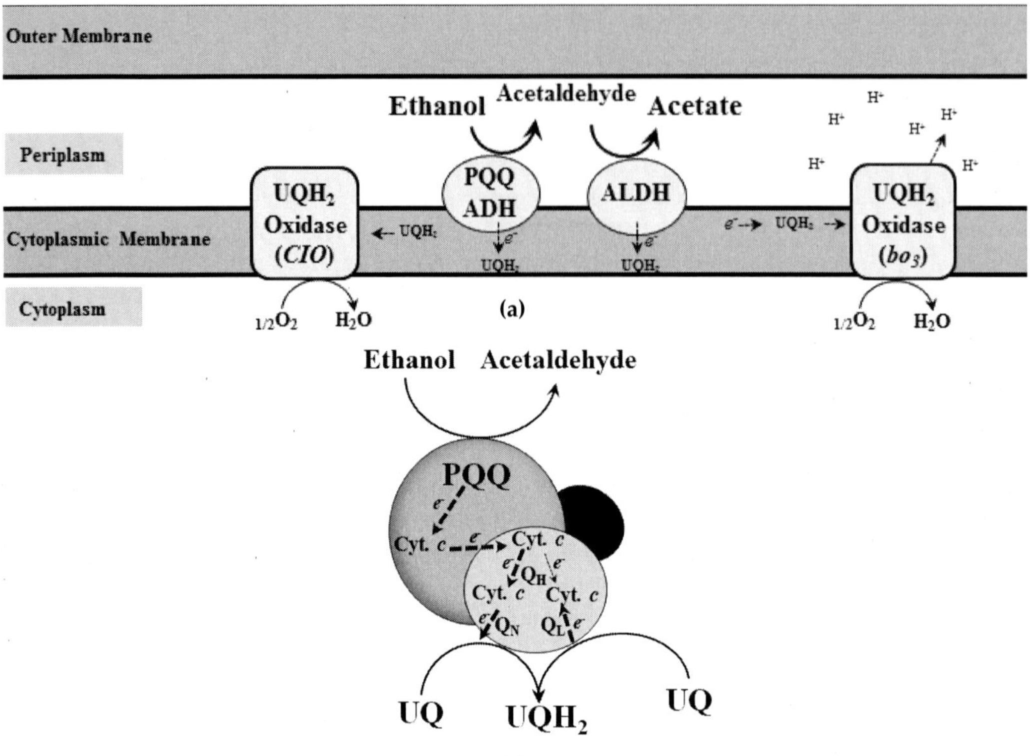

Figure 2: (a) Oxidative fermentation for acetic acid from ethanol linked to the respiratory chains. Electrons are transferred from substrate via UQH_2 to the terminal UQH_2 oxidases. (b) Proposed electron transfer mechanism of PQQ-ADH with three quinone binding sites (adapted from Adachi et al. 2007, Matsutani et al. 2014).

converting sugars to carbon dioxide and water by the TCA cycle, which is not functional in *Gluconobacter* (Mamlouk and Gullo 2013). However, some cellulose-synthesizing *Acetobacter* and *Komagataeibacter* strains use ED pathway since it is more active than the pentose phosphate pathway (Mamlouk and Gullo 2013).

AAB oxidize D-glucose through a pH- and glucose concentration-dependent way (Weenk et al. 1984, Qazi et al. 1991), firstly to D-glucono-δ-lactone by a PQQ-glucose dehydrogenase (PQQ-GDH), a primary dehydrogenase of AAB, and then to GA either spontaneously or by a gluconolactonase present in the membrane (Raspor and Goranovič 2008). PQQ-GDH consists of two subunits of which a large subunit contains PQQ and is responsible for D-glucose oxidation while the small subunit has an unknown function. Production of GA by AAB is important for industrial uses as GA is an important food additive in the food industry as well as Ca-GA, which is widely used as a metal polish in the lens industry (Adachi et al. 2007). GA is then converted to either 2KGA by an FAD-containing gluconate dehydrogenase (FAD-GADH) or 5-keto-D-gluconate (5KGA) by PQQ-glycerol dehydrogenase (PQQ-GLDH). In some AAB strains, 2KGA is later converted to 25KGA by FAD-containing 2KGA dehydrogenase (FAD-2KGA DH) (Fig. 3) (Matsushita et al. 1994, Singh and Kumar 2007, Toyama et al. 2007, Shinagawa et al. 2009). Strain type and the growth conditions affect the production ratio of 2KGA and 5KGA as production relies on the certain ratio of FAD-GADH and PQQ-GLDH activities (Ano et al. 2011, Saichana et al. 2015). Exclusive production of 5KGA by a wild-type *Gluconobacter* strain was achieved by controlling the pH of the medium at 3.5-4.0. It was found that the 5KGA production is

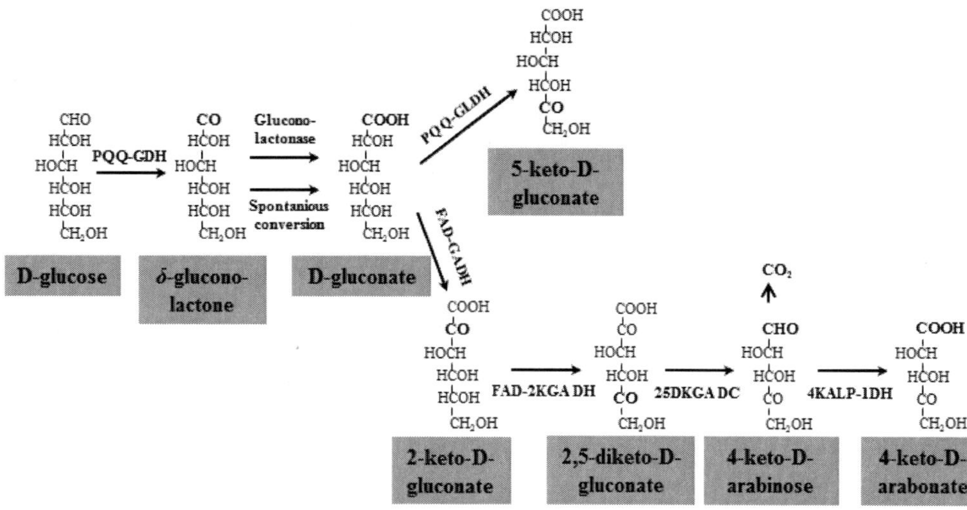

Figure 3: Oxidative fermentation for 2KGA and 5KGA formation from D-glucose with the conversion of 4KAB from 25KGA (adapted from Saichana et al. 2015).

correlated with the gluconate oxidizing activity (Ano et al. 2011). PQQ-GLDH responsible for 5KGA formation is a hetero-dimeric enzyme of which the large subunit contains a PQQ molecule with dehydrogenase function and the small subunit is believed to be involved in the attachment of the enzyme to the membrane (Fig. 4). The *sldBA* genes encoded for the small and large subunits of PQQ-GLDH have been reported and deletion of only *sldA* could diminish 5KGA formation (Hoshino et al. 2003). The small subunit of PQQ-GLDH is a hydrophobic protein with 5-membrane spanning regions, while the large subunit contains PQQ on the active site. Genes for FAD-GADH in *G. frateurii* NBRC 3271 have been reported as *gndSLC* encoding the small, large and cytochrome *c* subunits of the enzyme (Toyama et al. 2007). These genes showed moderate identities to the genes for FAD-GADH reported on the genome of *Gluconobacter* (Prust et al. 2005). The large subunit of FAD-GADH is the dehydrogenase subunit having FAD as the prosthetic group. The cytochrome *c* subunit contains three heme *c* which are believed to be involved in electron transfer from FAD in the dehydrogenase subunit to ubiquinone in the ubiquinone pool in the membrane while the function of the small subunit is unknown (Matsushita et al. 1994, Toyama et al. 2007). Disruption of large subunit gene for FAD-GADH resulted in the loss of 2KGA formation, except in the case of *G. frateurii* NBRC 3271, which was suggested to have two types of FAD-GADH (Elfari et al. 2005, Merfort et al. 2006, Toyama et al. 2007, Saichana et al. 2009). The genes *gndSLC* of *G. frateurii* NBRC 3271 are believed to be the genes for another type of FAD-GADH, which is different from FAD-GADH found in the genome of *G. oxydans* 621H (Toyama et al. 2007). Another set of genes for FAD-GADH were also reported from thermotolerant *Gluconobacter* strains THE42, THF55 and THG42 as *gndFGH*. The genes showed higher identities in the FAD-GADH genes on the genome of *Gluconobacter* than the *gndSLC* of *G. frateurii* NBRC 3271. FAD-GADHs encoded from *gndSLC* and *gndFGH* are believed to be different types of GADH with distinct physiological functions. 5KGA is a functional raw material to produce xylaric acid, tartaric acid and 4-hydroxy-5-methyl-dehydrofuranone-3, a valuable flavor compound (Salusjarvi et al. 2004). 5KGA is also appropriate for vitamin C production by Gray's method (Gray 1945a, b).

Soluble enzymes located in the cytoplasm are involved in gluconate metabolism in the assimilation of 2KGA or 5KGA as 2KGA and 5KGA produced from oxidation accumulate in the culture medium and both the ketogluconates are transported to the cytoplasm by

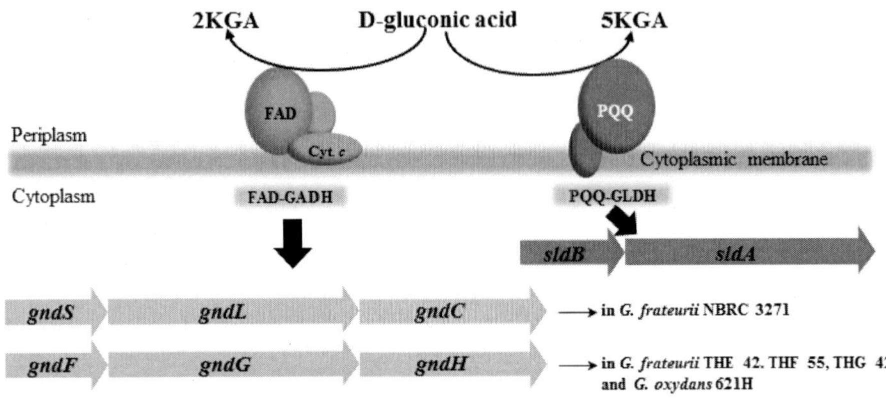

Figure 4: Oxidation of D-gluconic acid in AAB by two different D-gluconate oxidizing enzymes. Production of 2KGA and 5KGA are competitive reaction *in vivo.*

transporter proteins before being converted to D-gluconate by 2KGA reductase (2KGR) or 5KGA reductase (5KGR), respectively. The resultant products, which are assimilated by cytoplasmic reductases, are then subjected to pentose phosphate pathway in order to produce cell mass and the second growth phase is observed (Saichana et al. 2015). This pathway is known as the most important way for phosphorylative breakdown of sugars to carbon dioxide in *Gluconobacter* strains (Deppenmeier and Ehrenreich 2009, Mamlouk and Gullo 2013). However, this further oxidation to KGAs is undesirable during the production of gluconic acid by *Gluconobacter* strains. Therefore, use of high glucose concentrations and low pH values with a high aeration rate is recommended during wide-ranging production of gluconic acid by *Gluconobacter* strains (Mamlouk and Gullo 2013). *Acetobacter* strains can also produce high levels of GA, but not as high as *G. oxydans* since it can produce up to 120 g gluconic acid/L (Raspor and Goranovič 2008).

On the other hand, some *Gluconobacter* strains were found to convert 25KGA with a two-step biotranformation to 4-keto-D-arabonate (4KAB) as the end product. Within this two-step biotransformation, 25KGA is firstly converted to 4-keto-D-arabinose (4KAR) by a membrane-bound 2,5-diketo-D-gluconate decarboxylase (25DKGA DC) and then to 4KAB by 4-keto-D-aldopentose-1-dehydrogenase (4KALP 1DH) (Fig. 3) (Adachi et al. 2011).

Cleton-Jansen et al. (1991) demonstrated that a point mutation in the gene coding for PQQ-GDH of a *G. oxydans* strain resulted in the replacement of histidine at position 787 by asparagines. This resulted in conversion of maltose to maltobionic acid in addition to the oxidation of glucose, since it changed the substrate-specificity of PQQ-GDH (Kersters et al. 2006).

AAB oxidize D-fructose to 5-keto-D-fructose (5KF) by FAD-fructose dehydrogenase (FAD-FDH) (Fig. 5). This enzyme consisted of three subunits with the large dehydrogenase subunit containing FAD, the cytochrome *c* subunit and the small subunit with an unknown function (Adachi et al. 2007). The basis of this oxidative fermentation is that D-fructose oxidation by FAD-FDH continues until the D-fructose added initially in the reaction mixture is oxidized completely to the reaction product. There is no reaction equilibrium, unlike NAD(P)-dependent dehydrogenases. The reaction rate of D-fructose oxidation is not affected even if it is oxidized in the presence of the same concentration of substrate analogs, like D-glucose, D-fructose-6-phosphate, 5-keto-D-fructose, D-fructose-1,6-diphosphate, D-mannose, D-glucose-6-phosphate, GA, D-glucose-1-phosphate, 2KGA and 5KGA (Adachi et al. 2007).

Aldohexoses, aldopentoses, 5-keto-D-gluconate, 2-keto-D-gluconate, D-galactonate, 6-phospho-D-gluconate, D-mannonate, L-idonate, D-arabonate and D-xylonate are not oxidized by GADH. Some *Gluconobacter* spp. oxidize D-arabinose and D-arabonate to 4KAB as the final product. However, 4KAR is formed as an intermediate by D-aldopentose-4-dehydrogenase in D-arabinose conversion to 4KAB, but in the conversion of D-arabonate, the oxidation at C-4 position is catalyzed by D-pentonate-4-dehydrogenase (Fig. 6) (Adachi et al. 2011).

D-ribose is oxidized to 4-keto-D-ribose (4KRB) by D-aldopentose 4-dehydrogenase and then further oxidized to 4-Keto-D-ribonate (4KRN) by 4KALP 1DH (Fig. 7) (Adachi et al. 2013).

D-xylose oxidation is also catalyzed by PQQ-GDH since there is no specific D-xylose dehydrogenase other than PQQ-GDH (Adachi et al. 2007).

In addition to PQQ-GDH, *Acinetobacter calcoaceticus* contains soluble glucose dehydrogenase (s-GDH), which is believed to be the same enzymes or interconvertible

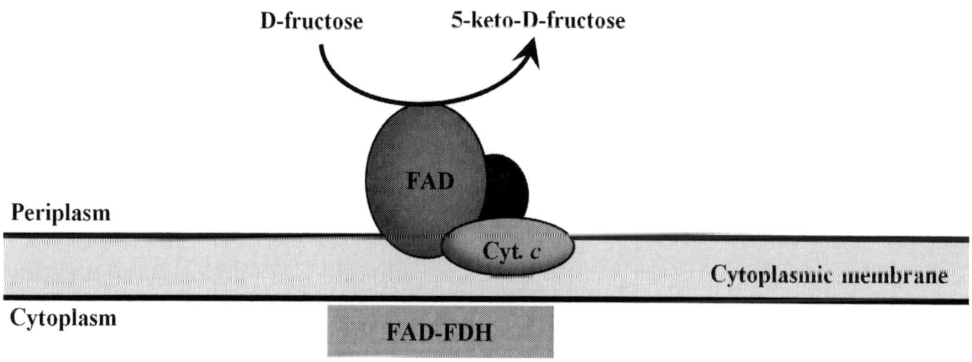

Figure 5: Oxidative fermentation for 5KF formation from D-fructose.

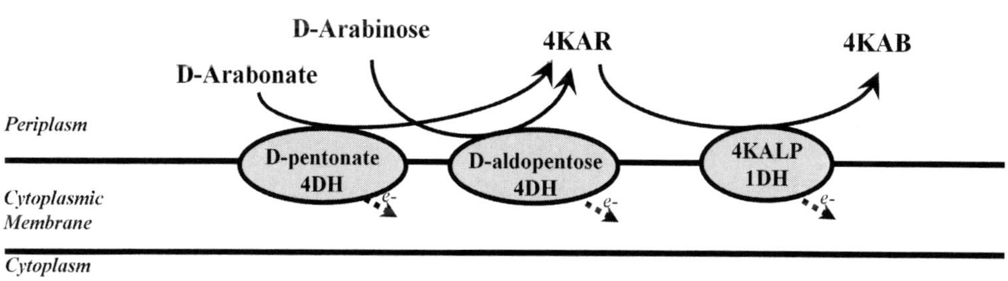

Figure 6: The conversion of 4KAB from D-arabinose and D-arabinate.

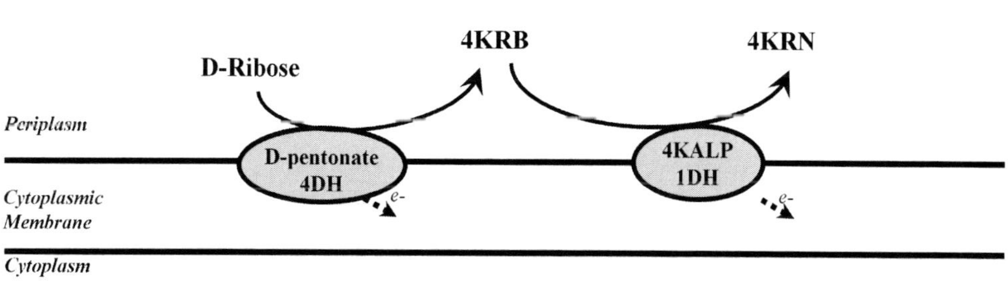

Figure 7: The conversion of KRN from D-ribose.

forms for many years. However, they are very different in all aspects, including optimum pH, kinetics, substrate specificity, U reactivity, immunoreactivity and molecular size (Matsushita et al. 1989). s-GDH can catalyze the oxidation of D-glucose (100 per cent), maltose (93 per cent), D-fucose (28 per cent), D-galactose (30 per cent), D-xylose (15 per cent) and lactose (72 per cent) while PQQ-GDH is less reactive with disaccharides: D-fucose (119 per cent), D-glucose (100 per cent), D-xylose (81 per cent), D-galactose (73 per cent), D-ribose (54 per cent), maltose (13 per cent) and lactose (5 per cent) (Adachi et al. 2007).

Sugar Alcohols Oxidation

Besides oxidizing ethanol to acetic acid, AAB oxidize various sugar alcohols with different carbon chain lengths, like D-sorbitol to L-sorbose, glycerol to dihydroxyacetone (DHA), D-mannitol to D-fructose, D-arabitol to D-xylulose, D- and *meso*-erythritol to L-erythrulose and ribitol to L-ribulose (Prust et al. 2005, Adachi et al. 2007, Mamlouk and Gullo 2013).

Gluconobacter strains oxidize various sugar alcohols and polyols according to the so-called 'rule of Bertrand-Hudson' since they have a great oxidative ability. This rule states, "Polyols with a *cis*-arrangement of two secondary hydroxyl groups in D-configuration to the adjacent primary alcohol group (D-erythro configuration) are oxidized regioselectively to the corresponding ketoses" (Kulhánek 1989, Deppenmeier et al. 2002). This particular mode of oxidation facilitates the synthesis of several new ketoses.

Gluconobacter strains oxidize many sugar alcohols incompletely to produce the corresponding organic acids, aldehydes and ketones, like D-gluconic acid, ketogluconic acids and dihydroxyacetone (DHA). For example, some *G. oxydans* strains regioselectively oxidize D-sorbitol to L-sorbose and to 2-keto-L-gluconic acid, which are both important intermediates in the production of L-ascorbic acid (Vitamin C), by PQQ-containing polyol dehydrogenase (known also as PQQ-glycerol dehydrogenase, PQQ-GLDH) in addition to FAD-containing D-sorbitol dehydrogenase (FAD-SLDH) (Fig. 8) (Matsushita et al. 2003, Adachi et al. 2007). Moreover, cyclitols and chemically modified pentitols and hexitols (deoxy-, deoxyamino-, deoxyhalogen-, deoxy-sulfuralditols, ω-deoxypolyols and ω-deoxy-acetyl polyols) are incompletely oxidized (Schedel 2000). Besides, many oxidoreductases also attack polyols and alcohols, utilize $NADH^+$ and do not use the Bertrand-Hudson rule (Kersters et al. 1965). Moonmangmee et al. (2000) also isolated and managed to apply a thermotolerant *Gluconobacter frateurii* CHM 54 to L-sorbose production. However, a strict temperature control is needed for this oxidative fermentation, especially in the hot summer days since a 2-3°C increase in temperature causes a highly probable failure in

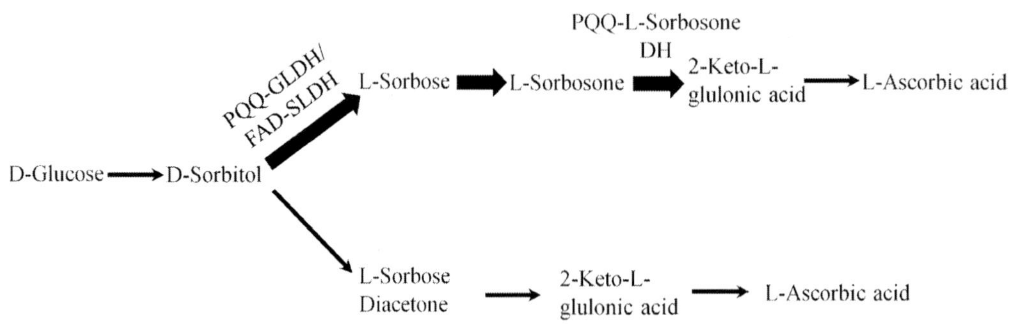

Figure 8: The conversion of L-sorbose and 2-keto-L-gulonic acid from D-sorbitol by *G. oxydans* in the production of L-ascorbic acid. Big arrows represent the transformations performed by different dehydrogenases of *G. oxydans*, while the little arrows represent the non-biological conversions (adapted from Kersters et al. 2006).

both fermentation rate and fermentation efficiency. Besides, in the last decade, industrial L-ascorbic acid production has revealed direct conversion of L-sorbosone to L-ascorbic acid by a quinoprotein L-sorbosone dehydrogenase (PQQ-L-sorbosone DH), which is purified from *Ketogluconicigenium vulgare* DSM 4025 (belongs to the *Rhodobacter*-lineage of the *Alphaproteobacteria*) (Kersters et al. 2006, Miyazaki et al. 2006, Adachi et al. 2007).

G. oxydans oxidizes glycerol to DHA, which is mostly used as a cosmetic tanning agent for pharmaceutical purposes and as an intermediate for the synthesis of various organic chemicals and surfactants, by PQQ-GLDH, according to the Bertrand–Hudson rule (Deppenmeier et al. 2002, Adachi et al. 2007). However, the inhibition of DHA formation due to increase at its concentration (feedback inhibition) and irreversible damage of the cells, which inhibits pentose cycle activity causes a great problem (Ohrem and Voß 1995).

Adachi et al. (2007) concluded from their studies that AAB oxidize different polyols (according to the Bertrand–Hudson rule) highly probable by an enzyme like GLDH, which is considered as the fundamental dehydrogenase of *G. oxydans* that exhibits a wide-range of substrate specificity (Fig. 9) (Moonmangmee et al. 2002, Adachi et al. 2003, Mamlouk and Gullo 2013). Besides, a quinoprotein D-arabitol dehydrogenase (PQQ-ARDH) from *G. thailandicus* NBRC 3257 (formerly known as *G. oxydans* IFO 3257) is reported to have the same wide substrate specificity as PQQ-GLDH (Saichana et al. 2015). *G. frateurii* CHM 43 catalyze *meso*-erythritol to L-erythrulose by PQQ-GLDH and NAD(P)-dependent enzymes have no function in L-erythrulose production (Adachi et al. 2007).

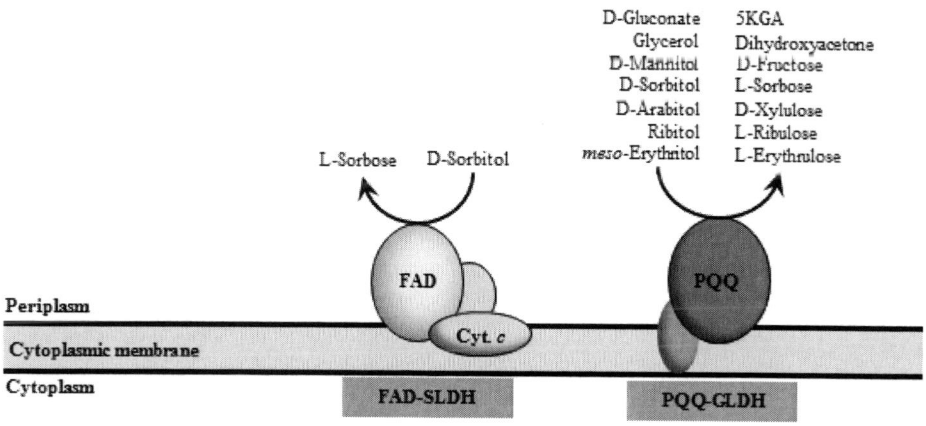

Figure 9: Conversion overview of sugar alcohols and polyols by AAB. FAD-SLDH is a specific enzyme for D-sorbitol oxidation to L-sorbose while PQQ-GLDH is a major polyol dehydrogenase which catalyzes the oxidations of many polyols, including D-sorbitol.

Organic Acids Oxidation

Most of the AAB (except strains of *Gluconobacter* since they do not have a functional TCA cycle) are also able to oxidize various organic acids [(acetic (AcOH), citric, fumaric, lactic, malic, pyruvic and succinic acids] to carbon dioxide and water through the TCA cycle (Matsushita et al. 2004, Raspor and Goranovič 2008). AAB oxidize organic acids at an optimum pH near to 6.0; however, there is evidence that it occurs at lower values (3.5-4.0) (Mamlouk and Gullo 2013).

Quinate oxidation: Some *Gluconobacter* strains, such as *G. oxydans* NBRC 3244, *G. oxydans* NBRC 3292 and *G. oxydans* NBRC 3294, produce 3-dehydroshikimate (DSA), which is the direct precursor for production of protocatechuic acid (PCA) from quinate. These strains have the membrane-bound quinate dehydrogenase (QDH) catalyzing the oxidation of quinate to 3-dehydroquinate (DQA). There are two types of enzymes responsible for the formation of DSA from DQA, which have been known as soluble 3-dehydroquinate dehydratase (sDQD) presented in the cytoplasm (Kleanthous et al. 1992) and the periplasmic DQD (pDQD) reported by Adachi et al. (2008). The activity of pDQD was 100-fold higher than the sDQD activity in the cytoplasm. The membrane-bound pDQD is believed to co-exist with membrane-bound QDH, oxidizing quinate to DSA via DQA (Adachi et al. 2008). The *qdh* gene of 2,475bp has been isolated from *G. oxydans* NBRC 3244 and the functional enzyme was expressed in *Pseudomonas putida* HK5 (Vangnai et al. 2010). In addition, *G. oxydans* NBRC 3244 oxidizes quinate to DSA in log phase and DSA accumulates in the medium which was then converted to PCA in the stationary phase. The enzyme in control for conversion of DSA to PCA is 3-dehydroshikimate dehydratase (DSD) (Fig. 10). The periplasmic pDQD was expressed in *G. oxydans* NBRC 3244, which showed a 100-fold higher activity than the original strain. This high expression level increased the rate of conversion of quinate to DSA in *G. oxydans* NBRC 3244 and almost 100 per cent conversion yield was achieved after adjusting the pH of the medium to 7.3 (Nishikura-Imamura et al. 2014).

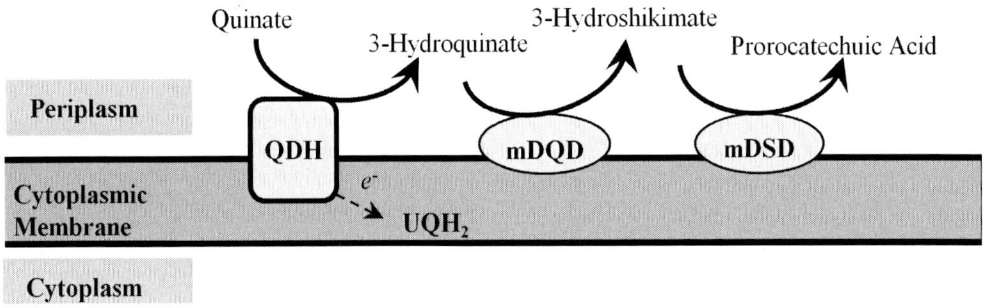

Figure 10: Conversion overview of quinate to PCA by membrane-bound enzymes in AAB.

It is reported that there are cytosolic 3-dehydroshikimate dehydratase (sDSD) and membrane-bound 3-dehydroshikimate dehydratase (mDSD) observed in *G. oxydans* (Shinagawa et al. 2010). PCA is an important antioxidant and anti-inflamatory compound. By using AAB, PCA is easily produced from quinate with the simple step of fermentation and the product is accumulated in the culture medium, allowing simple downstream processing.

Exopolysaccharide Production and Biofilm Formation of Acetic Acid Bacteria

AAB produce dextrans, levans and cellulose as the main exopolysaccharides (EPS) by glucose metabolism (Sievers and Swings 2005). However, cellulose, which is a polymer of β-1,4-linked glucose units, is the most common polysaccharide produced by AAB. It is synthesized by some strains of *Acetobacter* and *Komagataeibacter* (Matsutani et al. 2015). In the presence of oxygen and in a glucose-containing medium, cellulose production occurs, which can also occur during traditional vinegar production since the vinegar biofilm produced by AAB is composed of various EPS. In recent years, studies of cellulose production in AAB have centered on the composition and structure of the EPS biofilms produced by *K. medellinensis* strains (Valera et al. 2015).

The structural and mechanical characteristics of microbial cellulose are different from plant cellulose, which makes the microbial cellulose more valuable in the industry (Chawla et al. 2009). The size of the microbial cellulose fibrils is about 100 times smaller than that of plant cellulose, which gives it excellent property to hold a large amount of water, perfect elasticity, conformability and high wet strength. Especially cellulose produced by *K. medellinensis* has excellent tensile strength, transparency, fiber-binding ability, adaptability to the living body and a perfect biodegradability (Raspor and Goranovič 2008). The pathway of cellulose production from glucose has been well documented and it has four enzymatic steps (Fig. 11).

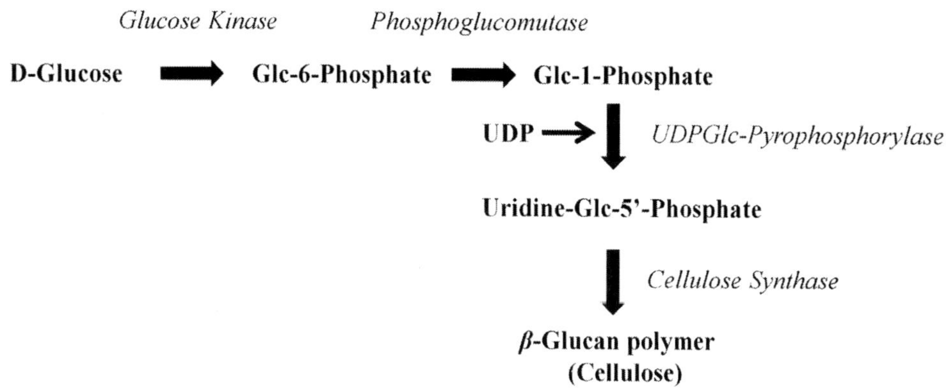

Figure 11: Reactions of cellulose synthesis.

Besides producing water-insoluble cellulose, most *K. medellinensis* and *K. intermedius* strains also produce acetan, which is a water-soluble polysaccharide containing glucose, mannose, glucuronic acid and rhamnose (Mamlouk and Gullo 2013).

Cellulose produced by AAB is used to manufacture wound dressings for patients with burns, chronic venous leg ulcers or other extensive loss of tissue (Alvarez et al. 2004, Raspor and Goranovič 2008).

Conclusion

AAB are obligately aerobic bacteria, which have simple respiratory chains linked to membrane-bound dehydrogenases. Electron transport chain in AAB was linked by the respiratory ubiquinones, either ubiquinone-9 or -10. The oxidative fermentation is an incomplete, but a rapid oxidation process that is a very useful feature in industrial applications as the products from the reaction are released into the culture medium, allowing simple downstream processing. The main well-known application of AAB is the production of acetic acid in the form of vinegar, which has a huge market, especially in Europe and some Asian countries. Moreover, AAB have been used for the production of several sugar derivatives, such as L-erytrulose, D-xylulose, L-ribulose, D-fructose, 2KGA and 5KGA, which are important precursors for the synthesis of several valuable compounds. Some strains of AAB also produce PCA, an antioxidant used in pharmaceutical applications. The nature of enzymes with enantio- and regioselectivities in *Gluconobacter* allow the applications for enantioselective oxidation of chiral and prochiral alcohol and diols, such as biotransformations of *N*-(2-hydroxyethyl) glucamine to produce 6-(2-hydroxyethyl) amino-6-deoxy-α-l-sorbofuranose, which is the main intermediate for synthesis of miglitol for type II diabetes treatment and production of (*S*)-2-methylbutanoic acid—an aroma compound found in fruits and used in the food industry (Keliang and Dongzhi 2006). Moreover, some AAB strains produce significant amounts of exopolysaccharides with various unique properties that can be useful in material science development.

References

Adachi, O., Moonmangmee, D., Toyama, H., Yamada, M., Shinagawa, E. and Matsushita, K. (2003). New developments in oxidative fermentation. *Applied Microbiology and Biotechnology* **60:** 643-653.

Adachi, O., Ano, Y., Toyama, H. and Matsushita, K. (2007). Biooxidation with PQQ- and FAD-dependent dehydrogenases. *In:* R.D. Schmid and V.B. Urlacher (Eds.). *Modern Biooxidation: Enzymes, Reactions and Applications*. Wiley-VCH Verlag GmbH & Co. KGaA, Weinheim, pp. 1-41.

Adachi O., Ano Y., Toyama H. and Matsushita K. (2008). A novel 3-dehydroquinate dehydratase catalyzing extracellular formation of 3-dehydroshikimate by oxidative fermentation of *Gluconobacter oxydans* IFO 3244. *Bioscience, Biotechnology and Biochemistry* **72:** 1475-1482.

Adachi, O., Hours, R.A., Shinagawa, E., Akakabe, Y., Yakushi, T. and Matsushita, K. (2011). Formation of 4-keto-D-aldopentoses and 4-pentulosonates (4-keto-D-pentonates) with unidentified membrane-bound enzymes from acetic acid bacteria. *Bioscience, Biotechnology and Biochemistry* **75:** 1801-1806.

Adachi, O., Hours, R.A., Akakabe, Y., Shinagawa, E., Ano, Y., Yakushi, T. and Matsushita, K. (2013). Pentose oxidation by acetic acid bacteria led to a finding of membrane-bound purine nucleosidase. *Bioscience, Biotechnology and Biochemistry* **77:** 1131-1133.

Alauzet, C., Teyssier, C., Jumas-Bilak, E., Gouby, A., Chiron, R., Rabaud, C., Counil, F., Lozniewski, A. and Marchandin, H. (2010). *Gluconobacter* as well as *Asaia* species, newly emerging opportunistic human pathogens among acetic acid bacteria. *Journal of Clinical Microbiology* **48:** 3935-3942.

Alvarez, O.M., Patel, M., Booker, J. and Markowitz, L. (2004). Effectiveness of a biocellulose wound dressing for the treatment of chronic venous leg ulcers: Results of a single center randomised study involving 24 patients. *Wounds* **16:** 224-233.

Ano, Y., Shinagawa, E., Adachi, O., Toyama, H., Yakushi, T. and Matsushita, K. (2011). Selective, high conversion of D-glucose to 5-keto-D-gluoconate by *Gluconobacter suboxydans. Bioscience, Biotechnology and Biochemistry* **75:** 586-589.

Baldrian, P. (2006). Fungal lactases. *FEMS Microbiology Reviews* **30:** 215-242.

Bartowsky, E.J. and Henschke, P.A. (2008). Acetic acid bacteria spoilage of bottled red wine—A review. *International Journal of Food Microbiology* **125:** 60-70.

Carr, J.G. (1968). Methods for identifying acetic acid bacteria. *In:* B.M. Gibbs and D.A. Shapton (Eds.). *Identification Methods for Microbiologists*, Academic Press, London, pp. 1-8.

Chawla, P.R., Bajaj, I.B., Survase, S.A. and Singhal, R.S. (2009). Microbial cellulose: Fermentative production and applications. *Food Technology and Biotechnology* **47:** 107-124.

Cirigliano, M.C. (1982). A selective medium for the isolation and differentiation of *Gluconobacter* and *Acetobacter*. *Journal of Food Science* **47:** 1038-1039.

Cleenwerck, I. and De Vos, P. (2008). Polyphasic taxonomy of acetic acid bacteria: An overview of the currently applied methodology. *International Journal of Food Microbiology* **125:** 2-14.

Cleton-Jansen, A.M., Dekker, S., van de Putte, P. and Goosen, N. (1991). A single amino acid substitution changes the substrate specificity of quinoprotein glucose dehydrogenase in *Gluconobacter oxydans*. *Molecular Genetics and Genomics* **229:** 206-212.

Czaja, W., Krystynowicz, A., Bielecki, S. and Brown, R.M. (2006). *Microbial cellulose—The natural power to heal wounds. Biomaterials* **27:** 145-151.

Czaja, W., Young, D., Kawecki, M. and Brown, R.M. (2007). The future prospects of microbial cellulose in biomedical applications. *Biomacromolecules* **8:** 1-12.

Deppenmeier, U., Hoffmeister, M. and Prust, C. (2002). Biochemistry and biotechnological applications of *Gluconobacter* strains. *Applied Microbiology and Biotechnology* **60:** 233-242.

Deppenmeier, U. and Ehrenreich, A. (2009). Physiology of acetic acid bacteria in light of the genome sequence of *Gluconobacter oxydans*. *Journal of Molecular Microbiology and Biotechnology* **16:** 69-80.

Desai, J.D. and Banat, I.M. (1997). Microbial production of surfactants and their commercial potential. *Microbiology and Molecular Biology Reviews* **61:** 47-64.

De Vero, L., Gullo, M. and Giudici, P. (2010) Acetic acid bacteria, biotechnological applications. *In:* M.C. Flickinger (Ed.). *Encyclopedia of Industrial Biotechnology: Bioprocess Bioseparation and Cell Technology*. Wiley, New York, pp. 9-25.

Drysdale, G.S. and Fleet, G.H. (1988). Acetic acid bacteria in winemaking: A review. *American Journal of Enology and Viticulture* **9:** 143-154.

Dutta, D. and Gachhui, R. (2006). Novel nitrogen-fixing *Acetobacter nitrogenifigens* sp. nov., isolated from Kombucha tea. *International Journal of Systematic and Evolutionary Microbiology* **56:** 1899-1903.

Du Toit, W.J. and Pretorius, I.J. (2002). The occurrence, control and esoteric effect of acetic acid bacteria in winemaking. *Annals of Microbiology* **52:** 155-179.

Elfari, M., Ha, S.-W., Bremus, C., Merfort, M., Khodaverdi, V., Herrmann, U., Sahm, H. and Gorisch, H. (2005). A *Gluconobacter oxydans* mutant converting glucose almost quantitatively to 5-keto-D-gluconic acid. *Applied Microbiology and Biotechnology* **66:** 668-674.

Fuentes-Ramirez, L.E., Bustillos-Cristales, R., Tapia-Hernandez, A., Jimenez-Salgado, T., Wang, E.T., Martinez-Romero, E. and Caballero-Mellado, J. (2001). Novel nitrogen-fixing acetic acid bacteria, *Gluconacetobacter johannae* sp nov and *Gluconacetobacter azotocaptans* sp nov, associated with coffee plants. *International Journal of Systematic and Evolutionary Microbiology* **51:** 1305-1314.

Gómez-Manzo, S., Chavez-Pacheco, J.L., Contreras-Zentella, M., Sosa-Torres, M.E., Arreguín-Espinosa, R., Pérez de la Mora, M., Membrillo-Hernández, J. and Escamilla, J.E. (2010). Molecular and catalytic properties of the aldehyde dehydrogenase of *Gluconacetobacter diazotrophicus*, a quinoheme protein containing pyrroloquinoline quinone, cytochrome *b*, and cytochrome *c*. *Journal of Bacteriology* **192:** 5718-5724.

González, A., Guillamón, J.M., Mas, A. and Poblet, M. (2006). Application of molecular methods for routine identification of acetic acid bacteria. *International Journal of Food Microbiology* **108:** 141-146.

Gouby, A., Teyssier, C., Vecina, F., Marchandin, H., Granolleras, C., Zorgniotti, I. and Jumas-Bilak, E. (2007). *Acetobacter cibinongensis* bacteremia in human. *Emerging Infectious Diseases* **13:** 784-785.

Gray, G.E. (1945a). *Fermentation of 2-ketogulonic Acid and Its Salts*. US patent 2,421,621.

Gray, G.E. (1945b). *Fermentation of 2-ketogulonic Acid and Its Salts*. US patent 2,421,612.

Greenberg, D.E., Porcella, S.F., Stock, F., Wong, A., Conville, P.S., Murray, P.R., Holland, S.M. and Zelazny, A.M. (2006). *Granulibacter bethesdensis* gen. nov., sp. nov., a distinctive pathogenic acetic acid bacterium in the family *Acetobacteraceae*. *International Journal of Systematic and Evolutionary Microbiology* **56:** 2609-2616.

Guillamón, J.M. and Mas, A. (2009). Acetic acid bacteria. *In:* H. König, G. Unden and J. Fröhlich (Eds.). *Biology of Microorganisms on Grapes, in Must and in Wine*. Springer, Verlag Berlin Heidelberg, pp. 31-46.

Gullo, M. and Giudici, P. (2008). Acetic acid bacteria in traditional balsamic vinegar: Phenotypic traits relevant for starter cultures selection. *International Journal of Food Microbiology* **125:** 46-53.

Gullo, M., Verzelloni, E. and Canonico, M. (2014). Aerobic submerged fermentation by acetic acid bacteria for vinegar production: Process and biotechnological aspects. *Process Biochemistry* **49:** 1571-1579.

Gupta, A., Singh, V.K., Qazi, G.N. and Kumar, A. (2001). *Gluconobacter oxydans*: Its biotechnological applications. *Journal of Molecular Microbiology and Biotechnology* **3:** 445-456.

Hattori, H., Yakushi, T., Matsutani, M., Moonmangmee, D., Toyama, H., Adachi, O. and Matsushita, K. (2012) High-temperature sorbose fermentation with thermotolerant *Gluconobacter frateurii* CHM43 and its mutant strain adapted to higher temperature. *Applied Microbiology and Biotechnology* **95:** 1531-1540.

Hermann, M., Petermeier, H. and Vogel, R.F. (2015). Development of novel sourdoughs with in situ formed exopolysaccharides from acetic acid bacteria. *European Food Research and Technology* **241:** 185-197.

Hölscher, T. and Görisch, H. (2006). Knockout and overexpression of pyrroloquinoline quinone biosynthetic genes in *Gluconobacter oxydans* 621H. *Journal of Bacteriology* **188:** 7668-7676.

Hoshino, T., Sugisawa, T., Shinjoh, M., Tomiyama, N. and Miyazaki, T. (2003). Membrane-bound D-sorbitol dehydrogenase of *Gluconobacter suboxydans* IFO 3255-enzymatic and genetic characterization. *Biochimica et Biophysica Acta-Bioenergetics* **1647:** 278-288.

Jakob, F., Steger, S. and Vogel, R.F. (2012). Influence of novel fructans produced by selected acetic acid bacteria on the volume and texture of wheat breads. *European Food Research and Technology* **234:** 493-499.

Joyeux, A., Lafon-Lafourcade, S. and Ribéreau-Gayon, P. (1984). Evolution of acetic acid bacteria during fermentation and storage of wine. *Applied and Environmental Microbiology* **48:** 153-156.

Kanchanarach, W., Theeragool, G., Yakushi, T., Toyama, H., Adachi, O. and Matsushita, K. (2010). Characterization of thermotolerant *Acetobacter pasteurianus* strains and their quinoprotein alcohol dehydrogenases. *Applied Microbiology and Biotechnology* **85:** 741-751.

Katsura, K., Kawasaki, H., Potacharoen, W., Saono, S., Seki, T., Yamada, Y., Uchimura, T. and Komagata, K. (2001). *Asaia siamensis* sp. nov., an acetic acid bacterium in the α-proteobacteria. *International Journal of Systematic and Evolutionary Microbiology* **51:** 559-563.

Keliang, G. and Dongzhi, W. (2006). Asymmetric oxidation by *Gluconobacter oxydans*. *Applied Microbiology and Biotechnology* **70:** 135-139.

Kersters, K., Wood, W.A. and De Ley, J. (1965). Polyol dehydrogenases of *Gluconobacter oxydans*. *Journal of Biological Chemistry* **240:** 965-974.

Kersters, K., Lisdiyanti, P., Komagata, K. and Swings, J. (2006). The family *Acetobacteraceae*: The genera *Acetobacter, Acidomonas, Asaia, Gluconacetobacter, Gluconobacter* and *Kozakia*. *In:* M. Dworkin, S. Falkow, E. Rosenberg, K-H. Schleifer and E. Stackebrandt (Eds.). *The Prokaryotes—A Handbook on the Biology of Bacteria*, 3rd edn, vol. 5. Springer-Verlag, New York, pp. 163-200.

Kleanthous, C., Deka, R., Davis, K., Kelly, S.M., Cooper, A., Harding, S.E., Price, N.C., Hawkings, A.R. and Coggins, J.R. (1992). A comparison of the enzymological and biophysical properties of two distinct classes of dehydroquinase enzymes. *Biochemistry Journal* **282:** 687-695.

Kösebalaban, F. and Özilgen, M. (1992). Kinetics of wine spoilage by acetic-acid bacteria. *Journal of Chemical Technology and Biotechnology* **55:** 59-63.

Kostner, D., Luchterhand, B., Junker, A., Volland, S., Daniel, R., Büchs, J., Liebl, W. and Ehrenreich, A. (2015). The consequence of an additional NADH dehydrogenase paralog on the growth of Gluconobacter oxydans DSM3504. *Applied Microbiology and Biotechnology* **99:** 375-386.

Kulhánek, M. (1989). Microbial dehydrogenations of monosaccharides. *Advances in Applied Microbiology* **34:** 141-181.

Li, L., Praet, J., Borremans, W., Nunes, O. C., Manaia, C. M., Cleenwerck, I., Meeus, I., Smagghe, G., De Vuyst, L. and Vandamme, P. (2015). *Bombella intestini* gen. nov., sp. nov., an acetic acid bacterium isolated from bumble bee crop. *International Journal of Systematic and Evolutionary Microbiology* **65:** 267-273.

Loganathan, P. and Nair, S. (2004). *Swaminathania salitolerans* gen. nov., sp. nov., a salt-tolerant, nitrogen-fixing and phosphate-solubilizing bacterium from wild rice (*Porteresia coarctata* Tateoka). *International Journal of Systematic and Evolutionary Microbiology* **54:** 1185-1190.

Malimas, T., Yukphan, P., Takahashi, M., Muramatsu, Y., Kaneyasu, M., Potacharoen, W., Tanasupawat, S., Nakagawa, Y., Tanticharoen, M. and Yamada, Y. (2009). *Gluconobacter japonicus* sp. nov., an acetic acid bacterium in the alphaproteobacteria. *International Journal of Systematic and Evolutionary Microbiology* **59:** 466-471.

Malimas, T., Chaipitakchonlatarn, W., Thi Lan Vu, H., Yukphan, P., Muramatsu, Y., Tanasupawat, S., Potacharoen, W., Nakagawa, Y., Tanticharoen, M. and Yamada, Y. (2013). *Swingsia samuiensis* gen. nov., sp. nov., an osmotolerant acetic acid bacterium in the α-Proteobacteria. *The Journal of General and Applied Microbiology* **59:** 375-384.

Mamlouk, D. and Gullo, M. (2013). Acetic acid bacteria: Physiology and carbon sources oxidation. *Indian Journal of Microbiology* **53:** 377-384.

Masud, U., Matsushita, K. and Theeragool, G. (2010). Cloning and functional analysis of *adhS* gene encoding quinoprotein alcohol dehydrogenase subunit III from *Acetobacter pasteurianus* SKU1108. *Interntional Journal of Food Microbiology* **138**: 39-49.

Matsushita, K., Shinagawa, E., Adachi, O. and Ameyama, M. (1987). Purification, characterization and reconstitution of cytochrome o-type oxidase from *Gluconobacter suboxydans*. *Biochimica et Biophysica Acta-Bioenergetics* **894:** 304-312.

Matsushita, K., Shinagawa, E., Adachi, O. and Ameyama, M. (1989). Quinoprotein D-glucose dehydrogenase of the *Acinetobacter calcoaceticus* respiratory chain: Membrane-bound and soluble forms are different molecular species. *Biochemistry* **28:** 6276-6280.

Matsushita, K., Shinagawa, E., Adachi, O. and Ameyama, M. (1990). Cytochrome a_1 of *Acetobacter aceti* is a cytochrome *ba* functioning as ubiquinol oxidase. *Proceedings of the National Academy of Sciences of the United States of America* **87:** 9863-9867.

Matsushita, K., Ebisuya, H., Ameyama, M. and Adachi, O. (1992). Change of the terminal oxidase from cytochrome a_1 in shaking cultures to cytochrome *o* in static cultures of *Acetobacter aceti*. *Journal of Bacteriology* **174:** 122-129.

Matsushita, K., Toyama, H. and Adachi, O. (1994). Respiratory chain and bioenergetics of acetic acid bacteria. *In:* A.H. Rose and D.W. Tempest (Eds.). *Advances in Microbial Physiology*. Academic Press, London, pp. 247-301.

Matsushita, K., Toyama, H. and Adachi, O. (2002). Quinoproteins: Structure, function, and biotechnological applications. *Applied Microbiology and Biotechnology* **58:** 13-22.

Matsushita, K., Fujii, Y., Ano, Y., Toyama, H., Shinjoh, M., Tomiyama, N., Miyazaki, T., Sugisawa, T., Hoshino, T. and Adachi O. (2003). 5-KetoD-gluconate production is catalyzed by a quinoprotein glycerol dehydrogenase, major polyol dehydrogenase in *Gluconobacter* species. *Applied and Environmental Microbiology* **69:** 1959-1966.

Matsushita, K., Toyama, H. and Adachi, O. (2004). Respiratory chains in acetic acid bacteria: Membrane bound periplasmic sugar and alcohol respirations. *In:* D. Zannoni (Ed.). *Respiration in Archaea and Bacteria—Advances in Photosynthesis and Respiration*. Springer, Dordrecht, pp. 81-99.

Matsushita, K., Kobayashi, Y., Mizuguchi, M., Toyama, H., Adachi, O., Sakamoto, K. and Miyoshi, H. (2008). A tightly bound quinone functions in the ubiquinone reaction sites of quinoprotein alcohol dehydrogenase of an acetic acid bacterium, *Gluconobacter suboxydans*. *Bioscience, Biotechnology and Biochemistry* **72:** 2723-2731.

Matsutani, M., Fukushima, K., Kayama, C., Arimitsu, M., Hirakawa, H., Toyama, H., Adachi, O., Yakushi, T. and Matsushita, K. (2014). Replacement of a terminal cytochrome *c* oxidase by ubiquinol oxidase during the evolution of acetic acid bacteria. *Biochimica et Biophysica Acta-Bioenergetics* **1837:** 1810-1820.

Matsutani, M., Ito, K., Azuma, Y., Ogino, H., Shirai, M., Yakushi, T. and Matsushita, K. (2015). Adaptive mutation related to cellulose producibility in *Komagataeibacter medellinensis* (*Gluconacetobacter xylinus*) NBRC 3288. *Applied Microbiology and Bitechnology* **99:** 7229-7240.

Merfort, M., Herrmann, U., Bringer-Meyer, S. and Sahm, H. (2006). High-yield 5-keto-D-gluconic acid formation is mediated by soluble and membrane-bound gluconate-5-dehydrogenases of *Gluconobacter oxydans*. *Applied Microbiology and Biotechnology* **73:** 443-451.

Miyazaki, T., Sugisawa, T. and Hoshino, T. (2006). Pyrroloquinoline quinone-dependent dehydrogenases from *Ketogulonicigenium vulgare* catalyze the direct conversion of L-sorbosone to L-ascorbic acid. *Applied Environmental Microbiology* **72:** 1487-1495.

Miura, H., Mogi, T., Ano, Y., Migita, C. T., Matsutani, M., Yakushi, T., Kita, K. and Matsushita, K. (2013). Cyanide-insensitive quinol oxidase (*CIO*) from *Gluconobacter oxydans* is a unique terminal oxidase subfamily of cytochrome *bd*. *Journal of Biochemistry* **153:** 535-545.

Moonmangmee, D., Adachi, O., Shinagawa, E., Toyama, H., Theeragool, G., Lotong, N. and Matsushita, K. (2002). L-erythrulose production by oxidative fermentation is catalyzed by PQQ-containing membrane-bound dehydrogenase. *Bioscience, Biotechnology and Biochemistry* **66:** 307-318.

Ndoye, B., Lebecque, S., Dubois-Dauphin, R., Tounkara, L., Guiro, T.-A., Kere, C., Diawara, B. and Thonart, P. (2006). Thermoresistant properties of acetic acid bacteria isolated from tropical products of Sub-Saharan Africa and destined to industrial vinegar. *Enzyme and Microbial Technology* **39:** 916-923.

Nishikura-Imamura, S., Matsutani, M., Insomphun, C., Vangnai, A. S., Toyama, H., Yakushi, T., Abe, T., Adachi, O. and Matsushita, K. (2014). Overexpression of a type II 3-dehydroquinate dehydratase enhances the biotransformation of quinate to 3-dehydroshikimate in *Gluconobacter oxydans*. *Applied Microbiology and Biotechnology* **98:** 2955-2963.

Ohrem, H.L. and Voß, H. (1995). Inhibitory effects of dihydroxyacetone on *Gluconobacter* cultures. *Biotechnology Letters* **17:** 981-984.

O'Toole, D.K. and Kun, L.Y. (2003). Fermented foods. *In:* L.Y. Kun (Ed.). *Microbial Biotechnology: Principles and Applications*, 2nd edn. World Scientific Publishing Co. Pte. Ltd., Singapore, pp. 201-256.

Pedraza, R.O. (2008). Recent advances in nitrogen-fixing acetic acid bacteria. *International Journal of Food Microbiology* **125:** 25-35.

Prust, C., Hoffmeister, M., Liesegang, H., Wiezer, A., Fricke, W.F., Ehrenreich, A., Gottschalk, G. and Deppenmeier, U. (2005). Complete genome sequence of the acetic acid bacterium *Gluconobacter oxydans*. *Nature Biotechnology* **23:** 195-200.

Qazi, G.N., Parshad, R.,Verma,V., Chopra, C.L., Buse, R., Träger, M. and Onken, U. (1991). Diketo-gluconate fermentation by *Gluconobacter oxydans*. *Enzyme and Microbial Technology* **13:** 504-507.

Qi, Z., Yang, H., Xia, X., Xin, Y., Zhang, L., Wang, W. and Yu, X. (2013). A protocol for optimization vinegar fermentation according to the ratio of oxygen consumption versus acid yield. *Journal of Food Engineering* **116:** 304-309.

Raspor, P. and Goranovič, D. (2008). Biotechnological applications of acetic acid bacteria. *Critical Reviews in Biotechnology* **28:** 101-124.

Richhardt, J., Luchterhand, B., Bringer, S., Buchs, J. and Bott, M. (2013). Evidence for a key role of cytochrome bo_3 oxidase in respiratory energy metabolism of *Gluconobacter oxydans*. *Journal of Bacteriology* **195:** 4210-4220.

Roh, S.W., Nam, Y.-D., Chang, H.-W., Kim, K.-H., Kim, M.-S., Ryu, J.-H., Kim, S.-H., Lee, W.-J. and Bae, J.-W. (2008). Phylogenetic characterization of two novel commensal bacteria involved with innate immune homeostasis in *Drosophila melanogaster*. *Applied and Environmental Microbiology* **74:** 6171-6177.

Salusjarvi, T., Povelainen, M., Hvorsley, N., Eneyskaya, E.V., Kulminskaya, A.A., Shabalin, K.A., Neustroev, K.N., Kalkkinen, N. and Miasnikov, A.N. (2004). Cloning of a gluconate/ polyol dehydrogenase gene from *Gluconobacter suboxydans* IFO 12528, characterization of the enzyme and its use for the production of 5-ketogluconate in a recombinant *Escherichia coli* strain. *Applied Microbiology and Biotechnology* **65:** 306-314.

Saichana, I., Moonmangmee, D., Adachi, O., Matsushita, K. and Toyama, H. (2009). Screening of thermotolerant *Gluconobacter* strains for production of 5-keto-D-gluconic acid and disruption of flavin adenine dinucleotide-containing D-gluconate dehydrogenase. *Applied and Environmental Microbiology* **75:** 4240-4247.

Saichana, N., Matsushita, K., Adachi, O., Frébort, I. and Frebortova, J. (2015). Acetic acid bacteria: A group of bacteria with versatile biotechnological applications. *Biotechnology Advances* **33:** 1260-1271.

Schedel, M. (2000). Regioselective oxidation of aminosorbitol with *Gluconobacter oxydans*, a key reaction in the industrial synthesis of 1-deoxynojirimycin. *In:* D.R. Kelly (Ed.). *Biotechnology: Biotransformations* II, 2nd edn, vol. 8b. Wiley-VCH Verlag GmbH & Co. KGaA, Weinheim, pp. 296-308.

Schlepütz, T., Gerhards, J.P. and Büchs J. (2013). Ensuring constant oxygen supply during inoculation is essential to obtain reproducible results with obligatory aerobic acetic acid bacteria in vinegar production. *Process Biochemistry* **48:** 398-405.

Sengun, I.Y. and Karabiyikli, S. (2011). Importance of acetic acid bacteria in food industry. *Food Control* **22:** 647-656.

Sengun, I.Y. (2015). Acetic acid bacteria in food fermentations. *In:* D. Montet and R.C. Ray (Eds.). *Fermented Foods: Part 1: Biochemistry and Biotechnology*. CRC Press, Boca Raton, USA, pp. 91-111.

Shinagawa, E., Ano, Y., Yakushi, T., Adachi, O. and Matsushita, K. (2009). Solubilization, purification, and properties of membrane-bound D-glucono-δ-lactone hydrolase from *Gluconobacter oxydans*. *Bioscience, Biotechnology and Biochemistry* **73:** 241-244.

Shinagawa, E., Adachi, O., Ano, Y., Yakushi, T. and Matsushita, K. (2010). Purification and characterization of membrane-bound 3-dehydroshikimate dehydratase from *Gluconobacter oxydans* IFO 3244, a new enzyme catalyzing extracellular protocatechuate formation. *Bioscience, Biotechnology and Biochemistry* **74:** 1084-1088.

Sievers, M. and Swings, J. (2005). Family II. Acetobacteraceae. *In:* D.J. Brenner, N.R. Krieg, J.T. Staley and G.M. Garrity, G.M. (Eds.). *Bergey's Manual of Systematic Bacteriology*, 2nd edn, vol. 2, Parts A, B and C. Springer-Verlag, New York, pp. 41-95.

Singh, O.V. and Kumar, R. (2007). Biotechnological production of gluconic acid: Future implications. *Applied Microbiology and Biotechnology* **75:** 713-772.

Snyder, R.W., Ruhe, J., Kobrin, S., Wasserstein, A., Doline, C., Nachamkin, I. and Lipschutz, J.H. (2004). *Asaia bogorensis* peritonitis identified by 16S ribosomal RNA sequence analysis in a patient receiving peritoneal dialysis. *American Journal of Kidney Diseases: The Official Journal of the National Kidney Foundation* **44:** e15-17.

Takemura, H., Tsuchida, T., Yoshinaga, F., Matsushita, K. and Adachi O. (1994). Prosthetic group of aldehyde dehydrogenase in acetic acid bacteria not pyrroloquinoline quinone. *Bioscience, Biotechnology and Biochemistry* **58:** 2082–2083.

Thi Lan Vu, H., Yukphan, P., Chaipitakchonlatarn, W., Malimas, T., Muramatsu, Y., Thi Tu Bui, U., Tanasupawat, S., Cong Duong, K., Nakagawa, Y., Thanh Pham, H. and Yamada, Y. (2013). *Nguyenibacter vanlangensis* gen. nov., sp. nov., an unusual acetic acid bacterium in the α-Proteobacteria. *The Journal of General and Applied Microbiology* **59:** 153-166.

Thurner, C., Vela, C., Thöny-Meyer, L., Meile, L. and Teuber, M. (1997). Biochemical and genetic characterization of the acetaldehyde dehydrogenase complex from *Acetobacter europaeus*. *Archives of Microbiology* **168:** 81-91.

Toyama, H., Furuya, N., Saichana, I., Ano, Y., Adachi, O. and Matsushita, K. (2007). Membrane-bound, 2-keto-D-gluconate-yielding D-gluconate dehydrogenase from *"Gluconobacter dioxyacetonicus"* IFO 3271: Molecular properties and gene disruption. *Applied and Environmental Microbiology* **273:** 6551-6556.

Trcek, J., Toyama, H., Czuba, J., Misiewicz, A. and Matsushita, K. (2006). Correlation between acetic acid resistance and characteristics of PQQ-dependent ADH in acetic acid bacteria. *Applied Microbiology and Biotechnology* **70:** 366-373.

Tuuminen, T., Heinäsmäki, T. and Kerttula, T. (2006). First report of bacteremia by *Asaia bogorensis*, in a patient with a history of intravenous-drug abuse. *Journal of Clinical Microbiology* **44:** 3048-3050.

Valera, M.J., Torija, M.J., Mas, A. and Mateo, E. (2015). Cellulose production and cellulose synthase gene detection in acetic acid bacteria. *Applied Microbiology and Biotechnology* **99:** 1349-1361.

Vangnai, A.S., Promden, W., De-Eknamkul, W., Matsushita, K. and Toyama, H. (2010). Molecular characterization and heterologous expression of quinate dehydrogenase gene from *Gluconobacter oxydans* IFO 3244. *Biochemistry* **75:** 452-459.

Yukphan, P., Malimas, T., Muramatsu, Y., Takahashi, M., Kaneyasu, M., Tanasupawat, S., Nakagawa, Y., Suzuki, K., Potacharoes, W. and Yamada, Y. (2008). *Tanticharoenia sakaeratensis* gen. nov., sp. nov., a new osmotolerant acetic acid bacterium in the α-proteobacteria. *Bioscience, Biotechnology and Biochemistry* **72:** 672-676.

Yukphan, P., Malimas, T., Muramatsu, Y., Potacharoen, W., Tanasupawat, S., Nakagawa, Y., Tanticharoen, M. and Yamada, Y. (2011). *Neokomagataea* gen. nov., with descriptions of *Neokomagataea thailandica* sp. nov. and *Neokomagataea tanensis* sp. nov., osmotolerant acetic acid bacteria of the α-proteobacteria. *Bioscience, Biotechnology and Biochemistry* **75:** 419-426.

Wang, B., Shao, Y. and Chen, F. (2015). Overview on mechanisms of acetic acid resistance in acetic acid bacteria. *World Journal of Microbiology and Biotechnology* **31:** 255-263.

Weenk, G., Olijve, W. and Harder, W. (1984). Ketogluconate formation by *Gluconobacter* species. *Applied Microbiology and Biotechnology* **20:** 400-405.

4

Acetic Acid Bacteria Strategies Contributing to Acetic Acid Resistance During Oxidative Fermentation

Cristina Andrés-Barrao[1] and François Barja[2,*]

[1] King Abdullah University of Science and Technology,
Division of Biological and Environmental Sciences and Engineering,
Center for Desert Agriculture, 4700 Thuwal 23955-6900, Saudi Arabia

[2] University of Geneva, Department of Botany and Plant Biology, Microbiology Unit,
Quai Ernest-Ansermet 30 (Sciences III), 1211 Geneva 4, Switzerland

Introduction

Vinegar was discovered fortuitously in ancient times when wine stored undisturbed in the open air turned spontaneously into vinegar. Since then, in addition to other benefits for humankind, vinegar has for long been used as a food preservative for its ability to kill the majority of microorganisms. The bacteriostatic and bactericidal effects of vinegar are due to acetic acid (4-6 per cent in commercial vinegars) (Solieri and Giudici 2009), a weak organic acid that is commonly added to food products to avoid microbial spoilage. Low concentration of acetic acid of 0.1-0.17 per cent has been able to inhibit the growth of important food-borne pathogens, such as *Proteus vulgaris*, *Acinetobacter baumannii*, *Pseudomonas aeruginosa* or the enterohemorrhagic *Escherichia coli* (EHEC) (Levine and Fellers 1940, Entani et al. 1998, Ryssel et al. 2009, Fraise et al. 2013). The capacity of biofilm-forming bacteria to resist antibiotics and other antimicrobial agents can cause persistence of chronic infections. Treatment with 0.5-1 per cent acetic acid has been proved to be effective in eradicating three-day-old flow chamber biofilm of *P. aeruginosa* and *Staphylococcus aureus* (Bjarnsholt et al. 2015). Disinfectant treatment using low percentage of acetic acid (vinegar) is also effective in killing *Mycobacterium tuberculosis* as well as other very resistant mycobacteria and can prove to be an effective alternative to the toxic and expensive mycobactericidal agents that are currently used (Cortesia et al. 2014).

The toxicity of acetic acid at low pH has been explained by different effects: uncoupling, intracellular pH regulation, anion accumulation, anion carriers, genetic regulation and resistance (Russell 1992). The protonated, uncharged form of this organic acid easily penetrates the microbial membrane and dissociates at the more alkaline bacterial cytoplasm, releasing products, particularly proton (H^+) but also acetate (CH_3COO^-), which are toxic for the cell (Russell 1992, Russell and Díez-González 1998) (Fig. 1).

The production of vinegar from wine or any other alcoholic source is the result of the microbiological transformation of ethanol into acetic acid, in a process called acetic acid fermentation (other accepted terminologies are oxidative fermentation, acetic fermentation

*Corresponding author: Francois.Barja@unige.ch

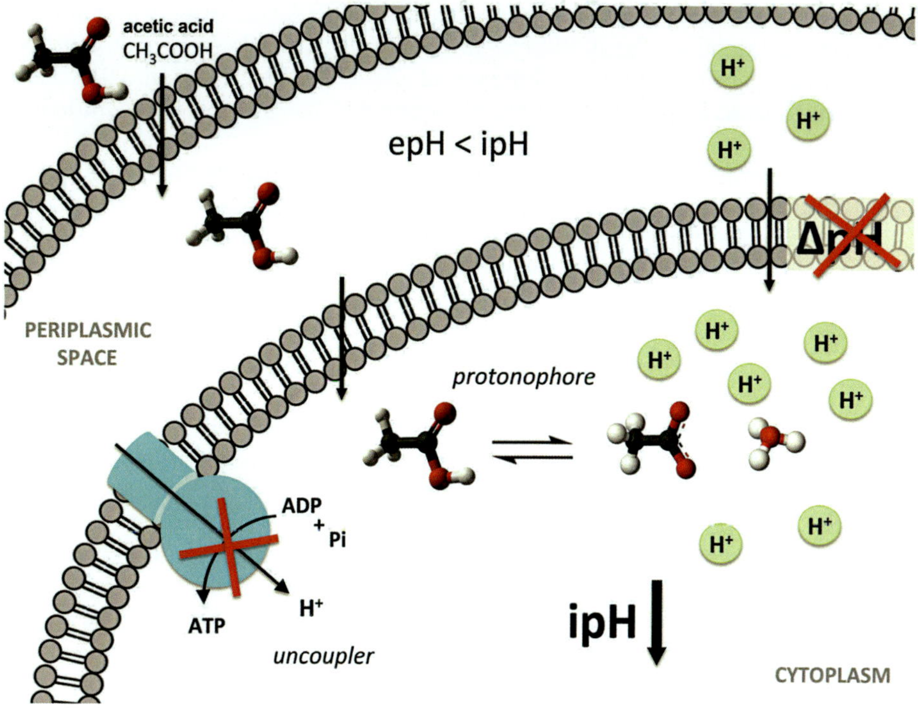

Figure 1: Cytotoxic effects of acetic acid. Inside the cell, the acetic acid molecule dissociates and releases a proton (H^+). The accumulation of H^+ leads to the acidification of the cytoplasm and the dissipation of the pH gradient (ΔpH) across the cytoplasmic membrane. Acetic acid acts as protonophore, carrying H^+ across the cell membrane, and causes the uncoupling between the electron transport and the oxidative phosphorylation, thus inhibiting the synthesis of ATP, ultimately leading to the bacterial cell death. epH and ipH stand for external and internal pH, respectively.

or acetification) and the microorganisms responsible for this transformation are known as acetic acid bacteria (AAB). These Gram-negative bacteria are strict aerobes and are characterized by their capability to oxidize sugars, alcohols and sugar alcohols into their corresponding organic acids.

AAB are taxonomically affiliated to the family *Acetobacteraceae* of the *Alphaproteobacteria*. The first recognized genera of AAB were *Acetobacter*, the type genus of the family (Beijerinck 1898), and *Gluconobacter* (Asai 1935). The genus *Acetobacter* was subsequently divided into two subgenera, *Acetobacter* and *Gluconoacetobacter*, on the basis of differing ubiquinones as the general electron acceptor. *Acetobacter* spp. contained ubiquinone-9, while *Gluconoacetobacter* spp. contained ubiquinone-10 (Yamada and Kondo 1985). So far, the presence of ubiquinone-9 is a unique characteristic of the genus *Acetobacter* among all currently recognized AAB. Later, the subgenus *Gluconoacetobacter* was elevated to the generic level and the new genus *Gluconacetobacter* was proposed (Yamada et al. 1997). Urakami et al. (1989) proposed the reclassification of the methylotrophic strain *Acetobacter methanolicus* as *Acidomonas methanolica* comb. nov., into the new genus *Acidomonas*. By the end of the 90's, only four genera of AAB were recognized: *Acetobacter*, *Gluconobacter*, *Gluconacetobacter* and *Acidomonas*. Since then, a great number of AAB species have been identified. Eighteen genera and 88 species are currently recognized in the group of AAB. *Acetobacter*, *Acidomonas*, *Ameyamaea*, *Asaia*, *Endobacter*, *Gluconacetobacter*, *Gluconobacter*, *Granulibacter*, *Komagataeibacter*, *Kozakia*, *Neoasaia*, *Neokomagataea*, *Saccharibacter*, *Swaminathania* and *Tanticharoenia* are the genera found in *The List of Prokaryotic Names with Standing in Nomenclature* (LPSN) (www.bacterio.net) (Parte 2014). Three additional monotypic genera, *Swingsia, Nguyenibacter* and *Bombella,* have been recently proposed but not yet included in the LPSN (Malimas et al. 2013, Thi Lan Vu et al. 2013, Li et al. 2015). The recently described *Acetobacter musti* (Ferrer et al. 2016) is also not included yet in the LPSN.

Most of the genera classified into the group of AAB are composed of a single species, and 73 out of the 88 recognized species are predominated by *Acetobacter* (26 spp.), *Asaia* (8 spp.), *Gluconacetobacter* (11 spp.), *Gluconobacter* (14 spp.) and *Komagataeibacter* (14 spp.). Until recently, the genus *Gluconacetobacter* consisted of two phylogenetically, phenotypically and ecologically distinct *Gluconacetobacter liquefaciens* and *Gluconacetobacter xylinus* groups. The division of this genus due to these distinctions was controversial until Yamada et al. (2012) proposed the reclassification of the species included into the *Gluconacetobacter xylinus* group into a new genus, *Komagataeibacter*.

In addition to the capability of AAB to oxidize sugars and alcohols into organic acids, some specialized strains of *Acetobacter* and *Komagataeibacter* spp. can transform ethanol into acetic acid with high efficiency, in an acidic environment. Therefore, these specialized strains are involved in the industrial production of vinegar. The general process involves a double oxidation from ethanol to acetaldehyde, and then to acetic acid, in a sequential reaction that occurs in the periplasmic space. This reaction is carried out by two enzymes located in the cytoplasmic membrane, the alcohol dehydrogenase (ADH) and the aldehyde dehydrogenase (ALDH) (Adachi et al. 1978, Ameyama et al. 1981) (Fig. 2).

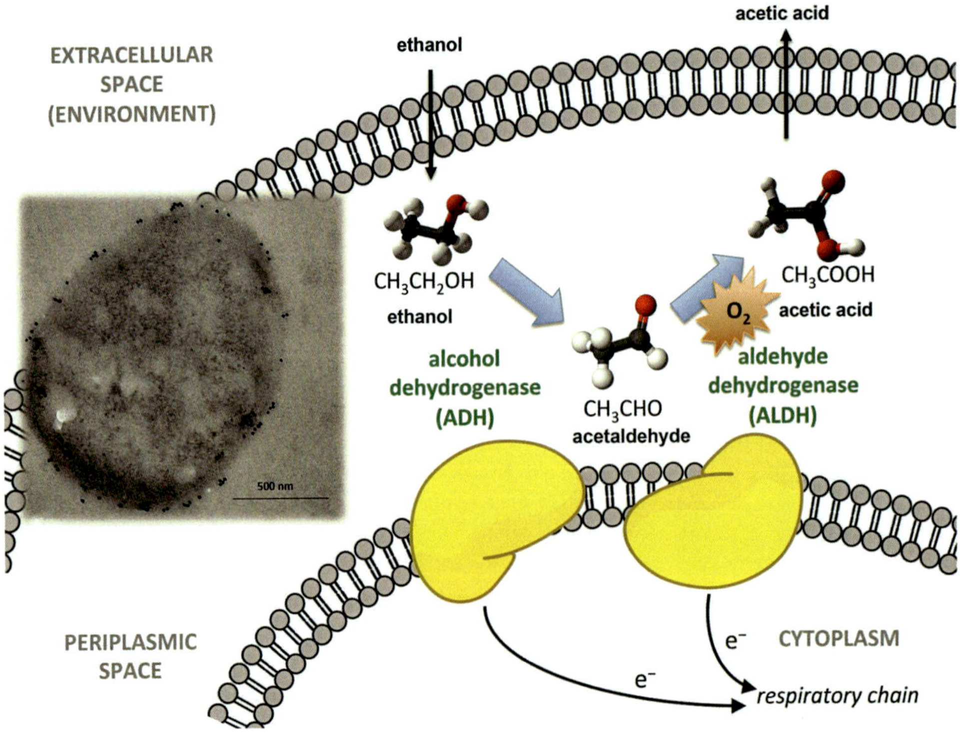

Figure 2: General overview of the acetic acid fermentation reaction. Ethanol reaches the periplasmic space of AAB and is sequentially transformed into acetic acid by action of two enzymes, alcohol dehydrogenase (ADH) and aldehyde dehydrogenase (ALDH). The acetic acid produced diffuses through the external membrane and the electrons generated during the oxidation are directed to the bacterial respiratory chain. Immunolocalization of the ADH enzyme on the bacterial membrane is highlighted at the left side of the figure.

The degree of resistance against acetic acid differs among AAB. *Acetobacter* spp., mostly *A. aceti*, *A. pasteurianus*, *A. pomorum*, *A. cerevisiae* and *A. malorum*, tolerate acidities of about 6-10 per cent and are involved in traditional or static acetic acid fermentations (Sokollek et al. 1998, Yamada 2003, Gullo et al. 2006, Andrés-Barrao et al. 2011a). *Komagataeibacter* spp. in contrast, though they can also be found in static procedures, are mainly involved in submerged industrial processes. Strains of *K. xylinus*, *K. hansenii* and *K. europaeus* have been found during static vinegar productions, showing resistance to acetic acid as high as 10-15 per cent (Yamada 2003, Fernández-Pérez et al. 2010). The highest tolerance has been shown by *K. europaeus*, *K. intermedius*, *K. oboediens* and *Gluconacetobacter entanii*, which have been isolated from submerged processes with maximum acidities of 15-20 per cent (Schüller et al. 2000, Emde 2006).

In research on the molecular mechanisms of acetic acid resistance of AAB, several strategies have been proposed. Among them, a stable core of proteins adapted to low intracellular pH, low pH membrane gradient, the production of acid stress and oxidative stress-induced proteins, the acetate metabolism by the tricarboxylic acid (TCA) cycle, the exportation of acetate by specialized efflux pumps systems, the modification in the lipid composition of the cytoplasmic membrane and the production of capsular polysaccharides (Menzel and Gottschalk 1985, Fukaya et al. 1990, Russell 1992, Steiner and Sauer 2001, Moonmangmee et al. 2002, Okamoto-Kainuma et al. 2002, 2004, Nakano et al. 2004, 2006, Deeraksa et al. 2005, Matsuhita et al. 2005b, Francois et al. 2006, Francois and Kappock 2007, Trček et al. 2007, Mullins et al. 2008, Kanchanarach et al. 2010a, Andrés-Barrao et al. 2012).

Research on the nature and metabolism of AAB and their resistance to acetic acid during the production of vinegar, a condiment of high economic value, is currently a subject of great interest, not only for industrials, but also for the general scientific community. The progress in high-throughput-sequencing technologies that have become available during the last years has resulted in an extraordinary development of genomic studies based on bacterial whole genomes. Prust et al. (2005) published the first completely sequenced genome of an AAB, *Gluconobacter oxydans* 621H. Since then, the number of AAB whole genome sequencing projects that have been completed or are in progress increases at an exponential rate; focusing on strains isolated from fruits, flowers or spoiled alcoholic beverages such as wine or beer, plant-growth promoters, industrially important biocatalyzers, thermotolerants, human opportunistic pathogens or insect symbionts (Greenberg et al. 2007, Roh et al. 2008, Ryu et al. 2008, Bertalan et al. 2009, Giongo et al. 2010, Iyer et al. 2010, Matsutani et al. 2011, Sakurai et al. 2011, Soemphol et al. 2011, Gao et al. 2012, Kim et al. 2012, Matsutani et al. 2012, Ge et al. 2013, Illeghems et al. 2013, Matsutani et al. 2013, Sato et al. 2013, Chouaia et al. 2014, Higashiura et al. 2014, Hung et al. 2014, Koike et al. 2014, Kubiak et al. 2014, Matsutani et al. 2014, Schmid et al. 2014, Sheng et al. 2014, Miao et al. 2015). The following draft genomes from strains isolated from vinegar have also been published: *A. pasteurianus* NBRC 3283, *A. pasteurianus* 3P3, *K. europaeus* LMG 18890^T, *K. europaeus* LMG 18494, *K. europaeus* 5P3, *K. oboediens* 174Bp2, *K. medellinensis* NBRC 3288 (formerly *Gluconacetobacter xylinus*) and *Komagataeibacter* sp. SXCC-1 (formerly *Gluconacetobacter* sp.) (Azuma et al. 2009, Andrés-Barrao et al. 2011b, Du et al. 2011, Ogino et al. 2011). The extraordinary high amount of information derived from sequencing projects and their comparative analysis provides new insights in the general metabolism of AAB and in the innate differences in acetic acid resistance among different AAB species.

Tolerance to Low pH: Acidophiles

The mechanisms known to maintain the pH homeostasis in acidophiles include a cell membrane highly impermeable to protons, membrane channels with a reduced pore size, the inhibition of the proton influx through creation of a chemiosmotic gradient, the secretion of exceeding protons by membrane pumps, a system of cytoplasmic buffering to maintain the intracellular pH, the proton uncoupling by organic acids, the action of chaperones to repair acid damage in DNA and proteins, the stabilization of intracellular enzymes by 'iron rivets', or specific adapted genes to facilitate the growth at low pH (Baker-Austin and Dopson 2007).

AAB are naturally adapted to grow at low pH, sharing with acidophiles the most part of their strategies to maintain the pH homeostasis. Additionally, AAB involved in vinegar production are also adapted to acetic acid, calling for more specialized strategies.

Intracellular pH Regulation and Homeostasis, and Adaptation of Cytoplasmic Components to Internal Acidification

During growth on ethanol broth (acetic acid fermentation medium), *Acetobacter aceti* DSM 2002 has been reported to decrease its internal pH as the external pH decreases, thus maintaining a low pH gradient (ΔpH) in the presence of high concentrations of acetic acid and keeping the inner cell slightly more alkaline than the external environment (ΔpH = 0.4). (Menzel and Gottschalk 1985). The intracellular pH in this strain can reach values as low as 3.9 (Menzel and Gottschalk 1985), which normally should inactivate the intracellular enzymes. The tolerance of AAB to acidic cytoplasms suggests their intracellular proteins are unusually stable in acidic environments. In this regard, it has been reported that N^5-carboxyaminoimidazole ribonucleotide mutase (PurE) of *A. aceti* 1023 (recently reclassified as *A. pasteurianus* by Hung et al. (2014)), an enzyme involved in the *de novo* synthesis of purine nucleotides, is structurally acid-stable and possesses catalytic properties of ordinary PurE (Constantine et al. 2006). One characteristic of the PurE structure that appears to contribute to its acid stability is a higher number of inter-subunit hydrogen bounds (Settembre et al. 2004). In addition, the presence of an increased number of arginine-containing salt bridges seems to be a general stress strategy that not only confers stability in the acid cytoplasm of AAB, but also contributes to thermotolerance (Settembre et al. 2004, Matsutani et al. 2011).

The purification and characterization of the citrate synthase (CS) from the highly acetate-tolerant *K. europaeus* DSM 6160^T (formerly *Acetobacter europaeus*) revealed that the enzyme is a hexamer (Sievers et al. 1997). The authors observed that CS activity was inhibited by ATP, enhanced by ADP and acetate and not affected by NADH or NADPH. This suggested its central role in the supply of sufficient ATP to protect the cells from acetic acid. *A. aceti* 1023 has also been reported to possess a CS that is a hexameric type II enzyme, not regulated by NADH, which is acid stable (Francois et al. 2006), as well as an alanine racemase (Alr) that is also stable in acidic conditions (Francois and Kappock 2007).

An enzyme that works in tandem with the oxalyl-CoA decarboxylase to consume protons and thus contributes to generate acid resistance, is formyl-coenzyme A (CoA): oxalate CoA-transferase (FCOCT/UctB). This enzyme has been described recently to remain folded at low cytoplasmic pH shown in *A. aceti* 1023 (Mullins et al. 2012). UctB show an apparent neutral surface in which charged surface residues are surrounded by compensating charges but do not form salt bridges. A pattern of residue substitution in UctB was also identified that could be the consequence of selection for protein stability by constant exposure to acetic acid. This surface charge pattern, which is a distinctive feature of *A. aceti* 1023 proteins, creates a stabilizing electrostatic network without stiffening the protein or compromising protein-solvent interactions (Mullins et al. 2012).

It is reasonable that the cytoplasmic enzymes of AAB are intrinsically resistant to inactivation by low pH, but also functionally adapted to low pH. The isolated PQQ-ADH enzymes of *A. pasteurianus* MSU10, SKU1108 and IFO 3191, for example, have been found to retain >90 per cent of their activity in a wide range of pH, from 8.0 to 2.3 (Kanchanarach et al. 2010b). The capacity to tolerate cytoplasmic acidification, which involves an unusual adaptation of cytoplasmic components, an intrinsic acid stability, may be a consequence of strong selective pressure to function at low pH (Francois et al. 2006).

The contribution of the intrinsic stability of intracellular proteins and general stress proteins, such as chaperones and chaperonins to the phenotypic expression of acetic acid resistance, has also been pointed out by Hanada et al. (2001).

Tolerance to Acetic Acid: Responses to Acetic Acid Stress

Bacterial strains of the genera *Acetobacter*, *Gluconacetobacter* and *Komagataeibacter* grow in ethanol broth following a typical diauxic/biphasic growth curve (Fig. 3). First, bacteria grow exponentially at the same time that the accumulation of acetic acid by incomplete oxidation of ethanol occurs. The following stationary phase is reached when ethanol is completely depleted, entering what is called the acetic acid resistance phase. After a certain time that varies according to the strain and culture condition, bacterial cells show rapid growth anew (second exponential phase), at the same time that acetic acid is internalized and completely oxidized to CO_2 and H_2O (overoxidation) through the TCA cycle (Saeki et al. 1997, Saeki et al. 1999, Matsushita et al. 2005a). The pH of the culture broth decreases during the ethanol oxidation phase from pH 5-6 to a minimum value of pH ~3 and increases again during acetate overoxidation (Saeki et al. 1999, Andrés-Barrao et al. 2012). Under

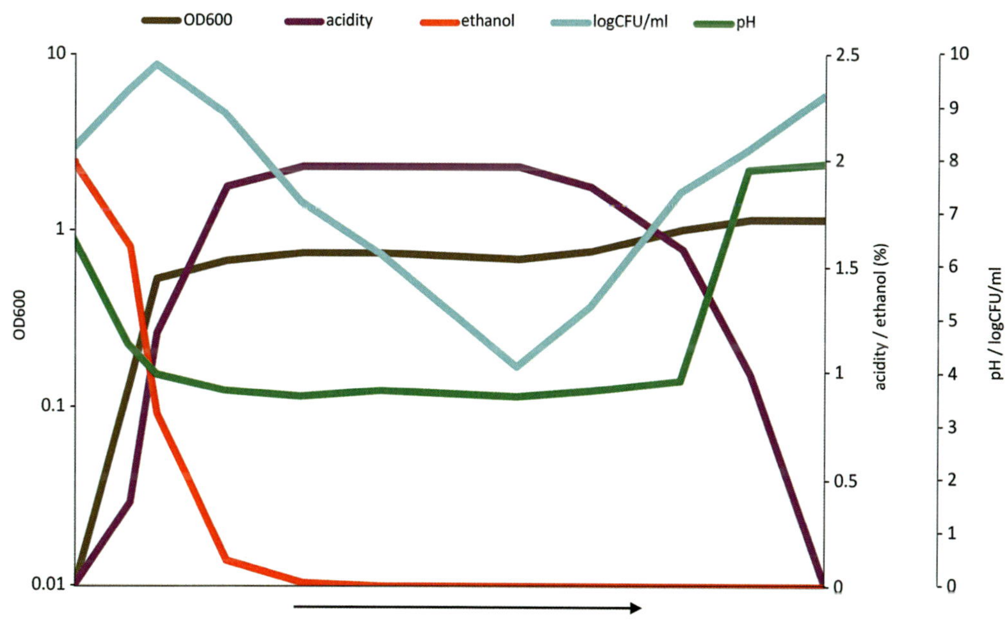

Figure 3: Physiology of AAB when growing in ethanol broth, showing a typical diauxic growth curve, with three phases. Ethanol oxidation to produce acetic acid (1st exponential growth phase) is followed by a stationary phase. Later, the overoxidation of the accumulated acetic acid occurs (2nd exponential growth phase).

such a dynamic and increasingly acidified environment, the bacteria metabolism must rapidly adapt and different molecular strategies must be active during each growth phase.

Regarding the elucidation of the molecular mechanisms conferring resistance to acetic acid stress, most studies have been performed in *A. aceti* (type species of the genus *Acetobacter*) and *A. pasteurianus*. Little work has been done so far which focuses on the physiological responses in *Komagataeibacter* spp.

Membrane-bound Dehydrogenases

The membrane-bound PQQ-ADH is considered to be the key enzyme in acetic acid fermentation due to its essential role in the oxidation of ethanol to acetaldehyde, the intermediate that will be oxidized later by the membrane-bound ALDH to acetic acid. The quino-hemeprotein nature of ADH has been thoroughly proved (Matsushita et al. 1994, Chinnawirotpisan et al. 2003). However, the nature of the prosthetic group of ALDH is still unclear. Although it was previously believed to be a pyrroloquinoline quinone (PQQ)-dependent protein, defective PQQ mutants of *Acetobacter* sp. BPR2001 showed a decrease in the activity of PQQ-glucose dehydrogenase (GDH) or PQQ-ADH, but the activity of ALDH remained unchanged; thus demonstrating PQQ was not the prosthetic group of ALDH. Molybdopterin cytosine dinucleotide (MCD) was suggested to be a cofactor instead (Takemura et al. 1994). The nature of MCD-dependent ALDH of *K. europaeus* DSM 6160^T (formerly *A. europaeus*) was later supported by Thurner et al. (1997). However, it has been demonstrated that the ALDH in *Ga. diazotrophicus* Pal5 contains a PQQ, a heme *B* and three hemes *C* (Gómez-Manzo et al. 2010). Moreover, the PQQ-ADH of *Ga. diazotrophicus* Pal5 has been recently characterized as an enzymatic complex showing double function, capable of oxidizing ethanol and acetaldehyde as alternate substrates, thus catalyzing the transformation of ethanol into acetic acid without accumulation of the toxic acetaldehyde into the media (Gómez-Manzo et al. 2015).

The chemical nature of the prosthetic group associated with the ALDH enzyme among AAB species seems thus to be heterogeneous, and more thorough studies are needed to clarify what type of prosthetic group is present in members of each different species.

A direct link between acetic acid resistance and the capability to oxidize ethanol during the growth of *A. aceti* 1023 has been established by Ohmori et al. (1982). Prolonged cultivation of the strain in ethanol broth (beyond the stationary phase) frequently led to the appearance of acetic-acid-sensitive mutants that were deficient in ethanol oxidation. Less frequently, strains with complete absence of acetic acid resistance and ethanol oxidizing ability were also isolated. These two characteristics of AAB were found to be rather variable and the cause of this phenomenon was suggested to be a plasmid or some extrachromosomal genetic element (Ohmori et al. 1982).

The key role of these membrane enzymes in the resistance mechanisms has been highlighted by additional findings. The defect in membrane-bound ADH solely or both ADH and ALDH has been also associated with the reduction in acetic acid resistance in *A. aceti* 10-8 and 1023, as well as in *A. pasteurianus* NCI1380 and SKU1108 (Okumura et al. 1985, Takemura et al. 1991, Chinnawirotpisan et al. 2003). The introduction of a plasmid containing the ALDH fragment in a mutant defective in both ADH and ALDH derived from *A. aceti* 10-8 led to the recovery of about 25 per cent of the activity of the parental strain. However, a 2-fold increase was observed upon introduction of the plasmid into a wild type *K. xylinus* NBI 2099 (formerly *A. aceti* subsp. *xylinum*). As expected, the elevation in membrane-bound ALDH activity by gene amplification improved the acetic acid fermentation in *K. xylinus* NBI 2099 by enhancing the bacterial growth and the concentration of acetic acid finally attained, from 68.4 g/L to 96.6 g/L (Fukaya et al. 1989). Although the authors did not comment upon the relationship with acetic acid resistance, the improved growth and acetic acid fermentation capability are clearly due to the physiological adaptation of the bacteria and better acetic acid resistance capability. In our understanding, higher ADH and ALDH activities might result in a bigger energy pool available for membrane-bound associated processes, such as the acetate/acetic acid export system, which may be involved in the resistance mechanisms to high acetic acid concentrations.

Differences in acetic acid resistance exist among vinegar-related AAB species, with *Komagataeibacter* spp. generally showing higher resistance than *Acetobacter* spp. These differences in acetic acid resistance suggest different adaptation mechanisms against acetic acid have been developed among AAB strains and species, and the direct relationship between the activity of membrane-bound dehydrogenases and acetic acid resistance likely involves different expression levels of the enzymes. Relative to this, Trček et al. (2006) compared the growth properties, acetic acid fermentation performance and characteristics of PQQ-ADH among three AAB strains isolated from vinegar: *A. pasteurianus* KKP 584, *K. europaeus* LMG 18494 (formerly *Ga. europaeus*) and *K. intermedius* DSM 13111 (formerly *Ga. intermedius*). *K. europaeus* exhibited the highest resistance to acetic acid (10 per cent), while the other two strains resisted up to 6 per cent. The authors reported a PQQ-ADH activity in *K. europaeus* and *K. intermedius* at least twice as high as that in *A. pasteurianus*, under the same growth conditions. The acid stability of the purified enzyme was also higher in *K. europaeus* and *K. intermedius* than in *A. pasteurianus*. These results suggested that high ADH activity in *K. europaeus* and high acetic acid stability of the purified enzyme represent two of the unique features that enable this species to grow and stay metabolically active at extremely high concentrations of acetic acid (Trček et al. 2006). In a later report, the same authors described the induction of the PQQ-ADH expression in all strains by an increase of acetic acid concentration from 1 to 2 per cent, although a regulation at the transcriptional level could not be demonstrated (Trček et al. 2007).

Despite the clear relationship between the dehydrogenase activity and acetic acid resistance at different phases of the bacterial growth, a recent report described that the expression levels of the genes coding for the PQQ-ADH (*adhA*), membrane-bound ALDH (*aldF*) and two NAD^+-ADH did not change significantly during the growth of *A. aceti* NBRC 14818 in ethanol broth (Sakurai et al. 2012). Additionally, contradictory reports have been published concerning the induction and activation of PQQ-ADH by ethanol (Takemura et al. 1993, Andrés-Barrao et al. 2012) and the data in literature support the regulation of the enzyme at a post-transcriptional level (Jucker and Ettlingler 1985, Matsushita et al. 1995, Trček et al. 2007, Quintero et al. 2009).

Protein Responsives to Acetic Acid Stress

Detoxification by Efficient Assimilation of Acetic Acid Incorporated into the Cell

There are currently 10 recognized genera of AAB that are able to oxidize ethanol to acetic acid. Among them, *Endobacter, Gluconobacter, Neoasaia, Neokomagataea, Saccharibacter, Swingsia* and *Tanticharoenia* are not able to further overoxidize acetate (Malimas et al. 2013, Mamlouk and Gullo 2013, Ramírez-Bahena et al. 2013, Thi Lan Vu et al. 2013, Li et al. 2015). The incapacity of *Gluconobacter oxydans* (formerly *A. suboxydans*) to overoxidize acetic acid has been found to be due to a non-functional tricarboxylic acid (TCA) cycle (the enzymatic activity of succinate dehydrogenase (SDH) was not detected) and the lack of the glyoxylate cycle (Greenfield and Claus 1972). Moreover, the analysis of the genome sequence of *G. oxydans* 621H confirmed the absence of the genes coding for the SDH and succinyl-CoA synthetase (SucCD) (Prust et al. 2005). These genera have been proved to be sensitive or less resistant to acetic acid, and although it has not been confirmed, the reason for this phenotype might be the incapacity to dissimilate the excess of cytosolic acetate through the TCA cycle.

In contrast to the former genera, *Acetobacter*, *Gluconacetobacter*, *Komagataeibacter*, as well as *Ameyamaea*, the methalotrophic *Acidomonas* and the recently described *Nguyenibacter*, are able to intensively oxidize acetate to CO_2 and H_2O (Malimas et al. 2013, Mamlouk and Gullo 2013, Thi Lan Vu et al. 2013). This distinctive phenotypic trait relies on the capacity of the bacterial cells to metabolize acetate through a fully functional TCA cycle and this reaction has been associated to the tolerance that strains of the genera *Acetobacter* and *Komagataeibacter* show against acetic acid during the industrial production of vinegar. The TCA cycle has so far been designated as the molecular mechanism responsible for acetic acid assimilation during the acetate overoxidation phase (diauxic growth curve) (Matsushita et al. 2005a).

Fukaya et al. (1990) provided the first clue to a better understanding of acetic acid resistance with the discovery of the first genes responsible for this phenomenon. Applying mutagenesis with *N*′-methyl-*N*′-nitro-*N*-nitrosoguanidine, authors obtained several acetic acid sensitive mutants of *A. aceti* 1023 and found that at least three genes were required to restore the acetic acid resistant phenotype: *aarA*, *aarB* and *aarC*. AarA was a hexameric type II citrate synthase (CS), an enzyme involved in the TCA cycle; NADH independent and specialized to be stable at low pH (Fukaya et al. 1990, Francois et al. 2006, Sievers et al. 1997). AarC was identified as an enzyme responsible for the assimilation of acetic acid, probably involved in the TCA cycle or the glyoxylate cycle (Fukaya et al. 1993) and was later characterized as succinyl-coenzyme A (CoA):acetate-CoA-transferase, which replaces SCS in a modified TCA cycle by Mullins et al. (2008). The same authors also identified AarB as SixA, a transcriptional activator of the TCA cycle flux. Additionally, they found that *A. aceti* 1023 harvested malate:quinone oxidoreductase (Mqo) replacing malate dehydrogenase (Mdh). Moreover, the draft genome of this strain lacked the genes coding for SCS, Mdh and the glyoxylate shunt (isocitrate lyase (ICL) and malate synthase (MS)) (Hung et al. 2014). Similar findings were revealed by analysis of the genome sequence of *A. pasteurianus* NBRC 3283 (=IFO 3283). The strain did not harvest the gene coding for succinyl-CoA synthetase and the process might be substituted with the acetic acid resistance gene product AarC (Azuma et al. 2009). Succinyl-coenzyme A (CoA): acetate-CoA-transferase (AarC) is not a simple substitute for SCS, which converts succinyl-CoA to succinate, but a functional bypass to the TCA cycle connecting with the direct conversion of acetate to acetyl-CoA. For this reason, cytosolic acetate is rapidly detoxified with the completion of the TCA cycle. This is considered as one of the characteristic and sophisticated strategies of these AAB for their acetic acid resistance properties.

The induction of many proteins in response to acetic acid has been described in *A. aceti*, although most of these proteins have not been further identified or characterized (Lasko et al. 1997, Steiner and Sauer 2001). The comparison of the protein expressed by *Acetobacter aceti* 10-8S2 (acetic acid-sensitive mutant strain derived from *A. aceti* 10-8) grown in the presence or absence of 1 per cent acetic acid by two-dimensional gel electrophoresis (2D-PAGE), showed a protein of 100 KDa (S100) whose expression was highly enhanced in acetic acid. The protein was identified as aconitase (AcnA), another enzyme involved in the TCA cycle (Nakano et al. 2004, Nakano and Fukaya 2008). The overexpression of the aconitase gene was found to increase the acetic acid resistance of the strain and to improve the productivity of acetic acid fermentation (Nakano et al. 2004). AcnA, as well as other proteins related with the TCA cycle, such as 2-oxoglutarate dehydrogenase (α-ketoglutarate DH) or isocitrate dehydrogenase, have also been observed by our group to be up-regulated during acetic acid fermentation (Andrés-Barrao et al. 2012).

A recent study of the transcriptome changes in *Acetobacter aceti* NBRC 14818 during growth on ethanol has shown the induction of genes coding for the whole set of TCA cycle enzymes at the end of the ethanol oxidation, specially citrate synthase (*gltA*, *aarA*), aconitase (*acnA*), isocitrate DH (*icd1*) and succinate DH (*sdhC*) (Sakurai et al. 2012).

Acetyl-CoA synthetase (ACS) and phosphotransacetylase (PTA), together with phospoenolpyruvate carboxylase (PEPC), are enzymes which play an important role in the flow of acetate to the TCA cycle. Their activities tended to increase in *A. racens* SKU 111 during the acetate overoxidation phase, while their activity after ethanol oxidation was very low or not detected. When cells were growing in glycerol containing medium (D-glucose omitted), the activity of isocitrate dehydrogenase (IDH) was also observed to increase up to ~6-fold during acetate overoxidation; so did the activity of fumarase, but to a lesser extent. Pyruvate kinase (PK) and malate synthase (MS), in contrast, exhibited decreased activity during the same phase. The decrease of MS activity, together with a low and stable activity of isocitrate lyase (ICL), maintained a low supply of succinate through the glyoxylate pathway during acetate overoxidation. This pathway, in contrast, seems to be rather important during the previous phases (ethanol oxidation and acetic acid resistance phases), until acetate oxidation began (Saeki et al. 1999). The TCA cycle has been proved to be the typical route for acetate catabolism in acetic acid bacteria, but a relatively long lag-phase when cells are grown in acetic acid containing medium indicates that ATP or other forms of energy are required to initiate the process (Saeki et al. 1997).

During the industrial production of vinegar by surface or submerged methodologies, repetitive cycles of ethanol oxidation occur and the fermentation process is terminated when the acetic acid concentration reaches the highest level, before the complete depletion of ethanol. This procedure prevents bacteria from reaching the transition phase and entering the undesirable acetate overoxidation phase, with the concomitant productivity and economic cost losses for vinegar manufacturers. Thus, among the molecular mechanisms allowing bacteria to resist up to 15-20 per cent acetic acid during acetic acid fermentation, while acetate overoxidation is prevented, other strategies than the assimilation of acetic acid by the TCA cycle must be involved.

Detoxification by Efficient Exportation of the Intracellular Acetic Acid

Preliminary observations that *A. aceti* DSMZ 2002 adapted to high levels of acetic acid exhibited lower intracellular acetate concentrations than non-adapted strains, suggested an efflux system as a plausible acetic acid resistance mechanism during acetate fermentation (Steiner and Sauer 2003). Such a mechanism was confirmed some years later, when Matsushita et al. (2005b) examined the first system for acetic acid exportation in *A. pasteurianus* IFO 3283 (formerly *A. aceti*) during the ethanol oxidation phase. The authors studied the acetic acid and acetate transport and accumulation by measuring the membrane potential ($\Delta\Psi$) generation and the pH gradient (ΔpH) in intact cells and in membrane vesicles. The measurements indicated that the system was an efflux pump specific for acetic acid that was driven by a proton motive force but not by ATP, in the cytoplasmic membrane.

By comparing 2D-PAGE of the membrane protein fraction of the acetate sensitive *A. aceti* 10-8S2, grown in the presence or absence of 1 per cent acetic acid, Nakano et al. (2006) discovered a membrane protein of 60 KDa (AatA) that was induced in the presence of 1 per cent acetic acid. The protein was characterized as a putative ABC transporter and homologs were found in *Acetobacter* spp. and *Komagataeibacter* spp. This transporter was found to confer resistance to acetic acid, but also to other short-chain organic acids, such as formic, propionic and lactic acids (Nakano et al. 2006, Nakano and Fukaya 2008). The suggested function of the transporter was an efflux pump for acetic acid. As AatA did not show hydrophobic apparent transmembrane domains, it was suggested to be tightly bound to a membrane protein (Nakano et al. 2006).

Authors have hypothesized a link between the two detoxification mechanisms, assimilation and exportation of acetate. The enhancement of the cytosolic enzyme activity to assimilate acetate through the TCA cycle would lead to the production of more ATP, which would then be used by an ATP-binding cassette (ABC) transporter (Nakano et al. 2006). Additionally, the overexpression of *acnA* and *aatA* resulted in an improved growth of *A. aceti* 10-8S2 under high concentrations of acetic acid and in a higher final yield of acetic acid fermentation. This result was probably due to the maintenance of a low concentration of intracellular acetate (Nakano et al. 2006). Interestingly, the transcription levels of *aatA* have not been found to change significantly during ethanol oxidation, and a probable post-transcriptional regulation of the transporter has been suggested (Sakurai et al. 2012).

The transporters described so far have been proposed to be one of the main acetic acid resistance mechanisms during ethanol oxidation phase (Trček et al. 2007). This phase is especially important for industrial vinegar production in contrast to the overoxidation one, when the catabolic conversion of acetate into CO_2 and H_2O occurs. The acetate exportation is of more ecological significance and probably favors the survival of AAB in nature (Matsushita et al. 2005a).

General Stress Proteins

The heat shock systems, GroESL and DnaKJ, are among the more abundant cytosolic proteins in prokaryotes and eukaryotes. These chaperones and chaperonins are general stress proteins that protect other proteins from denaturation and aggregation caused by heat but also environmental stresses as oxidative, acid, salt, starvation or organic solvent stresses. The two systems are known to be differently induced in different microorganisms, by various stresses resulting from environmental changes (Hartl and Hayer-Hartl 2002, Susin et al. 2006).

When looking at the available literature on stress response of AAB against acetic acid or ethanol, it is very important to differentiate the response occurring under two different conditions: a) shock-like conditions and b) continuous acid stress. Under the first condition, bacteria react against a sudden addition of a certain amount of exogenous stressor (acetic-acid). Similar response is triggered when unadapted cells are grown in liquid media with high concentrations of acetic acid. On the other hand, under the second condition (acetic acid fermentation), bacteria growing in ethanol broth are exposed to minute gradients of acetic acid; a consequence of the gradual increase in endogenous acetic acid produced by metabolic oxidation of ethanol.

Response against Acid Shock-like Conditions

One of the first observations was the induction of GroEL in *A. aceti* DSMZ 2002 under increasing concentrations of acetate (0-2 per cent) in the growing medium, in batch cultures (Steiner and Sauer 2001). The important role of the GroESL and DnaKJ systems during acetic acid fermentation in *A. pasteurianus* NBRC 3283 (formerly *A. aceti*) was suggested when researchers applied Northern-blot hybridization and observed an increase in the *groELS*, *grpE* and *dnaKJ* transcript levels as a response to heat shock as well as 4 per cent ethanol and 3 per cent acetic acid stresses, until at least 30 minutes after the addition of the stressor to the growing medium (Matsushita et al. 1994, Okamoto-Kainuma et al. 2002, 2004, Matsushita et al. 2005a, Ishikawa et al. 2010a).

Additionally, the overexpression of the *groESL* operon was found to provide resistance to heat shock as well as 5 per cent ethanol and 1 per cent acetic acid, as proved by the delayed growth of the wild-type strain compared with the overexpressing mutants, under the three stress conditions (Okamoto-Kainuma et al. 2002). Similar results were obtained for the *dnaKJ* operon overexpressing mutant, which grew nearly normally in the presence of 5 per cent ethanol, while the growth of the wild-type was clearly delayed. But in contrast, while the *groESL* overexpressing mutant grew normally, the *dnaKJ* overexpressing and the wild-type strains could not grow in the presence of 1 per cent acetic acid (Okamoto-Kainuma et al. 2004). It is known that the DnaK co-chaperone GrpE acts as a nucleotide exchange factor, essential for the proper chaperone activity of the DnaKJ system (Harrison 2003). So the growth behavior under stress conditions was suggested to be a consequence of the lack of *grpE* in the *dnaKJ* overexpressing strain (Okamoto-Kainuma et al. 2004).

To elucidate the effect of the co-chaperone in the function of DnaKJ, researchers repeated the same set of experiments using the *grpE, dnaK-dnaJ* and *grpE-dnaK-dnaJ* overexpressing mutants. Results showed that the overexpression of *grpE,* alone or with *dnaKJ,* improved the growth under heat shock and 5 per cent ethanol, but could not increase the resistance against 1 per cent acetic acid (Ishikawa et al. 2010a).

The relative transcription levels of *grpE, dnaK* and *dnaJ* genes after the addition of acetic acid stressors were investigated by qRT-PCR. Consistent with the previous results, the level of the *grpE* transcript showed a peak of 10 minutes after the addition of 4 per cent ethanol, but no significant change was observed as response to 3 per cent acetic acid. The growth comparison among *grpE* and *dnaKJ* overexpressing mutants and the wild-type was similar to the previous results and in accordance with the qRT-PCR analysis (Ishikawa et al. 2010a).

ClpB is the other molecular chaperone whose relationship with acetic acid fermentation stressors has been investigated. The expression level of *clpB* in the same *A. pasteurianus* strain was studied by using qRT-PCR, showing an increase to up to 4.6-fold after addition of 4 per cent EtOH, but no marked effect after exposure to 3 per cent acetic acid (Ishikawa et al. 2010b). These results were similar to those previously described for *dnaKJ* and *grpE* (Ishikawa et al. 2010a). However, the bacterial growth of a *clpB* disruptant, compared with that of the wild-type, was not affected by ethanol or acetic acid. The same tendency was observed for a *clpB* overexpressing strain growing in 5 per cent ethanol (Ishikawa et al. 2010b).

Recently, RpoH, a RNA polymerase sigma factor (σ^{32}) and regulation factor of the heat shock response was also investigated by Okamoto-Kainuma et al. (2011). The disruption of *rpoH* repressed chaperonin genes, *groEL, dnaK, dnaJ, grpE* and *clpB,* and demonstrated that RpoH is indispensable for a proper heat shock response under heat shock conditions. In this study, a *rpoH* disruptant showed a decreased resistance under ethanol (5 per cent) and acetic acid (1 per cent) stress conditions. Although the biomass of wild-type and mutant strains reached the same final yield, the growth of the disruptant was remarkably delayed (Okamoto-Kainuma et al. 2011). While ClpB was finally related with only heat shock stress, these results suggested an important role for RpoH in acetic acid fermentation.

Response under Continuous Stress Conditions

The expression levels of *groESL, dnaKJ* and *grpE/grpE-dnaKJ* were previously found to be induced during growth in 4 per cent ethanol (Okamoto-Kainuma et al. 2002, 2004, Ishikawa et al. 2010a). In accordance with these results, a recent study from our group showed that the protein expression levels of the whole chaperone system in *A. pasteurianus* LMG1262^{T}, including GroESL, DnaKJ and GrpE, increased during growth in Reinforced Acetic Acid Ethanol (RAE) medium supplemented with 4 per cent EtOH (Andrés-Barrao et al. 2012). These results support the importance of the roles of these heat shock proteins during industrial acetic acid fermentation.

It is important to highlight that, in the experiments performed by Okamoto-Kainuma et al. (2002), although the growth in 5 per cent EtOH of the wild-type compared to the *groELS* and *dnaKJ* overexpressing mutants was delayed, both strains reached similar optical density (OD600) levels. Under this condition, bacteria were able to grow in the presence of a maximum concentration of 2 per cent acetic acid (reached after 36 h culture) that was produced gradually from ethanol oxidation. In contrast, the wild-type showed little or no growth in a medium containing 1 per cent acetic acid. These results demonstrate the different mechanisms that AAB must bring into play to overcome both kinds of acid stresses.

In accordance, a recent study reported the transcription of *dnaJ* and *clpB* genes in *A. aceti* NBRC 14818 to be up-regulated during ethanol oxidation (1 per cent EtOH-containing broth), showing a maximum level of gene expression that coincided with the maximum in the acetic acid accumulation in the medium (Sakurai et al. 2012). In a previous study, the transcription level of *clpB* was also observed to be 4.8-fold up-regulated during growth in 4 per cent EtOH and the *clpB* disruptant strain could not grow properly nor perform acetic acid fermentation (Ishikawa et al. 2010b).

Cytoplasmic Membrane Modifications

Lipid Components

The outer membrane of the AAB has characteristic lipid components that are limited to some Gram-negative bacteria. Those are the main phospholipids: phosphatidylcholine (PC), phosphatidylethanolamine (PE) and phosphatidylglycerol (PG), as well as alkali-stable lipids (ASL) (Fukami et al. 2010). ASL includes hopanoids (terpenoids compounds), sphingolipids (dihydroceramine and sphinganine), aminolipids and free fatty acids (*cis*-vaccenic acid) (Tahara et al. 1976, Fujino et al. 1978, Tahara et al. 1986, Simonin et al. 1994). Although the highly specialized composition of the outer membrane of AAB has been extensively studied, few studies have sought to elucidate the role of membrane lipids in acetic acid resistance.

A role of PC in acetic acid resistance was first pointed out by Hanada et al. (2001). Working with *A. pasteurianus* IFO 3283 (formerly *A. aceti*), the authors observed that the disruption of phosphatidylethanolamine N-methyltransferase gene (*pmt*), coding for an enzyme that catalyzes the methylation of PE to PC, led to a mutant defective in the synthesis of PC that accumulated 10-fold PE and 2-fold PG. The *pmt* disruptant showed slower growth than the wild-type in the media supplemented with 170 mM sodium acetate or 170 mM acetic acid; but interestingly, no difference in growth were observed when strains were grown in media supplemented with 170 mM ethanol. Although the PC content increased with the increase in acidity during acetic acid fermentation, PC appeared to be dispensable for growth in ethanol medium (Hanada et al. 2001). PC deficiency had no effect in acetic acid production and consistently with this result, the *pmt* gene expression has been found not to change significantly during the growth of *A. aceti* IFO 14818 in ethanol (Sakurai et al. 2012).

The *K. europaeus* V3 strain (formerly *Ga. europaeus*), isolated by Trček et al. (2000) from high-acid submerged vinegar, was further used to investigate the modifications on the lipid composition as other adaptation strategy for acetic acid tolerance (Trček et al. 2007). The authors compared the lipid content of this strain grown in a glucose broth with no ethanol or acetic acid (RAE 0a/0e), with 3 per cent ethanol (RAE 0a/3e) or with 3 per cent ethanol and 3 per cent acetic acid (RAE 3a/3e). Gas chromatography analysis of total fatty acids methyl esters (FAMEs) showed that, under the three conditions studied, the major fatty acid (FA) was *cis*-vaccenic acid (>60 per cent).

In accordance with the usual composition established by Goldfine (1984) for Gram-negative bacteria with high proportions of unsaturated fatty acids (>65 per cent), the main phospholipids identified in *K. europaeus* V3 were PC, PE, PG and cardiolipin (CL).

In the presence of 3 per cent ethanol and 3 per cent acetic acid (RAE 3a/3e), the total lipids decreased about 14 per cent in comparison to the control without acetic acid (RAE 0a/3e). While neutral lipids and glycolipids slightly increased, the phospholipid portion decreased to a great extent. The ratio of individual phospholipids also changed: PC and PG increased while PE and CL decreased. Being PE and PG intermediates in the synthesis of PC and CL respectively, the modification observed indicated that the synthesis of PC was favored to enhance the membrane stability, while the synthesis of CL was inhibited. The authors speculated the involvement of PG in transport of acetic acid across the membrane (Trček et al. 2007). The decrease in total phospholipids might cause a decrease in the area available for passive transport of small lipophilic molecules, such as acetic acid, due to a diminution of the phospholipids/proteins ratio and more rigid membranes. The slight increase in glycolipids might reinforce the hydrophobic barrier of the cell. Unexpectedly, in the presence of 3 per cent acetic acid, the unsaturated *cis*-vaccenic acid increased to a great extent, leading to a more fluid membrane. Due to *cis*-vaccenic acid being the major FA in PG in almost all AAB, representing up to 80 per cent (Tahara et al. 1976, Yamada et al. 1981, Franke et al. 1999, Jojima et al. 2004, Logaonathan and Nair 2004, Greenberg et al. 2006), the authors assessed its increase might have resulted from the substantial increase in the relative content of PG observed (Trček et al. 2007).

Sphingolipids are lipid components widely distributed in eukaryots, but rare in bacterial microorganisms. Nevertheless, a highly pure 2-hydroxypalmitoyl-sphinganine (dihydroceramide) has been found in *Acetobacter* and *Komagataeibacter* strains (Fujino et al. 1978, Tahara et al. 1986, Fukami et al. 2010). Recently, Ogawa et al. (2010) observed the induction of ceramide during acetic acid production in several AAB isolated from various fermented foods, and the sphingolipid was suggested to play a role in tolerance to acetic acid and low pH. The strain *A. malorum* S24 accumulated the highest yield of ceramide and was selected for further study. The stress produced by addition of 0.3 M acetic acid in the growth medium when this strain reached the stationary phase also showed an induction of 1.5-fold in ceramide production. An additional role proposed by the authors for ceramide during acetic acid fermentation was the stabilization of the PQQ-ADH (unpublished data). The idea that this sphingolipid is involved in resistance to acetic acid in AAB was reinforced by the findings of Goto and Nakano (2008). These authors observed that the acetic acid tolerance of *A. aceti* 1023 was enhanced by overexpression of palmitoyltransferase, an essential enzyme that catalyzes the synthesis of sphinganine, one of the components of ceramide.

The outer plasma membrane of AAB was found to be rich in hopanoids (Fukami et al. 2010). Hopanoids are natural-occurring pentacyclic compounds, belonging to the group of triterpenes, whose main function in the bacterial membrane is to improve their strength and rigidity (Rohmer et al. 1984). In many bacteria, these lipid components may play important roles in the adjustment of cell membrane permeability and adaptation to extreme environmental conditions. Hopanoids have also been reported in the bacterial membrane of the ethanol-fermenting *Zymomonas mobilis,* where they have a role in adaptation of cell membranes to ethanol accumulation and to temperature changes, which influence membrane functions (Madigan and Martinko 2006). Similarly, in AAB they might have a role in adaptation to acetic acid accumulation.

Surface Polysaccharides

Lipopolysaccharides (LPSs) are the main components of the outer membrane of Gram-negative bacteria. They are located exclusively in the outer leaflet and confer the membrane highly impermeable characteristics (Matsushita et al. 1985, Nikaido 2003). LPS consist of a polysaccharide portion that is anchored to the membrane through attachment to a lipid-A portion. The polysaccharide portion is at the same time composed by a core polysaccharide (Kdo) that is connected to an *O*-polysaccharide (*O*-antigen), that is facing the extracellular space and whose composition varies from strain to strain. The classification of LPS in rough-type (rLPS) or smooth-type (sLPS) is determined by the presence or absence, respectively, of the *O*-antigen (Nikaido 2003). Gram-negative bacteria are also able to synthesize membrane-bound capsular polysaccharides (CPSs) that are physically attached or remain tightly associated with the cell surface. CPSs form a thick, mucus-like layer surrounding and protecting the cell from a harmful environment and usually consist of acidic water-soluble polysaccharides (Tayama et al. 1986, Grimmecke et al. 1994, Guo et al. 2008).

Additionally, Gram-negative bacteria are able to synthetize exopolysaccharides (EPSs), which are secreted and released in the outer media to form a protective biofilm. AAB are able to synthetize cellulose or cellulose-like EPS that allow them to float on the surface of growth liquid media. This capability is tremendously important during the traditional (static) production of acetic acid, where, to perform the strictly aerobic process of ethanol oxidation, bacteria need to be in contact with the oxygen in the air (MacCormick et al. 1996, Kornmann et al. 2003, Ali et al. 2011, Rani et al. 2011).

The capacity of AAB to synthetize surface polysaccharides, especially CPSs and EPSs, has been traditionally correlated with acetic acid resistance by acting as a barrier to prevent the acetic acid entrance into the cell (Deeraksa et al. 2005).

Differences in the synthesis of CPSs in *Acetobacter* spp. have given rise to the formation of two colony types on agar plates: a rough-surfaced colony (R strain) and a mucoid smooth-surfaced colony (S strain). Although R strains have been observed to be predominant in static cultures and produce a CPSs pellicle that allows them to float on the medium surface, they can also grow in shaking cultures. In contrast, S strains are predominant in shaking cultures but are not able to grow in static cultures due to the lack of pellicle formation (Matsushita et al. 1992, Moonmangmee et al. 2002, Deeraksa et al. 2005, Kanchanarach et al. 2010a). No differences in the ADH nor ALDH activities have been observed between the two colony types in *A. lovaniensis* IFO 3284 (formerly *A. aceti*). Nevertheless, the transition between the R and S phenotypes has been related to a change in the terminal oxidase from cytochrome *o* (static) to cytochrome *a1* (shaking) (Matsushita et al. 1992). The inactivation of glycerol kinase (SNP G144>t) has been pointed out as being correlated with the transition from R to S phenotypes in *A. pasteurianus* IFO 3283-01 growing in glycerol medium and evidenced by a remarkable loss in the production of neutral polysaccharides (Azuma et al. 2009). Although the S phenotype was maintained, the authors observed an increase in the amount of neutral polysaccharides produced in the liquid medium, accompanied with the disruption of the protein translocase *secB2* gene (Azuma et al. 2009).

The R strain of *A. tropicalis* SKU1100 showed higher resistance against acetic acid stress. They were able to grow on 1 per cent acetic acid containing agar plates. In contrast, the growth of the S strain and two defective mutants in pellicle formation (Pel^- and $\Delta polB$) was inhibited on agar plates containing 0.5 per cent acetic acid (Deeraksa et al. 2005).

No other report dealing with pellicle formation and acetic acid resistance in AAB was published until a direct relationship was established by Kanchanarach et al. (2010a). In this study, three strains of *A. pasteurianus* (IFO 3283, SKU1108 and MSU10) were cultured in ethanol broth and they produced a polysaccharide pellicle and aggregate during stationary growth (acetic acid resistance phase), after ethanol was completely oxidized. The aggregation decreased during the second exponential growth (acetic acid overoxidation phase). The two colony types described, R and S, were also isolated from each of the three strains and their acetic acid fermentation performance was evaluated. All the three R strains (pellicle forming) showed a better acetification activity, higher biomass yield and acetic acid overoxidation; while the acetification activity of the S strains (pellicle non-forming) was very poor, with very low biomass yield and no acetic acid overoxidation. The authors concluded that S strains do not tolerate the acetic acid produced gradually during acetic acid fermentation. Additionally, the acetate transport was studied by the supplementation of 0.4 mM (pH = 6.5) or 80 mM (pH = 3.9) [1-^{14}C] acetate to the culture medium, and all S strains showed a higher uptake of marked acetate than the R strains (Kanchanarach et al. 2010a).

Interestingly, the observation of Kanchanarach et al. (2010a) of a polysaccharide amorphous material covering the aggregated cells during the stationary phase appears to contradict the previous finding that S strains (pellicle non-forming) were predominant in shacking cultures (Matsushita et al. 1992). Deeraksa et al. (2006) identified these differences by means of electron microscopy. Scanning electron micrographs revealed R strains showing a rough surface, while S strains were covered by an amorphous EPS material. In the light of these results, we have to consider the cell aggregation as a result of the production of EPS, which entrap the cells (biofilm), rather than CPS.

The former discoveries led to the acceptance of the general thought that the CPS pellicle formed by AAB has an active role in the resistance against acetic acid. Nevertheless, a recent study of our group exposed contradictory results. In this study, we reported the down-regulation of the dTDP-4-dehydrorhamnose reductase (RfbD), an essential enzyme involved in the synthesis of dTDP-rhamnose, during acetic acid fermentation of *A. pasteurianus* LMG 1262^T (Andrés-Barrao et al. 2012). The enzyme down-regulation, which could be related with a lower rhamnose synthesis, was consistent with a decrease in the polysaccharide layer surrounding the cell that was observed by PATAg specific staining of transmission electron micrographs (Andrés-Barrao et al. 2012). Rhamnose was found to be a sugar component of heteropolysaccharides produced by *Acetobacter* spp. and *Komagataeibacter* spp. (Savidge and Colvin 1985, Tayama et al. 1985, 1986, Jansson et al. 1993, Grimmecke et al. 1994, Moonmangmee et al. 2002, Deeraksa et al. 2005, Mungdee et al. 2006), but it can be also involved in the formation of *O*-antigen in other bacteria (Stevenson et al. 1994, Mitchison et al. 1997, Mullane et al. 2008). This novel result led us to suggest that a rhamnose-containing CPSs or LPSs in *A. pasteurianus* LMG 1262^T was not involved directly in resistance against the acetic acid gradually produced during the acetic acid fermentation.

This is not the first time that contradictory reports have been published related to the involvement of surface polysaccharides to organic acid resistance. For example, the full expression of sLPS (including the *O*-antigen) and enterobacterial common antigen (ECA) have been reported to be indispensable for the resistance of Shiga Toxin-producing *E. coli* O157:H7 and *Salmonella* against acetic acid and other short-chain fatty acids (Barua et al. 2002). Although the lack of LPS *O*-antigen appeared to reduce the resistance of *E. coli* against organic acids, the same lack was found to render *Bradyrhizobium japonicum* 61A101C more resistant to them (Oh et al. 2004).

Conclusion

AAB are known to show a high mutation rate, which is a consequence of genetic instability due to a very high number in transposases and insertion sequences contained in their genome sequences (Shimwell and Carr 1964, Azuma et al. 2009). This great plasticity has been advantageous in adaptation to their natural extreme environment.

AAB have evolved to acquire a very complex machinery that allows them to resist high concentrations of acetic acid at a low pH. The main strategies are (a) acid-stable cytoplasmic enzymes, (b) efficient chaperones and chaperonins to protect DNA and proteins from denaturation, (c) a low membrane ΔpH, (d) a specialized TCA cycle to assimilate acetate, (e) the exportation of acetate excess by ABC-type transporters or efflux pumps, (f) a highly impermeable outer membrane rich in PC and PG, with *cis*-vaccenic acid as the major FA, and (g) the production of CPS pellicle as a barrier in *Acetobacter* strains (Fig. 4). We must yet point out that many of the studies on acetic acid resistance have used moderately tolerant *Acetobacter* strains, whereas the highest resistance to acetic acid is shown by industrially adapted strains of the genus *Komagataeibacter*.

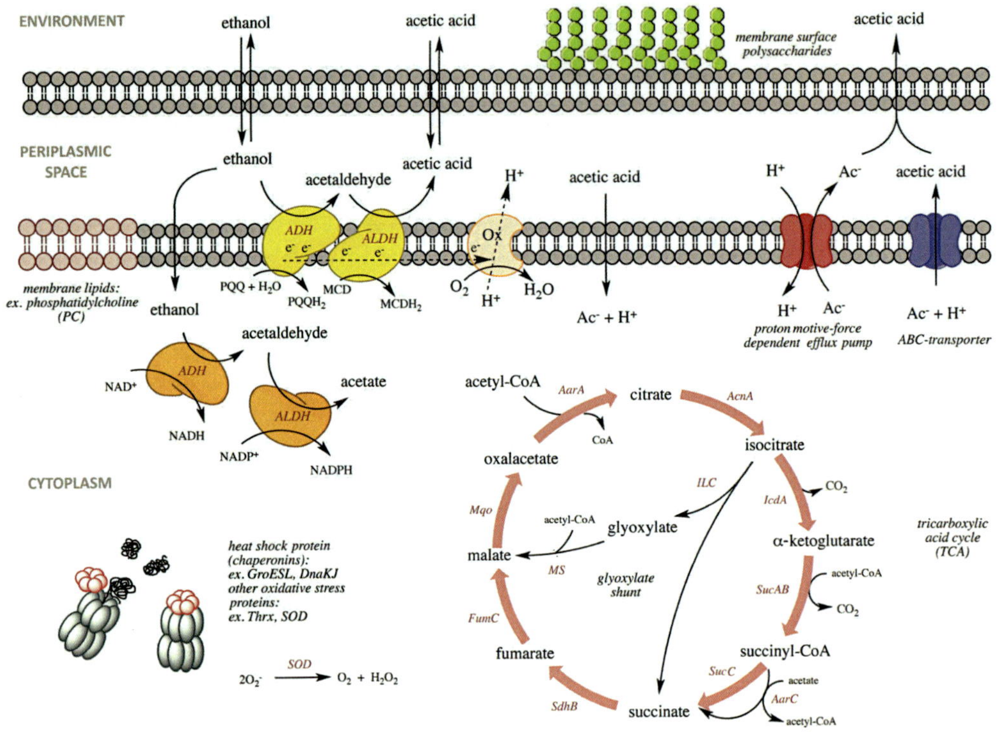

Figure 4: Main mechanisms that have been described to contribute to acetic acid resistance in AAB (adapted from Matsushita et al. 2005b, Nakano et al. 2006, Mullins et al. 2008, Nakano and Fukaya 2008). Ac^- = ion acetate, Ox = cytochrome *c* oxidase (last complex of the bacterial respiratory chain), ADH = alcohol dehydrogenase, ALDH = aldehyde dehydrogenase, AarA = citrate synthase (CS), AcnA = aconitase/aconitate hydratase, IcdA = isocitrate dehydrogenase, SucAB = α-ketoglutarate dehydrogenase, SucC = succinyl-CoA ligase, AarC = succinyl-coenzyme A (CoA): acetate CoA-transferase, SdhB = succinate dehydrogenase, FumC = fumarate hydratase, Mqo = malate: quinone oxidoreductase, ICL = isocitrate lyase, MS = malate synthetase, SOD = superoxide dismutase.

The traditional approach used in the study of acetic acid resistance in AAB led to the observation of the cell response after the addition of a certain amount of acetic acid in the growing media. The results obtained in such studies are related with the response of unadapted bacteria to the stress produced by the acid added (acid shock). Although this approach may provide insights on strategies leading to the survival of bacteria in nature, such as the acetate oxidation through the TCA cycle, it is clear that it is very far from the natural process in which vinegar industrials are interested. The acetic acid that is gradually produced during acetic acid fermentation does not produce a shock response, but a progressive adaptation through a gradual modification of bacterial metabolism. The bacteria used in the production of vinegar are already adapted to the extreme environment provided by acetic acid and low pH. But, although this adaptation must have been acquired as a survival strategy during evolution, it seems not to be an innate trait and is very quickly lost when bacteria grow out of the fermentation broth, in a non-aggressive environment (Ohmori et al. 1982).

The pellicle polysaccharide produced by R type *Acetobacter* strains is suggested to provide cells higher acetic acid resistance than that showed by S type strains (Kanchanarach et al. 2010a). However, recent findings of our group contrasted this suggestion. Whereas *A. pasteurianus* grown on RAE plates supplemented with 1 per cent acetic acid and 2 per cent ethanol (1a/2e) showed a surrounding polysaccharide layer relatively thick, the pellicle was absent in members of the genus *Komagataeibacte*r grown under the same conditions. Furthermore, *Komagataeibacter* spp. directly harvested from running vinegar at the maximum acidity of 14 per cent showed the same pellicle absence (Andrés-Barrao et al. 2016) (Fig. 5). *Acetobacter* spp. are less resistant to acetic acid than *Komagataeibacter* spp. and although it has been proved that the CPS pellicle present in *Acetobacter* spp. contributes to acetic acid resistance, strains of the genus *Komagataeibacter* must show some other additional and specific strategies. In that sense, the genome comparison of several *Acetobacter* and *Komagataeibacter* strains by our group (Andrés-Barrao 2012), showed that the genome of *Komagataeibacter* spp. was about 1 Mb bigger (4.0 Mb) than that of *Acetobacter* spp. (2.9 Mb).

Both AAB genera involved in the production of vinegar must share some mechanisms to resist moderate to high concentrations of acetic acid, but the genus *Komagataeibacter*, that is able to resist the extremely high concentrations of acetic acid produced in industrial fermentors as well as high oxidative stress due to a high aeration rate, must have acquired additional and specialized strategies to adapt and survive in its unique environment. The molecular strategies discovered in *Acetobacter* spp. or *Komagataeibacter* spp. cannot be generalized and comparative studies between the two genera are necessary to elucidate the specific mechanisms conferring resistance to both moderate to extremely high concentrations of acetic acid.

The integration of the recent advances in the development of genomic, transcriptomic and proteomic high-throughput technologies will facilitate the discovery of new genes or metabolic functions that are unique to each of the two genera of AAB and specially adapted to grow under medium to high concentrations of acetic acid and low pH.

Prospectives

Taking advantage of the high number of prokaryotic genomes that are currently available on public databases, the comparative genome analysis of acetic acid resistant and sensitive bacteria would allow the identification of potential DNA adaptations to growth in the presence of organic acids at low pH.

Although the analysis of controlled processes under laboratory conditions is always a complementary approach, the main prospective would be the metagenomic study of spontaneous acetic acid fermentation processes; specially, the application of transcriptomics to find the genes that are active during the oxidative fermentation process. This approach would give insights on the metabolic modifications and genome adaptations occurring in the natural process.

Another future recommended research direction should focus on the study of the AAB membrane characteristics, being subjects of special interest the membrane potential, the nature and function of membrane channels and pumps and the modification of the

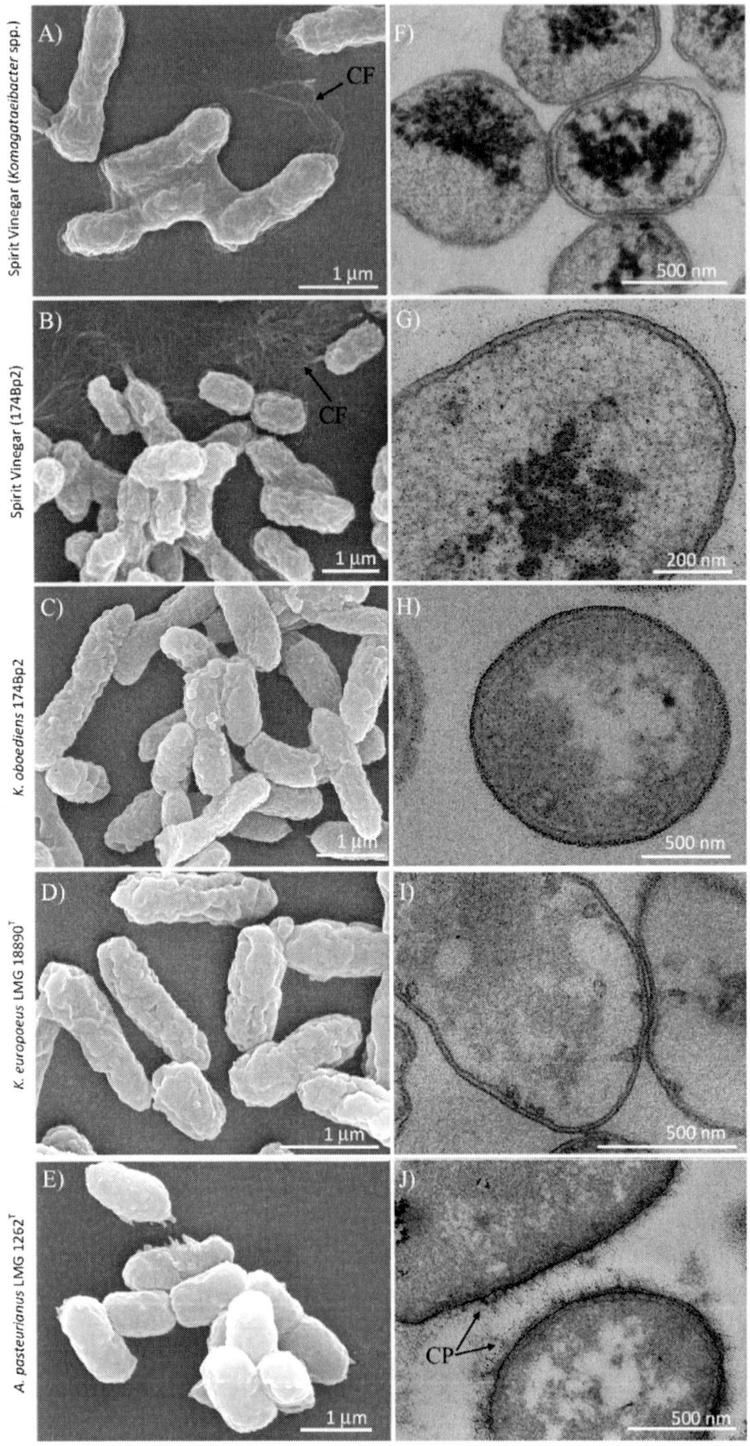

Figure 5: Scanning electron micrographs (*left*) and transmission electron microscopy sections (*right*) of AAB from high-acid spirit vinegar (A-B, F-G) and synthetic ethanol broth (C-E, H-J). Only *Acetobacter pasteurianus* LMG 1262^T growing in synthetic medium shows a layer of capsular polysaccharides (CP) surrounding the cells. Cellulose-like fibers (CF) are observed in spirit vinegar samples. [Reproduced with permission from Andrés-Barrao et al. (2016), copyright © 2016 Elsevier].

lipid content of the high impermeable membrane during the acetic acid fermentation. The unequivocal role of membrane polysaccharides in *Acetobacter* and *Komagataeibacter* must also be determined.

References

Adachi, O., Miyagawa, E., Shinagawa, E., Matsushita, K. and Ameyama, M. (1978). Purification and properties of particulate alcohol dehydrogenase from *Acetobacter aceti*. *Agricultural and Biological Chemistry* **42:** 2331-2340.

Ali, A.I.A., Akakabe, Y., Moonmangmee, S., Deeraksa, A., Matsutani, M., Yakushi, T., Yamada, M. and Matsushita, K. (2011). Structural characterization of pellicle polysaccharide of *Acetobacter tropicalis* SKU1100 wild type and mutant strains. *Carbohydrate Polymers* **86:** 1000-1006.

Ameyama, M., Osada, K., Shinagawa, E., Matsushita, K. and Adachi, O. (1981). Purification and characterization of aldehyde dehydrogenase of *Acetobacter aceti*. *Agricultural and Biological Chemistry* **45:** 1889-1890.

Andrés-Barrao, C., Weber, A., Chappuis, M.L., Theiler, G. and Barja, F. (2011a). Acetic acid bacteria population dynamics and natural imposition of *Gluconacetobacter europaeus* during submerged vinegar production. *Archives des Sciences* **64:** 99-114.

Andrés-Barrao, C., Falquet, L., Calderon-Copete, S.P., Descombes, P., Ortega Pérez, R. and Barja, F. (2011b). Genome sequences of the high-acetic acid-resistant bacteria *Gluconacetobacter europaeus* LMG 18890^T and *G. europaeus* LMG 18494 (reference strains), *G. europaeus* 5P3, and *Gluconacetobacter oboediens* 174Bp2 (isolated from vinegar). *Journal of Bacteriology* **193:** 2670-2671.

Andrés-Barrao, C. (2012). Characterization of Acetic Acid Bacteria and Study of the Molecular Strategies Involved in the Resistance to Acetic Acid during Oxidative Fermentation. Ph.D. Thesis, Université de Genève, Genève, Switzerland.

Andrés-Barrao, C., Saad, M.M., Chappuis, M.L., Boffa, M., Perret, X., Ortega Pérez, R. and Barja, F. (2012). Proteome analysis of *Acetobacter pasteurianus* during acetic acid fermentation. *Journal of Proteomics* **75:** 1701-1717.

Andrés-Barrao, C., Saad, M.M., Cabello Ferrete, E., Bravo, D., Chappuis, M.L., Ortega Pérez, R., Junier, P., Perret, X. and Barja, F. (2016). Metaproteomics and ultrastructure characterization of *Komagataeibacter* spp. involved in high-acid spirit vinegar production. *Food Microbiology* **55:** 112-122.

Asai, T. (1935). Taxonomic studies on acetic aicd bacteria and allied oxidative bacteria isolated from fruits: A new classification of the oxidative bacteria. *Journal of the Agricultural Chemical Society of Japan* **11:** 674-708.

Azuma, Y., Hosoyama, A., Matsutani, M., Furuya, N., Horikawa, H., Harada, T., Hirakawa, H., Kuhara, S., Matsushita, K., Fujita, N. and Shirai, M. (2009). Whole-genome analyses reveal genetic instability of *Acetobacter pasteurianus*. *Nucleic Acids Research* **37:** 5768-5783.

Baker-Austin, C. and Dopson, M. (2007). Life in acid: pH homeostasis in acidophiles. *Trends in Microbiology* **15:** 165-171.

Barua, S., Yamashino, T., Hasegawa, T., Yokoyama, K., Torii, K. and Ohta, M. (2002). Involvement of surface polysaccharides in the organic acid resistance of Shiga Toxin-producing *Escherichia coli* O157:H7. *Molecular Microbiology* **43:** 629-640.

Beijerinck, M.W. (1898). Uber die arten der Essigbakterien. Zentralblatt fur Bakteriologie, Parasitnkunde. *Infektionskrankheiten und Hygiene Abeteilung* II **4:** 209-216.

Bertalan, M., Albano, R., de Pádua, V., Rouws, L., Rojas, C., Hemerly, A., Teixeira, K., Schwab, S., Araujo, J., Oliveira, A., França, L., Magalhães, V., Alquéres, S., Cardoso, A., Almeida, W., Loureiro, M.M., Nogueira, E., Cidade, D., Oliveira, D., Simão, T., Macedo, J., Valadão, A., Dreschsel, M., Freitas, F., Vidal, M., Guedes, H., Rodrigues, E., Meneses, C., Brioso, P., Pozzer, L., Figueiredo, D., Montano, H., Junior, J., de Souza Filho, G., Martin Quintana Flores, V., Ferreira, B., Branco, A., González, P., Guillobel, H., Lemos, M., Seibel, L., Macedo, J., Alves-Ferreira, M., Sachetto-Martins, G., Coelho, A., Santos, E., Amaral, G., Neves, A., Pacheco, A.B., Carvalho, D., Lery, L., Bisch, P., Rössle, S.C., Ürményi, T., Rael, Pereira, A., Silva, R., Rondinelli, E., von Krüger, W., Martins, O., Baldani, J.I. and Ferreira, P.C. (2009). Complete genome sequence of the sugarcane nitrogen-fixing endophyte *Gluconacetobacter diazotrophicus* Pal5. *BMC Genomics* **10:** 450.

Bjarnsholt, T., Alhede, M., Jensen, P.Ø., Nielsen, A.K., Johansen, H.K., Homøe, P., Høiby, N., Givskov, M. and Kirketerp-Møller, K. (2015). Antibiofilm properties of acetic acid. *Advances in Wound Care (New Rochelle)* **4:** 363-372.

Chinnawirotpisan, P., Theeragool, G., Limtong, S., Toyama, H., Adachi, O. and Matsushita, K. (2003). Quinoprotein alcohol dehydrogenase is involved in catabolic acetate production, while NAD-dependent alcohol dehydrogenase in ethanol assimilation in *Acetobacter pasteurianus* SKU1108. *Journal of Bioscience and Bioengineering* **96:** 564-571.

Chouaia, B., Gaiarsa, S., Crotti, E., Comandatore, F., Degli Esposti, M., Ricchi, I., Alma, A., Favia, G., Bandi, C. and Daffonchio, D. (2014). Acetic acid bacteria genomes reveal functional traits for adaptation to life in insect guts. *Genome Biology and Evolution* **6:** 912-920.

Constantine, C.Z., Starks, C.M., Mill, C.P., Ransome, A.E., Karpowicz, S.J., Francois, J.A., Goodman, R.A. and Kappock, T.J. (2006). Biochemical and structural studies of N^5-carboxyaminoimidazole ribonucleotide mutase from the acidophilic bacterium *Acetobacter aceti*. *Biochemistry* **45:** 8193-8208.

Cortesia, C., Vilchèze, C., Bernut, A., Contreras, W., Gómez, K., de Waard, J., Jacobs, W.R. Jr, Kremer, L. and Takiff, H. (2014). Acetic acid, the active component of vinegar, is an effective tuberculocidal disinfectant. *mBio* **5:** e00013-14.

Deeraksa, A., Moonmangmee, S., Toyama, H., Yamada, M., Adachi, O. and Matsushita, K. (2005). Characterization and spontaneous mutation of a novel gene, polE, involved in pellicle formation in *Acetobacter tropicalis* SKU1100. *Microbiology* **151:** 4111-4120.

Deeraksa, A., Moonmangmee, S., Toyama, H., Adachi, O. and Matsushita, K. (2006). Conversion of capsular polysaccharide, involved in pellicle formation, to extracellular polysaccharide by *galE* deletion in *Acetobacter tropicalis*. *Bioscience, Biotechnology and Biochemistry* **70:** 2536-2539.

Du, X.J., Jia, S.R., Yang, Y. and Wang, S. (2011). Genome sequence of *Gluconacetobacter* sp. strain SXCC-1, isolated from Chinese vinegar fermentation starter. *Journal of Bacteriology* **193:** 3395-3396.

Emde, F. (2006). State of the art technologies in submersible vinegar production. *In:* I. García-García (Ed.). *Proceedings of the Second Symposium of R+D+I for Vinegar Production*. Servicio de Publicaciones de la Universidad de Córdoba, Cordoba, Spain, pp. 101-109.

Entani, E., Asai, M., Tsujihata, S., Tsukamoto, Y. and Ohta, M. (1998). Antibacterial action of vinegar against food-borne pathogenic bacteria, including *Escherichia coli* O157:H7. *Journal of Food Protection* **61:** 953-959.

Fernández-Pérez, R., Torres, S., Sanz, S. and Ruiz-Larrea, F. (2010). Rapid molecular methods for enumeration and taxonomical identification of acetic acid bacteria responsible for submerged vinegar production. *European Food Research and Technology* **231:** 813-819.

Ferrer, S., Mañes-Lázaro, R., Benavent-Gil, Y., Yépez, A. and Pardo, I. (2016). *Acetobacter musti* sp. nov. isolated from Bobal grape must. *International Journal of Systematic Bacteriology* **66:** 957-961.

Fraise, A.P., Wilkinson, M.A.C., Bradley, C.R., Oppenheim, B. and Moiemen, N. (2013). The antibacterial activity and stability of acetic acid. *Journal of Hospital Infection* **84:** 329-331.

Francois, J.A., Starks, C.M., Sivanuntakorn, S., Jiang, H., Ransome, A.E., Nam, J.W., Constantine, C.Z. and Kappock, T.J. (2006). Structure of a NADH-insensitive hexameric citrate synthase that resists acid inactivation. *Biochemistry* **45:** 13487-13499.

Francois, J.A. and Kappock, T.J. (2007). Alanine racemase from the acidophile *Acetobacter aceti*. *Protein Expression and Purification* **51:** 39-48.

Franke, I.H., Fegan, M., Hayward, C., Leonard, G., Stackebrandt, E. and Sly, L.I. (1999). Description of *Gluconacetobacter sacchari* sp. nov., a new species of acetic acid bacterium isolated from the leaf sheath of sugar cane and from the pink sugar-cane mealy bug. *International Journal of Systematic Bacteriology* **49:** 1681-1693.

Fujino, Y., Ito, S., Masai, H. and Fujimori, M. (1978). Studies on lipid components in *Acetobacter*. *Research Bulletin of Obihiro University* **10:** 917-925.

Fukami, H., Tachimoto, H., Kishi, M., Kaga, T. and Tanaka, Y. (2010). Acetic acid bacterial lipids improve cognitive function in dementia model rats. *Journal of Agricultural and Food Chemistry* **58:** 4084-4090.

Fukaya, M., Tayama, K., Tamaki, T., Tagami, H., Okumura, H., Kawamura, Y. and Beppu, T. (1989). Cloning of the membrane-bound aldehyde dehydrogenase gene of *Acetobacter polyoxogenes* and improvement of acetic acid production by use of the cloned gene. *Applied and Environmental Microbiology* **55:** 171-176.

Fukaya, M., Takemura, H., Okumura, H., Kawamura, Y., Horinouchi, S. and Beppu, T. (1990). Cloning of genes responsible for acetic acid resistance in *Acetobacter aceti*. *Journal of Bacteriology* **172:** 2096-2104.

Fukaya, M., Takemura, H., Tayama, K., Okumura, H., Kawamura, Y., Horinouchi, S. and Beppu, T. (1993). The *aarC* gene responsible for acetic acid assimilation confers acetic acid resistance on *Acetobacter aceti*. *Journal of Bioscience and Bioengineering* **76:** 270-275.

Gao, L., Zhou, J., Liu, J., Du, G. and Chen, J. (2012). Draft genome sequence of *Gluconobacter oxydans* WSH-003, a strain that is extremely tolerant of saccharides and alditols. *Journal of Bacteriology* **194:** 4455-4456.

Ge, X., Zhao, Y., Hou, W., Zhang, W., Chen, W., Wang, J., Zhao, N., Lin, J., Wang, W., Chen, M., Wand, Q., Jiao, Y., Yuan, Z. and Xiong, X. (2013). Complete genome sequence of the industrial strain *Gluconobacter oxydans* H24. *Genome Announcements* **1:** pii:e0003-13.

Giongo, A., Tyler, H.L., Zipperer, U.N. and Triplett, E.W. (2010) Two genome sequences of the same bacterial strain, *Gluconacetobacter diazotrophicus* PAl 5, suggest a new standard in genome sequence submission. *Standards in Genomic Sciences* **2:** 309-317.

Goldfine, H. (1984). Bacterial membranes and lipid packing theory. *The Journal of Lipid Research* **25:** 1501-1507.

Gómez-Manzo, S., Chavez-Pacheco, J.L., Contreras-Centella, M., Sosa-Torres, E., Arreguín-Espinosa, R., de la Mora, M.P., Membrillo-Hernández, J. and Escamilla, J.E. (2010). Molecular and catalytic properties of the aldehyde dehydrogenase of *Gluconacetobacter diazotrophicus*, a quinoheme protein containing pyrroloquinoline quinone, cytochrome b, and cytochrome c. *Journal of Bacteriology* **192:** 5718-5724.

Gómez-Manzo, S., Escamilla, J.E., González-Valdez, A., López-Velázquez, G., Vanoye-Carlo, A., Marcial-Quino, J., de la Mora, I., Garcia-Torres, I., Enríquez-Flores, S., Contreras-Zentella, M.L., Arreguín-Espinosa, R., Kroneck, P.M. and Sosa-Torres, M.E. (2015). The oxidative fermentation of ethanol in *Gluconacetobacter diazotrophicus* is a two-step pathway catalyzed by a single enzyme: Alcohol-aldehyde dehydrogenase (ADHa). *International Journal of Molecular Sciences* **16:** 1293-1311.

Goto, H. and Nakano, S. (2008). Gene participating in acetic acid tolerance, acetic acid bacteria bred using the gene, and process for producing vinegar with the use of the acetic acid bacteria. United States Mitsukan Group Corporation (Handa-shi, JP). US Patent No. 7446192 (W/O03/078635). Available from: http://www.freepatentsonline.com/7446192.html.

Greenberg, D.E., Porcella, S.F., Stock, F., Wong, A., Conville, P.S, Murray, P.R., Holland, S.M. and Zelazny, A.M. (2006). *Granulibacter bethesdensis* gen. nov., sp. nov., a distinctive pathogenic acetic acid bacterium in the family *Acetobacteraceae*. *International Journal of Systematic and Evolutionary Microbiology* **56:** 2609-2616.

Greenberg, D.E., Porcella, S.F., Zelazny, A.M., Virtaneva, K., Sturdevant, D.E., Kupko, J.J., Barbian, K.D., Babar, A., Dorward, D.W. and Holland, S.M. (2007). Genome sequence analysis of the emerging human pathogenic acetic acid bacterium *Granulibacter bethesdensis*. *Journal of Bacteriology* **189:** 8727-8736.

Greenfield, S. and Claus, G.W. (1972). Nonfunctional tricarboxylic acid cycle and the mechanism of glutamate biosynthesis in *Acetobacter suboxydans*. *Journal of Bacteriology* **112:** 1295-1301.

Grimmecke, H.D., Knirel, Y.A., Kiesel, B., Voges, M. and Rietschel, E.T. (1994). Structure of the *Acetobacter methanolicus* MB 129 capsular polysaccharide and of oligosaccharides resulting from degradation by bacteriophage Acm7. *Carbohydrates Research* **259:** 45-58.

Guo, H., Yi, W., Song, J.K. and Wang, P.G. (2008). Current understanding on biosynthesis of microbial polysaccharides. *Current Topics in Medicinal Chemistry* **8:** 141-151.

Gullo, M., Caggia, C., De Vero, L. and Giudici, P. (2006). Characterization of acetic acid bacteria in 'traditional balsamic vinegar'. *International Journal Food Microbiology* **106:** 209-212.

Hanada, T., Kashima, Y., Kosugi, A., Koizumi, Y., Yanagida, F. and Udaka, S. (2001). A gene encoding phosphatidylethanolamine N-methyltransferase from *Acetobacter aceti* and some properties of its disruptant. *Bioscience, Biotechnology and Biochemistry* **65:** 2741-2748.

Harrison, C. (2003). GrpE, a nucleotide exchange factor for DnaK. *Cell Stress Chaperones* **8:** 218-224.

Hartl, F.U. and Hayer-Hartl, M. (2002). Molecular chaperones in the cytosol: From nascent chain to folded protein. *Science* **295:** 1852-1858.

Higashiura, N., Hadano, H., Hirakawa, H., Matsutani, M., Takebe, S., Matsushita, K. and Azuma, Y. (2014). Draft genomic DNA sequence of the facultatively methylotrophic bacterium *Acidomonas methanolica* type strain MB58. *FEMS Microbiology Letters* **351:** 9-13.

Hung, J.E., Mill, C.P., Clifton, S.W., Magrini, V., Bhide, K., Francois, J.A., Ransome, A.E., Fulton, L., Thimmapuram, J., Wilson, R.K. and Kappock, T.J. (2014). Draft genome sequence of *Acetobacter aceti* strain 1023, a vinegar factory isolate. *Genome Announcements* **2:** e00550-14.

Illeghems, K., De Vuyst, L. and Weckx, S. (2013). Complete genome sequence and comparative analysis of *Acetobacter pasteurianus* 386B, a strain well-adapted to the cocoa bean fermentation ecosystem. *MBC Genomics* **14:** 526.

Ishikawa, M., Okamoto-Kainuma, A., Jochi, T., Suzuki, I., Matsui, K., Kaga, T. and Koizumi, Y. (2010a). Cloning and characterization of *grpE* in *Acetobacter pasteurianus* NBRC 3283. *Journal of Bioscience and Bioengineering* **109:** 25-31.

Ishikawa, M., Okamoto-Kainuma, A., Matsui, K., Takigishi, A., Kaga, T. and Koizumi, Y. (2010b). Cloning and characterization of *clpB* in *Acetobacter pasteurianus* NBRC 3283. *Journal of Bioscience and Bioengineering* **110:** 69-71.

Iyer, P.R., Geib, S.M., Catchmark, J., Kao, T.H. and Tien, M. (2010). Genome sequence of a cellulose-producing bacterium, *Gluconacetobacter hansenii* ATCC 23769. *Journal of Bacteriology* **192:** 4256-4257.

Jansson, P.E., Lindgerb, J., Swarna Wimalasiri, K.M. and Dankert, M.A. (1993). Structural studies of acetan, an exopolysaccharide elaborated by *Acetobacter xylinum*. *Carbohydrate Research* **245:** 303-310.

Jojima, Y., Mihara, Y., Suzuki, S., Yokozeki, K., Yamanaka, S. and Fudou, R. (2004). *Saccharibacter floricola* gen. nov., sp. nov., a novel osmophilic acetic acid bacterium isolated from pollen. *International Journal of Systematic and Evolutionary Microbiology* **54:** 2263-2267.

Jucker, W. and Ettlinger, L. (1985). The inhibition of acetate oxidation by ethanol in *Acetobacter aceti*. *Archives of Microbiology* **143:** 283-289.

Kanchanarach, W., Theeragool, G., Inoue, T., Yakushi, T., Adachi, O. and Matsushita, K. (2010a). Acetic acid fermentation of *Acetobacter pasteurianus*: Relationship between acetic acid resistance and pellicle polysaccharide formation. *Bioscience, Biotechnology and Biochemistry* **74:** 1591-1597.

Kanchanarach, W., Theeragool, G., Yakushi, T., Toyama, H., Adachi, O. and Matsushita, K. (2010b). Characterization of thermotolerant *Acetobacter pasteurianus* strains and their quinoprotein alcohol dehydrogenases. *Applied Microbiology and Biotechnology* **85:** 741-751.

Kim, E.K., Kim, S.H., Nam, H.J., Choi, M.K., Lee, K.A., Choi, S.H., Seo, Y.Y., You, H., Kim, B. and Lee, WJ. (2012). Draft genome sequence of *Gluconobacter morbifer* G707T, a pathogenic gut bacterium isolated from *Drosophila melanogaster* intestine. *Journal of Bacteriology* **194:** 1245.

Koike, H., Sato, S., Morita, T., Fukuoka, T. and Habe, H. (2014). Draft genome sequence of *Acetobacter tropicalis* type strain NBRC16470, a producer of optically pure D-glyceric acid. *Genome Announcements* **2:** e01329-14.

Kubiak, K., Kurzawa, M., Jedrzejczak-Krzepkowska, M., Ludwicka, K., Krawczyk, M., Migdalski, A., Kacprzak, M.M., Loska, D., Krystynowicz, A. and Bielecki, S. (2014). Complete genome sequence of *Gluconacetobacter xylinus* E25 strain—valuable and effective producer of bacterial nanocellulose. *Journal of Biotechnology* **176:** 18-19.

Kornmann, H., Duboc, P., Marison, I. and von Stockar, U. (2003). Influence of nutritional factors on the nature, yield, and composition of exopolysaccharides produced by *Gluconacetobacter xylinus* I-2281. *Applied and Environmental Microbiology* **69:** 6091-6098.

Lasko, D.R., Schwerdel, C., Bailey, J.E. and Sauer, U. (1997). Acetate-specific stress response in acetate-resistant bacteria: An analysis of protein patterns. *Biotechnology Progress* **13:** 519-523.

Levine, A.S. and Fellers, C.R. (1940). Inhibiting effect of acetic acid upon microorganisms in the presence of sodium chloride and sucrose. *Journal of Bacteriology* **40:** 255-269.

Li, L., Praet, J., Borremans, W., Nunes, O.C., Manaia, C.M., Cleenwerck, I., Meeus, I., Smagghe, G., De Vuyst, L. and Vandamme, P. (2015). *Bombella intestini* gen. nov., sp. nov., an acetic acid bacterium isolated from bumble bee crop. *International Journal of Systematic and Evolutionary Microbiology* **65:** 267-273.

Loganathan, P. and Nair, S. (2004). *Swaminathania salitolerans* gen. nov., sp. nov., a salt-tolerant, nitrogen-fixing and phosphate-solubilizing bacterium from wild rice (*Porteresia coarctata Tateoka*). *International Journal of Systematic and Evolutionary Microbiology* **54:** 1185-1190.

MacCormick, C.A., Harris, J.E., Jay, A.J., Ridout, M.J., Colquhoun, I.J. and Morris, V.J. (1996). Isolation and characterization of a new extracellular polysaccharide from an *Acetobacter* species. *Journal of Applied Biotechnology* **81:** 419-424.

Madigan, M., Martinko, J. and Brock T.D. (2006). *Brock Biology of Microorganisms*, 11th edn., Upper Saddle River, NJ, Prentice Hall/Pearson Education.

Malimas. T., Chaipitakchonlatarn, W., Thi Lan Vu, H., Yukphan, P., Muramatsu, Y., Tanasupawat, S., Potacharoen, W., Nakagawa, Y., Tanticharoen, M. and Yamada, Y. (2013). *Swingsia samuiensis* gen. nov., sp. nov., an osmotolerant acetic acid bacterium in the α-Proteobacteria. *Journal of General and Applied Microbiology* **59:** 375-384.

Mamlouk, D. and Gullo, M. (2013). Acetic acid bacteria: Physiology and carbon source oxidation. *Indian Journal of Microbiology* **53:** 377-384.

Matsushita, K., Nonobe, M., Shinagawa, E., Adachi, O. and Ameyana, M. (1985). Isolation and characterization of outer and cytoplasmic membranes from spherolasts of *Acetobacter aceti*. *Agricultural and Biological Chemistry* **40:** 3519-3526.

Matsushita, K., Ebisuya, H., Ameyama, M. and Adachi, O. (1992). Change of the terminal oxidase from cytochrome a1 in shaking cultures to cytochrome o in static cultures of *Acetobacter aceti*. *Journal of Bacteriology* **174:** 122-129.

Matsushita, K., Toyama, H. and Adachi, O. (1994). Respiratory chains and bioenergetics of acetic acid bacteria. *Advances in Microbial Physiology* **36:** 247-301.

Matsushita, K., Yakushi, T., Takaki, Y., Toyama, H. and Adachi, O. (1995). Generation mechanism and purification of an inactive form convertible in vivo to the active form of quinoprotein alcohol dehydrogenase in *Gluconobacter suboxydans*. *Journal of Bacteriology* **177:** 6552-6559.

Matsushita, K., Inoue, T., Theeragol, G., Trček, J., Toyama, H. and Adachi, O. (2005a). Acetic acid production in acetic acid bacteria leading to their 'death' and survival. *In:* M. Yamada (Ed.). *Survival and Death in Bacteria*. Research Singpost, Kerala, India, pp. 169-181.

Matsushita, K., Inoue, T., Adachi, O. and Toyama, H. (2005b). *Acetobacter aceti* possesses a proton motive force-dependent efflux system for acetic acid. *Journal of Bacteriology* **187:** 4346-4352.

Matsutani, M., Hirakawa, H., Nishikura, M., Soemphol, W., Ali, I.A., Yakushi, T. and Matsushita, K. (2011). Increased number of Arginine-based salt bridges contributes to the thermotolerance of thermotolerant acetic acid bacteria, *Acetobacter tropicalis* SKU1100. *Biochemical and Biophysical Research Communications* **409:** 120-124.

Matsutani, M., Hirakawa, H., Saichana, N., Soemphol, W., Yakushi, T. and Matsushita, K. (2012). Genome-wide phylogenetic analysis of differences in thermotolerance among closely related *Acetobacter pasteurianus* strains. *Microbiology* **158:** 229-239.

Matsutani, M., Kawajiri, E., Yakushi, T., Adachi, O. and Matsushita, K. (2013). Draft genome sequence of dihydroxyacetone-producing *Gluconobacter thailandicus* strain NBRC 3255. *Genome Announcements* **1:** e0011813.

Matsutani, M., Suzuki, H., Yakushi, T. and Matsushita, K. (2014). Draft genome sequence of *Gluconobacter thailandicus* NBRC 3257. *Standards in Genomic Sciences* **9:** 614-623.

Menzel, U. and Gottschalk, G. (1985). The internal pH of *Acetobacterium wieringae* and *Acetobacter aceti* during growth and production of acetic acid. *Journal of Bacteriology* **143:** 47-51.

Miao, Y., Zhou, X., Xu, Y. and Yu, S. (2015). Draft genome sequence of *Gluconobacter oxydans* NL71, a strain that efficiently biocatalyzes xylose to xylonic acid at a high concentration. *Genome Annoncements* **3:** e00615-15.

Mitchison, M., Bulach, D.M., Vinh, T., Rajakumar, K., Faine, S. and Adler, B. (1997). Identification and characterization of the dTDP-rhamnose biosynthesis and transfer genes of the lipopolysaccharide-related *rfb* locus in *Leptospira interrogans* serovar Copenhageni. *Journal of Bacteriology* **179:** 1262-1267.

Moonmangmee, S., Kawabata, K., Tanaka, S., Toyama, H., Adachi, O. and Matsushita, K. (2002). A novel polysaccharide involved in the pellicle formation of *Acetobacter aceti*. *Journal of Bioscience and Bioengineering* **93:** 192-200.

Mullane, N., O'Gaora, P., Nally, J.E., Iversen. C., Whyte, P., Wall, P.G. and Fanning, S. (2008). Molecular analysis of the *Enterobacter sakazakii* O-antigen gene locus. *Applied and Environmental Microbiology* **74:** 3783-3794.

Mullins, E.A., Francois, J.A. and Kappock, T.J. (2008). A specialized citric acid cycle requiring succinyl-coenzyme A (CoA):acetate CoA-transferase (AarC) confers acetic acid resistance on the acidophile *Acetobacter aceti*. *Journal of Bacteriology* **190:** 4933-4940.

Mullins, E.A., Starks, C.M., Francois, J.A., Sael, L., Kihara, D. and Kappock, T.J. (2012). Formyl-coenzyme A (CoA):oxalate CoA-transferase from the acidophile *Acetobacter aceti* has a distinctive electrostatic surface and inherent acid stability. *Protein Science* **21:** 686-696.

Mungdee, S., Moonmangmee, D. and Moonmangmee, S. (2006). Purification and characterization of pellicle polysaccharide from *Acetobacter aceti* IFO 3284 R strain. Proceedings of 44th Kasetsart University Annual Conference 'Agricultural Science: Carrying Forward the Royal Bio-Energy Initiative', Bangkok, Thailand, pp. 451-458.

Nakano, S., Fukaya, M. and Horinouchi, S. (2004). Enhanced expression of aconitase raises acetic acid resistance in *Acetobacter aceti*. *FEMS Microbiology Letters* **234:** 315-322.

Nakano, S., Fukaya ,M. and Horinouchi, S. (2006). Putative ABC transporter responsible for acetic acid resistance in *Acetobacter aceti*. *Applied and Environmental Microbiology* **72:** 497-505.

Nakano, S. and Fukaya, M. (2008). Analysis of proteins responsive to acetic acid in *Acetobacter*: Molecular mechanisms conferring acetic acid resistance in acetic acid bacteria. *International Journal of Food Microbiology* **125:** 54-59.

Nikaido, H. (2003). Molecular basis of bacterial outer membrane permeability revisited. *Microbiology and Molecular Biology Reviews* **67:** 593-656.

Ogawa, S., Tachimoto, H. and Kaga, T. (2010). Elevation of ceramide in *Acetobacter malorum* S24 by low pH stress and high temperature stress. *Journal of Bioscience and Bioengineering* **109:** 32-36.

Ogino, H., Azuma, Y., Hosoyama, A., Nakazawa, H., Matsutani, M., Hasegawa, A., Otsuyama, K., Matsushita, K., Fujita, N. and Shirai, M. (2011). Complete genome sequence of NBRC 3288, a unique cellulose-nonproducing strain of *Gluconacetobacter xylinus* isolated from vinegar. *Journal of Bacteriology* **193:** 6997-6998.

Oh, E.T., Ju, Y.J., Koh, S.C., Kim, Y., Kim, J.S. and So, J.S. (2004). Lack of *O*-polyssacharide renders *Bradyrhizobium japonicum* more resistant to organic acid stress. *Journal of Microbiology and Biotechnology* **14:** 1324-1326.

Ohmori, S., Uozumi, T. and Beppu, T. (1982). Loss of acetic acid resistance and ethanol oxidizing ability in an *Acetobacter* strain. *Agricultural and Biological Chemistry* **46:** 381-389.

Okamoto-Kainuma, A., Yan, W., Kadono, S., Tayama, K., Koizumi, Y. and Yanagida, F. (2002). Cloning and characterization of *groESL* operon in *Acetobacter aceti*. *Journal of Bioscience and Bioengineering* **94:** 140-147.

Okamoto-Kainuma, A., Yan, W., Fukaya, M., Tukamoto, Y., Ishikawa, M. and Koizumi, Y. (2004). Cloning and characterization of the *dnaKJ* operon in *Acetobacter aceti*. *Journal of Bioscience and Bioengineering* **97:** 339-342.

Okamoto-Kainuma, A., Ishikawa, M., Nakamura, H., Fukazawa, S., Tanaka, N., Yamagami, K. and Koizumi, Y. (2011). Characterization of rpoH in *Acetobacter pasteurianus* NBRC3283. *Journal of Bioscience and Bioengineering* **111:** 429-432.

Okumura, H., Uozumi, T. and Beppu, T. (1985). Biochemical characteristics of spontaneous mutants of *Acetobacter aceti* deficient in ethanol oxidation. *Agricultural and Biological Chemistry* **49:** 2485-2487.

Parte, A.C. (2014). LPSN--list of prokaryotic names with standing in nomenclature. *Nucleic Acids Research* **42:** D613-D616.

Prust, C., Hoffmeister, M., Liesegang, H., Wiezer, A., Fricke, W.F., Ehrenreich, A., Gottschalk, G. and Deppenmeier, U. (2005). Complete genome sequence of the acetic acid bacterium *Gluconobacter oxydans*. *Nature Biotechnology* **23:** 195-200.

Quintero, Y., Poblet, M., Guillamón, J.M. and Mas, A. (2009). Quantification of the expression of reference and alcohol dehydrogenase genes of some acetic acid bacteria in different growth conditions. *Journal of Applied Microbiology* **106:** 666-674.

Ramírez-Bahena, M.H., Tejedor, C., Martín, I., Velázquez, E. and Peix, A. (2013). *Endobacter medicaginis* gen. nov., sp. nov., isolated from alfalfa nodules in an acidic soil. *International Journal of Systematic and Evolutionary Microbiology* **63:** 1760-1765.

Rani, M.U., Rastogi, N.K. and Appaiah, K.A. (2011). Statistical optimization of medium composition for bacterial cellulose production by *Gluconacetobacter hansenii* UAC09 using coffee cherry husk extract—An agro-industry waste. *Journal of Microbiology and Biotechnology* **21:** 739-745.

Roh, S.W., Nam, Y.D., Chang, H.W., Kim, K.H., Kim, M.S., Ryu, J.H., Kim, S.H., Lee, W.J. and Bae, J.W. (2008). Phylogenetic characterization of two novel commensal bacteria involved with innate immune homeostasis in *Drosophila melanogaster*. *Applied and Environmental Microbiology* **74:** 6171-6177.

Rohmer, M., Bouvier-Navez, P. and Ourisson, G. (1984). Distribution of hopanoid triterpenes in prokaryotes. *Journal of General Microbiology* **130:** 1137-1150.

Russell, J.B. (1992). Another explanation for the toxicity of fermentation acids at low pH: Anion accumulation versus uncoupling. *Journal of Applied Microbiology* **73:** 363-370.

Russell, J.B. and Díez-González, F. (1998). The effects of fermentation acids on bacterial growth. *Advances in Microbial Physiology* **39:** 205-234.

Ryssel, H., Kloeters, O., Germann, G., Schäfer, Th., Wiedemann, G. and Oehlbauer, M. (2009). The antimicrobial effect of acetic acid—An alternative to common local antiseptics. *Burns* **35:** 695-700.

Ryu, J.H., Kim, S.H., Lee, H.Y., Bai, J.Y., Nam, Y.D., Bae, J.W., Lee, D.G., Shin, S.C., Ha, E.M. and Lee, W.J. (2008). Innate immune homeostasis by the homeobox gene caudal and commensal-gut mutualism in *Drosophila*. *Science* **319:** 777-782.

Saeki, A., Taniguchi, M., Matsushita, K., Toyama, H., Theeragool, G., Lotong, N. and Adachi, O. (1997). Microbial aspects of acetate oxidation by acetic acid bacteria, unfavourable phenomena in vinegar fermentation. *Bioscience, Biotechnology and Biochemistry* **61:** 317-323.

Saeki, A., Matsushita, K., Takeno, S., Taniguchi, M., Toyama, H., Theeragool, G., Lotong, N. and Adachi, O. (1999). Enzymes responsible for acetate oxidation by acetic acid bacteria. *Bioscience, Biotechnology and Biochemistry* **63:** 2102-2109.

Sato, S., Umemura, M., Koike, H. and Habe, H. (2013). Draft genome sequence of *Gluconobacter frateurii* NBRC 103465, a glyceric acid-producing strain. *Genome Announcements* **1:** e00369-13.

Sakurai, K., Arai, H., Ishii, M. and Igarashi, Y. (2011). Transcriptome response to different carbon sources in *Acetobacter aceti*. *Microbiology* **157:** 899-910.

Sakurai, K., Arai, H., Ishii, M. and Igarashi, Y. (2012). Changes in the gene expression profile of *Acetobacter aceti* during growth on ethanol. *Journal of Bioscience and Bioengineering* **113:** 343-348.

Savidge, R. and Colvin, J. (1985). Production of cellulose and soluble polysaccharides by *Acetobacter xylinum*. *Canadian Journal of Microbiology* **31:** 1019-1025.

Schmid, J., Koenig, S., Pick, A., Steffler, F., Yoshida, S., Miyamoto, K. and Sieber, V. (2014). Draft genome sequence of *Kozakia baliensis* SR-745, the first sequenced *Kozakia* strain from the family *Acetobacteraceae*. *Genome Announcements* **2:** e00594-14.

Schüller, G., Hertel, C. and Hammes, W.P. (2000). *Gluconacetobacter entanii* sp. nov., isolated from submerged high-acid industrial vinegar fermentations. *International Journal of Systematic and Evolutionary Microbiology* **50:** 2013-2020.

Settembre, E.C., Chittuluru, J.R., Mill, C.P., Kappock, T.J. and Ealick, S.E. (2004). Acidophilic adaptations in the structure of *Acetobacter aceti* N^5-carboxyaminoimidazole ribonucleotide mutase (PurE). *Acta Crystallographica Section D: Biological Crystallography* **60:** 1753-1760.

Sheng, B., Ni, J., Gao, C., Ma, C. and Xu, P. (2014). Draft genome sequence of the *Gluconobacter oxydans* strain DSM 2003, an important biocatalyst for industrial use. *Genome Announcements* **2:** e00417-14.

Shimwell, J.L. and Carr, J.G. (1964). Mutant frequency in *Acetobacter*. *Nature* **201:** 1051-1052.

Sievers, M., Stöckli, M. and Teuber, M. (1997). Purification and properties of citrate synthase from *Acetobacter europaeus*. *FEMS Microbiology Letters* **146:** 53-58.

Simonin, P., Tindall, B. and Rohmer, M. (1994). Structure elucidation and biosynthesis of 31-methylhopanoids from *Acetobacter europaeus*. Studies on a new series of bacterial triterpenoids. *European Journal of Biochemistry* **225:** 765-771.

Soemphol, W., Deeraksa, A., Matsutani, M., Yakushi, T., Toyama, H., Adachi, O., Yamada, M. and Matsushita, K. (2011). Global analysis of the genes involved in the thermotolerance mechanisms of thermotolerant *Acetobacter tropicalis* SKU1100. *Bioscience, Biotechnology and Biochemistry* **75:** 1921-1928.

Sokollek, S.J., Hertel, C. and Hammes, W.P. (1998). Description of *Acetobacter oboediens* sp. nov. and *Acetobacter pomorum* sp. nov., two new species isolated from industrial vinegar fermentations. *International Journal of Systematic Bacteriology* **48:** 935-940.

Solieri, L. and Giudici, P. (2009). Vinegars of the world. *In*: L. Solieri and P. Giudici (Eds.). *Vinegars of the World*. Springer-Verlag, Italy. Pp 1-16.

Steiner, P. and Sauer, U. (2001). Proteins induced during adaptation of *Acetobacter aceti* to high acetate concentrations. *Applied and Environmental Microbiology* **67:** 5474-5481.

Steiner, P. and Sauer, U. (2003). Long-term continuous evolution of acetate resistant *Acetobacter aceti*. *Biotechnology and Bioengineering* **84:** 40-44.

Stevenson, G., Neal, B., Liu, D., Hobbs, M., Packer, N.H., Batley, M., Redmond, J.W., Lindquist, L. and Reeves, P. (1994). Structure of the O antigen of *Escherichia coli* K-12 and the sequence of its rfb gene cluster. *Journal of Bacteriology* **176:** 4144-4156.

Susin, M.F., Baldini, R.L., Gueiros-Filho, F. and Gomes, S.L. (2006). GroES/GroEL and DnaK/DnaJ have distinct roles in stress responses and during cell cycle progression in *Caulobacter crescentus*. *Journal of Bacteriology* **188:** 8044-8053.

Tahara, Y., Kameda, M., Yamada, Y. and Kondo, K. (1976). Phospholipid composition of *Gluconobacter cerinus*. *Agricultural and Biological Chemistry* **40:** 2355-2360.

Tahara, Y., Nakagawa, A. and Yamada, Y. (1986). Occurrence of free ceramide in *Acetobacter xylinum*. *Agricultural and Biological Chemistry* **50:** 2949-2950.

Tayama, K., Minakami, H., Entani, E., Fujiyama, S. and Masai, H. (1985). Structure of an acidic polysaccharide from *Acetobacter* sp. NBI 1022. *Agricultural and Biological Chemistry* **49:** 959-966.

Tayama, K., Minakami, H., Fujiyama, S., Masai, H. and Misaki, A. (1986). Structure of an acidic polysaccharide elaborated by *Acetobacter* sp. NBI 1005. *Agricultural and Biological Chemistry* **50:** 1271-1278.

Takemura, H., Horinouchi, S. and Beppu, T. (1991). Novel insertion sequence IS1380 from *Acetobacter pasteurianus* is involved in loss of ethanol-oxidizing ability. *Journal of Bacteriology* **173:** 7070-7076.

Takemura, H., Kondo, K., Horinouchi, S. and Beppu, T. (1993). Induction by ethanol of alcohol dehydrogenase activity in *Acetobacter pasteurianus*. *Journal of Bacteriology* **175:** 6857-6866.

Takemura, H., Tsuchida, T., Yoshinaga, F., Matsushita, K. and Adachi, O. (1994). Prosthetic group of aldehyde dehydrogenase in acetic acid bacteria not pyrroloquinoline quinone. *Bioscience, Biotechnology and Biochemistry* **58:** 2082-2083.

Thi Lan Vu, H., Yukphan, P., Chaipitakchonlatarn, W., Malimas, T., Muramatsu, Y., Thi Tu Bui, U., Tanasupawat, S., Cong Duong, K., Nakagawa, Y., Thanh Pham, H. and Yamada, Y. (2013). *Nguyenibacter vanlangensis* gen. nov., sp. nov., an unusual acetic acid bacterium in the α-Proteobacteria. *Journal of General and Applied Microbiology* **59:** 153-166.

Thurner, C., Vela, C., Thöny-Meyer, L., Meile, L. and Teuber, M. (1997). Biochemical and genetic characterization of the acetaldehyde dehydrogenase complex from *Acetobacter europaeus*. *Archives of Microbiology* **168:** 81-91.

Trček, J., Raspor, P. and Teuber, M. (2000). Molecular identification of *Acetobacter* isolates from submerged vinegar production, sequence analysis of plasmid pJK2-1 and application in the development of a cloning vector. *Applied Microbiology and Biotechnology* **53:** 289-295.

Trček, J., Toyama, H., Czuba, J., Misiewicz, A. and Matsushita, K. (2006). Correlation between acetic acid resistance and characteristics of PQQ-dependent ADH in acetic acid bacteria. *Applied Microbiology and Biotechnology* **70:** 366-373.

Trček, J., Jernejc, K. and Matsushita, K. (2007). The highly tolerant acetic acid bacterium *Gluconacetobacter europaeus* adapts to the presence of acetic acid by changes in lipid composition, morphological properties and PQQ-dependent ADH expression. *Extremophiles* **11:** 627-635.

Urakami, T., Tamaoka, J., Suzuki, K.I. and Komagata, K. (1989). *Acidomonas* gen. nov., incorporating *Acetobacter methanolicus* as *Acidomonas methanolica* comb. nov. *International Journal of Systematic Bacteriology* **39:** 50-55.

Yamada, Y., Nunoda, M., Ishikawa, T. and Tahara, Y. (1981). The cellular fatty acid composition in acetic acid bacteria. *Journal of General and Applied Microbiology* **27:** 405-417.

Yamada, Y. and Kondo, K. (1985). *Gluconoacetobacter*, a new subgenus comprising the acetate-oxidizing acetic acid bacteria with ubiquinone-10 in the genus *Acetobacter*. *Journal of General and Applied Microbiology* **30:** 297-303.

Yamada, Y., Hoshino, K. and Ishikawa, T. (1997). The phylogeny of acetic acid bacteria based on the partial sequences of 16S ribosomal RNA: The elevation of the subgenus *Gluconoacetobacter* to the generic level. *Bioscience, Biotechnology and Biochemistry* **61:** 1244-1251.

Yamada, Y. (2003). Taxonomy of acetic acid bacteria utilized for vinegar fermentation. Proceedings of the 1st International Symposium and Workshop on 'Insight into the World of Indigenous Fermented Foods for Technology Development and Food Safety', n.p.: Ministry of University Affairs, Bangkok (Thailand). Division of International Cooperation. Available from: http://www.agriqua.doae.go.th/worldfermentedfood/I_8_Yamada.pdf (Accessed: June 2016).

Yamada, Y., Yukphan, P., Thi Lan Vi, H., Muramatsu, Y., Ochaikui, D., Tanasupawat, S. and Nakagawa, Y. (2012). Description of *Komgataeibacter* gen. nov., with proposals of new combinations (*Acetobacteraceae*). *Journal of General and Applied Microbiology* **58:** 397-404.

5

Exopolysaccharide Production of Acetic Acid Bacteria

Seval Dağbağlı[1,*] and Yekta Göksungur[2]

[1] Celal Bayar University, Faculty of Engineering, Department of Food Engineering, 45140, Muradiye Manisa, Turkey

[2] Ege University, Faculty of Engineering, Department of Food Engineering, 35100, Bornova, Izmir, Turkey

Introduction

Polysaccharides derived from Gram-positive and Gram-negative bacteria are divided into two groups; capsular polysaccharides (CPS) and exopolysaccharides (EPS) (Kumar et al. 2007, Ali et al. 2011). CPS is permanently attached to the outer surface of the cells and EPS is secreted into the growth medium (Ali et al. 2011, Perumpuli et al. 2014). These polysaccharides are either homopolysaccharides containing a single type of monosaccharide or heteropolysaccharides containing several types of monosaccharides (Donot et al. 2012, Ahmad et al. 2015).

Bacterial extracellular polysaccharides (EPS) have high molecular weight and incomparable rheological properties because of their high purity and regular structure (Kornman et al. 2003, Roca et al. 2015). Depending on their chemical and physical properties, they are commercially used in various industrial sectors, such as pharmaceuticals, food, cosmetics, oil drilling and paper manufacturing (Kumar et al. 2007, Roca et al. 2015). Despite this potential, bacterial EPS is not used in large quantities because of high production costs, except in very specific small markets, where the production cost does not become a deterrent in their commercialization (Roca et al. 2015).

Many strains of acetic acid bacteria (AAB) synthesize high amounts of exopolysaccharides (Kornmann et al. 2003, Jakob et al. 2013). They have the ability to float on the surface of a static culture by producing pellicles, which consist of cells and a self-produced matrix of cell-attached polysaccharide (Ali et al. 2011, Perumpuli et al. 2014). Some strains of *Acetobacter* and *Komagataeibacter* produce extracellular homopolysaccharides, such as cellulose or heteropolysaccharides like acetan (Ojinnaka et al. 1996, Perumpuli et al. 2014, Matsutani et al. 2015).

Cellulose

Chemical Structure and Properties of Cellulose

Cellulose is the most abundant organic polymer on earth and is mostly produced by vascular plants (Khajavi et al. 2011, Dayal et al. 2013, Lin et al. 2013). Some bacterial strains

*Corresponding author: seval.dagbagli@cbu.edu.tr

also produce cellulose with the same chemical structure of linear β-1,4 glucan chains as plant-derived cellulose. However, the chemical and physical properties of the bacterial polymers are different from plant polymers. Bacterial cellulose (BC) is free of other polymers while plant-derived cellulose chains are closely associated with hemicellulose, lignin and pectin (Sulaeva et al. 2015).

Bacterial cellulose is an unbranched polymer with nanofibrils and these linear glucan chains form highly regular intra- and inter-molecular hydrogen bonds. In the process of forming these fibers, polymerization and crystallization occur together to include both characteristics. These nanofibrils have cross-sectional dimensions in the nanometer range, which can aggregate to form microfibrils with a width of 50-80 nm and a thickness of 3-8 nm. These can then form a 3D network structure (Fig. 1). This fine structure makes BC different from other microbial polysaccharides (Shi et al. 2014).

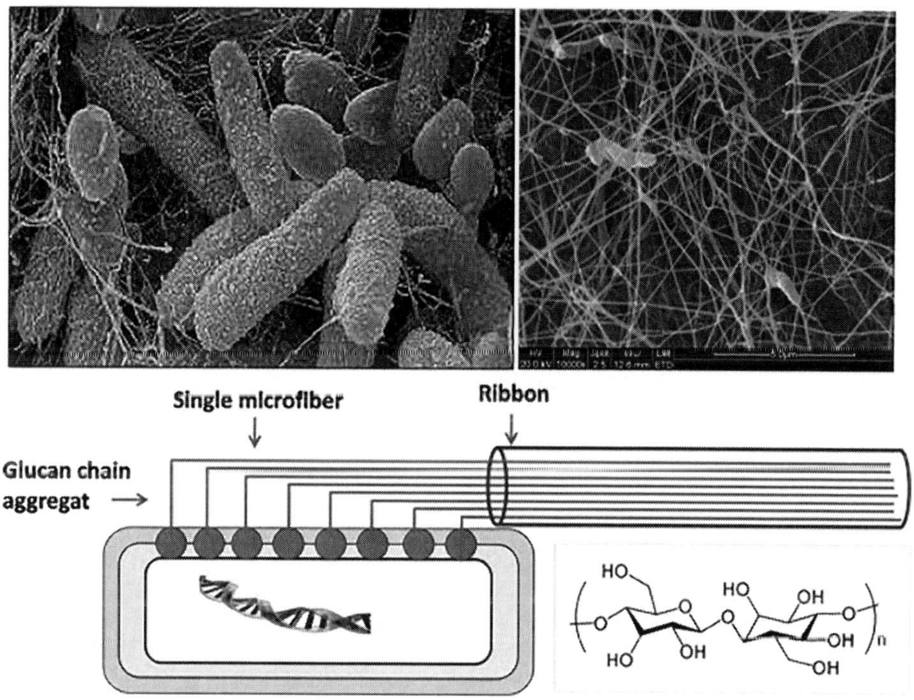

Figure 1: SEM images of *Acetobacter xylinus* and formation of bacterial cellulose (Shi et al. 2014).

Cellulose is produced in two forms by *Komagataeibacter* species as cellulose I and cellulose II (Chawla et al. 2009). Cellulose I is a ribbon-like polymer in predominant native form, while cellulose II is a thermodynamically more stable amorphous form of the polymer. Cellulose I composed of parallel β-1,4 glucan chains arranged uniaxially with van der Waals forces while β-1,4 glucan chains of cellulose II are arranged in a random manner. They are mostly antiparallel high numbers of hydrogen bonds resulting in a more stable form (Yu and Atalla 1996, Mohite and Patil 2014a, b). In Nature, Cellulose I structure is found in allomorphic forms I_{α} or I_{β}, depending on the arrangement of the chains between each other. Cellulose belonging to plant cell walls shows a higher percentage of I_{β} structure as compared with cellulose from algae and bacteria. This shows a higher percentage of I_{α} structure and seems to be a less stable displacement (Mohite and Patil 2014a). There are differences in accumulation of cellulose I and cellulose II outside the cytoplasmic membrane (Fig. 2).

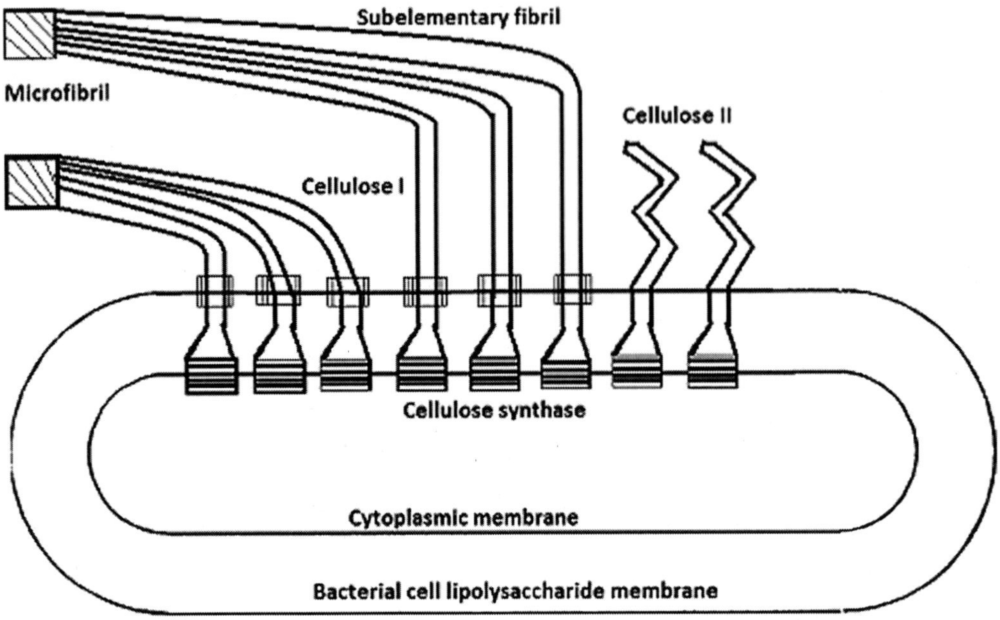

Figure 2: Schematic representation of synthesis of BC at the cell surface (Mohite and Patil 2014b).

The biosynthesis of bacterial cellulose occurs between the outer and cytoplasmic membranes in three main steps. In the first step, glucose residues are polymerized in β-1,4 glucan after which 10–15 of these chains are arranged in parallel structure to form a 1.5 nm wide protofibril (sub-elementary fibrils). In the second step, several sub-elementary fibrils are assembled into 2–4 nm wide microfibrils. In the third step, microfibrils are bundled into a ribbon-shape (20-100 nm) after which cellulose is secreted (Iguchi et al. 2000, Klemm et al. 2001, Castro et al. 2011, 2015, Jozala et al. 2016).

A. xylinum produces cellulose from glucose in the presence of oxygen. Production pathway of cellulose by this bacterium is shown in Fig. 3. Cellulose biosynthesis consists of a cycle of lipid-phosphate-carbohydrate intermediates. As a result of the glucose reaction, glucose-6-phosphate, glucose-1-phosphate, uridine diphosphoglucose (UDP-glc) and finally cellulose synthesis occur (Lin et al. 2013). The biosynthesis of bacterial cellulose is controlled by glucokinase, isomerase, phosphoglucomutase, UDPG-pyrophosphorylase and cellulose synthase enzymes. This pathway consists of the branched hexose monophosphate pathway (HMP) and tricarboxylic acid cycle (Lin et al. 2013).

Cellulose Producing AAB

The vinegar biofilm produced by AAB is composed of various exopolysaccharides (EPS), including cellulose. Several AAB species belonging to the genera *Gluconacetobacter (Ga.)*, *Komagataeibacter (K.)*, *Acetobacter (A.)* or *Gluconobacter (G.)* have been reported as cellulose producers and especially the species *K. xylinus* is considered to be the model microorganism for BC production (Valera et al. 2015). Only *Komagataeibacter* species can produce cellulose at commercial levels (Lin et al. 2013), especially *K. xylinus* and *K. hansenii* are the most potential producers compared to other strains (Rangaswamy et al. 2015). BC-producing strains can be isolated from fruits, flowers, fermented foods, beverages and vinegar (Lin et al. 2016). Rangaswamy et al. (2015) aimed to isolate cellulose-producing bacteria from rotten fruits and rotten vegetables and isolated thirty different cellulose producers from natural sources. The isolate RV28 from rotten pomegranate showed a better cellulose

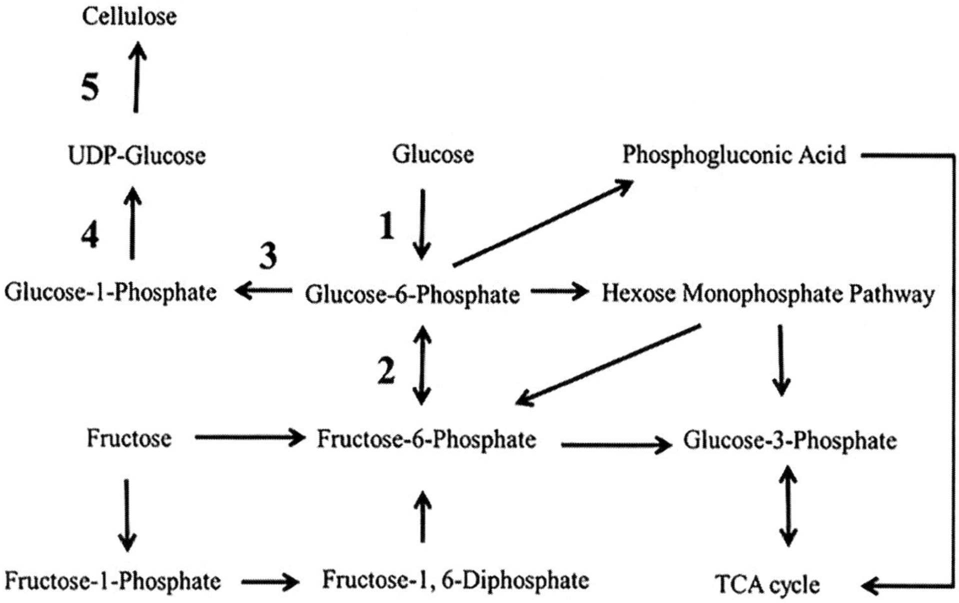

Figure 3: Pathway of cellulose synthesis by *Acetobacter xylinum*. 1 glucokinase, 2 isomerase, 3 phosphoglucomutase, 4 UDPG-pyrophosphorylase, 5 cellulose synthase (Lin et al. 2013).

production capacity as compared to other isolates and was identified as *Gluconacetobacter* sp. RV28 on the basis of its biochemical characteristics.

Cellulose is synthesized by AAB as a primary metabolite. The aerobic bacteria move to the oxygen-rich area due to its cellulose synthesis mechanism (Castro et al. 2011).

Production Conditions

Cellulose produced by microbial fermentation offers several advantages when compared to plant celluloses. BC production is not dependent on regional or climatic conditions. Microbial growth rates can be controlled to produce BC in the required quantities and time period. Cellulose produced by microorganisms contains no lignin or other contaminants. Therefore, the energy utilised in the purification of plant cellulose can be eliminated and microorganisms are genetically modified to produce BC with desired properties (Shi et al. 2014).

In recent years, developments on cellulose production are mainly focusing on improvement of the properties of cellulose, selection of new carbon sources to reduce the production cost and isolation of new strains. To reduce the cost of BC production, researchers have been evaluating the use of agroindustrial wastes rich in fructose, sucrose, along with nitrogen and vitamins which are required for cellulose biosynthesis (Castro et al. 2011).

BC is biosynthesized from sugar and organic acids by AAB (Serrato et al. 2013). It is produced from a variety of synthetic (Shezad et al. 2009, Shah et al. 2010, Ul-Islam et al. 2012) and non-synthetic media, such as molasses (El-Saied et al. 2008, Çakar et al. 2014a, b), pineapple juice (Kurosumi et al. 2009, Castro et al. 2011) and orange juice (Kurosumi et al. 2009) by ABB.

Different carbon sources are used for BC production. It is reported that the main difference is in the productivity factor. With the exception of two sugar alcohols, arabitol and mannitol, all the sources are less active than glucose. Cellulose production in a medium containing arabitol and mannitol is respectively 6.2 and 3.8 times higher than the medium containing glucose. Also, the efficiency of cellulose production from monosaccharide (glucose) is higher than disaccharides (lactose and sucrose) (Khajavi et al. 2011).

Rangaswamy et al. (2015) investigated the effects of carbon and nitrogen sources, pH, temperature and inoculum density for cellulose production by *Gluconacetobacter* sp. RV28, which is isolated from rotten pomegranate. They used maltose, mannitol, mannose, sucrose, lactose, glucose and fructose as carbon source at 2 per cent (w/v) concentration in standard Hestrin-Schramm medium. They found that *Gluconacetobacter* sp. RV28 produced 4.7 g/L of cellulose at optimum growth conditions of temperature (30°C), pH (6.0), sucrose (2 per cent), peptone (0.5 per cent) and inoculum density (5 per cent).

Lin et al. (2014) investigated the production of BC from waste beer yeast (WBY) by *Ga. hansenii* CGMCC 3917 (reclassified as *K. hansenii*). They designed a two-step pre-treatment and found that ultrasonication combined with mild acid hydrolysis is an effective pre-treatment to improve the reducing sugar yield of WBY and BC. WBY hydrolyzates containing 3 per cent of final sugar concentration gave the highest BC yield (7.02 g/L), almost six times higher than that produced from untreated WBY (1.21 g/L). Furthermore, the properties and microstructure of BC produced by WBY hydrolyzates were as good as those obtained from the conventional chemical media (Lin et al. 2014).

The current methods of BC production are the static culture, which is the most commonly used method (El-Saied et al. 2008, Castro et al. 2011) and agitated culture (Yan et al. 2008). The agitated culture of *Ga. hansenii* (presently *K. xylinus*) strain produces cellulose in the form of characteristic spheres (Fig. 4). The mechanism of cellulose production of agitated cultures has not been fully resolved. The cells in fresh liquid medium attach to the surface of air bubbles and reproduce, forming a cellulose ribbon resulting in a compact structure in the form of spheres (Mohite and Patil 2014a).

Figure 4: BC spheres produced under the agitated culture condition (Mohite and Patil 2014a).

Large scale, semi-continuous and continuous fermentation modes are expected to meet the commercial demand in BC production. In all cases, the main objective is to achieve maximum production of BC with optimum form and suitable properties for the intended applications (Lin et al. 2013).

The results of a study by Çakar et al. (2014a) show that BC can be produced in high quantities by *Ga. xylinus* (presently *K. xylinus*) in molasses with a static semi-continuous process than a batch process. Çakar et al. (2014a) declared that molasses can be used for industrial-scale BC productions as a low-cost medium in static culture with semi-continuous mode of operation.

Concentration of carbon, nitrogen source and other minor nutrients influence the growth of the microorganisms as well as the production of microbial cellulose. BC productions from different microorganisms using different carbon sources are summarized in Table 1.

Different types of bioreactors, such as stirred-tank, air-lift, rotation disk, rotary biofilm contactor, aerosol bioreactor, membrane bioreactor, horizontal lift reactor (HoLiR) and modified air-lift or gas-lift bioreactors have been tested for BC production (Fig. 5) (Sani and Dahman 2010, Shi et al. 2014). Stirred-tank and the air-lift reactors are the most commonly used bioreactors (Sani and Dahman 2010). Rotary disc reactors can achieve 86.78 per cent higher yield of cellulose than the traditional static fermentation (Shi et al. 2014).

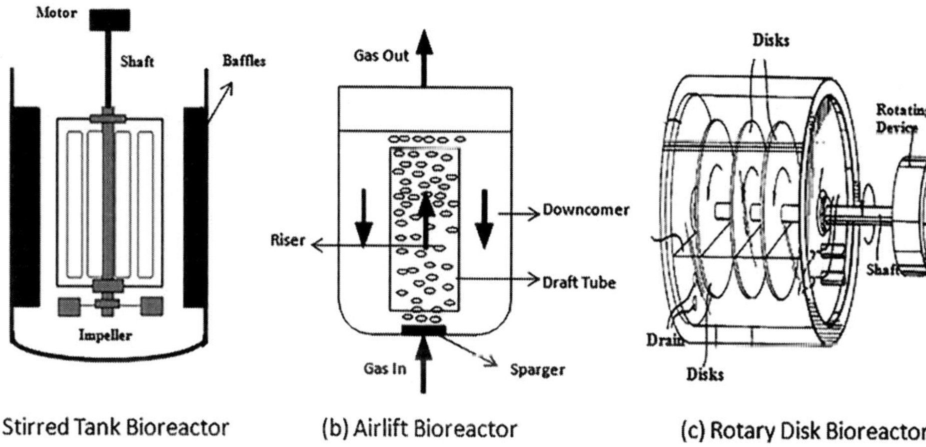

Figure 5: Different types of bioreactors used to produce BC (Sani and Dahman 2010).

Characterization of Cellulose

BC is a water-insoluble homopolysaccharide (Dayal et al. 2013, Lin et al. 2013). It has exceptional properties such as high purity, free of lignin and hemi-cellulose, high crystallinity, water absorption, tensile strength, high degree of polymerization and high mechanical strength. Because of these properties, using BC in the manufacture of cellulose-based products may be a better choice (Dayal et al. 2013, Mohite and Patil 2014a, Çakar et al. 2014b).

Dimensions of BC fibrils are 3-8 nm thick and 50-80 nm wide. Due to these dimensions, BC fibrils are about 100 times thinner than plant cellulose. Nano-fibrils of BC have a very high surface area and highly porous matrix (Khajavi et al. 2011, Ul-Islam et al. 2012, Mohite and Patil 2014b). Nevertheless, these properties change according to the activity of the producing organism, the composition of culture media, the cultivation technique and the variation in carbon sources used for its production (Sani and Dahman 2010, Ul-Islam et al. 2012).

Microfibrillar structure of BC is responsible for most of its properties, such as a high crystallinity index, high tensile strength and higher degree of polymerization (Mohite and Patil 2014a).

Rozenberga et al. (2016) produced BC samples using *K. rhaeticus* and *K. hansenii*. They obtained nanopaper from BC and determined its nanofiber structure and mechanical properties. They found that water retention values of BC were 86–95 per cent and the

Table 1: Studies of BC production by using different strains of AAB

Strain	*Carbon source*	*BC production (g/L)*	*Condition (significance)*	*References*
Ga. swingsii sp.	Pine apple peel juice	2.8	In static culture at 28°C for 13 days	Castro et al. 2011
	Hestrin and Schramm's medium	2.1		
Ga. xylinus ATCC 10245	Acid treated molasses solution (17 per cent w/v) and corn steep liquor (8 per cent v/v)	4.7	In static culture	El-Saied et al. 2008
Ga. hansenii PJK	Glucose	245.41 (wet weight)	Fed-batch cultivation in static conditions	Shezad et al. 2009
Ga. hansenii PJK (KCTC 10505BP)	Waste from beer fermentation broth	13.95	In static conditions	Ha et al. 2008
Acetobacter sp. 4B-2	Replacement of glucose in modified HS medium with sucrose as carbon source	11.98	In static conditions	Pourramezan et al. 2009
A. aceti subsp. *xylinus* ATCC 23 770	Wheat straw hydrolysates	15.4	In static conditions	Hong et al. 2011
A. xylinum BPR 2001	Maple syrup	1.53	Shaken flask	Zeng et al. 2011
A. xylinum subsp. *sucrofermentans* BPR2001	Corn steep liquor-fructose	8.7	Airlift reactor	Chao et al. 2001
		6.9	Stirred tank reactor	
Gluconacetobacter sp. RKY5	Hestrin and Schramm's medium	6.17	Rotary biofilm contactor	Kim et al. 2007
A. xylinum KJ1	Saccharified food wastes	5.8	Modified airlift-type bubble column bioreactor	Song et al. 2009
Ga. hansenii PJK (KCTC 10505BP)	Glucose	5.472	Surface modified reactors	Shah et al. 2010
A. xylinum KJ1	Saccharified food waste	7.37	Modified bubble column bioreactor	Li et al. 2011

**Acetobacter* and *Gluconacetobacter* spp. mentioned in the references reclassified as *Komagataeibacter*.

highest degree of polymerisation determined was 2540. Nanofibers displayed were 80–120 nm wide and 600–1200 nm long. They concluded that BC was highly crystalline (up to 15 per cent) and had strong nanofibers.

Applications: Cellulose in Industry

BC is a type of dietary fiber and classified as 'Generally Recognized as Safe' (GRAS) by the USA Food and Drug Administration in 1992 (Shi et al. 2014). BC can improve the rheology of foods, produce low-calorie food ingredients and produce low-cholesterol products (Shi et al. 2014). BC applications in the food industry are summarized in Table 2.

Table 2: Application of modified bacterial cellulose (BC) and its composites in food industry (Esa et al. 2014)

Materials	*Function*	*Types of food*
BC/nisin	Antimicrobial food packaging	Meat
BC/polylysine	Biodegradable food packaging	Sausage
BC	Emulsifier	Surimi
Carboxymethycellulose	Regulate dough rheology	Flour dough
Hydroxypropyl methyl cellulose	Texture enhancer	Whipped cream
Methyl cellulose	Enhancing shelf life	Egg
Methyl cellulose	Enhance bioavailability	Vitamin C

Packaging with bio-based materials is an important method for the protection and preservation of foods. Nowadays bio-based materials in the packaging industry are preferred due to their biodegradable and high water-resistant performances. Although BC has no antimicrobial and antioxidant properties, it has been identified as one of the suitable materials (Shah et al. 2013, Esa et al. 2014).

Polymer composites are synthesized through numerous methods depending on the nature of the polymer and the combining partner. The synthesis strategy also varies with the required applications. In general, there are two basic composite synthesis approaches, *in situ* and *ex situ*, which are shown in Fig. 6a, b. *In situ* method utilizes the addition of reinforcement material to the polymer during its synthesis, which then becomes part of the polymer structure. In *ex situ* synthesis, the polymer matrix is impregnated with reinforcement materials to produce composites. These techniques can further be improved based on the nature of the combining agents and the development strategy. BC is a biopolymer that can be subjected to all of the aforementioned techniques for polymer synthesis (Shah et al. 2013).

Levan

Chemical Structure and Properties of Levan

Levan is a homopolymer of fructose derived from plants and microorganisms (Ahmed et al. 2014, Srikanth et al. 2015a, b, Semjonovs et al. 2016). Microbial levan, known as microbial extracellular fructan, is more advantageous than plant-derived levan and is widely used in many different fields (Srikanth et al. 2015a, b). This homopolysaccharide is one of two main types of fructans. Levan composed of five-member fructofuranosyl residues is linked by β-2,6 linkage and β-2,1 linkage in core and branch chain, respectively (Fig. 7) (Arvidson et al. 2006, Ahmed et al. 2014, Srikanth et al. 2015a, b). In contrast to inulin, which is the other essential type of fructan, the main chain is formed by β-2,1 linkages (Arvidson et al. 2006).

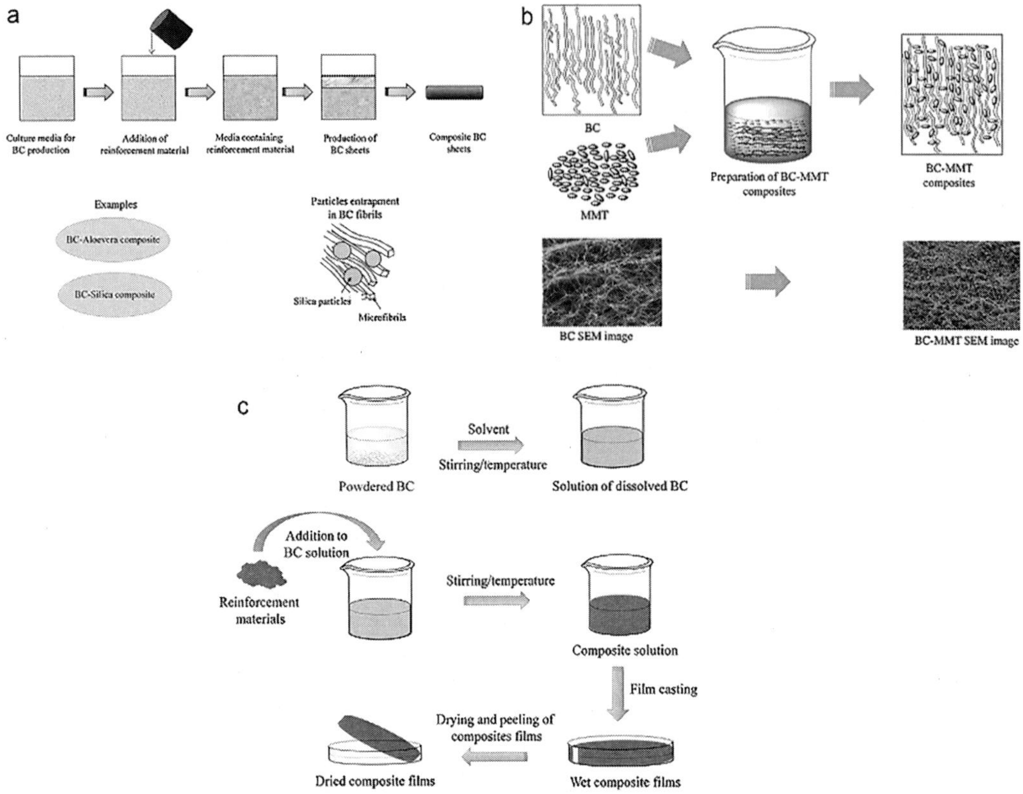

Figure 6: (a) Schematic representation of BC composites synthesized through an *in situ* synthetic strategy. Particles entrapped between growing BC fibrils are visible in the figure. (b) Schematic representation of BC composites synthesized through an *ex situ* synthetic strategy. The example illustrates the penetration of particles in the BC matrix. (c) Schematic representation of BC composites synthesized from dissolved BC solutions. The composite solution was casted to prepare BC films (Shah et al. 2013).

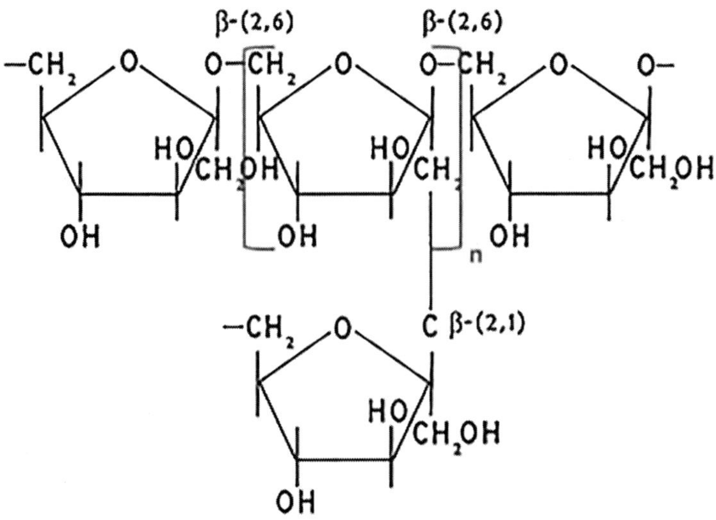

Figure 7: Levan structure showing the fructane backbone along with β-(2,6) linkages and β-(2,1) branching (Srikanth et al. 2015a).

Levans contain more than 11 monosaccharide units (Molinari and Boiardi 2013). Bacterial levans are much larger than those produced by plants. They have multiple branches and their molecular weights range from 2 to 100 million Da. Molecular weight and degree of branching of levan depends on the microorganisms, the reaction conditions and sucrose concentration (Arvidson et al. 2006, Wu et al. 2013).

Plant fructans with molecular weights ranging from 2,000 to 33,000 Da are available. Many microorganisms, like *A. xylinum,* also produce extracellular levan of high molecular weight when grown on sucrose (Arvidson et al. 2006, Rairakhwada 2007).

Levan-producing AAB

Microbial levan is produced by trans-fructosylation reaction of levansucrase (beta-2,6 fructose; d-glucose-fructosyl transferase EC2.4.1.10) secreted by organisms from various genera, including *Zymomonas, Streptococcus, Xanthomonas, Bacillus, Halomonas, Sachharomyces, Erwinia* and *Pseudomonas* (Srikanth et al. 2015b, Semjonovs et al. 2016). Also, AAB, particularly certain strains of *Komagataeibacter,* can produce extracellular fructans from sucrose-based media (Molinari and Boiardi 2013, Semjonovs et al. 2016). Fructans are isolated from the fructan-overproducing AAB strains *Gluconobacter cerinus, Gluconobacter frateurii, Kozakia baliensis* and *Neoasaia chiangmaiensis* (Jakob et al. 2012, Jakob et al. 2013).

Levansucrase, known as sucrose 6-fructosyltransferase carries out levan biosynthesis (Hernandez et al. 1999). Gram-negative bacteria, such as AAB, produce levansucrase by a signal peptide-independent pathway (Hernandez et al. 1999, Srikanth et al. 2015a). Levan is synthesized in five stages (Fig. 8). In the first stage, levansucrase is synthesized in the cell and accumulated in the periplasmic space where it adopts its final confirmation (Stage 2). In stage 3, levansucrase is excreted from the cell into the surrounding environment either by the cleavage of signal peptide followed by protein folding or by the mechanism initiated through a signal peptide pathway itself. In the fourth stage, the synthesis of levan starts by active levansucrase enzyme catalysing transfrucosylation reactions on substrates. In stage 5, fructose subunits by levansucrase are added on the growing levan polysaccharide chain so that the synthesis of levan is completed (Srikanth et al. 2015a).

Production Conditions

Figure 9 presents a levan production scheme with a different production strategy. Microbial exopolysaccharides, such as levan, are mostly produced using submerged fermentation. After several days of growth at specific pH, temperature, aeration and agitation, the culture medium is centrifuged to remove the bacterial cells. The levan is precipitated with different strategies such as alcohol precipitation, isoelectric point precipitation, precipitation by salting out or flocculation by polyelectrolytes. Then the collected precipitate is either freeze-dried or vacuum-dried (Srikanth et al. 2015a).

Cultural conditions, such as temperature, pH, aeration and agitation required for the growth varies for different microorganisms and thus the production of levan is unique for each and every microorganism. Hence, for increased yields, these conditions must be established and optimized for each organism. The production of levan can also be done *in vitro* using the enzyme levansucrase, which shows specificity for sucrose (Han 1990, Abdel-Fattah et al. 2005). The concentration of carbon, nitrogen and other minor nutrients significantly affects the growth of the microorganism and the production of microbial levan (Srikanth et al. 2015a). A major problem in levan production is that the activity of levansucrases is strongly inhibited by glucose generated from sucrose hydrolysis (Molinari and Baiardi 2013).

In recent years, researchers are focusing on the production strategies, such as different organisms and production conditions specific for each organism for increased yields.

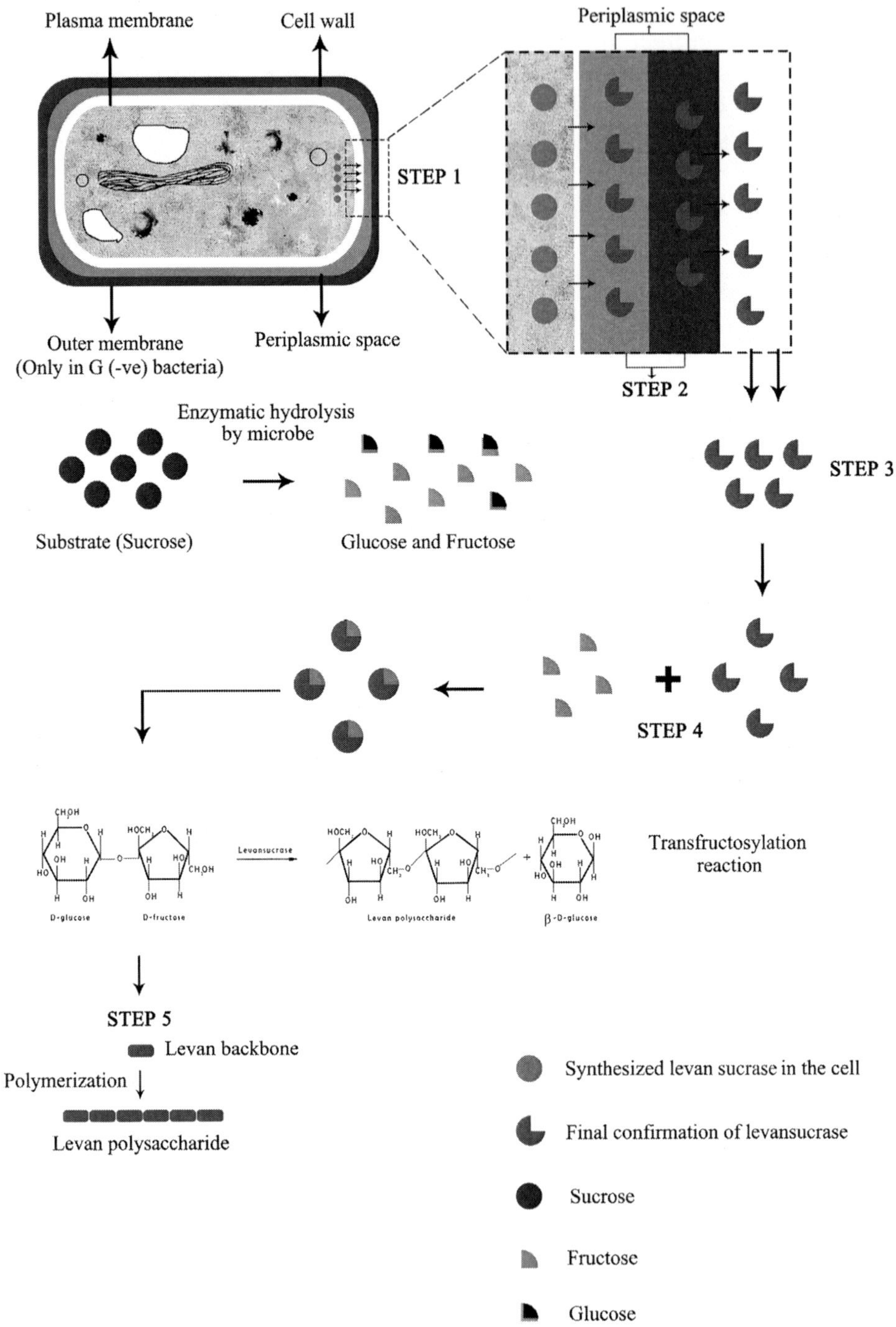

Figure 8: Schematic representation of biosynthesis of levan from sucrose (Srikanth et al. 2015a).

Idogawa et al. (2014) investigated the synthesis of levan by *Ga. diazotrophicus* Pal5 on a solid medium containing a high concentration of phosphate. They found that synthesis of levan by *Ga. diazotrophicus* Pal5 increased on this medium. Therefore, the production medium with a high concentration of phosphate may be important for industrial production of levan.

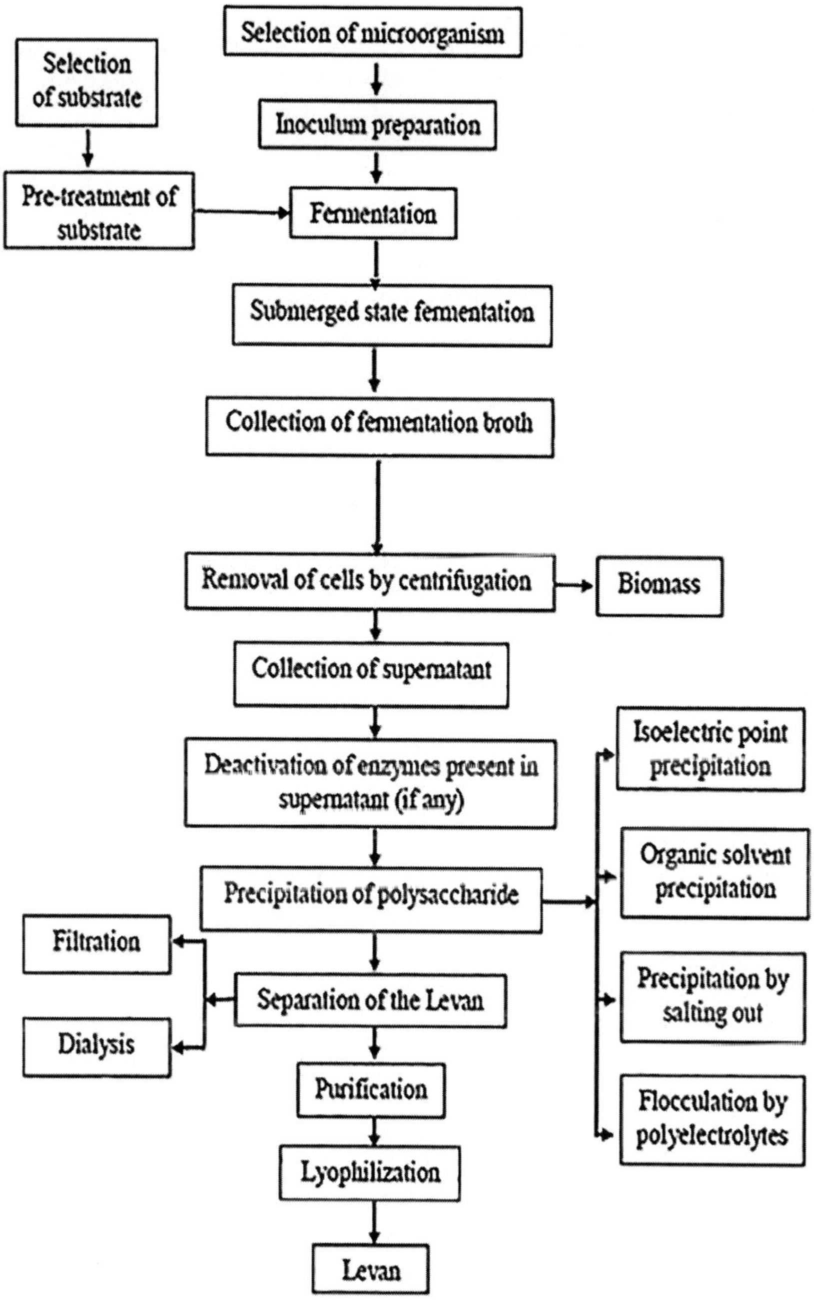

Figure 9: Schematic representation of microbial levan production (Srikanth et al. 2015a).

In a study by Jakob et al. (2012), 21 AAB strains of the genera *Acetobacter*, *Gluconobacter*, *Gluconacetobacter*, *Neoasaia* and *Kozakia* were screened for their ability to produce polysaccharides on sucrose-containing medium. Total EPS amounts by *G. frateurii* TMW 2.767, *G. cerinus* DSM 9533, *Neoasaia chiangmaiensis* NBRC 101099 and *Kozakia baliensis* DSM 14400 were 59.5, 31.5, 36.5 and 39 g, respectively. The HPLC analysis showed that the isolated exopolysaccharide was a homopolysaccharide consisting of fructose (fructan, levan).

Molinari and Boiardi (2013) investigated the production of levans, gluconic acid and keto gluconic acids in cultures with different sucrose concentration and nitrogen sources. In their study, levan production was observed in all sucrose concentrations, varying from 8 per cent to 25 per cent. Maximum levan production was 24.70 g/L in cultures under BNF (biological N_2-fixation) with 100 g/L of sucrose. However, the highest levan production was 14 g/L in cultures under NoBFN with 80 g/L of sucrose. Sugar conversion into levan under BNF and NoBNF in cultures with 20 g/L of sucrose were 17 per cent and 10 per cent, respectively. In cultures using yeast extract and tryptone as complex nitrogen sources, approximately 15 per cent sugar conversion into levan was determined.

In a study by Semjonovs et al. (2016), the productivity (Qp) and yield (Yp/s) of fructan (levan) production is determined for different cultivation conditions. Qp and Yp/s values ranged from 0.774 to 1.244 g/Lh and from 0.181 to 0.436 g/g, respectively. It was concluded that *G. nephelii* P1464 can be used as a promising strain for commercial production of levan.

Characterization of Levan

After fermentation, the determination of physical and chemical properties of the product is a very important step. Chemical and physical properties, molecular mass, molecular structure, morphology and mechanical, thermal and few other chemical properties are determined by characterization techniques (Srikanth et al. 2015a).

Dry levan powder is white and nonhydroscopic (Kang et al. 2009) and is a carbohydrate polymer, soluble both in water and oil due to its β-2,6 linkage (Srikanth et al. 2015a). However, it is insoluble in many organic solvents, such as ethanol, methanol, acetone, isopropanol, n-propanol, methylethylketone, toluene, acetic anhydride and furfuryl alcohol (Srikanth et al. 2015a).

Molecular weight of levan is highly dependent on the microorganism and the fermentation parameters used for its production. Molecular mass estimations showed that fructose polymer synthesized by *G. nephelii* P1464 has a relatively low molecular weight and a sizeable polydispersity when compared to other AAB (Table 3). Molecular weight can influence the physiological activity of polymers, for example, fructans with lower molecular weights can be more effective prebiotics (Semjonovs et al. 2016).

Table 3: Average molecular weight and rheological indexes of exopolysaccharide (levan) synthesized by *Gluconobacter nephelii* P1464 (Semjonovs et al. 2016)

Characteristics	*A numeric value ± SD*
Number average molecular weight, Mn	51,985 ± 5480.7
Weight average molecular weight, Mw (kDa)	1122.939 ± 153.453
Polydispersity, Mw/Mn	21.57 ± 1.60
Intrinsic viscosity, η (cL g^{-1})	0.179 ± 0.004
Huggins' constant	0.582 ± 0.108

Determination of molecular structure of the obtained sample can be done with Fourier Transform Infrared Spectroscopy (FTIR). Characterization of levan by FTIR shows hydroxyl stretching vibration, C–H stretching vibration, carbonyl C=O stretching vibration, glycosidic bond (C–O–C) stretching of the polysaccharide and the presence of the furanoid ring of the sugar units (Srikanth et al. 2015a) as well as the specific 'fingerprint' absorption of the major chemical groups (Semjonovs et al. 2015).

Levan has many advantageous properties, such as low viscosity, high solubility in water and oil, compatibility with salts and surfactants, film formation, stability to heat (25-70°C), acid and alkali, high holding capacity for water and chemicals and good biocompatibility (Küçükaşik et al. 2011, Bekers et al. 2005).

Applications: Levan in Industry

Levan is an exopolysaccharide with a wide variety of applications. It can be used in medicine as plasma substitute, drug activity prolongator, radio-protector, antitumor, hypo-cholesterol, immune modulator and antiinflammatory (Oliveira et al. 2007, Esawya et al. 2013). Di-fructofuranoses, fructose and fructooligosaccharides are formed by acid hydrolysis of levan in the stomach. Therefore, levan is used in foods as a source of these components. These components have prominent prebiotic effects and hypocholesterolaemic effects (Oliveira et al. 2007, Srikanth et al. 2015b).

Levan polysaccharides have the potential for applications in the food industry as emulsifiers, stabilizers, coloring and flavoring agents, fat substitute and coating materials (Han 1990, Molinari and Boiardi 2013, Srikanth et al. 2015a, b).

Levan has a benign nature and contains higher number of hydroxyl and reducing keto sugar units; therefore, levan is an attractive and promising biopolymer for nanotechnology (Ahmed et al. 2014).

Acetan

Chemical Structure and Properties of Acetan

Acetan, also referred to as xylinan, is a microbial exopolysaccharide that contains glucose, mannose, glucuronic acid and rhamnose in a 4:1:1:1 molar ratio (Semino and Dankert 1993, Berth et al. 1996). It is an anionic heteropolysaccharide and the chemical structure of acetan consists of a cellulosic backbone substituted on alternate glucose residues with a charged pentasaccharide sidechain (Ojinnaka et al. 1996). The structure proposed for the repeating unit of acetan is shown in Fig. 10.

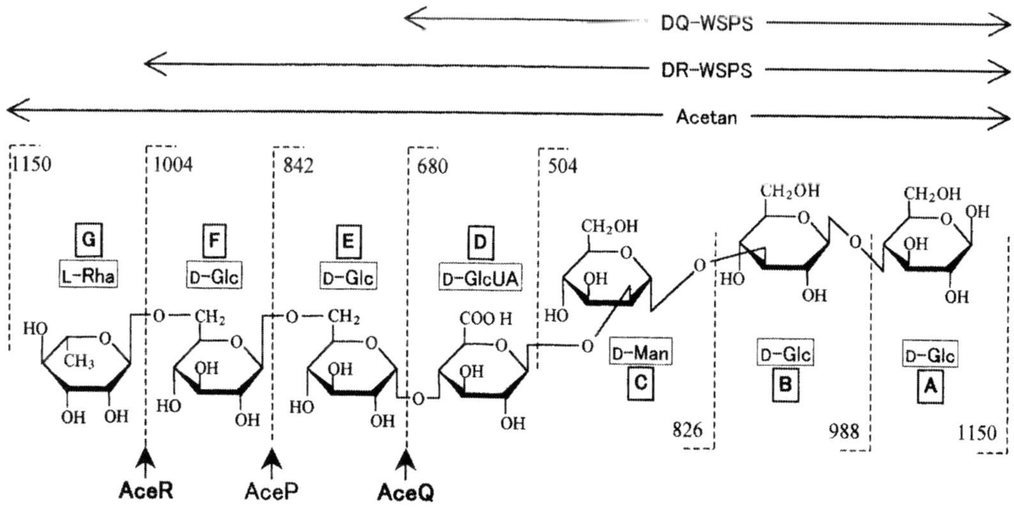

Figure 10: Structure proposed for the repeat unit of acetan (Ishida et al. 2002).

Acetan-producing AAB

Some of AAB, such as *Ga. entanii* and *A. xylinum* (= *A. xylinus* = *Ga. xylinus* = *K. xylinus*) produce acetan (MacCormick et al. 1993, Moonnangmee et al. 2002, Kornmann et al. 2003, Velasco-Bedran and Lopez-Isunza 2007).

Semino and Dankert (1993) investigated *in vitro* biosynthesis of acetan using electroporated *A. xylinum* cells as enzyme preparations. They showed and proposed schemes for the reactions in the biosynthesis of acetan and the other saccharides mentioned in their work.

Velasco-Bedran and Lopez-Isunza (2007) reported the metabolic pathway of *Ga. entanii*, particularly for the synthesis of bacterial cellulose and acetan, which is a polymer of industrial interest.

In Fig. 11, the routes to acetan and cellulose synthesis of *Ga. xylinus* are shown. Acetan, which is a water-soluble polysaccharide, is synthesized from glucose 6P using rhamnose 6 phosphate and glucuronic acid phosphate, with mannose 1 phosphate coming from F6P (Velasco-Bedran and Lopez-Isunza 2007).

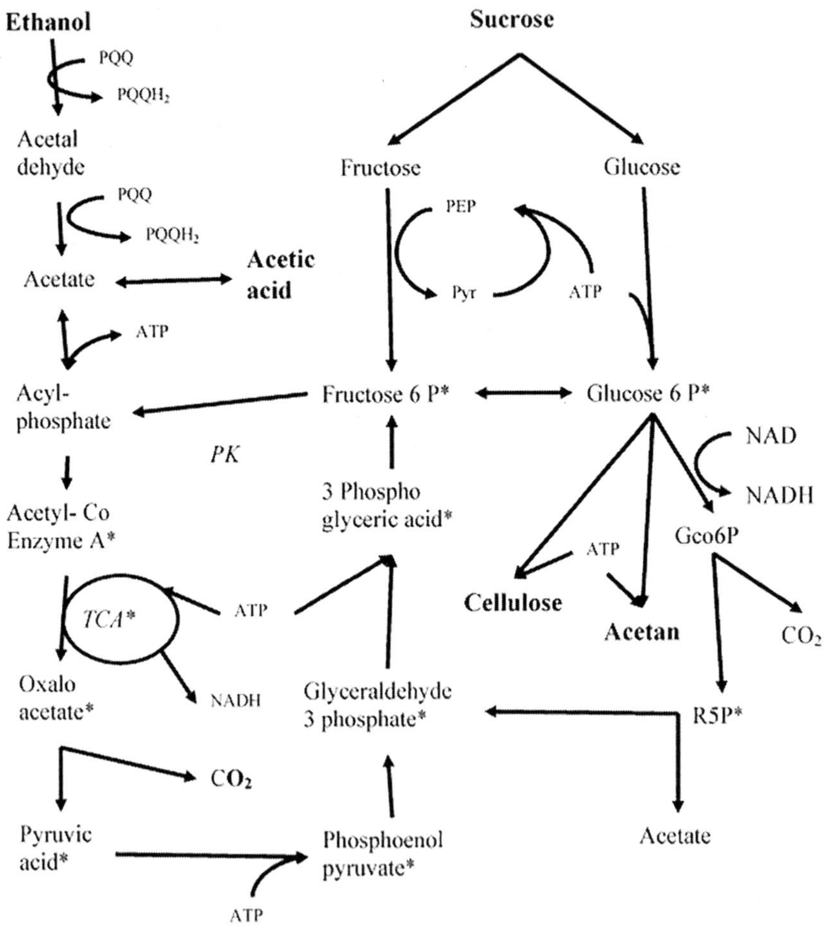

Figure 11: Schematic of the structured metabolism of *Gluconacetobacter xylinus*, showing the links between the ethanol and glucose/fructose catabolisms with the central metabolic pathways. TCA* stands for the glyoxilate shunt of the tricarboxylic acid cycle. Gco6P stands for gluconic acid 6-phosphate and R5P, for ribose 5 phosphate (Velasco-Bedran and Lopez-Isunza 2007).

Production Conditions

Acetan is produced from different carbon sources, such as sucrose, glucose and ethanol. *A. xylinum* BPR2001 produces the water-soluble polysaccharide acetan, in corn steep liquor fructose medium (Ishida et al. 2003).

Velasco-Bedran and Lopez-Isunza (2007) investigated the production of acetan with different media compositions in the continuous culture by *Ga. entanii.* They found that acetan concentrations were 4.63, 2.3, 3.13 and 2.1 g/L, on standard (acetate, glucose and ethanol) media, at 0.08, 0.078, 0.06 and 0.02 dilution rates (h^{-1}), respectively. They also found that the highest concentration of acetan was 5.4 g/L in a growth medium containing sucrose at diluton rate of 0.028 h^{-1}.

Characterization and Applications of Acetan

The polysaccharides after fermentation were precipitated with two volumes of ethanol and the acetan polysaccharide was purified by selective precipitation with hexadecyl trimethyl ammonium bromide (MacCormick et al. 1993, Ojinnaka et al. 1996).

Acetan, which consists of a structure similar to xanthan, is widely used in various industries as an emulsifier, viscosifier, etc. (Ishida et al. 2002).

Other Exopolysaccharides

AAB, such as *Ga. xylinus* (presently *K. xylinus*), has the ability to produce different types of water-soluble exopolysaccharides (EPS): free EPS and hard to extract EPS (HE-EPS) (Fang and Catchmark 2014, 2015) and a variety of polymers, such as neutral mannan and gluconacetan (MacCormick et al. 1993, Chandrasekaran et al. 2003, Kornmann et al. 2004). Free and HE-EPS were purified using the process shown in Fig. 12.

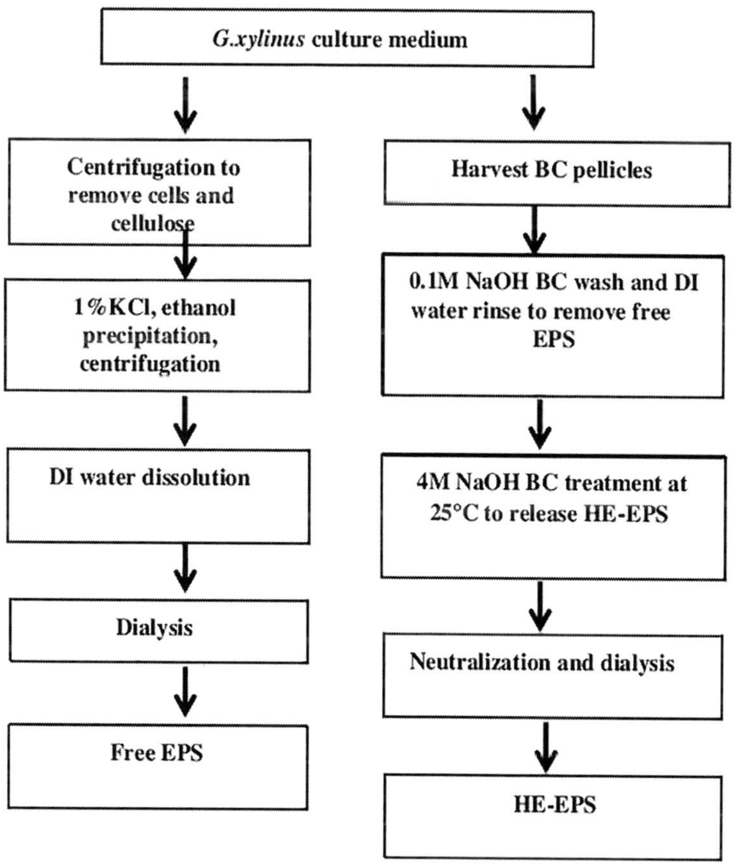

Figure 12: Process for isolation of free EPS and HE-EPS from the cultivation medium (Fang and Catchmark 2014).

The yields of exopolysaccharides were dependent on both the carbon source and the *Gluconacetobacter* strain. The composition of the free EPS was also influenced by the carbon substrate (Fang and Catchmark 2015).

One of the important water-soluble polymers produced by AAB has been reported to be the neutral mannan polymer (MacCormick et al. 1993, Fang and Catchmark 2015). MacCormick et al. (1993) reported that *A. xylinum* NRRL B42, which is the wild type strain and *A. xylinum* CR1/4, the mutant type strain, produced mannan in shake flask cultures. Yields (g polysaccharide g^{-1} cell dry weight) of mannan by *A. xylinum* NRRL B42 and *A. xylinum* CR1/4 were found to be 0.3 and 0.9, respectively.

ABB, like *Ga. xylinus*, produces gluconacetan, also referred to as konjac mannan (Chandrasekaran et al. 2003), on the medium containing ethanol (Kornmann et al. 2004). The metabolic states of *Ga. xylinus* are shown in Fig. 13. In the first metabolic state, ethanol was oxidized to acetate. In the second metabolic state, oxidation of acetate resulted in biomass production. In the third metabolic state, gluconacetan was produced by a concentrated fructose solution (Kornmann et al. 2004).

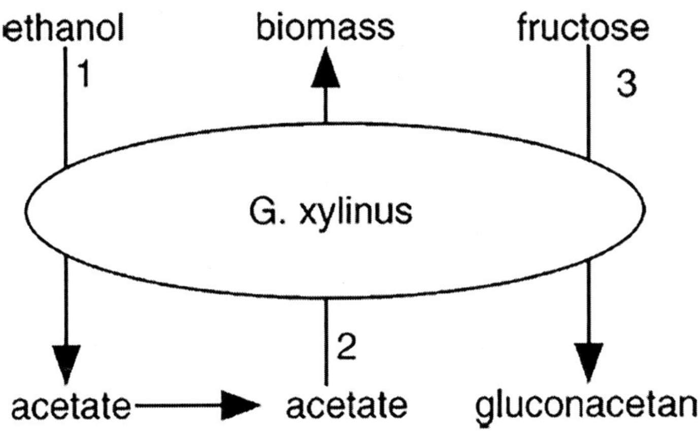

Figure 13: Metabolic states for biomass and gluconacetan production (Kornmann et al. 2004).

Kornmann et al. (2004) investigated the production of biomass and gluconacetan on ethanol and fructose, respectively in batch and fed-batch fermentation of *Ga. xylinus*. Productivity of gluconacetan in batch and fed-batch processes were 1.8×10^{-3} [C-mol/C-mol substrates/h] and 2.9×10^{-3} [C-mol/C-mol substrates/h], respectively.

Gluconacetan is a food-grade EPS and shows industrial interest as it can be used as a texturing agent in food products (Kornmann et al. 2004). Polysaccharides, such as acetan and glucomannan, are good thickeners that do not form gels under normal conditions (Chandrasekaran et al. 2003).

Genetic modifications for ABB by adaptation and genetic engineering were also applied to improve their properties, such as productivity and heat resistance (Saichana et al. 2015).

A. tropicalis SKU1100 produces a pellice-forming capsular polysaccharide, made of galactose, glucose and rhamnose as reported by Deeraksa et al. (2006). They cloned the galE gene from *A. tropicalis* SKU1100 by PCR. They found that the ΔgalE mutant of *A. tropicalis* SKU1100 secreted an EPS, but did not produce a capsular polysaccharide and the molecular size of the purified EPS was found to be more than 400 kDa (Deeraksa et al. 2006).

Conclusion

Microbial polysaccharides are used in many industrial areas and it is no doubt that the use of these microbial polysaccharides will increase in the future. Microbial polysaccharides have some advantages over polysaccharides from other sources. Their production is not limited to the production fields and seasonal variations. They have controlled physicochemical properties and production processes and, hence, the production wastes are more balanced. AAB produce different polysaccharides and the most well-known polysaccharide produced by AAB is cellulose. The cellulose produced by plant sources that contain impurities, such as lignin and hemicellulose, can be replaced by pure cellulose produced by AAB. The development of genetic engineering tools also exhibits great potential for improvement of the properties of AAB suitable for industrial exopolysaccharide production. The most important disadvantage of microbial polysaccharides is the high production costs compared with the production costs of polysaccharides from other sources. Today, academics and engineers with different backgrounds are trying to optimize the production of microbial polysaccharides and working hard to lower the production costs. They are investigating agricultural wastes and industrial by-products as cost-effective carbon sources in EPS production. Optimization studies of microbial polysaccharide fermentations with low production costs will enable widespread use of these polysaccharides in the industry. It seems that there is a great potential for exploiting AAB in the production of microbial polysaccharides, which is used in biomedical, industrial, food and pharmaceutical fields.

Acknowledgements

We thank Mrs. Huriye Göksungur for editing the manuscript.

References

Abdel-Fattah, A.F., Mahmoud, D.A. and Esawy, M.A. (2005). Production of levan-sucrase from *Bacillus subtilis* NRC 33a and enzymic synthesis of levan and fructo-oligosaccharides. *Current Microbiology* **51:** 402-407.

Ahmad, N.H., Mustafa, S. and Man, Y.B.C. (2015). Microbial Polysaccharides and Their Modification Approaches: A Review. *International Journal of Food Properties* **18:** 332-347.

Ahmed, K.B.A., Kalla, D., Uppulurib, K.B. and Anbazhagan, V. (2014). Green synthesis of silver and gold nanoparticles employing levan, a biopolymer from *Acetobacter xylinum* NCIM 2526, as a reducing agent and capping agent. *Carbohydrate Polymers* **112:** 539-545.

Ali, I.A.I., Akakabe, Y., Moonmangmee, S., Deeraksa, A., Matsutani, M., Yakushi, T., Yamada, M. and Matsushita, K. (2011). Structural characterization of pellicle polysaccharides of *Acetobacter tropicalis* SKU1100 wild type and mutant strains. *Carbohydrate Polymers* **86:** 1000-1006.

Arvidson, S.A., Rinehart, B.T. and Gadala-Maria F. (2006). Concentration regimes of solutions of levan polysaccharide from *Bacillus* sp. *Carbohydrate Polymers* **65:** 144-149.

Bekers, M., Upite, D. and Kaminska, E. (2005). Stability of levan produced by *Zymomonas mobilis*. *Process Biochemistry* **40:** 1535-1539.

Berth, G., Dautzenberg, H., Christensen, B.E., Rother, G. and Smidsrsd, O. (1996). Physicochemical studies on xylinan (acetan) I. Characterization by gel permeation chromatography on sepharose cı-2b coupled with static light scattering and viscometry. *Biopolymers* **39:** 709-719.

Castro, C., Zuluaga, R., Putaux, J.L., Caro, G., Mondragon, I. and Ganan, P. (2011). Structural characterization of bacterial cellulose produced by *Gluconacetobacter swingsii* sp. from Colombian agroindustrial wastes. *Carbohydrate Polymers* **84:** 96-102.

Castro, C., Cordeiro, N., Faria, M., Zuluaga, R., Putaux, J.L., Filpponen, I., Velez, L., Rojas, O.J. and Ganán P. (2015). *In-situ* glyoxalization during biosynthesis of bacterial cellulose. *Carbohydrate Polymers* **126:** 32-39.

Chandrasekaran, R., Janaswamy, S. and Morris, J.V. (2003). Acetan: Glucomannan interactions—A molecular modeling study. *Carbohydrate Research* **338:** 2889-2898.

Chao, Y., Mitarai, M., Sugano, Y. and Shoda, M. (2001). Effect of addition of water-soluble polysaccharides on bacterial cellulose production in a 50-L airlift reactor. *Biotechnology Progress* **17:** 781-785.

Chawla, P.R., Bajaj, I.B., Survase, S.A. and Singhal, R.S. (2009). Microbial cellulose: Fermentative production and applications. *Food Technology and Biotechnology* **47:** 107-124.

Çakar, F., Özer, I., Aytekin, A.Ö. and Şahin, F. (2014a). Improvement production of bacterial cellulose by semi-continuous process in molasses medium. *Carbohydrate Polymers* **106:** 7-13.

Çakar, F., Katı, A., Özer, I., Demirbağ, D.D., Şahin, F. and Aytekin, A.Ö. (2014b). Newly developed medium and strategy for bacterial cellulose production. *Biochemical Engineering Journal* **92:** 35-40.

Dayal, M.S., Goswami, N., Sahai, A., Jain, V., Mathur, G. and Mathur, A. (2013). Effect of media components on cell growth and bacterial cellulose production from *Acetobacter aceti* MTCC 2623. *Carbohydrate Polymers* **94:** 12-16.

Deeraksa, A., Moonmangmee, S., Toyama, H., Adachi, O. and Matsushita, K. (2006). Conversion of capsular polysaccharide, involved in pellicle formation, to extracellular polysaccharide by *galE* deletion in *Acetobacter tropicalis*. *Bioscience, Biotechnology and Biochemistry* **70:** 2536-2539.

Donot, F., Fontana, A., Baccou, J.C. and Schorr-Galindo, S. (2012). Microbial exopolysaccharides: Main examples of synthesis, excretion, genetics and extraction. *Carbohydrate Polymers* **87:** 951-962.

El-Saied, H., El-Diwany, A.I., Basta, A.H., Atwa, N.A. and El-Ghwas, D.E. (2008). Production and characterization of economical bacterial cellulose. *BioResources* **3:** 1196-1217.

Esa, F., Tasirin, S.M. and Rahman, N.A. (2014). Overview of bacterial cellulose production and application. *Agriculture and Agricultural Science Procedia* **2:** 113-119.

Esawya, M.A., Amer, H., Gamal-Eldeenc, A.M., El Enshasy, H.A., Helmy, W.A., Abo-Zeid, M.A.M., Malek, R., Ahmed, E.F. and Awad, G.E.A. (2013). Scaling up, characterization of levan and its inhibitory role in carcinogenesis initiation stage. *Carbohydrate Polymers* **95:** 578-587.

Fang, L. and Catchmark, J.M. (2014). Characterization of water-soluble exopolysaccharides from *Gluconacetobacter xylinus* and their impacts on bacterial cellulose crystallization and ribbon assembly. *Cellulose* **21:** 3965-3978.

Fang, L. and Catchmark, J.M. (2015). Characterization of cellulose and other exopolysaccharides produced from *Gluconacetobacter* strains. *Carbohydrate Polymers* **115:** 663-669.

Ha, J., Shehzad, O., Khan, S., Lee, S., Park, J., Khan, T. and Park, J. (2008). Production of bacterial cellulose by a static cultivation using the waste from beer culture broth. *Korean Journal of Chemical Engineering* **25:** 812-815.

Han, Y.W. (1990). Microbial levan. *Advances in Applied Microbiology* **35:** 171-194.

Hernandez, L., Arrieta, J., Betancourt, L., Falcon, V., Madrazo, J., Coego, A. and Menendez, C. (1999). Levansucrase from *Acetobacter diazotrophicus* SRT4 is secreted via periplasm by a signal-peptide-dependent pathway. *Current Microbiology* **39:** 146-152.

Hong, F., Zhu, Y.X., Yang, G. and Yang, X.X. (2011). Wheat straw acid hydrolysate as a potential cost-effective feedstock for production of bacterial cellulose. *Journal of Chemical Technology and Biotechnology* **86:** 675-680.

Idogawa, N., Amamoto, R., Murata, K. and Kawai, S. (2014). Phosphate enhances levan production in the endophytic bacterium *Gluconacetobacter diazotrophicus* Pal5. *Bioengineered* **5:** 173-179.

Iguchi, M., Yamanaka, S. and Budhiono, A. (2000). Bacterial cellulose – a masterpiece of nature's arts. *Journal of Materials Science* **35:** 261-270.

Ishida, T., Sugano, Y. and Shoda, M. (2002). Novel glycosyltransferase genes involved in the acetan biosynthesis of *Acetobacter xylinum*. *Biochemical and Biophysical Research Communications* **295:** 230-235.

Ishida, T., Mitarai, M., Sugano, Y. and Shoda, M.A. (2003). Role of water-soluble polysaccharides in bacterial cellulose production. *Biotechnology and Bioengineering* **83:** 474-478.

Jakob, F., Steger, S. and Vogel, R.F. (2012). Influence of novel fructans produced by selected acetic acid bacteria on the volume and texture of wheat breads. *European Food Research and Technology* **234:** 493-499.

Jakob, F., Pfaff, A., Novoa-Carballal, R., Rübsam, H., Becker, T. and Vogel, R.F. (2013). Structural analysis of fructans produced by acetic acid bacteria reveals a relation to hydrocolloid function. *Carbohydrate Polymers* **92:** 1234-1242.

Jozala, A.F., de Lencastre-Novaes, L.C., Lopes, A.M., Santos-Ebinuma, V.D., Mazzola, P.G., Pessoa, A., Grotto, D., Gerenutti, M. and Chaud, M.V. (2016). Bacterial nanocellulose production and application: A 10-year overview. *Applied Microbiology and Biotechnology* **100:** 2063-2072.

Kang, S.A., Jang, K.H., Seo, J.W., Kim, K.H., Kim, Y.H., Rairakhwada, D., Seo, M.Y., Lee, J.O., Ha, S.D., Kim, C.H. and Rhee, S.K. (2009). Levan: Applications and perspectives. *In:* Rehm, B.H.A. (Ed.). *Microbial Production of Biopolymers and Polymer Precursors*. Caister Academic Press, Norfolk, pp. 145-162.

Khajavi, R., Esfahani, E.J., Sattari, M. and Univer, I.A. (2011). Crystalline structure of microbial cellulose compared with native and regenerated cellulose. *International Journal of Polymeric Materials* **60:** 1178-1192.

Kim, Y.J., Kim, J.N., Wee, Y.J., Park, D.H. and Ryu, H.W. (2007). Bacterial cellulose production by *Gluconacetobacter* sp. RKY5 in a rotary biofilm contactor. *Applied Biochemistry and Biotechnology* **137:** 529-537.

Klemm, D., Schumann, D., Udhardt, U. and Marsch, S. (2001). Bacterial synthesized cellulose—Artificial blood vessels for microsurgery. *Progress in Polymer Science* **26:** 1561-1603.

Kornmann, H., Duboc, P., Marison, I. and von Stockar, U. (2003). Influence of nutritional factors on the nature, yield, and composition of exopolysaccharides produced by *Gluconacetobacter xylinus* I-2281. *Applied and Environmental Microbiology* **69:** 6091-6098.

Kornmann, H., Valentinotti, S., Duboc, P., Marison, I. and von Stockar, U. (2004). Monitoring and control of *Gluconacetobacter xylinus* fed-batch cultures using in situ mid-IR spectroscopy. *Journal of Biotechnology* **113:** 231-245.

Kumar, A.S., Mody, K. and Jha, B. (2007). Bacterial exopolysaccharides—A perception. *Journal of Basic Microbiology* **47:** 103-117.

Kurosumi, A., Sasaki, C., Yamashita, Y. and Nakamura, Y. (2009). Utilization of various fruit juices as carbon source for production of bacterial cellulose by *Acetobacter xylinum* NBRC 13693. *Carbohydrate Polymers* **76:** 333-335.

Küçükaşik, F., Kazak, H., Güney, D., Finore, I., Poli, A., Yenigün, O., Nicolaus, B. and Öner E.T. (2011). Molasses as fermentation substrate for levan production by *Halomonas* sp. *Applied Biochemistry and Biotechnology* **89:** 1729-1740.

Li, H., Kim, S.J., Lee, Y.W., Kee, C. and Oh, I. (2011). Determination of the stoichiometry and critical oxygen tension in the production culture of bacterial cellulose using saccharified food wastes. *Korean Journal of Chemical Engineering* **28:** 2306-2311.

Lin, S.P., Calvar, I.L., Catchmark, J.M., Liu, J.R., Demirci, A. and Cheng, K.C. (2013). Biosynthesis, production and applications of bacterial cellulose. *Cellulose* **20:** 2191-2219.

Lin, D., Lopez-Sanche, P., Li, R. and Li, Z. (2014). Production of bacterial cellulose by *Gluconacetobacter hansenii* CGMCC 3917 using only waste beer yeast as nutrient source. *Bioresource Technology* **151:** 113-119.

Lin, S.P., Huang, Y.H., Hsu, K-D., Lai Y.J., Chen, Y.K. and Cheng, K.C. (2016). Isolation and identification of cellulose-producing strain *Komagataeibacter intermedius* from fermented fruit juice. *Carbohydrate Polymers* (in press).

MacCormick, C.A., Harris, J.E., Gunning, A.P. and Morris, V.J. (1993). Characterization of a variant of the polysaccharide acetan produced by a mutant of *Acetobacter xylinum* strain CR1/4. *Journal of Applied Bacteriology* **74:** 196-199.

Matsutani, M., Ito, K., Azuma, Y., Ogino, H., Shirai, M., Yakushi, T. and Matsushita, K. (2015). Adaptive mutation related to cellulose producibility in *Komagataeibacter medellinensis* (*Gluconacetobacter xylinus*) NBRC 3288. *Applied Microbiology and Biotechnology* **99:** 7229-7240.

Mohite, B.V. and Patil, S.V. (2014a). Physical, structural, mechanical and thermal characterization of bacterial cellulose by *G. hansenii* NCIM 2529. *Carbohydrate Polymers* **106:** 132-141.

Mohite, B.V. and Patil, S.V. (2014b). A novel biomaterial: bacterial cellulose and its new era applications. *Biotechnology and Applied Biochemistry* **61:** 101-110.

Molinari, M.L. and Boiardi, J.L. (2013). Levans production by *Gluconacetobacter diazotrophicus*. *Electronic Journal of Biotechnology* **16:** 3. Available from: http://dx.doi.org/10.2225/vol16-issue3-fulltext-9 (Accessed: July 2016).

Moonmangmee, S., Kawabata, K., Tanaka, S., Toyama, H., Adachi, O. and Matsushita, K. (2002). A novel polysaccharide involved in the pellicle formation of *Acetobacter aceti*. *Journal of Bioscience and Bioengineering* **93:**192-200.

Ojinnaka, C., Jay, A.J., Colquhoun, I.J., Brownsey, G.J., Morris, E.R. and Morris, V.J. (1996). Structure and conformation of acetan polysaccharide. *International Journal of Biological Macromolecules* **19:** 149-156.

Oliveira, M.R., Silva, R.S.S.F., Buzato, J.B. and Celligoi, M.A.P.C. (2007). Study of levan production by *Zymomonas mobilis* using regional low-cost carbohydrate source. *Biochemical Engineering Journal* **37:** 177-183.

Perumpuli, P.A.B.N., Watanabe, T. and Toyama, H. (2014). Pellicle of thermotolerant *Acetobacter pasteurianus* strains: Characterization of the polysaccharides and of the induction patterns. *Journal of Bioscience and Bioengineering* **118:** 134-138.

Pourramezan, G., Roayaei, A. and Qezelbash, Q. (2009). Optimization of culture conditions for bacterial cellulose production by *Acetobacter* sp. 4B-2. *Biotechnology* **8:** 150-154.

Rairakhwada, D., Pal, A.K., Bhathena, Z.P., Sahu, N.P., Jha, A. and Mukherjee, S.C. (2007). Dietary microbial levan enhances cellular non-specific immunity and survival of common carp (*Cyprinus carpio*) juveniles. *Fish & Shellfish Immunology* **22:** 477–486.

Rangaswamy, B.E., Vanitha, K.P. and Hungund, B.S. (2015). Microbial cellulose production from bacteria isolated from rotten fruit. *International Journal of Polymer Science* Article ID **280784:** 1-8.

Roca, C., Alves, V.D., Freitas, F. and Reis, M.A.M. (2015). Exopolysaccharides enriched in rare sugars: Bacterial sources, production, and applications. *Frontiers in Microbiology* **6:** 288.

Rozenberga, L., Skute, M., Belkova, L., Sable, I., Vikelea, L., Semjonovs, P., Saka, M., Ruklisha, M. and Paegle, L. (2016). Characterisation of films and nanopaper obtained from cellulose synthesised by acetic acid bacteria. *Carbohydrate Polymers* **144:** 33-40.

Saichana, N., Matsushita, K., Adachi, O., Frébort, I. and Frebortova, J. (2015). Acetic acid bacteria: A group of bacteria with versatile biotechnological applications. *Biotechnology Advances* **33:** 1260-1271.

Sani, A. and Dahman, Y. (2010). Improvements in the production of bacterial synthesized biocellulose nanofibres using different culture methods. *Journal of Chemical Technology and Biotechnology* **85:** 151-164.

Semino., C.E. and Dankert, M.A. (1993). *In vitvo* biosynthesis of acetan using electroporated *Acetobacter xylinum* cells as enzyme preparations. *Journal of General Microbiology* **139:** 2745-2756.

Semjonovs, P., Shakirova, L., Treimane, R., Shvirksts, K., Auzina, L., Cleenwerck, I. and Zikmanis, P. (2016). Production of extracellular fructans by *Gluconobacter nephelii* P1464. *Letters in Applied Microbiology* **62:** 145-152.

Serrato, R.V., Meneses, C.H.S.G., Vidal, M.S., Santana-Filho, A.P., Iacomini, M., Sassaki, G.L. and Baldani, J.I. (2013). Structural studies of an exopolysaccharide produced by *Gluconacetobacter diazotrophicus* Pal5. *Carbohydrate Polymers* **98:** 1153-1159.

Shah, N., Ha, J.H. and Park, J.K. (2010). Effect of reactor surface on production of bacterial cellulose and water soluble oligosaccharides by *Gluconacetobacter hansenii* PJK. *Biotechnology and Bioprocess Engineering* **15:** 110-118.

Shah, N. Ul-Islam, M., Khattak, W.A. and Park, J.K. (2013). Overview of bacterial cellulose composites: A multipurpose advanced material. *Carbohydrate Polymers* **98:** 1585-1598.

Shezad, O., Khan, S., Khan, T. and Park, J.K. (2009). Production of bacterial cellulose in static conditions by a simple fed-batch cultivation strategy. *Korean Journal of Chemical Engeneering* **26:** 1689-1692.

Shi, Z., Zhang, Y., Phillips, G.O. and Yang, G. (2014). Utilization of bacterial cellulose in food. *Food Hydrocolloids* **35:** 539-545.

Song, H.J., Li, H., Seo, J.H., Kim, M.J. and Kim, S.J. (2009). Pilot-scale production of bacterial cellulose by a spherical type bubble column bioreactor using saccharified food wastes. *Korean Journal of Chemical Engineering* **26:** 141-146.

Srikanth, R., Reddy, C.H.S.S.S., Siddartha, G., Ramaiah, M.J. and Uppuluri, K.B. (2015a). Review on production, characterization and applications of microbial levan. *Carbohydrate Polymers* **120:** 102-114.

Srikanth, R., Siddartha, G., Reddy, C.H.S.S.S., Harish, B.S., Ramaiah, M.J. and Uppuluri, K.B. (2015b). Antioxidant and anti-inflammatory levan produced from *Acetobacter xylinum* NCIM2526 and its statistical optimization. *Carbohydrate Polymers* **123:** 8-16.

Sulaeva, I., Henniges, U., Rosenau, T. and Potthast, A. (2015). Bacterial cellulose as a material for wound treatment: Properties and modifications. A review. *Biotechnology Advances* **33:** 1547-1571.

Ul-Islam, M., Khana, T. and Park, J.K. (2012). Water holding and release properties of bacterial cellulose obtained by *in situ* and *ex situ* modification. *Carbohydrate Polymers* **88:** 596-603.

Valera, M.J., Torija, M.J., Mas, A. and Mateo, E. (2015). Cellulose production and cellulose synthase gene detection in acetic acid bacteria. *Applied Genetics and Molecular Biotechnology* **99:** 1349-1361.

Velasco-Bedran, H. and Lopez-Isunza, F. (2007). The unified metabolism of *Gluconacetobacter entanii* in continuous and batch processes. *Process Biochemistry* **42:** 1180-1190.

Wu, F.C., Chou, S.Z. and Shih, I.L. (2013). Factors affecting the production and molecular weight of levan of *Bacillus subtilis* natto in batch and fed-batch culture in fermenter. *Journal of the Taiwan Institute of Chemical Engineers* **44:** 846-853.

Yan, Z., Chen, S., Wang, H., Wang, B. and Jiang, J. (2008). Biosynthesis of bacterial cellulose/multi-walled carbon nanotubes in agitated culture. *Carbohydrate Polymers* **74:** 659-665.

Yu, X. and Atalla, R.H. (1996). Production preliminary communication of cellulose II by *Acetobacter xylinum* in the presence of 2, 6-dichlorobenzonitrile. *International Journal of Biological Macromolecules* **19:** 145-146.

Zeng, X., Small, D.P. and Wan, W. (2011). Statistical optimization of culture conditions for bacterial cellulose production by *Acetobacter xylinum* BPR 2001 from maple syrup. *Carbohydrate Polymers* **85:** 506-513.

6

Improvements of Acetic Acid Bacterial Strains: Thermotolerant Properties of Acetic Acid Bacteria and Genetic Modification for Strain Development

Natsaran Saichana[1,*]

[1] Mae Fah Luang University, School of Science, Chiang Rai – 57100, Thailand

Introduction

Acetic acid bacteria (AAB) are natural oxidizers for ethanol, sugars, sugar alcohols and sugar acids (Matsushita et al. 1994, Adachi et al. 2007). They rapidly oxidize the substrates, accumulating the oxidized products in the culture media. This property is preferable for industrial dowstream processing as the products can be separated easily from the bacterial cells. This characteristic is suitable for biotransformation of various kinds of substrates. However, in order to apply some valuable AAB into industries, the selected strains need to be improved to meet the requirements of industrial criteria. Thermotolerant ability of some AAB is considered as a preferable property for industrial application as thermotolerant strains are more resistant to environmental changes during the course of fermentation. AAB are mesophilic bacteria with the optimum temperature for normal growth ranging around 30°C and the growth will be dramatically affected by increasing the growth temperature from the optimal temperature. The thermotolerant AAB strains, however, show normal growth at around 30°C and continue to grow very well at higher temperature (37°C or above). Application of mesophilic AAB in industrial production of vinegar and other products needs strict temperature control at 30°C during the fermentation period as heat generated from the biological activity increases the temperature in the fermentation vessel and, therefore, a cooling system is required to maintain the optimum temperature. The growth and production of mesophillic AAB will be significantly affected by the failure in temperature control and there is usually neither growth nor production at temperatures over 34°C (Saichana et al. 2015). The discovery of thermotolerant AAB has induced the possibility of reducing the cost for cooling system and allows flexible operation during fermentation. Most thermotolerant AAB can grow and perform oxidative fermentation at 37°C and some strains can even grow at temperatures going up to 42°C (Soemphol et al. 2011). Genome-wide analysis of closely related *Acetobacter pasteurianus* strains revealed that the thermotolerant strains are closer to each other than to thermotolersensitive ones (Matsutani et al. 2012). AAB with thermotolerant ability are more suitable for industrial

*Corresponding author: natsaran.sai@mfu.ac.th

applications and have been already applied in industrial scales. Thermotolerant AAB are considered as improved strains for industrial applications, but most thermotolerant strains still require further improvements to be more suitable for industrial uses. This calls for improvement in genetic level to create suitable strains for industrial applications.

In this chapter, thermotolerant properties of AAB and the applications of thermotolerant strains in industrial uses will be discussed. Furthermore, improvement of AAB strains by genetic modification will also be discussed.

Sources of Thermotolerant AAB and Their Industrial Applications

Thermotolerant AAB have been isolated from several places around the world, where the average temperature throughout the year is considerably higher than other regions and the properties of the useful AAB isolates were studied intensively for decades. The usual sources of isolations are tropical fruits or flowers, fermented foods and some alcoholic beverages, among which presence of high amounts of sugars and alcohols are the preferred carbon sources for AAB. Most of thermotolerant AAB isolated so far are from sources in the countries around the equator, such as Thailand, Indonesia and some countries in Africa and the Middle-East. In addition, some AAB can also be isolated from soils, woods from fruit trees, cocoa heap fermentation and also some spoiled foods. Moreover, some thermotolerant AAB strains were also reported from countries far above the equator (Chen et al. 2016). Isolation of thermotolerant AAB may require an enrichment step to exclude other contaminating microorganisms by addition of ethanol or acetic acid to suppress growth of some competitive microorganisms. Temperature for isolation can be 30°C in the enrichment step to ensure a healthy growth of all AAB and then selection for thermotolerant strains is performed at 37°C in the medium containing ethanol or D-glucose or some sugar alcohols as the sole carbon sources (Saeki et al. 1997, Lu et al. 1999, Moonmangmee et al. 2000, Chen et al. 2016). AAB can be easily grown in a medium containing peptone and yeast extract as nitrogen and vitamin sources with ethanol as a carbon source and $CaCO_3$ as selection indicator. The isolates that produce acid from ethanol will show a clear halo on the ethanol medium supplemented with $CaCO_3$. D-Glucose or D-sorbitol might also be used as the carbon sources as well as ethanol. However, some other contaminating microorganisms can also grow in this medium. The most common microorganisms co-isolated with AAB are the lactic acid bacteria (LAB) and some yeasts if D-glucose is used as the carbon source.

Thermotolerant AAB for Vinegar Production

Vinegar is widely consumed as a major seasoning and preserving ingredient for foods around the world. Production of vinegar, a commercial and consumable form of acetic acid, is usually performed at 30°C which is the optimum temperature for mesophilic AAB strains. An increase of 2-3°C in the fermentation process can cause a marked decrease in both the fermentation rate and the yield (Ohmori et al. 1980). The first report on isolation of thermotolerant AAB was published in 1980 by Ohmori et al. Several AAB were isolated from vinegar mash from surface vinegar fermentation, soil from vinegar factory and spoiled fruits and were tested for their ability to produce acetic acid at higher temperatures. The isolate No. 1023 from rice vinegar mash from surface fermentation was identified as *Acetobacter aceti* that showed a normal growth and fermentation ability at 35°C. The strain retained 45 per cent of the fermentation activity when it was grown at 38°C. In 1997, thermotolerant AAB strains isolated from fruit samples from all over Thailand, a tropical country, were reported by Saeki et al. Several thermotolerant AAB isolates were obtained and majority of them were classified as *Acetobacter rancens* subsp. *pasteurianus, A. lovaniensis* subsp. *lovaniensis*, *A. aceti* subsp. *liquefaciens*, and *A. xylinum* subsp. *xylinum* (re-classified as *Komagataeibacter xylinus*). All of the thermotolerant strains oxidized ethanol up to 9 per cent (v/v) and produced acetic acid at 38 to 40°C, which is an advantage over other mesophilic strains. Fermentation efficiency for vinegar production of thermotolerant strains was found to be similar to mesophilic acetic acid producers grown at 30°C. Among these thermotolerant strains, *A. lovaniensis* subsp. *lovaniensis* was selected as a good candidate of thermotolerant AAB in vinegar production and was later reclassified as *A. pasteurianus* SKU 1108 (Kanchanarach et al. 2010a, b). Lu et al. (1999) isolated *Acetobacter* sp. I14-2 possessing thermotolerant ability with ethanol resistance and high acetic acid productivity from a spoiled banana in Taiwan. This strain showed a comparable yield at 37°C with that of *A. pasteurianus* SKU 1108 reported previously by Saeki et al. (1997). The SKU 1108 and I14-2 strains were proposed as suitable strains for vinegar production at higher temperature. In 2006, Ndoye et al. reported new thermotolerant *A. tropicalis* CWBI-B418 (re-classified as *A. senegalensis* by Ndoye et al. (2007) as a novel species), isolated from mango wine in Senegal and *A. pasteurianus* CWBI-B419, isolated from 'Dolo', a local beer made of cereals in Burkina Faso. *A. tropicalis* CWBI-B418 exhibited an optimum growth at 35°C while *A. pasteurianus* CWBIB419 has the optimum temperature for growth at 38°C. These two strains did not show any noticeable lag time at their optimum temperatures for growth and showed high acetic acid fermentation ability at the temperatures, which are considered as superior properties when compared to mesophilic strains. From the dominant characteristics of *A. pasteurianus* CWBI-B419 as a thermoresistant and acidoresistant strain, it was proposed as a strain for development of vinegar production in Sub-Saharan Africa (Ndoye et al. 2006).

In recent years, the search for new and better thermotolerant AAB is still posing a challenge. Several groups of researchers have successfully isolated new thermotolerant AAB strains for vinegar production from different sources and the new strains can be used as alternative strains in specific types of vinegar as will be described below. New thermotolerant AAB, *A. pasteurianus* SL13E-2, SL13E-3, and SL13E-4 were isolated from coconut water vinegar from Sri Lanka by Perumpuli et al. (2014a). These strains showed comparable acetic acid production ability with the thermotolerant acetic acid producer *A. pasteurianus* SKU 1108 at 37 to 40°C (Perumpuli et al. 2014a). They could grow at a temperature up to 42°C as well as the SKU 1108 did on potato agar plates. These thermotolerant strains were then tested for suitability in the production of coconut water vinegar and they were found to be effective strains for the production in a simultaneous system, of which the starters of yeast and AAB were added to vinegar substrate simultaneously. Similarly, *A. pasteurianus* SKU 1108 and the three thermotolerant strains grew without lag time and produced comparable amounts of acetic acid. Strains SL13E-2 and SL13E-3 produced up to 4 per cent (w/v) and 2.5 per cent (w/v) of acetic acid at 37 and 40°C, respectively, which is slightly higher than that produced by *A. pasteurianus* SKU 1108 (Perumpuli et al. 2014a).

Konate et al. (2014) isolated five thermotolerant AAB strains from an Ivorian palm wine and were classified as members of *A. pasteurianus*. The isolated strains showed the production of acetic acid at above 30 g/L, which is considerably high, with acceptable acid and ethanol tolerance. Two strains, called S3 and S32, were subsequently used for production of vinegar from overripe banana juice at 37°C in semi-continuous fermentation (Konate et al. 2015). The two strains showed very high efficiencies for acetic acid production which are suitable for industrial production of vinegar.

Cocoa heap fermentation has been known to be a source for isolation of AAB. The temperature during the fermentation usually rises over 40°C and thermotolerant AAB are expected to be present in this process. Yao et al. (2014) isolated 86 strains of thermotolerant AAB which are able to grow at temperatures up to 45°C from cocoa fermentation in Côte d'Ivoire. From the 86 strains which are able to grow in 10 per cent ethanol, 53 of them could normally grow in the presence of 15 per cent ethanol at temperatures up to 40°C. All the isolates did not show acid formation at 45°C, however, acid formation of the isolated strains at the temperature from 40-45°C has not been studied in detail and it is possible that there might be some strains that produce acetic acid at temperatures between 41-44°C.

Seven AAB that are tolerant to temperatures above 40°C and ethanol concentrations above 10 per cent (v/v) were isolated from Chinese vinegar *Pei* by Chen et al. (2016). All isolates are classified to *A. pasteurianus* and the isolate AAB4 is the best strain that is resistant to 12 per cent ethanol and still grew at 43°C. This strain also produced the highest acid yield of 61.2 g/L at 30°C in a medium with 10 per cent ethanol in the shaking culture. Most of thermotolerant strains reported so far produced about 30 g/L acetic acid at 40°C. However, AAB4 strain produced 38 g/L of the acid at the same condition and still produced 30 g/L of the acid at 42°C (Chen et al. 2016). Production of acetic acid in 10-L fermenter in the medium with 10 per cent ethanol at 37°C, which is a condition that AAB4 faced two stresses at the same time, was found to be 42.0 g/L (Chen et al. 2016). These results clearly indicated that the AAB4 strain is a very excellent strain for industrial vinegar production.

It has been reported that the flavor of vinegar relies on raw materials, such as vinegar substrates and starter cultures (Natera Marín et al. 2002, Gullo et al. 2009). Production of gluconic acid and acetic acid during the course of vinegar fermentation was proposed as the factors affecting vinegar sensory quality (Falcone et al. 2007). Two thermotolerant strains AF01 and CV01, which are members of *A. pasteurianus*, were isolated from apple and cactus fruits by Mounir et al. (2016). The two strains were cultured in semi-continuous fermentation system to evaluate the acetic acid and gluconic acid fermentation abilities. Simultaneous production of the two acids was observed in strain CV01, whereas the AF01 strain mainly produced acetic acid at 41°C. The strain CV01 produced 10.08 per cent (w/v) acetic acid and showed less sensitivity to ethanol depletion while the strain AF01 reached 7.64 per cent (w/v) acetic acid at the end of the fermentation (Mounir et al. 2016). The ability of strain CV01 to produce acetic acid and gluconic acid simultaneously is very important as this property is rarely found in most AAB and production of gluconic acid improves vinegar sensory quality.

Thermotolerant AAB for Polyol and Sugar Oxidations

Most of *Acetobacter* spp. are known for their ability to oxidize ethanol to acetic acid, *Gluconobacter* spp., on the other hand, have the outstanding oxidative fermentation ability for sugars, sugar alcohols and sugar acids (Moonmangmee et al. 2000, Gupta et al. 2001, Mamlouk and Gullo 2013). It is believed that all *Gluconobacter* spp. possess a general polyol dehydrogenase, called the pyrroloquinoline quinone-dependent glycerol dehydrogenase (PQQ-GLDH), which is a broad substrate specificity enzyme that catalyzes the oxidation of glycerol, *meso*-erythritol, D-arabitol, ribitol, D-sorbitol, D-mannitol and D-gluconate and yields dihydroxyacetone, L-erytrulose, D-xylulose, L-ribulose, L-sorbose, D-fructose and 5-keto-D-gluconic acid (5KGA), respectively (see Fig. 1) (Moonmangmee et al. 2002a, Matsushita et al. 2003, Keliang and Dongzhi 2006, Adachi et al. 2007). The affinity of the enzyme for each substrate varies from strain to strain, with each strain producing the oxidized products at different levels. This enzyme was found to be a primary dehydrogenase that is active during growth at higher temperatures of several thermotolerant *Gluconobacter* strains, which is very important as PQQ-GLDH is a major enzyme for oxidation of polyols.

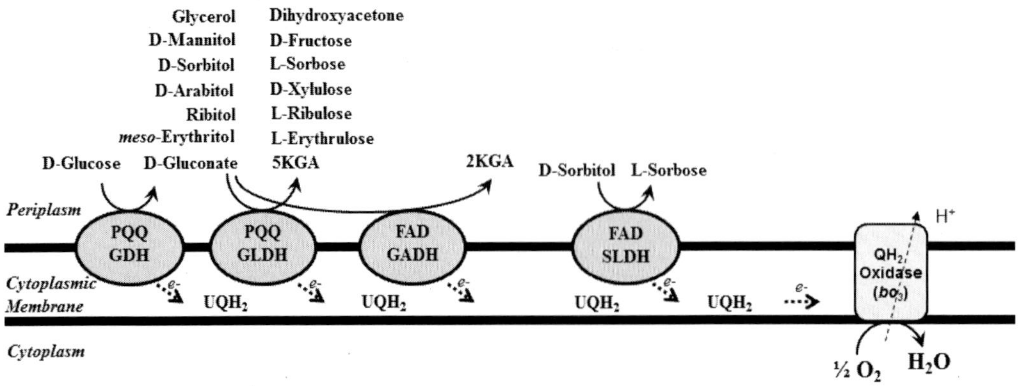

Figure 1: Oxidation of various sugars, sugar alcohols, and sugar acid in *Gluconobacter* strains.

Moonmangmee et al. (2000) isolated eight thermotolerant *Gluconobacter* strains called 'CHM' strains from various kinds of fruits and flowers in Thailand. The CHM strains grew at pH 3.0 with temperature ranging from 10 to 37°C, the optimum temperature being around 30-33°C. Among the newly isolated strains, *Gluconobacter frateurii* CHM 16 produced D-fructose from D-mannitol at 37°C within 24 hours. This is an interesting property as none of the mesophilic strains could grow or produce any product at this temperature (Moonmangmee et al. 2000). The same authors also reported that *G. frateurii* CHM 54 efficiently produced L-sorbose at 37°C with almost 100 per cent conversion yield reached after 24-36 hours of fermentation without any lag time. They also later reported another isolate, *G. frateurii* CHM 43, which produced a high amount of L-erythrulose from *meso*-erythritol (Moonmangmee et al. 2002a).

L-Sorbose production in *Gluconobacter* spp. is known to be catalyzed by two membrane-bound enzymes; flavin adenine dinucleotide-containing D-sorbitol dehydrogenase (FAD-SLDH) (Shinagawa et al. 1982, Toyama et al. 2005, Soemphol et al. 2008) and the PQQ-GLDH (Ameyama et al. 1985, Matsushita et al. 2003). The two enzymes were found to have different physiological functions as PQQ-GLDH is a primary enzyme that oxidizes D-sorbitol which is believed to be present in all *Gluconobacter* strains while FAD-SLDH is believed to exist only in certain strains. In the strain having both D-sorbitol oxidizing enzymes, such as *G. frateurii* THD 32, the excess D-sorbitol and accumulated L-sorbose, primarily produced by PQQ-GLDH, are the factors that induced FAD-SLDH to facilitate faster oxidation of D-sorbitol (Soemphol et al. 2008). It was also found that PQQ-GLDH preferably linked to a cytochrome bo_3 terminal oxidase while FAD-SLDH preferred the cyanide-insensitive oxidase (*CIO*) as an electron accepter (Soemphol et al. 2008). FAD-SLDH is a dominant enzyme for L-sorbose production in *G. frateurii* THD 32 (Soemphol et al. 2008), while PQQ-GLDH is responsible for production of L-sorbose in *G. frateurii* CHM 54 and *G. frateurii* CHM 43 strains (Hattori et al. 2012).

Most *Gluconobacter* strains produce 2-keto-D-gluconic acid (2KGA) and 5KGA when D-glucose is used as a carbon source. The ratio of 2KGA and 5KGA varies from strain to strain and culture conditions as some strains are 2KGA major producers or 5KGA major producers, but natural strains, that exclusively produce 2KGA or 5KGA, are rare. Screening of thermotolerant *Gluconobacter* strains isolated from Thailand with the ability to produce 2KGA or 5KGA has been reported (Saichana et al. 2009). Similar to other mesohilic strains, the thermotolerant *Gluconobacter* strains oxidize D-glucose to D-gluconic acid (GA), which resulted in the decrease of pH of the medium and GA is then further converted to 2KGA and 5KGA. As mentioned earlier, PQQ-GLDH oxidizes GA to 5KGA, while 2KGA formation is catalyzed by a flavin adenine dinucleotide-containing D-gluconate dehydrogenase (FAD-

GADH). 2KGA is a dominant product in all thermotolerant strains isolated from Thailand since FAD-GADH seemed to have higher affinity to D-gluconate than PQQ-GLDH did. *Gluconobacter* sp. THE 42, THF 55, and THG 42, which showed high 5KGA production at 37°C, were selected (Saichana et al. 2009). These new strains showed strong ability to grow and produce 2KGA/5KGA at 37°C; however, further improvement of the strains to produce only 2KGA or 5KGA is required.

Perumpuli et al. (2014a) isolated four thermotolerant *Gluconobacter* strains (SL13-5, SL13-6, SL13-7, and SL13-8) from Sri Lankan coconut water vinegar. They were all classified as members of *G. frateurii*, which showed growth at 37°C similar to several thermotolerant strains previously isolated. Although these new strains have not been tested for their ability to produce any typical product, the fact that they are new strains of thermotolerant *Gluconobacter* is interesting as they can serve as alternative strains to be selected for industrial applications.

Polysaccharide Production by Thermotolerant AAB

Bacterial cellulose is a very well-known polysaccharide produced by AAB that showed some superior properties than plant cellulose. Besides producing cellulose, most thermotolerant AAB produce various types of polysaccharides; extra cellular polysaccharides (EPS) or capsular polysaccharides (CPS). CPS is known as pellicles which are capsules that cover cells of AAB and help the cells to float on the medium surface in order to take up enough O_2 for cellular activities. Pellicles are also involved in acetic acid resistance in AAB as a protective barrier for cells (Deeraksa et al. 2005, Kanchanarach et al. 2010a). Pellicles from thermotolerant *A. tropicalis* SKU 1100 were reported to be a novel type of polysaccharide composed of glucose, galactose, and rhamnose in the molar ratio of 1:1:1 (Moonmangmee et al. 2002b, Deeraksa et al. 2005, 2006). Further studies revealed that the structure of this novel polysaccharide comprises a branched hexasaccharide repeating unit with two moles of 2,3-α-L-rhamnopyranosyl, and one mole each of 6-β-D-galactopyranosyl and 2-α-D glucopyranosyl residues with branches at the rhamnosyl residues and its chains terminating with terminal-β-D-galactofuranosyl and α-D-glucopyranosyl residues (Ali et al. 2011). The *A. tropicalis* SKU 1100 produced more pellicles at higher temperatures reaching up to 40°C in static culture and the polysaccharide has the molecular mass of 120 kDa (Deeraksa et al. 2005). Thermotolerant *A. pasteurianus* SL13E-2, SL13E-3, and SL13E-4 from Sri Lanka were also reported to produce new types of CPS (Perumpuli et al. 2014b). The CPS from strain SL13E-2 composed of one rhamnose with eight glucoses, whereas the strain SL13E-4 produced CPS with rhamnose, glucose and xylose in the ratio of 1:5:2. CPS from strains SL13E-3 and SKU 1108 comprising rhamnose, glucose and galactose in the ratio of 2:2:1 and 1:5:2.5, respectively (Perumpuli et al. 2014b). The varieties of polysaccharides produced by various thermotolerant strains are considered as new sources for material science that can be applied in food, cosmetic and other industries.

Thermotolerant Mechanisms of AAB

Thermotolerant mechanisms of AAB have not been well understood. The genes relating to heat-shock responses, such as *groELS*, *grpE*, *dnaKJ*, and *clpB* in *Acetobacter* strains, have been isolated in *Acetobacter* (Okamoto-Kainuma et al. 2002, 2004, Ishikawa et al. 2010a, b). The *groESL* operon of *A. aceti* contains a heat-shock promoter and a highly conserved inverted repeat. Transcription of *groELS* operon in this strain was induced by heat-shock as well as by exposure to ethanol and acetic acid during ethanol fermentation (Okamoto-Kainuma et al. 2002). Overexpression of *groES*/*groEL* resulted in increased resistance to stressors, such as heat, ethanol, or acetic acid, which suggested the association of GroES and GroEL in acetic acid fermentation. Genes on *dnaKJ* operon, which response to temperature shift and exposure to ethanol and acetic acid, were also reported from *A. aceti* (Okamoto-Kainuma et al. 2004). Growth of the *dnaKJ* overexpressed strain under high temperature and ethanol improved. However, resistance to acetic acid did not increase. The *grpE* gene was found at the upstream sequence of *dnaKJ* of *A. pasteurianus* NBRC 3283. The expression of *grpE* was induced by heat-shock or ethanol and the relative transcription levels of *grpE, dnaK*, and *dnaJ* mRNA were in the ratio of approximately 1:2:0.1, and the genes were transcribed as *grpE-dnaK, dnaK*, and *dnaJ* (Ishikawa et al. 2010a). Co-overexpression of GrpE with DnaK/J improved growth as compared to the single overexpression of DnaK/J in high temperature or ethanol-containing conditions. The findings of GroESL and GrpE/DnaKJ involvement in responses to high temperature and ethanol and acetic acid resistances could lead to the development of new strains with higher tolerance to stresses, such as high temperature and ethanol/acetic acid. It was also found that *groEL, grpE, dnaKJ*, and *clpB* which are major heat-shock responsible genes in *Alphaproteobacteria* have the putative recognition sequences for RpoH (σ^{32}) (Okamoto-Kainuma et al. 2011).

Transposon Tn*10* mutagenesis was employed to identify genes required for growth at high temperatures in *A. tropicalis* SKU 1100 by Soemphol et al. (2011). There were 32 transconjugants with deficient growth at high temperature and they were selected for further investigation. Among 24 genes identified in the mutants (from 32 transconjugants), a putative small heat-shock protein HspA (ATPR_3088) was only a heat-shock protein found to involve in the thermotolerance of *A. tropicalis* SKU 1100. Many other genes possibly involved in stress response, cell division, cell wall or membrane biosynthesis and membrane transport also played a part in thermotolerance and/or acetic acid resistance of *A. tropicalis* SKU 1100 (Soemphol et al. 2011). In order to adapt to higher temperature conditions, it was expected that the adaptation for thermotolerance requires changes (mutation and/or acquisition) in many genes that have functionally diverged and not only the genes of heat-shock proteins (Matsutani et al. 2013). Thus, adaptive mutation of some *Acetobacter* strains was carried out and the resulted strains were found to have several mutations in the genome.

The genome of an adapted strain of *A. pasteurianus* NBRC 3283, namely IFO3283-01/42C, was compared with its parental strain. It was found that a 92 kb truncation of the genome was the most critical mutation in the thermotolerant IFO3283-01/42C strain (Matsushita et al. 2016). Higher expression of the genes for oxidoreduction and stress response was found in the IFO3283-01/42C strain grown at 42°C as compared to the parental strain grown at 30 and 37°C. In addition to these genes, several genes encoding transcriptional factors, including heat shock sigma factor RpoH and transcriptional regulator for starvation/low temperature, were also shown to be highly expressed in IFO3283-01/42C grown at 42°C than in the parental ones at 30 and 37°C. These results suggested that the thermotolerant phenotype of the IFO3283-01/42C resulted from the partial decrement of the replication burden together with higher expression of the oxidoreductase and stress responsive genes (Matsushita et al. 2016).

A. pasteurianus SKU 1108 was recognized as one of the best thermotolerant acetic acid producers. Adaptive mutation of this strain resulted in the generation of TI and TH-3 strain which showed higher thermotolerance than the SKU 1108 wild-type strain. The whole genome mapping of the wild-type SKU 1108 and the adapted TI and TH-3 strains was carried out and intensive comparison of the genomes revealed mutations at 6 and 11 positions on the genomes of TI and TH-3, respectively (Matsutani et al. 2013). Among the mutated genes identified, mutations of amino acid transporter (APT_1698), transcriptional regulator MarR (APT_2081) and C4-dicarboxylate transporter (APT_2237) were found in both TI and TH-3 strains. Disruption studies of *marR* and *APT1698* were carried out and the growth characteristic of the disrupted strains was compared with the wild-type and the TI and TH-3 strains. The *marR* disruptant strain showed similar growth characteristics under high temperature with higher acetic acid resistance and faster ethanol oxidation compared to the wild-type (Matsutani et al. 2013). The disruption of *APT1698* (amino acid transporter) resulted in increased growth at 39°C, at which the wild-type could not grow. This mutant strain also showed a delayed acetic acid overoxidation, indicating the involvement of *APT1698* in the maintenance of prolonged acetic acid resistance phase. It was suggested that all mutated genes discovered in the adapted strains may not necessarily be involved in thermotolerance, but the combined effects of the mutations could result in the thermotolerant and other relevant traits (Matsushita et al. 2016).

Development of AAB Strains by Genetic Modifications

Genetic Adaptation of AAB

Many bacterial strains have fluid genetic traits that can be present or absent according to suitable growth environments. Some AAB possess high mutability during growth in certain conditions that result from temporal acclimation or heritable adaptation (Deeraksa et al. 2005, Azuma et al. 2009). Some *Acetobacter* strains acquire the resistant ability to acetic acid after long-term passage during traditional vinegar fermentation of which the seed culture is from previous batch of fermentation, called the 'Mother of Vinegar'. This acetic acid resistance can be lost if the strains were grown under non-stressed conditions for some generations. This genetic instability has caused the loss of various useful phenotypic characteristics of AAB, such as ethanol oxidation, acetic acid resistance, pellicle formation and cellulose biosynthesis (Matsutani et al. 2013). The genetic properties of these instabilities were not well understood, but phenotypic modifications by transposon insertion were reported in ethanol oxidation, acetic acid resistance and cellulose formation (Matsushita et al. 2016). Whole genome analyses of *A. pasteurianus* NBRC 3283 revealed more than 280 transposons that covered about 9 per cent of the total genes found in the genome (Azuma et al. 2009). This strain also contains six plasmids that might be the factor involving genetic flexibility as three smaller plasmids were removable from cells. In addition to having large numbers of transposons, hyper-mutable tendem repeats (HTRs) were also found in the genome of *A. pasteurianus* NBRC 3283. For example, the $(AGGAC)_n$ repeats were found to be shorter $(AGGAC)_{n-1}$ or longer $(AGGAC)_{n+1}$ in some varieties of cells mutated from the original strain. This finding suggested the possibility that the repetitive numbers (*n*) can quickly alter the coding frames or amino acid sequences following the HTRs and thus influence its functions (Matsushita et al. 2016).

Some groups of researchers have performed adaptive mutagenesis in AAB. This mutagenesis is an *in vitro* evolution that forced the bacterial cells to survive under stressful (high temperature) conditions. It has been shown that *A. pasteurianus* NBRC 3283 could be adapted to become a thermotolerant bacterium by repeated cultivation at high temperatures (Azuma et al. 2009). The obtained thermotolerant strain, IFO3283-01/42C, could grow at 42°C after 27 days of cultivation at 40°C and the thermotolerant ability was found to be stable (Matsushita et al. 2016). This finding has opened the possibility that AAB can be adapted to acquire mutations in the genome in order to survive at higher temperatures and that mutation(s) is inheritable. In addition, adaptive mutation of *G. frateurii* CHM 43, an outstanding thermotolerant strain for L-sorbose fermentation at 37°C, was achieved (Hattori et al. 2012). The original CHM 43 strain was repeatedly grown at 38.5°C for several generations until improved growth was observed. By this adaptation procedure, the obtained thermoadapted strain, CHM43AD, showed improved thermotolerance as it could grow and produce L-sorbose at 38.5 to 40°C. It was found that the CHM43AD strain produced more PQQ than the CHM 43 strain, suggesting that the adapted strain has more

PQQ to keep GLDH, the enzyme responsible for L-sorbose formation, in holo-form that resulted in maintaining the level of L-sorbose production at higher temperature (Hattori et al. 2012).

Mutagenesis of *A. pasteurianus* SKU 1108 was performed by repeated cultivation at 39°C or step-wise increase in the adaptive cultivation temperature from 38.5 to 40°C (*see* Fig. 2) (Matsutani et al. 2013). The wild-type strain of *A. pasteurianus* SKU 1108 has been known to produce high amounts of acetic acid at high temperatures reaching up to 38°C. By repeated cultivation at 39°C for 1150 hours (11 passages to fresh medium), the TI strain was obtained as a thermoadapted strain. The TH-3 strain was obtained by step-wise adaptation of overall 2,500 hours (total 38 passages to fresh medium). TI and TH-3 strains exhibited the improved growth characteristics as they grew normally at 39°C and 40°C. TH-3 has acquired ethanol oxidation and acetic acid resistant abilities about 1.6-fold higher than the wild-type and TI strains (Matsutani et al. 2013). Furthermore, the TH-3 strain was adapted to higher ethanol concentrations to acquire ethanol-tolerant strain. The TH-3 strain was cultivated in 5 per cent ethanol and subsequently introduced into 6 per cent and 7 per cent ethanol. The resultant adapted strain, namely 7E-13, was obtained. This strain exhibited good fermentation ability at 7 per cent ethanol and high temperature similar to the TH-3 strain (Matsushita et al. 2016).

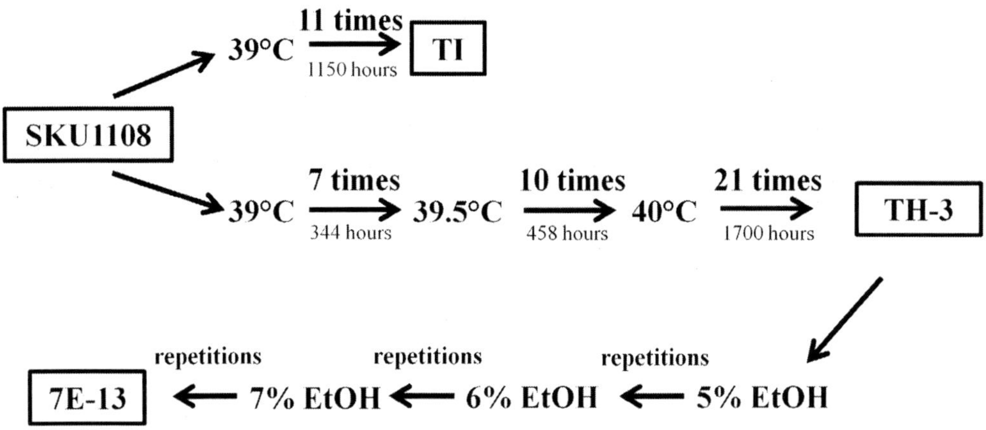

Figure 2: Diagram of adaptive mutagenesis of *A. pasteurianus* SKU1108 to obtain TI, TH-3 and 7E-13 strains (adapted from Matsutani et al. 2013).

In addition, a thermotolerant *A. pasteurianus* T24 was obtained through adaptive mutagenesis of the parental HN 101 strain (Wei et al. 2015). The adaptation was performed by repeating cultivation at 40°C for 48 hours in 12 cycles. The resultant adapted strain T24 exhibited better growth at 40°C in solid medium and under the condition of low ethanol concentration.

The adapted strain showed a large number of mutations in highly diversified genes which could be categorized into groups related to cell surface functions, ion or amino acid transporters and some transcriptional factors. Generally, cells grown under higher temperatures produced more reactive oxygen species (ROS) that could lead to damage of the cells. However, the thermal adaptation could lead to decreased ROS generation in cells. Thus, it is suggested that thermally adapted cells could become robust and resistant to many stressors and thus be useful for high-temperature fermentations (Matsushita et al. 2016).

Genetic Engineering in AAB

Transposon Mutagenesis for Mutation Studies in AAB

Mutation study is used when there is no information on genes responsible for specific traits of microorganisms. Transposable elements (Tn), such as Tn*5* or Tn*10*, have been used as tools for mutation studies and in locating the genes of interest. The Tn introduced into a bacterial cell will be inserted randomly into the chromosome and, as a result, the gene that is separated by the insertion of Tn will be inactivated. This gene disruption will result in changes in phenotypes of the bacteria. For example, if the gene of interest is responsible for thermotolerance in the bacteria, insertion of Tn into the gene will result in loss of thermotolerance. The cell showing expected phenotypes will be selected for further studies and the gene that was inactivated by Tn insertion will be identified through several molecular techniques, such as *in vitro* cloning strategy or thermal asymmetric interlaced polymerase chain reaction (TAIL-PCR). Mutation studies using plasmid pSUP2021 carrying Tn*5* have been used in several Gram-negative bacteria (Simon et al. 1983). Transfer of pSUP2021 to recipient AAB through conjugation requires an *Escherichia coli* donor strain having the transfer genes of the broad host range IncP–type plasmid RP4 integrated in its chromosome (such as *E. coli* S17-1). Transposon mutagenesis in *G. oxydans* ATCC 9937 was carried out to obtain the mutant deficient in glucose dehydrogenase activity by conjugating with *E. coli* S17-1 as the donor strain (Gupta et al. 1997). In addition, identification of genes for pellicle polysaccharide biosynthesis in thermotolerant *A. tropicalis* SKU 1100 was carried out by Tn*10* mutagenesis study. The mobilized plasmid pSUP2021 having Tn*10* was constructed by inserting Tn*10* from the transposon vehicle phage λNK1323 into pSUP2021 (Deeraksa et al. 2005). The mutant deficient in pellicle formation was then selected and the location of Tn*10* was identified on *polE*. The sequence of the whole *polABCDE* operon was determined by *in vitro* cloning strategy. Furthermore, Tn*10* mutagenesis was also used in identifying genes involved in thermotolerance in *A. tropicalis* SKU 1100 by TAIL-PCR technique (Soemphol et al. 2011). As described above, transposon mutagenesis is a useful tool for mutation study and identification of genes of interest that can be applied to studies in thermotolerant AAB.

Shuttle Vectors for AAB

Metabolic engineering of AAB requires a good expression system for the control and enhancement of specific gene expression. Development in transformation conditions that facilitated the use of several expression plasmids in *Gluconobacter oxydans* subsp. *suboxydans* by electroporation method was first described by Creaven et al. (1994). DNA transfer in AAB can also be facilitated by the conjugation method if the plasmid to be transferred contains conjugal transferring factors. Several plasmids, such as pSA19 (Tonouchi et al. 1994), pFG15A (Saito et al. 1997), pJT2 (Trcek et al. 2000), and pSG6 (Tonouchi et al. 2003), have been constructed as shuttle vectors for AAB. They were derived from conventional *E. coli* cloning plasmids and some cryptic plasmids of AAB. These plasmids contain origins of replications of both *E. coli* and AAB which allow them to replicate in the two bacterial groups. The reason that the vectors should be able to replicate in *E. coli* is that it is easier to produce and engineer the plasmids using *E. coli* as a host strain than using AAB. After the plasmids are engineered in *E. coli*, it will be transformed into AAB as working hosts.

Several broad-host-range vectors of the pCM-series for cloning and expression in methylotrophs and Gram-negative bacteria were constructed (Schmidhauser and Helinski 1985, Marx and Lidstrom 2001). Among several derivatives of pCM-series vectors, pCM62 has been used for expression in some AAB, such as the expression of genes in thermotolerant *A. tropicalis* SKU 1100 and *A. pasteurianus* SKU 1108 (Deeraksa et al. 2005, Soemphol et al. 2011, Tasanapak et al. 2013). The pCM-series plasmids are broad-host-range plasmids that can be applied for expression in *Alpha-*, *Beta-*, and *Gamma-proteobacteria*. However, the expression level was weak or moderate in most strains. In addition, a shuttle vector, pZL1, was constructed from a cryptic plasmid pGOX3 of *G. oxydans* DSM 2003 and an *E. coli* cloning vector pUC18 to be used in *G. oxydans* DSM 2003 (Zhang et al. 2010).

Besides, several expression plasmids have been constructed for AAB as described above, the plasmids that have been widely used are the derivatives of the broad-host-range plasmid pBBR1MCS, which are conjugal transferable plasmid derivatives (Kovach et al. 1994). Other than using electroporation, the pBBR1MCS derivatives can also be transferred into the host cells by tri-parental conjugation, using an *E. coli* donor strain (carrying the pBBR1MCS plasmid) with a helping strain *E. coli* HB 101 harboring pRK2013 and a AAB strain as a recipient strain. Originally, this vector contains *lac* promoter for gene expression with a chloramphenicol resistance gene as a selectable marker. Selection of the insert for pBBR1MCS is also facilitated by blue-write screening. The original pBBR1MCS with chloramphenicol resistant gene has limitation of use as many Gram-negative bacteria are known to be resistant to chloramphenicol. The new derivatives of pBBR1MCS carrying different antibiotic resistance genes as selectable markers were constructed for flexible use, namely pBBR1MCS2 (kanamycin), pBBR1MCS3 (tetracycline), pBBR1MCS4 (ampicillin) and pBBR1MCS5 (gentamycin) (Kovach et al. 1995). These modified pBBR1MCS plasmids are still transferable by conjugation which is applicable for some AAB strains that DNA transfer by electroporation is known to be ineffective. Furthermore, new pBBR1MCS derivatives, originating from the four original derivatives, have been constructed to improve the flexibility and enhance the expression level in AAB. Obranić et al. (2013) improved the plasmids by introducing an *Nde*I site to the start codon of *lacZα* gene on the vectors in order to introduce the DNA fragment for expression, of which the start codon is fused with *Nde*I recognition sequence, close to the *lac* promoter on the plasmids that resulted in generations of pBBR1MCS2_START, pBBR1MCS3_START, pBBR1MCS4_START and pBBR1MCS5_START. These new plasmids are improved for the ease of use which facilitate the cloning process and improve the promoter efficeincy.

Active Promoters for Expression in AAB

The original promoter for expression in pBBR1MCS is a *lac* promoter which is a universally active promoter for several bacteria. However, this promoter is considered as a weak promoter in AAB. In order to gain higher expression level in AAB, new effective promoters are required. Several *E. coli* promoters have been tested for expression of genes in AAB and it was found that the promoter of elongation factor TU (P_{tufB}), P_{tac}, and P_{PL} were strong promoters for AAB (Saito et al. 1997). Merfort et al. (2006a) constructed the derivatives of pBBR1MCS5 plasmids by introducing a P_{tufB} of *Gluconobacter* and a glucose dehydrogenase promoter (P_{gdh}) into the plasmids as both the promoters were constitutive promoters for primary proteins in *Gluconobacter* strains. The two promoters showed great improvement in expression and higher activity of the gene product when they were used for expression of a PQQ-GLDH for 5KGA production. Therefore, P_{tufB} and P_{gdh} were considered as strong promoters which are suitable for overexpression in *Gluconobacter* strains (Merfort et al. 2006a).

Schleyer et al. (2008, 2009) constructed plasmid pEXGOX-G from pBBR1MCS5 having P_{tufB} of *Gluconobacter* for expression. The plasmid has gentamycin-resistant cassette as the selective marker and it was then engineered to have unique restriction sites. A *Swa*I site was introduced to generate the blunt-ended vector for ligation to the blunt-ended PCR product starting with a start codon ATG. The *Swa*I site was placed at a proper distance from the promoter to ensure strong expression. The pEXGOX-G was then modified into pEXGOX-A and pEX-GOX-K by replacing gentamycin-resistant cassette with ampicillin and kanamycin-resistance cassettes, respectively. These new pEXGOX-series plasmids were tested for overexpression of several *Gluconobacter* ORFs and were found to be very effective expression vectors for *Gluconobacter* strains (Schleyer et al. 2008).

Zhang et al. (2010) constructed a shuttle vector, pZL1, from a cryptic plasmid pGOX3 of *G. oxydans* DSM 2003 and *E. coli* cloning vector pUC18. This vector was engineered to have a *tufB* promoter and was found to be equally efficient with pBBR1MCS5 carrying P_{tufB}. This shuttle vector is stably maintained in *G. oxydans* DSM 2003 with much higher copy number than pBBR1MCS5 in *E. coli* (Zhang et al. 2010). The plasmid was successfully used for expression in *G. oxydans* DSM 2003 and can be used as an alternative expression vector for *Gluconobacter* strains in addition to the pBBR1MCS derivatives.

Besides the P_{tufB} that has been widely used as a promoter for expression in *Gluconobacter* strains, promoter of PQQ-alcohol dehydrogenase (PQQ-ADH) is also considered a strong promoter. P_{adh} was introduced into pBBR1MCS4 and tested for expression in *G. oxydans* (Kawai et al. 2013). The result showed that P_{adh} is a strong promoter that can be applied for overexpression in *Gluconobacter* strains.

Finding new strong promoters for expression in AAB is a challenging task and many researchers have reported new developments in the past few years. The promoter regions of *gox0264* and *gox0452*, the genes encoding ribosomal proteins L35 and L13, respectively, were introduced into pBBR1MCS2 to create a plasmid with stronger promoters for gene expression in *Gluconobacter* strains (Kallnik et al. 2010). The resultant plasmids, pBBR1p264 and pBBR1p452, were tested for promoter activity and it was found that the newly introduced promoters were strong and moderate promoters while the original *lac* promoter on pBBR1MCS2 was classified as a weak promoter. Expression of PQQ-containing D-glucose dehydrogenase (PQQ-GDH) by pBBR1p452 vector in *G. oxydans* showed five-fold increase of GDH activity in the membrane fraction that resulted in 70 per cent increase of D-gluconate production (Meyer et al. 2013). These new expression vectors are useful for protein expression, of which different expression levels might be required due to the toxicity of the expressing proteins to the cells of the *Gluconobacter* host strains. In addition, the promoter region of *gox0169*, which was a hypothetical protein, was isolated and introduced into an expression plasmid, pBBR1p*gHp0169*. The $P_{gox0169}$ was found to be stronger than P_{tufB} that has been known as a strong promoter for expression in *Gluconobacter* strains (Shi et al. 2014).

Hu et al. (2015) performed a screening of strong promoters by proteomic analysis. From 3D-gel electrophoresis, a spot of protein which is most abundant in all phases of growth was selected for further identification. The protein was later identified as a hypothetical protein B932_2000 and the promoter region of the corresponding gene was cloned into pBBR1MCS2. The expression of a green fluorescence protein (*gfp*) expressed under P_{B932_2000} was found to be higher than *gfp* expressed under *E. coli* P_{tufB} and *G. oxydans* P_{tufB}. When P_{B932_2000} was used for expression of a D-sorbitol dehydrogenase enzyme, productivity increased by 33 per cent (Hu et al. 2015).

Shi et al. (2015) performed a number of mutations into the -10 and -35 regions of the original promoter and ribosomal binding region of pBBR1MCS5, which resulted in generations of several derivatives of pBBR1MCS5. The plasmid pBBR-3510 has mutations at the -10 and -35 regions, of which the nucleotide sequences were changed from CATAAT to TATAAT and from TTGACT to TTGACA, respectively. These mutations of pBBR-3510 resulted in the highest enhancement of promoter activity compared to other mutations. This plasmid was then tested for expression of FAD-GADH enzyme that is responsible for 2KGA production. The production was greatly improved as 0.63 g/g CWW/h of 2KGA was produced by the GADH-expressing strain under pBBR-3510, while the original strain produced 0.40 g/g CWW/h under pBBR1MCS5 original *lac* promoter. Compared to the parental wild-type strain, pBBR-3510 increased 186 per cent of the 2KGA productivity (Shi et al. 2015).

Metabolic Engineering in AAB

Gene disruption and overexpression are the prime techniques used in metabolic engineering to manipulate the flow of substrate into the desired routes for better yield of the final product. The disruption of genes requires a homologous recombination mechanism to facilitate the integration of the introduced mutated gene fragment into the cells while overexpression requires efficient promoter and necessary components in the expression plasmids. Some AAB strains produced less amounts of desirable products due to the competition in metabolic pathways. In *Gluconobacter*, production of 2KGA and 5KGA has been known to be a competitive reaction *in vivo*. The ratio of 2KGA and 5KGA production is influenced by growth conditions and it varies from strain to strain (Shinagawa et al. 1983, Ano et al. 2011). Production of 2KGA is catalyzed by a membrane-bound FAD-GADH whereas 5KGA is catalyzed by a membrane-bound PQQ-GLDH (Moonmangmee et al. 2002a, Matsushita et al. 2003, Keliang and Dongzhi 2006, Adachi et al. 2007). These two enzymes are believed to be present in all *Gluconobacter* strains and are responsible for the formation of 2KGA and 5KGA during growth in the D-glucose-containing medium. This property is not preferred as production of 2KGA will have 5KGA as a by-product and *vice versa*. In order to produce exclusive 2KGA or 5KGA, the genes for PQQ-GLDH or FAD-GADH should be disrupted, respectively. For the production of 5KGA, mutant deficient in production of FAD-GADH enzyme has been constructed (Elfari et al. 2005). The FAD-GADH-deficient mutant of *G. oxydans* DSM 2343 (identical to ATCC 621H) produced 5KGA at the yield of 84 per cent without any traceable 2KGA as the co-product. To acquire faster D-gluconic acid formation, the PQQ-GDH and PQQ-GLDH enzymes were overproduced (Merfort et al. 2006b). The resulting strain produced higher amounts of 5KGA and the strain was further improved by overexpressing PQQ-GLDH in the *Gluconobacter* strain under the control of P_{gdh} and P_{tufB} that resulted in the increase of 5KGA production (Merfort et al. 2006a). In the study, although GLDH activity of the overexpressed strain under P_{gdh} was lower than that of the P_{tufB}, production of 5KGA improved in the comparable level.

Production of 5KGA by thermotolerant *Gluconobacter* strains was reported by Saichana et al. (2009). From several thermotolerant *Gluconobacter* spp., *Gluconobacter* sp. THE 42, THF 55, and THG 42 were selected for the ability to produce high amounts of 5KGA at 37°C, although 2KGA was the major product. These strains were further improved by disrupting the *gndG* gene-encoding large subunit of FAD-GADH to diminish 2KGA formation. As a result, the mutants of all the three strains exclusively produced 5KGA (Saichana et al. 2009). These three mutant strains are thermotolerant with the ability to produce high amounts of 5KGA without any by-product. It was also found that production of 5KGA at 37°C could be improved by addition of 5 mM $CaCl_2$ into the culture medium (Saichana et al. 2009).

On the other hand, 2KGA production was improved in *G. oxydans* DSM 2003 by Shi et al. (2014). Overexpression of FAD-GADH genes in *G. oxydans* DSM 2003 under $P_{gox0169}$ was carried out and the production of 2KGA catalyzed by the overexpressed FAD-GADH could reach the yield of 98.3 per cent (Shi et al. 2014).

Production of dihydroxyacetone (DHA) in *Gluconobacter* strains is known as a product of glycerol oxidation by PQQ-GLDH enzyme (Adachi et al. 2007). Production of DHA in PQQ-ADH-deficient mutant was found to be higher than the wild-type and it was further enhanced by overexpression of PQQ-GLDH in this strain (Li et al. 2010). This improved strain was used for biotransformation of glycerol to DHA and it was found that the production increased 2.4-fold in resting cell system. Furthermore, production of DHA improved by combination of metabolic engineering and adaptive evolution as described by Lu et al. (2012). The PQQ-GDH-deficient mutant of *G. oxydans* was adapted to grow under D-glucose as a sole carbon source until the adapted strain, which grows very well in D-glucose, was obtained (Zhu et al. 2011). This mutant strain was then mutated to have PQQ-ADH deficiency to prevent accumulation of glyceric acid. The strain was then transformed with two different PQQ-GLDH-overexpressing plasmids of which the expression was under P_{lac} or P_{tufB}. The purpose of this study is to obtain high biomass of *Gluconobacter* cells grown in D-glucose, which is a cheap carbon source, while obtaining an enhanced strain that produces higher PQQ-GLDH activity from the D-glucose-grown cells. The overexpressed strains showed improved catalytic capacity for DHA production which is suitable for further improvement in industrial biotransformation of glycerol to DHA.

Some *Gluconobacter* strains could produce 3-dehydroshikimate (DSA), the direct precursor for production of protocatechuic acid (PCA), which is an important antioxidant and anti-inflamatory compound, from quinate (Adachi et al. 2008). Overexpression of a periplasmic 3-dehydroquinate dehydratase (pDQD) responsible for the formation of DSA was carried out in *G. oxydans* NBRC 3244 (Nishikura-Imamura et al. 2014). The activity of pDQD in the overexpressed strain has achieved more than 100-fold higher level than in the original strain. The conversion rate of quinate to DSA in *G. oxydans* NBRC 3244 increased and nearly 100 per cent conversion yield after amending the pH of the medium to 7.3 was obtained.

Production of 5-keto-D-fructose was enhanced by genetic engineering of the overexpression plasmid carrying genes for D-fructose dehydrogenase (FDH). The enzyme has three subunits of a small subunit, a large dehydrogenase subunit and a cytochrome *c* subunit. Kawai et al. (2013) constructed pBBR1MCS4 derivatives carrying P_{adh} with a change in the translational initiation codon of the gene for the small subunit from TTG to ATG. The designated $_{TTG}$FDH and $_{ATG}$FDH derivatives were transformed into *G. japonicus* NBRC 3260. Membrane fractions of the recombinant $_{TTG}$FDH- and $_{ATG}$FDH-producing cells exhibited approximately 20-fold and 100-fold higher specific activity than the parental strain, respectively. These results showed that there are several ways of improving the expression of desired proteins other than selecting the strong promoter for the expression.

2-keto-L-gulonic acid (2-KLG) is a direct precursor of vitamin C that can be produced by the two-step bacterial fermentation from D-sorbitol that involved three microorganisms. D-sorbitol is firstly oxidized to L-sorbose by PQQ-GLDH of *G. oxydans*. The second step requires the mix-culture of *Ketogulonicigenium vulgare* and *Bacillus megaterium* which converted L-sorbose to L-sorbosone using L-sorbose dehydrogenases (SDH) and then L-sorbosone to 2-KLG using L-sorbosone dehydrogenases (SNDH). Production of 2-KLG with a single step of fermentation is needed in order to establish a simpler process for industrial application. Different combinations of five L-sorbose dehydrogenases (SDH) and two L-sorbosone dehydrogenases (SNDH) of *Kg. vulgare* WSH-001 were introduced into *G. oxydans* WSH-003 resulted in the optimum combination that produced 4.9 g/L of 2-KLG (Gao et al. 2014). In addition, 10 different linker peptides were tested for expression of SDH fused with SNDH in *G. oxydans*. *G. oxydans* harboring pGUC-*k*0203-GS-*k*0095 was found to be the best strain that produced 32.4 g/L of 2-KLG after 168 h. Furthermore, enhancement of biosynthesis of PQQ, which is a cofactor of the two dehydrogenases, was carried out to improve 2-KLG production. This stepwise metabolic engineering of *G. oxydans* gave the final 2-KLG production at 39.2 g/L, which is eight-fold higher than that obtained from independent expression of the dehydrogenases (Gao et al. 2014).

Conclusion

Thermotolerant AAB are important for industrial application as they are resistant to changes during fermentation. It was demonstrated that the AAB that acquired thermotolerant ability also exhibited resistance to other stressors (Matsushita et al. 2016). The *A. pasteurianus* TH-3 strain was adapted to acquire higher resistance to ethanol and finally the 7E-13 strain which tolerates 7 per cent of ethanol was obtained. Acetic acid fermentation without temperature control was carried out by U. Masud-Tippayasak (unpublished) in a 100L fermentor where the temperature reached 40°C and the fermentation finished within 21 hours with a yield of 6.6 per cent acetic acid (Matsushita et al. 2016). The adapted thermotolerant strain of *G. frateurii* CHM43AD which sustained the PQQ-GLDH activity for oxidation of glycerol, *meso*-erythritol, D-arabitol, ribitol, D-sorbitol, D-mannitol and D-gluconate to produce dihydroxyacetone, L-erytrulose, D-xylulose, L-ribulose, L-sorbose, D-fructose and 5KGA, respectively, at higher temperatures was also constructed (Hattori et al. 2012). The production of L-sorbose from D-sorbitol in a fermentor without temperature control demonstrated that the temperature could reach 41.5°C at which CHM43AD could continue to grow and produce L-sorbose. This is a very interesting example of the application of the adapted thermotolerant AAB in industrial production of vinegar and L-sorbose as the system does not require any cooling apparatus which will greatly reduce the cost of production.

Thermotolerant AAB isolated from natural sources possess the thermotolerant ability with moderate productivity of the desired products, leading to the need for development to acquire better strains for industrial application. As mentioned in the previous section, adaptation of AAB could potentially improve thermotolerance of several AAB strains. Metabolic engineering of specific pathways, however, requires genetic engineering to specifically target the desired genes. Tools for genetic manipulations of AAB have been developed to facilitate research in this field. Introduction of DNA fragments into cells of AAB requires efficient DNA transformation techniques which have been developed in the past decade. Transferring DNA into many *Acetobacter* strains that produced CPS to cover the cell surface have been known to be difficult. This problem could be solved by using conjugation technique for DNA transfer with *E. coli* as donor cells (Deeraksa et al. 2005, Soemphol et al. 2011).

As described in this chapter, developments for expression in AAB are a promising approach to control and enhance metabolic pathways for the production of several desired products. Development in protein production is useful for development of metabolic engineering in AAB. Together with the discoveries of new thermotolerant AAB, the applications of AAB in industries are more welcome. These recent developments enhance the ability of AAB to withstand several stresses during fermentation and enhance productivity.

References

Adachi, O., Ano, Y., Toyama, H. and Matsushita, K. (2007). Biooxidation with PQQ- and FAD-dependent dehydrogenases. *In:* R.D. Schmid and V.B. Urlacher (Eds.). *Modern Biooxidation: Enzymes, Reactions and Applications*. Wiley-VCH Verlag GmbH & Co. KGaA, Weinheim, pp. 1-41.

Adachi, O., Ano, Y., Toyama, H. and Matsushita, K. (2008). A novel 3-dehydroquinate dehydratase catalyzing extracellular formation of 3-dehydroshikimate by oxidative fermentation of *Gluconobacter oxydans* IFO 3244. *Bioscience, Biotechnology and Biochemistry* **72:** 1475-1482.

Ali, I.A.I., Akakabe, Y., Moonmangmee, S., Deeraksa, A., Matsutani, M., Yakushi, T., Yamada, M. and Matsushita, K. (2011). Structural characterization of pellicle polysaccharides of *Acetobacter tropicalis* SKU1100 wild type and mutant strains. *Carbohydrate Polymers* **86:** 1000-1006.

Ameyama, M., Shinagawa, E., Matsushita, K. and Adachi, O. (1985). Solubilization, purification and properties of membrane-bound glycerol dehydrogenase from *Gluconobacter industrius. Agricultural Biological Chemistry* **49:** 1001-1010.

Ano, Y., Shinagawa, E., Adachi, O., Toyama, H., Yakushi, T. and Matsushita, K. (2011). Selective, high conversion of D-glucose to 5-keto-D-gluoconate by *Gluconobacter suboxydans. Bioscience, Biotechnology and Biochemistry* **75:** 586-589.

Azuma, Y., Hosoyama, A., Matsutani, M., Furuya, N., Horikawa, H., Harada, T., Hirakawa, H., Kuhara, S., Matsushita, K., Fujita, N. and Shirai, M. (2009) . Whole-genome analyses reveal genetic instability of *Acetobacter pasteurianus. Nucleic Acids Research* **37:** 5768-5783.

Chen, Y., Bai, Y., Li, D., Wang, C., Xu, N. and Hu, Y. (2016). Screening and characterization of ethanol-tolerant and thermotolerant acetic acid bacteria from Chinese vinegar *Pei. World Journal of Microbiology and Biotechnology* **32:** 14.

Creaven, M., Fitzgerald, R.J. and O'Gara, F. (1994). Transformation of *Gluconobacter oxydans* subsp. *suboxydans* by electroporation. *Canadian Journal of Microbiology* **40:** 491-494.

Deeraksa, A., Moonmangmee, S., Toyama, H., Yamada, M., Adachi, O. and Matsushita, K. (2005). Characterization and spontaneous mutation of a novel gene, *polE*, involved in pellicle formation in *Acetobacter tropicalis* SKU1100. *Microbiology* **151:** 4111-4120.

Deeraksa, A., Moonmangmee, S., Toyama, H., Adachi, O. and Matsushita, K. (2006). Conversion of capsular polysaccharide, involved in pellicle formation, to extracellular polysaccharide by *galE* deletion in *Acetobacter tropicalis. Bioscience, Biotechnology and Biochemistry* **70:** 2536-2539.

Elfari, M., Ha, S.-W., Bremus, C., Merfort, M., Khodaverdi, V., Herrmann, U., Sahm, H. and Görisch, H. (2005). A *Gluconobacter oxydans* mutant converting glucose almost quantitatively to 5-keto-D-gluconic acid. *Applied Microbiology and Biotechnology* **66:** 668-674.

Falcone, P.M., Chillo, S., Giudici, P. and Del Nobile, M.A. (2007). Measuring rheological properties for applications in quality assessment of traditional balsamic vinegar: Description and preliminary evaluation of a model. *Journal of Food Engineering* **80:** 234-240.

Gao, L., Hu, Y., Liu, J., Du, G., Zhou, J. and Chen, J. (2014). Stepwise metabolic engineering of *Gluconobacter oxydans* WSH-003 for the direct production of 2-keto-L-gulonic acid from D-sorbitol. *Metabolic Engineering* **24:** 30-37.

Gullo, M., De Vero, L. and Giudici, P. (2009). Succession of selected strains of *Acetobacter pasteurianus* and other acetic acid bacteria in traditional balsamic vinegar. *Applied and Environmental Microbiology* **75:** 2585-2589.

Gupta, A., Verma, V. and Qazi, G.N. (1997). Transposon induced mutation in *Gluconobacter oxydans* with special reference to its direct-glucose oxidation metabolism. *FEMS Microbiology Letters* **147:** 181-188.

Gupta, A., Singh, V.K., Qazi, G.N. and Kumar, A. (2001). *Gluconobacter oxydans*: Its biotechnological applications. *Journal of Molecular Microbiology and Biotechnology* **3:** 445-456.

Hattori, H., Yakushi, T., Matsutani, M., Moonmangmee, D., Toyama, H., Adachi, O. and Matsushita, K. (2012). High-temperature sorbose fermentation with thermotolerant *Gluconobacter frateurii* CHM43 and its mutant strain adapted to higher temperature. *Applied Microbiology and Biotechnology* **95:** 1531-1540.

Hu, Y., Wan, H., Li, J. and Zhou, J. (2015). Enhanced production of l-sorbose in an industrial *Gluconobacter oxydans* strain by identification of a strong promoter based on proteomics analysis. *Journal of Industrial Microbiology & Biotechnology* **42:** 1039-1047.

Ishikawa, M., Okamoto-Kainuma, A., Jochi, T., Suzuki, I., Matsui, K., Kaga, T. and Koizumi, Y. (2010a). Cloning and characterization of *grpE* in *Acetobacter pasteurianus* NBRC 3283. *Journal of Bioscience and Bioengineering* **109:** 25-31.

Ishikawa, M., Okamoto-Kainuma, A., Matsui, K., Takigishi, A., Kaga, T. and Koizumi, Y. (2010b). Cloning and characterization of *clpB* in *Acetobacter pasteurianus* NBRC 3283. *Journal of Bioscience and Bioengineering* **110:** 69-71.

Kallnik, V., Meyer, M., Deppenmeier, U. and Schweiger, P. (2010). Construction of expression vectors for protein production in *Gluconobacter oxydans. Journal of Biotechnology* **150:** 460-465.

Kanchanarach, W., Theeragool, G., Inoue, T., Yakushi, T., Adachi, O. and Matsushita, K. (2010a). Acetic acid fermentation of *Acetobacter pasteurianus*: Relationship between acetic acid resistance and pellicle polysaccharide formation. *Bioscience, Biotechnology and Biochemistry* **74:** 1591-1597.

Kanchanarach, W., Theeragool, G., Yakushi, T., Toyama, H., Adachi, O. and Matsushita, K. (2010b). Characterization of thermotolerant *Acetobacter pasteurianus* strains and their quinoprotein alcohol dehydrogenases. *Applied Microbiology and Biotechnology* **85:** 741-751.

Kawai, S., Goda-Tsutsumi, M., Yakushi, T., Kano, K. and Matsushita, K. (2013). Heterologous overexpression and characterization of a flavoprotein-cytochrome *c* complex fructose dehydrogenase of *Gluconobacter japonicus* NBRC3260. *Applied and Environmental Microbiology* **79:** 1654-1660.

Keliang, G. and Dongzhi, W. (2006). Asymmetric oxidation by *Gluconobacter oxydans*. *Applied Microbiology and Biotechnology* **70:** 135-139.

Konate, M., Akpa, E.E., Koffi, L.B., Kra, K.A.S., Megnanou, R.-M. and Niamke, S. (2014). Isolation of thermotolerant and high acetic acid-producing *Acetobacter pasteurianus* from Ivorian palm wine. *Emirates Journal of Food and Agriculture* **23:** 773-785.

Konate, M., Akpa, E.E., Bernadette, G.G., Koffi, L.B., Honore, O.G. and Niamke, S. (2015). Banana vinegars production using thermotolerant *Acetobacter pasteurianus* isolated from Ivorian palm wine. *Journal of Food Research* **4:** 92-103.

Kovach, M.E., Phillips, R.W., Elzer, P.H., Roop, R.M. and Peterson, K.M. (1994). pBBR1MCS: A broad-host-range cloning vector. *BioTechniques* **16:** 800-802.

Kovach, M.E., Elzer, P.H., Steven Hill, D., Robertson, G.T., Farris, M.A., Roop, R.M. and Peterson, K.M. (1995). Four new derivatives of the broad-host-range cloning vector pBBR1MCS, carrying different antibiotic-resistance cassettes. *Gene* **166:** 175-176.

Li, M.-H., Wu, J., Liu, X., Lin, J.-P., Wei, D.-Z. and Chen, H. (2010). Enhanced production of dihydroxyacetone from glycerol by overexpression of glycerol dehydrogenase in an alcohol dehydrogenase-deficient mutant of *Gluconobacter oxydans*. *Bioresource Technology* **101:** 8294-8299.

Lu, S.-F., Lee, F.-L. and Chen, H.-K. (1999). A thermotolerant and high acetic acid-producing bacterium *Acetobacter* sp. I14-2. *Journal of Applied Microbiology* **86:** 55-62.

Lu, L., Wei, L., Zhu, K., Wei, D. and Hua, Q. (2012). Combining metabolic engineering and adaptive evolution to enhance the production of dihydroxyacetone from glycerol by *Gluconobacter oxydans* in a low-cost way. *Bioresource Technology* **117:** 317-324.

Mamlouk, D. and Gullo, M. (2013). Acetic acid bacteria: Physiology and carbon sources oxidation. *Indian Journal of Microbiology* **53:** 377-384.

Marx, C.J. and Lidstrom, M.E. (2001). Development of improved versatile broad-host-range vectors for use in methylotrophs and other Gram-negative bacteria. *Microbiology* (Reading, England) **147:** 2065-2075.

Matsushita, K., Toyama, H. and Adachi, O. (1994). Respiratory chain and bioenergetics of acetic acid bacteria. *In:* A.H. Rose and D.W. Tempest (Eds.). *Advances in Microbial Physiology*. Academic Press, London, pp. 247-301.

Matsushita, K., Fujii, Y., Ano, Y., Toyama, H., Shinjoh, M., Tomiyama, N., Miyazaki, T., Sugisawa, T., Hoshino, T. and Adachi, O. (2003). 5-keto-D-gluconate production is catalyzed by a quinoprotein glycerol dehydrogenase, major polyol dehydrogenase, in *Gluconobacter* species. *Applied and Environmental Microbiology* **69:** 1959-1966.

Matsushita, K., Azuma, Y., Kosaka, T., Yakushi, T., Hoshida, H., Akada, R. and Yamada, M. (2016). Genomic analyses of thermotolerant microorganisms used for high-temperature fermentations. *Bioscience, Biotechnology and Biochemistry* **80:** 655-668.

Matsutani, M., Hirakawa, H., Saichana, N., Soemphol, W., Yakushi, T. and Matsushita, K. (2012). Genome-wide phylogenetic analysis of differences in thermotolerance among closely related *Acetobacter pasteurianus* strains. *Microbiology* **158:** 229-239.

Matsutani, M., Nishikura, M., Saichana, N., Hatano, T., Masud-Tippayasak, U., Theergool, G., Yakushi, T. and Matsushita, K. (2013). Adaptive mutation of *Acetobacter pasteurianus* SKU1108 enhances acetic acid fermentation ability at high temperature. *Journal of Biotechnology* **165:** 109-119.

Merfort, M., Herrmann, U., Bringer-Meyer, S. and Sahm, H. (2006a). High-yield 5-keto-D-gluconic acid formation is mediated by soluble and membrane-bound gluconate-5-dehydrogenases of *Gluconobacter oxydans*. *Applied Microbiology and Biotechnology* **73:** 443-451.

Merfort, M., Herrmann, U., Ha, S.-W., Elfari, M., Bringer-Meyer, S., Görisch, H. and Sahm, H. (2006b). Modification of the membrane-bound glucose oxidation system in *Gluconobacter oxydans* significantly increases gluconate and 5-keto-D-gluconic acid accumulation. *Biotechnology Journal* **1:** 556-563.

Meyer, M., Schweiger, P. and Deppenmeier, U. (2013). Effects of membrane-bound glucose dehydrogenase overproduction on the respiratory chain of *Gluconobacter oxydans*. *Applied Microbiology and Biotechnology* **97:** 3457-3466.

Moonmangmee, D., Adachi, O., Ano, Y., Shinagawa, E., Toyama, H., Theeragool, G., Lotong, N. and Matsushita, K. (2000). Isolation and characterization of thermotolerant *Gluconobacter* strains catalyzing oxidative fermentation at higher temperatures. *Bioscience, Biotechnology and Biochemistry* **64:** 2306-2315.

Moonmangmee, D., Adachi, O., Shinagawa, E., Toyama, H., Theeragool, G., Lotong, N. and Matsushita, K. (2002a). L-Erythrulose production by oxidative fermentation is catalyzed by PQQ-containing membrane-bound dehydrogenase. *Bioscience, Biotechnology and Biochemistry* **66:** 307-318.

Moonmangmee, S., Toyama, H., Adachi, O., Theeragool, G., Lotong, N. and Matsushit, K. (2002b). Purification and characterization of a novel polysaccharide involved in the pellicle produced by a thermotolerant *Acetobacter* strain. *Bioscience, Biotechnology and Biochemistry* **66:** 777-783.

Mounir, M., Shafiei, R., Zarmehrkhorshid, R., Hamouda, A., Ismaili Alaoui, M. and Thonart, P. (2016). Simultaneous production of acetic and gluconic acids by a thermotolerant Acetobacter strain during acetous fermentation in a bioreactor. *Journal of Bioscience and Bioengineering* **121:** 166-171.

Natera Marín, R., Castro Mejias, R., Garcia Moreno, M.V., Garcia Rowe, F. and Garcia Barroso, C. (2002). Headspace solid-phase microextraction analysis of aroma compounds in vinegar. *Journal of Chromatography* **A967:** 261-267.

Ndoye, B., Lebecque, S., Dubois-Dauphin, R., Tounkara, L., Guiro, A-T., Kere, C., Diawara, B. and Thonart, P. (2006). Thermoresistant properties of acetic acids bacteria isolated from tropical products of Sub-Saharan Africa and destined to industrial vinegar. *Enzyme and Microbial Technology* **39:** 916-923.

Ndoye, B., Cleenwerck, I., Engelbeen, K., Dubois-Dauphin, R., Guiro, A.T., Van Trappen, S., Willems, A. and Thonart, P. (2007). Acetobacter senegalensis sp. nov., a thermotolerant acetic acid bacterium isolated in Senegal (sub-Saharan Africa) from mango fruit (*Mangifera indica* L.). *International Journal of Systematic and Evolutionary Microbiology* **57:** 1576-1581.

Nishikura-Imamura, S., Matsutani, M., Insomphun, C., Vangnai, A.S., Toyama, H., Yakushi, T., Abe, T., Adachi, O. and Matsushita, K. (2014). Overexpression of a type II 3-dehydroquinate dehydratase enhances the biotransformation of quinate to 3-dehydroshikimate in *Gluconobacter oxydans*. *Applied Microbiology and Biotechnology* **98:** 2955-2963.

Obranić, S., Babić, F. and Maravić-Vlahoviček, G. (2013). Improvement of pBBR1MCS plasmids, a very useful series of broad-host range cloning vectors. *Plasmid* **70:** 263-267.

Ohmori, S., Masai, H., Arima, K. and Beppu, T. (1980). Isolation and identification of acetic acid bacteria for submerged acetic acid fermentation at high temperature. *Agricultural and Biological Chemistry* **44:** 2901-2906.

Okamoto-Kainuma, A., Yan, W., Kadono, S., Tayama, K., Koizumi, Y. and Yanagida, F. (2002). Cloning and characterization of *groESL* operon in *Acetobacter aceti*. *Journal of Bioscience and Bioengineering* **94:** 140-147.

Okamoto-Kainuma, A., Yan, W., Fukaya, M., Tukamoto, Y., Ishikawa, M. and Koizumi, Y. (2004). Cloning and characterization of the *dnaKJ* operon in *Acetobacter aceti*. *Journal of Bioscience and Bioengineering* **97:** 339-342.

Okamoto-Kainuma, A., Ishikawa, M., Nakamura, H., Fukazawa, S., Tanaka, N., Yamagami, K. and Koizumi, Y. (2011). Characterization of rpoH in Acetobacter pasteurianus NBRC3283. *Journal of Bioscience and Bioengineering* **111:** 429-432.

Perumpuli, P.A.B.N., Watanabe, T. and Toyama, H. (2014a). Identification and characterization of thermotolerant acetic acid bacteria strains isolated from coconut water vinegar in Sri Lanka. *Bioscience, Biotechnology and Biochemistry* **78:** 533-541.

Perumpuli, P.A.B.N., Watanabe, T. and Toyama, H. (2014b). Pellicle of thermotolerant *Acetobacter pasteurianus* strains: Characterization of the polysaccharides and of the induction patterns. *Journal of Bioscience and Bioengineering* **118:** 134-138.

Saeki, A., Theeragool, G., Matsushita, K., Toyama, H., Lothong, N. and Adachi, O. (1997). Development of thermotolerant acetic acid bacteria useful for vinegar fermentation at higher temperatures. *Bioscience, Biotechnology and Biochemistry* **61:** 138-145.

Saichana, I., Moonmangmee, D., Adachi, O., Matsushita, K. and Toyama, H. (2009). Screening of thermotolerant *Gluconobacter* strains for production of 5-keto-D-gluconic acid and disruption of flavin adenine dinucleotide-containing D-gluconate dehydrogenase. *Applied and Environmental Microbiology* **75:** 4240-4247.

Saichana, N., Matsushita, K., Adachi, O., Frébort, I. and Frebortova, J. (2015). Acetic acid bacteria: A group of bacteria with versatile biotechnological applications. *Biotechnology Advances* **33:** 1260-1271.

Saito, Y., Ishii, Y., Hayashi, H., Imao, Y., Akashi, T., Yoshikawa, K., Noguchi, Y., Soeda, S., Yoshida, M., Niwa, M., Hosoda, J. and Shimomura, K. (1997). Cloning of genes coding for L-sorbose and L-sorbosone dehydrogenases from *Gluconobacter oxydans* and microbial production of 2-keto-L-gulonate, a precursor of L-ascorbic acid, in a recombinant *G. oxydans* strain. *Applied and Environmental Microbiology* **63:** 454-460.

Schleyer, U., Bringer-Meyer, S. and Sahm, H. (2008). An easy cloning and expression vector system for *Gluconobacter oxydans*. *International Journal of Food Microbiology* **125:** 91-95.

Schleyer, U., Bringer-Meyer, S. and Sahm, H. (2009). Corrigendum to 'An easy cloning and expression vector system for *Gluconobacter oxydan*' [*Int. J. of Food Microbiol.* **125** (2008) 91-95]. *International Journal of Food Microbiology* **130:** 76.

Schmidhauser, T.J. and Helinski, D.R. (1985). Regions of broad-host-range plasmid RK2 involved in replication and stable maintenance in nine species of gram-negative bacteria. *Journal of Bacteriology* **164:** 446-455.

Shi, L., Li, K., Zhang, H., Liu, X., Lin, J. and Wei, D. (2014). Identification of a novel promoter *gHp0169* for gene expression in *Gluconobacter oxydans*. *Journal of Biotechnology* **175:** 69-74.

Shi, Y.-Y., Li, K.-F., Lin, J.-P., Yang, S.-L. and Wei, D.-Z. (2015). Engineered expression vectors significantly enhanced the production of 2-keto-d-gluconic acid by *Gluconobacter oxidans*. *Journal of Agricultural and Food Chemistry* **63:** 5492-5498.

Shinagawa, E., Matsushita, K., Adachi, O. and Ameyama, M. (1982). Purification and characterization of D-sorbitol dehydrogenase from membrane of *Gluconobacter suboxydans* var. α. *Agricultural and Biological Chemistry* **46:** 135-141.

Shinagawa, E., Matsushita, K., Adachi, O. and Ameyama, M. (1983). Selective production of 5-keto-D-gluconate by *Gluconobacter* strains. *Journal of Fermentation Technology* **61:** 359-363.

Simon, R., Priefer, U. and Pühler, A. (1983). A broad host range mobilization system for *in vivo* genetic engineering: Transposon mutagenesis in Gram negative bacteria. *Bio/Technology* **1:** 784-791.

Soemphol, W., Adachi, O., Matsushita, K. and Toyama, H. (2008). Distinct physiological roles of two membrane-bound dehydrogenases responsible for D-sorbitol oxidation in *Gluconobacter frateurii*. *Bioscience, Biotechnology and Biochemistry* **72:** 842-850.

Soemphol, W., Deeraksa, A., Matsutani, M., Yakushi, T., Toyama, H., Adachi, O., Yamada, M. and Matsushita, K. (2011). Global analysis of the genes involved in the thermotolerance mechanism of thermotolerant *Acetobacter tropicalis* SKU1100. *Bioscience, Biotechnology and Biochemistry* **75:** 1921-1928.

Tasanapak, K., Masud-Tippayasak, U., Matsushita, K., Yongmanitchai, W. and Theeragool, G. (2013). Influence of *Acetobacter pasteurianus* SKU1108 *aspS* gene expression on *Escherichia coli* morphology. *Journal of Microbiology* **51:** 783-790.

Tonouchi, N., Tsuchida, T., Yoshinaga, F., Horinouchi, S. and Beppu, T. (1994). A host-vector system for a cellulose-producing *Acetobacter* strain. *Bioscience, Biotechnology and Biochemistry* **58:** 1899-1901.

Tonouchi, N., Sugiyama, M. and Yokozeki, K. (2003). Construction of a vector plasmid for use in *Gluconobacter oxydans*. *Bioscience, Biotechnology and Biochemistry* **67:** 211-213.

Toyama, H., Soemphol, W., Moonmangmee, D., Adachi, O. and Matsushita, K. (2005). Molecular properties of membrane-bound FAD-containing D-sorbitol dehydrogenase from thermotolerant *Gluconobacter frateurii* isolated from Thailand. *Biscience, Biotechnology and Biochemistry* **69:** 1120-1129.

Trcek, J., Raspor, P. and Teuber, M. (2000). Molecular identification of *Acetobacter* isolates from submerged vinegar production, sequence analysis of plasmid pJK2-1 and application in the development of a cloning vector. *Applied Microbiology and Biotechnology* **53:** 289-295.
Wei, Y., Wu, X., Xu, Z., Tan, Z. and Jia, S. (2015). A thermotolerant *Acetobacter pasteurianus* T24 achieving acetic acid fermentation at high temperature in self-adaption experiment. *In:* T.-C. Zhang and M. Nakajima (Eds.). *Advances in Applied Biotechnology*. Springer, Berlin Heidelberg, pp. 287-293.
Yao, W., Ouattara, H., Goualie, B., Soumahoro, S. and Niamke, S. (2014). Analysis of some functional properties of acetic acid bacteria involved in Ivorian cocoa fermentation. *Journal of Applied Biosciences* **75:** 6282-6290.
Zhang, L., Lin, J., Ma, Y., Wei, D. and Sun, M. (2010). Construction of a novel shuttle vector for use in *Gluconobacter oxydans*. *Molecular Biotechnology* **46:** 227-233.
Zhu, K., Lu, L., Wei, L., Wei, D., Imanaka, T. and Hua, Q. (2011). Modification and evolution of *Gluconobacter oxydans* for enhanced growth and biotransformation capabilities at low glucose concentration. *Molecular Biotechnology* **49:** 56-64.

7

Identification Techniques of Acetic Acid Bacteria: Comparison between MALDI-TOF MS and Molecular Biology Techniques

Cristina Andrés-Barrao[1], François Barja[2*], Ruben Ortega Pérez[2], Marie-Louise Chappuis[2], Sarah Braito[2] and Ana Hospital Bravo[2]

[1] King Abdullah University of Science and Technology,
Division of Biological and Environmental Sciences and Engineering,
Center for Desert Agriculture, 4700 Thuwal 23955-6900, Saudi Arabia

[2] University of Geneva, Department of Botany and Plant Biology, Microbiology Unit,
Quai Ernest-Ansermet 30 (Sciences III), 1211 Geneva 4, Switzerland

Introduction

Acetic acid bacteria (AAB) are widespread microorganisms in nature, commonly found on different types of tropical flowers and fruits, in warm and humid climates and in rhizosphere soils (Yukphan et al. 2004, Trček and Barja 2015, Li et al. 2015). Some of them have established a symbiotic relationship with insects, which disseminate them throughout the environment (Crotti et al. 2010, Chouaia et al. 2014, Li et al. 2015) while some are considered opportunistic human pathogens (Greenberg et al. 2006, 2007, Alauzet et al. 2010). Furthermore, AAB present a great interest for industries due to their unique metabolic characteristics. Among other biotechnological applications, AAB play the main role in the natural oxidative fermentation of alcoholic broths; hence are used in industrial production of fermented food and beverages. Among these products, vinegar has one of the highest added value.

Despite the importance of vinegar production in the overall economy of industrialized countries (Solieri and Giudici 2009), the process is not yet sufficiently controlled, mostly due to a lack of knowledge concerning the microbial population that takes part in it.

As mentioned before, AAB are widely disseminated in nature and are involved in many biotechnological processes; however, they can also be spoiling microbiota in wine, beer, juices, fruits, etc (Bartowsky and Henschke 2008, Kregiel 2015). Given their relevance in industry, medicine and agriculture, the isolation and quick identification of these microorganisms is required to obtain a better understanding of oxidative fermentations.

Different methods have been used to identify AAB from fermented products, such as vinegar, wine, beer or fermented cocoa beans (first step in the manufacture of chocolate), and their current classification is the result of a polyphasic taxonomic approach, which combines phenotypic, genotypic and chemotaxonomic characterization (Cleenwerck and De Vos 2008). This polyphasic approach has yielded an array of molecular biology methods targeting different taxonomic markers. However, these methods are still labor-intensive, expensive and inaccurate. They lack the rapidity and simplicity needed for

*Corresponding author: François.Barja@unige.ch

routine analyses in industry, since they deal with a high number of samples on a daily basis and also underscore the need for faster, cost-effective and accurate methods.

In this regard, matrix-assisted laser desorption/ionization time-of-flight mass spectrometry (MALDI-TOF MS) has proved useful in categorization and identification of microbes, especially for routine identification of environmental and clinical microorganisms (Bizzini et al. 2010, Benagli et al. 2011), and was recently considered as a good alternative for quick and accurate identification of AAB. The accuracy of bacterial identification, using MALDI-TOF MS, largely depends on the robustness of the reference database and therefore, the construction and validation of such a reference database for the identification of AAB is crucial.

Identification of Acetic Acid Bacteria, an Historical Approach

The first recorded mention of vinegar dates back to 5000 B.C., when the Babylonians used the fruit and sap of date palm as raw material to produce alcoholic beverages, which in turn are naturally transformed into vinegar if left in contact with air (Bourgeois and Barja 2009). For centuries, vinegar has been obtained from alcohol-containing beverages or solutions through spontaneous oxidation without understanding the mechanism of the process. Progressively it was recognized that vinegar was produced by wine souring due to living organisms in contact with air and forming a film on the liquid surface. This gelatinous film was called 'mother of vinegar' (Boerhaave 1732). According to Asai (1968) and Bourgeois and Barja (2009), the first scientist to suggest that the mother of vinegar was composed of a 'vegetal substance' or 'flower' was Boerhaave (1668-1738). He also highlighted the importance of air in the production of vinegar and is given credit for the construction of the first packed generator for quick production of vinegar (Conner and Allgeier 1976). One century later, a botanist named Persoon (1822) was the first to attribute vinegar production to the viscous film formed on the surface of wine and beer left open to the air. He was convinced that this film was a fungus and he proposed the name *Mycoderma*. Another scientist, Desmazières (1826), distinguished between the viscous material formed on the top of acidified wine and on the surface of beer, calling it *Mycoderma vini* and *Mycoderma cerevisiae*, respectively. Some years later, Kützing (1837) named the surface film formed during transformation of alcohol into acetic acid as *Ulvina aceti*. Thomson (1852) suggested a change of the name to *Mycoderma aceti*. And in 1879, Hansen (1879) argued that the microbial flora involved in the transformation of wine into vinegar was composed of various bacterial species. In the meantime, Döbereiner (1823) published the now well-known equation: ethanol + oxygen = acetic acid + water. Finally, the real mechanism of vinegar elaboration was understood only when Pasteur published, in 1864, his famous memoir on acetous fermentation. Louis Pasteur (1924) identified scientifically the five criteria indispensable for production of vinegar: (1) alcohol: in wine, cider or other alcoholic beverages, (2) oxygen: that of air, (3) the ferment: *Mycoderma aceti*, (4) nutrients: such as sugar and proteins naturally present in wine, enhancing the growth of the bacteria, and (5) temperature: between 20 and 35°C. Nevertheless, Pasteur was unable to identify *Mycoderma* as a bacterium. It was finally recognized as such by von Knieriem and Mayer (1873) and also by Cohn (1872).

The first trials of classification of these microorganisms were due to Hansen (1894), based on the presence of a film in liquid media and its reaction with iodine. Hansen classified them into three kinds of acetic acid bacteria: *Bacterium aceti*, *Bacterium pasteurianum* and *Bacterium kützingianum*. He also affirmed that acetification of beer and wine was produced by the action of these bacteria. Beijerinck, in 1898, was the first to call acetic acid bacteria with the generic name *Acetobacter*. He mentioned four species of AAB: *Bacterium aceti*, *Bacterium rancens*, *Bacterium pasteurianum* and *Bacterium xylinum*.

Hooft F. (1925) proposed a classification based on biochemical and physiological criteria. Some years later, Asai (1934) isolated from fruits about 38 strains of bacteria, some of which were similar to those previously described as AAB, whilst others had a somewhat similar oxidative metabolism. He proposed to divide these bacteria under two genera, *Acetobacter* and *Gluconobacter*. The genus *Acetobacter* was further separated in two subgenera, that is to say, *Eucetobacter*, which oxidized ethanol, but not glucose, and *Acetogluconobacter*, which oxidized ethanol to acetic acid, weakly glucose to gluconic acid and mannitol to fructose (Carr and Shimwell 1961). The genus *Gluconobacter* was subdivided in *Eugluconobacter*, which oxidized glucose, but not ethanol, and *Gluconoacetobacter*, which oxidized glucose to gluconic acid, and weakly ethanol to acetic acid (Carr and Shimwell 1961).

Frateur (1950) had the merit of greatly reducing the number of species using five criteria: catalase, over-oxidation of acetic acid to CO_2 and H_2O, oxidation of lactose to CO_2 and H_2O, oxidation of glycerol to dihydroxyacetone, and production of gluconic acid from glucose. He proposed the subdivision of *Acetobacter* into four biochemical groups: *Peroxydans*, *Oxydans*, *Mesoxydans* and *Suboxydans*. Later Leifson (1954) made a distinction of AAB in two entirely different types. According his observations, he concluded that the *Acetobacter* genus should be divided into two genera: *Acetobacter* and *Acetomonas* gen. nov. The re-defined *Acetobacter* thus would include only peritrichously or non-flagellated species of like physiological characteristics, while the new genus *Acetomonas* would include polarly flagellated or non-motile species of like physiological characteristics. *Acetomonas* coincides with the *Gluconobacter* genus, with respect not only to flagellation but also to physiological and biochemical properties (Leifson 1954, Carr and Shimwell 1961, Asai 1968). Carr and Shimwell (1961) proposed to replace *Acetomonas* by *Gluconobacter*. Since then, a number of different species has been added to both the genera, and 18 genera are currently included in the group of AAB (*see* Chapter 1). The characteristics and classification for the AAB have been collected in *Bergey's Manual of Systematic Bacteriology*. The classification and taxonomy of these microorganisms, initially based on morphological and physiological criteria, have been subjected to continuous reorientations and remodeling. During the past 20 years, the description of new taxa and reclassification of previously described strains and species of AAB have increased exponentially, basically due to the application of molecular techniques. Moreover, the amount of complete bacterial whole-genome sequences that are currently available is due to the improvement of next-generation sequencing (NGS) technologies and affordable costs. This information will enable a whole revision of the current taxonomy of microorganisms in general and AAB in particular in the near future. In the meantime, the more recent taxonomic revision of AAB has been done by various researchers in this field, principally Yamada.

Molecular Techniques Used for Identification of Acetic Acid Bacteria

Cowan (1955) said: "In the biological sciences, classification is a compromise between the dynamics of evolution and the stasis of the present; with microbes there is less apparent stasis, for, with rapid multiplication, there are greater opportunities for observing evolution in action. All classifications are subjective... A classification consists of two elements: the objects to be classified and the subject who feels the urge to classify them."

Since the existence of the first microscopic organism was discovered by Hooke and van Leeuwenhoek in 1665-1683 (Gest 2004), the classification of microorganisms of environmental, industrial or clinical interest, initially based on morphological and physiological characteristics, has been subjected to continuous changes. AAB strains isolated from natural sources have, similarly like other microorganisms, suffered a high number of redefinitions and reclassifications along the history. The number of AAB new taxa have increased exponentially since the application of molecular techniques to replace the phenotypic methods traditionally used.

Among AAB, it is well known that different species show different capabilities to perform the transformation of ethanol into acetic acid during acetic acid fermentation. Hence, their classification is of paramount importance from the industrial point of view as well as for fundamental microbiology. In this context, ecological studies to analyze the bacterial population involved in acetic acid fermentation processes, at the level of strain and species, require the use of identification methods accurate and reproducible.

For the identification and typing of AAB, a polyphasic approach has been generally used (Cleenwerck and De Vos 2008). Several DNA-based methodologies were developed and have been extensively applied during the last 20 years. First, methods based on the analysis of Restriction Fragment Length Polymorphisms (RFLP) or direct sequencing of the 16S rRNA gene or the 16S-23S intergenic spacer region (ITS) were used to assess the population dynamics of AAB and lactic acid bacteria (LAB) during alcoholic and acetic fermentations (in the production of wine and vinegar) (Ruíz et al. 2000, González et al. 2005, González et al. 2006a, Andrés-Barrao et al. 2011). Although these techniques were extensively used for some time, they started to lose accuracy when the number of described AAB species began to increase. The discovery of novel AAB species has shot up during the last years due to ecological studies on yet unexplored niches, with the concomitant need to develop alternative methods for its genotyping and identification, such as amplified fragment-length polymorphism (AFLP), the Repetitive Element PCR fingerprinting (REP-PCR) by using Enterobacterial Repetitive Intergenic Consensus primers (ERIC-PCR) or the $(GTG)_5$ primer ($(GTG)_5$-rep-PCR) (Nanda et al. 2001, González et al. 2005, De Vuyst et al. 2008, Cleenwerck et al. 2009). As the 16S rRNA gene is sometimes unable to discriminate among closely related strains, alternative genomic markers have been developed; for instance the DNA repair protein RecA (*recA*), the first subunit of the pyrroloquinoline quinone (PQQ)-dependent alcohol dehydrogenase (*adhA*), the chaperones DnaJ and GroEL (*dnaJ*, *groEL*), the β subunit of bacterial RNA polymerase (*rpoB*) (Trček 2005, Cleenwerck et al. 2010), or the elongation factor Tu (*tuf*) (Yetiman and Kesmen 2015). An exhaustive list of these methods is compiled in Table 1.

Identification of Acetic Acid Bacteria at the Level of Species

Restriction Fragment Length Polymorphism Analysis of PCR-Amplified Fragments (PCR-RFLP)

PCR- RFLP-based analysis of the 16S rRNA gene was a very popular technique for genetic analysis of AAB and was later replaced by direct sequencing and comparison against public databases. It has been applied for detection of interspecies as well as intraspecies variations. The first step in a PCR-RFLP method consists of the amplification of the targeted genomic region, the 16S rRNA gene in this case, followed by the digestion of the amplified products with an appropriate set of restriction enzymes and analysis of the resulting fragments by polyacrylamide or agarose gel electrophoresis (Rasmussen 2012). The identification of bacterial species by using this method relays in the comparison of the restriction fragment patterns with that of well-characterized reference strains (Fajardo et al. 2009). The 16S-RFLP was extensively used to study the bacterial population dynamics of traditional acetic acid fermentations during the early 2000's. The restriction by using *TaqI* enzyme was sufficient to discriminate among the genera *Acetobacter*, *Gluconobacter* and *Gluconacetobacter* (Poblet et al. 2000). At this time, only three genera and 15 species of AAB were described and most of them could be differentiated using a set of up to four endonucleases: *TaqI*, *RsaI*, *AluI* and *MspI* (Poblet et al. 2000, Ruíz et al. 2000).

Table 1: Molecular biology methods used for the identification and typing of AAB, and their maximum level of discrimination (modified from Sengun and Karabiyikli 2011 and Trček and Barja 2015)

Method	*Level*[#]	*References*
PCR-RFLP profiling:		
16S rRNA gene	G/Sp	Poblet et al. 2000, Ruíz et al. 2000, González et al. 2006[a]
16S-23S rRNA genes	Sp/St	Andrés-Barrao et al. 2011
16S-23S internal transcribed spacer (ITS)	Sp/St	Trček and Teuber 2002, Trček 2005, González and Mas 2011
16S-23S-5S rRNA operon	G/Sp	Gullo et al. 2006
PCR and direct sequencing:		
16S rRNA gene	G/Sp	González et al. 2006a, Vegas et al. 2010, Andrés-Barrao et al. 2016
16S-23S internal transcribed spacer (ITS)	G/Sp	Greenberg et al. 2006, Andrés-Barrao et al. 2011, González and Mas 2011
First subunit of the PQQ-alcohol dehydrogenase (*adhA*)	Sp	Trček 2005
DNA repair protein RecA (*recA*)	Sp	Greenberg et al. 2006
Housekeeping genes (*dnaK*, *groEL*, *rpoB*)	Sp	Cleenwerck et al. 2010, Li et al. 2014, Andrés-Barrao et al. 2016
nifH, *nifD*[*]	Sp	Loganathan and Nair 2004, Dutta and Gachhui 2006, 2007
Elongation factor Tu (*tuf*)	Sp	Huang et al. 2014, Yetiman and Kesmen 2015
Genotyping by DNA profiling:		
Amplified Fragment Length Polymorphism (AFLP) fingerprinting	St	Cleenwerck et al. 2009
Random Amplified Polymorphic DNA (RAPD)	St	Nanda et al. 2001, Bartowsky et al. 2003
Enterobacterial Repetitive Intergenic Consensus-PCR (ERIC-PCR)	St	Nanda et al. 2001, González et al. 2004, González et al. 2005, Vegas et al. 2010, Sechi et al. 1998
Repetitive Extragenic Palindromic-PCR (REP-PCR)	St	González et al. 2004, González et al. 2005
Amplification of Repetitive Elements using the $(GTG)_5$ primer ($(GTG)_5$-rep-PCR)	St	De Vuyst et al. 2008, Papalexandratou et al. 2009, Vegas et al. 2010, Andrés-Barrao et al. 2016

(*Contd.*)

Culture-independent:		
Denaturing Gradient Gel Electrophoresis (DGGE)	G/Sp	Andorrà et al. 2008, De Vero et al. 2006, De Vero and Giudici 2008, Haruta et al. 2006, Yetiman and Kesmen 2015, Andrés-Barrao et al. 2016
Temporal Temperature Gradient Gel Electrophoresis (TTGE)	G/Sp	Ilabaca et al. 2008
$(GTG)_5$-rep-PCR fingerprinting	St	Yetiman and Kesmen 2015, Andrés-Barrao et al. 2016
Hybridization:		
Whole-genome DNA-DNA hybridization	Sp/St	Boesch et al. 1998, Fuentes-Ramírez et al. 2001, Cleenwerck et al. 2002, Lisdiyanti et al. 2001, González et al. 2006a
Fluorescence in situ hybridization (FISH) targeting the 16S rRNA gene sequence	Sp	Franke-Whittle et al. 2005
Others:		
Nested PCR[$]	G/Sp	González et al. 2006b
Real-Time PCR (RT-PCR)[$]	G/Sp	González et al. 2006b, Andorrà et al. 2008
RT-PCR with TaqMan–MGB probe[$]	G/Sp	Torija et al. 2010, Valera et al. 2013
Direct Epifluorescent Filter Technique (DEFT)[$$]	-	Du Toit et al. 2005, Mesa et al. 2003
Plasmid profiling	St	Teuber et al. 1987, Mariette et al. 1991, Trček 2015
Pulse Field Gel Electrophoreiss	St	Nanda et al. 2001
MALDI-TOF mass spectrometry	Sp/St	Andrés-Barrao et al. 2013, Wieme et al. 2014a, b, Spitaels et al. 2015

[#]G = genus, Sp = species, St = Strain
[*]Set of genes analyzed for the identification of a specific set of nitrogen-fixing AAB species
[$]Detection and quantification of AAB at low concentrations
[$$]Evaluation of the viability of AAB during acetic fermentation

However, due to the high conservation of 16S rRNA gene sequence among closely related strains, the use of others genomic regions is often needed to differentiate them (Rojas et al. 2009). To overcome the problem of high conservation of the 16S rRNA gene, PCR-RFLP of the spacer region between the 16S and 23S rRNA genes (ITS) was developed. Intergenic sequences are known to have higher variability than functional sequences, thus being able to resolve closely related species (Barry et al. 1991, Navarro et al. 1992). For instance, *Gluconacetobacter xylinus* and *Gluconacetobacter europaeus* (currently *Komagataeibacter xylinus* and *Komagataeibacter europaeus*) share 99 per cent of their respective 16S rRNA gene sequence (only five nucleotides difference) and could not be differentiated unless restriction of 16S-23S ITS with *PvuII* (González et al. 2006a). Nevertheless, due to the frequent interspecific variations of intergenic sequences among strains of the same species, PCR-RFLP of the 16S-23S ITS was proposed to be an adequate method to detect intraspecific variability, though not useful for the identification at species level of AAB isolates (Ruíz et al. 2000). PCR-RFLP of the entire 16S-23S-5S rRNA operon using *RsaI* enzyme was also used for identification of isolated AAB from Traditional Balsamic Vinegar (TBV) (Gullo et al. 2006). Although this method could not confidently identify bacterial isolates to the species level, the authors were able to ascribe the most widespread AAB in TBV as probably *Ga. xylinus* (currently *K. xylinus*).

Fingerprinting RFLP methods are indirect, but rapid and simple techniques that do not require state-of-art instruments and can be implemented in most laboratories. In addition, the design of primers and PCR-RFLP analyses are generally easy and can be accomplished by using public available programs. Further, the training of laboratory staff does not pose any serious difficulty.

Disadvantages include requirement of specific endonucleases and difficulties in identifying the exact small insertion or deletion. Moreover, some restriction enzymes are expensive and the exact genotyping cannot be reached when there is more than one nucleotide variation in a recognition site of a specific restriction enzyme. It is relatively time-consuming and is not suitable for high-throughput analysis (Rasmussen 2012). But the main drawback of these techniques is that in most cases they provide differentiation information. Different restriction patterns indicate different bacterial species, whereas identical patterns do not necessarily characterize identical taxa; they might be generated by different but closely related species.

DNA Amplification and Direct Sequencing

With the number of AAB species rapidly increasing and the decreasing costs of sequencing, PCR-RFLP analysis has become obsolete for the identification of novel isolates. Instead, the direct sequencing of specific genomic regions and their comparison with sequences in public databases was developed.

16S rRNA gene is the gold standard molecular target for phylogenetic analysis and for identification of novel taxa because of its high conservation among prokaryotes, with a conserved function over time that suggests random sequence changes are a more accurate measure of time (evolution); its large size (1450 bp), enough for informatics purposes; and the availability of a large number of sequences in public databases (Janda and Abbott 2007). Although the meaning of similarity scores >97 per cent is not as clear, gene sequences showing similarity scores <97 per cent are considered to belong to different species (Petty 2007). Although the amplification and sequencing of 16S rRNA gene has been used extensively for identification of AAB, phylogenetic analyses based on alternative taxonomic targets have been developed to differentiate among closely related species.

Despite the high intraspecies variability that this region might show, researchers suggested the usefulness of including the 16S-23S ITS as a part of the polyphasic approach required for further classification of AAB (González and Mas 2011). Phylogenetic analysis of this region was used extensively for the taxonomy and classification of members of the genus *Gluconobacter*. Similarly, Greenberg et al. (2006) include the phylogeny based on 16S-23S ITS in the study leading to characterization of *Granulibacter bethesdensis* sp. nov., the first human opportunistic pathogenic AAB that was described and isolated from a patient of chronic granulomatous disease (CGD). Another group of AAB very closely related, *Acetobacter cerevisiae* and *Acetobacter malorum*, share 99.9 per cent of their 16S rRNA gene sequences, with a difference of only two nucleotides; so discrimination between them was not possible unless an alternative taxonomical marker was used. In this case, the phylogenetic analysis by using 16S-23S ITS allowed differentiation of both the species, which are grouped in separate clusters with bootstrap values higher than 90 per cent (Valera et al. 2011).

So far, the 16S-23S ITS was applied as a suitable approach for a quick affiliation of AAB to a distinct group of restriction types and also for quick identification of novel isolates. However, with the exception of two conserved genes coding for tRNA-Ile and tRNA-Ala, and the antitermination boxB element, this genomic region is highly divergent among AAB species, and for this reason, Trček (2005) evaluated the gene coding for the subunit I of the PQQ-dependent alcohol dehydrogenase (*adhA*) as an alternative target among representative strains of genera *Acetobacter*, *Gluconobacter* and *Gluconacetobacter* (Trček 2005). Phylogenetic analysis based on the 16S-23S ITS region and the *adhA* gene were also used by our group to confirm the identification of vinegar isolates as *A. pasteurianus* and *Ga. europaeus* (currently *K. europaeus*) (Andrés-Barrao et al. 2011, Andrés-Barrao 2012).

The RecA protein, in addition to the 16S rRNA gene and the ITS region, has been used in multilocus sequence analysis for taxonomic studies of novel bacterial isolates of *Alphaproteobacteria* (Eisen 1995). It was also used by Greenberg et al. (2006) as a complementary approach for the proposal of *Granulibacter bethesdensis* as a new taxon in the family *Acetobacteraceae*. The use of RecA for phylogenetic analysis was possible because at that time sequences of the coding gene *recA* were available in public databases for members of genera *Acetobacter*, *Gluconobacter* and *Gluconacetobacter*. Focusing on a region of 152 amino acids encoded in 456 conserved nucleotides in the *recA* gene of the new isolate, the authors obtained similar results to those from ITS and 16S rRNA gene regions (Greenberg et al. 2006).

For the characterization of the novel AAB taxa *Swaminathania salitolerans*, *Acetobacter nitrogenifigens* and *Gluconacetobacter kombuchae*, isolated from plant-associated bacteria, the sequencing of *nifH* and *nifD*-specific PCR amplification products was used, at the same time as they served as confirmation of the capability of the isolates to fix atmospheric nitrogen (Loganathan and Nair 2004, Dutta and Gachhui 2006, 2007). Additionally, the elongation factor Tu (*tuf*) gene was recently proved to show higher discriminatory power than 16S rRNA and similar to that of 16S-23S ITS, for identification of *Acetobacter* species (Huang et al. 2014, Yetiman and Kesmen 2015).

Another method for genotypic characterization of prokaryotes at an intra-specific level is Multilocus Sequence Alignment (MLSA). This method is based on the phylogenetic analysis of concatenated sequences from single-copy ubiquitous protein-coding genes, which evolve faster than rRNA (Gevers et al. 2005, 2007). Sequences of housekeeping genes are superior to 16S rRNA gene sequences to resolve phylogenetic relationships of closely related species (Naser et al. 2005, Brady et al. 2008). To avoid uncovered horizontal gene transfer or recombinations in a single gene resulting in a misleading phylogenetic signal, it has been recommended to examine a minimum of three housekeeping genes (Konstantinidis et al. 2006).

Cleenwerck et al. (2010) evaluated a set of three single-copy housekeeping genes (*dnaK, rpoB, groEL*) for the construction of a MLSA framework for AAB. Phylogenetic trees based on concatenated sequences of the three housekeeping genes showed consistency with the topology showed by 16S rRNA gene and delineated all AAB species with high bootstrap values (≥99 per cent). Moreover, this technique allowed the delineation of closely related species of the *Ga. liquefaciens* and *Ga. xylinus* groups (species from the latter group are currently reclassified as independent genus *Komagataeibacter*) and the identification of *Ga. sucrofermentans* (currently *K. sucrofermentans*) as new taxon. The trees based on individual *dnaK, groEL* and *rpoB* sequences showed similar topology to that of the tree based on the concatenated gene sequences.

The convenience of this technique to identify bacterial isolates has been also exploited for the characterization of the new AAB species *Acetobacter siceare* (Li et al. 2014). Recently, our group characterized a set of spirit vinegar samples from a running acetic acid fermentation in a laboratory-scale fermentor by using a culture-independent approach. In this study, phylogenetic analysis, based on 16S rRNA gene and MLSA using the former three housekeeping genes, led to the identification of the bacterial species involved as a distinct member of the genus *Komagataeibacter* and probably a new species (Andrés-Barrao et al. 2016).

A massive application of MLSA that is gaining ground in general taxonomy of microorganisms is the phylogenetic analysis based on whole-genome sequences. In this case, a selection of all single-copy genes that are present in all bacterial species are studied. Matsutani et al. (2011) evaluated the suitability of genome-wide phylogenetic analysis in the framework of AAB. The authors selected five reference species whose complete genome sequence was available and identified a group of 753 orthologous genes that were present in all the five strains as single-copy genes. Phylogenetic analysis based on the concatenated sequence of these 753 genes strongly suggested that *Gluconobacter* was the first AAB genus to diverge from the common ancestor of *Gluconobacter*, *Acetobacter* and *Gluconacetobacter*, a relationship that is in agreement with the physiologies and habitats of these genera but contradicts what was previously shown based on 16S rRNA gene. Phylogenetic analysis using 16S rRNA gene sequences showed that *Gluconacetobacter* diverged first from the ancestor of these three genera (Matsutani et al. 2011, 2012).

Denaturing Gradient Gel Electrophoresis (DGGE)

Denaturing gradient gel electrophoresis (DGGE) is a common molecular fingerprinting technique used for separating PCR-generated DNA fragments according to their mobility under increasingly denaturing conditions (usually increasing formamide/urea concentrations) (Muyzer et al. 1993, 1998). The polymerase chain reaction of environmental samples or heterogeneous bacterial cultures can generate templates of differing DNA sequences that represent many of the dominant microorganisms present. However, since PCR products from different bacterial strains might be of similar size, conventional separation by agarose gel electrophoresis results only in a single DNA band that is largely non-descriptive (Myers et al. 1987). DGGE overcomes this limitation by separating PCR products based on sequence differences that result in differential denaturing characteristics of the DNA. During DGGE, PCR products encounter increasingly higher concentrations of a denaturant agent, formamide, as they migrate through a polyacrylamide gel electrophoresis. Upon reaching a certain threshold of denaturant concentration, the weaker melting domains of the double-stranded PCR product begins to denature and its migration slows dramatically. Differing sequences of DNA denature at different denaturant concentrations, resulting in a pattern of bands, with each band theoretically representing a different bacterial population present in the community (Myers et al. 1987).

The primers pair, WBAC1 and WBAC2^{GC}, has been extensively used to apply DGGE as a culture-independent approach to study the population dynamics of AAB during acetic acid fermentation. The primers were designed to specifically amplify both LAB and AAB in wine, covering the V7-V8 region of the 16S rRNA gene (López et al. 2003). A further evaluation of this technique as a rapid and cost-effective method for the screening of AAB in Traditional Balsamic Vinegar (TBV) was done. Although the authors suggested DGGE analysis as a suitable tool for the rapid and cost-effective screening of AAB, they recognized that the distinction of different groups of AAB could not be achieved (De Vero et al. 2006). Despite the capability of this technique to clearly group bacteria of the genera *Acetobacter*, *Gluconobacter* and *Gluconacetobacter*, it failed to discriminate among several groups of closely related species. For instance, the amplification products of the type strains of *Ga. xylinus*, *Ga. europaeus* and *Ga. intermedius* (currently *K. xylinus*, *K. europaeus* and *K. intermedius*) showed the same electrophoretic mobility; hence the group of species could not be resolved (De Vero et al. 2006, De Vero and Giudici 2008, Ilabaca et al. 2008).

The microbial succession during rice vinegar fermentation was also studied, and showed the imposition of *Lactobacillus acetotolerans* and *Acetobacter pasteurianus* at the stage at which acetic acid started to accumulate (Haruta et al. 2006). Recently, Yetiman and Kesmen (2015) applied DGGE to investigate the bacterial population of different traditionally-produced vinegars and mothers of vinegar. It showed that *K. hansenii* and the group *K. europaeus*/*K. xylinus* were the predominant species in all the samples, and the results were supported by $(GTG)_5$-rep-PCR as well as 16S rRNA, 16S-23S ITS and *tuf* gene sequencing (Yetiman and Kesmen 2015, Kathleen et al. 2014).

This technique is very sensitive to variations in DNA sequence and allows simultaneous analysis of multiple samples. On the other hand, it has the same handicaps as all PCR-based techniques for the analysis of microbial communities, including biases from DNA extraction and amplification. Additionally, the variation in 16S rRNA gene copy number in different microorganisms makes this technique only 'semi-quantitative'. Micro-heterogeneity in rRNA encoding genes present in some species may result in multiple bands for a single species and subsequently to an overestimation of bacterial diversity. This technique works well only with short fragments (< 600 bp), thus limiting phylogenetic characterization. In some gels, complex communities may look smeared due to the large number of bands and the results difficult to reproduce between gels. Additionally, concerning AAB, the WBAC primers set, which was originally developed for the 16S rRNA gene sequence of *Lactobacillus plantarum*, allows the differentiation of some of the AAB species. Therefore, authors suggest that a new set of primers targeting only AAB is required to obtain a higher resolution (Yetiman and Kesmen 2015).

Real-Time PCR (qPCR)

This molecular technique is a technical upgrade of the standard Polymerase Chain Reaction (PCR). It is based on the detection and quantification of the amplification product by monitoring the fluorescence emitted by dyes or probes during each cycle of the PCR reaction. The emission is directly proportional to the quantity of amplification product formed, and the calculation of the initial number of DNA molecules present in the sample is done assuming that the amplification efficiency is close to doubling the number of molecules per amplification cycle (Kubista et al. 2006). Typical uses of quantitative real-time PCR (qPCR) include, among others, detection of pathogens and analysis of gene expression, single nucleotide polymorphisms (SNPs) or chromosome aberrations (Kubista et al. 2006).

The application of qPCR for a fast and accurate detection and enumeration of AAB in wine and vinegar has been evaluated by González et al. (2006b). The technique was initially applied to reference strains and its suitability for routine analysis was validated on industrial samples. Andorrà et al. (2008) assessed the microbial population during different wine fermentations by using several culture-independent approaches. In this study, the specificity of the qPCR designed primers allowed an accurate quantification of minor microbial groups. On the contrary, DGGE could detect only a low diversity of yeasts and bacteria, corresponding with the more abundant species: *A. aceti* and *Ga. hansenii* (currently *K. hansenii*), which were detected during the entire fermentation process in all samples. Although the main bacterial species remained constant during the process, qPCR results revealed a reduction of AAB in must during alcoholic fermentation (Andorrà et al. 2008).

One of the primary advantages of real-time PCR is the capability not only to identify but also to enumerate specific PCR products by monitoring the accumulation of amplified fragments during each cycle. With the highly efficient detection chemistry, sensitive instrumentation and optimized assays that are available today, the number of DNA molecules of a particular sequence in a complex sample can be determined with unprecedented accuracy and sensitivity sufficient to detect a single molecule (Kubista et al. 2006). It is a fast and reliable technique for the identification and quantification of bacteria in environmental samples. However, despite its accuracy, qPCR is an expensive and complex technique (Higuchi et al. 1992, González et al. 2006b).

Identification of Acetic Acid Bacteria to the Level of Strain (Genotyping)

Random Amplification of Polymorphic DNA (RAPD)

The principle is that a single and short oligonucleotide primer, which binds to many different loci, is used to amplify random sequences from a complex DNA template. The principle of this technique is based on the random amplification of genomic DNA with a single and short oligonucleotide primer (10 nucleotides of length) that is able to hybridize with high affinity to chromosomal DNA sequences at low annealing temperature. The amplified products (of up to 3.0 kb) are usually separated by agarose gel electrophoresis (Kumar and Gurusubramanian 2011, Williams et al. 1990).

RAPD has found a wide range of applications in gene mapping, population genetics, molecular evolutionary genetics, plant and animal breeding, plant- and animal-microbe interaction and pesticide herbicide resistance. In contrast to the PCR amplification methods previously described, the amplification of repetitive regions does not require designing of gene-specific primers. A high number of fragments are produced during the PCR reaction and arbitrary primers are easily purchased. Moreover, the technique is quick, simple and efficient. Nevertheless, RAPD markers, observed as different-sized DNA segments amplified from the same locus, are detected very rarely. As PCR is an enzymatic reaction, the quality and concentration of template DNA, concentrations of PCR components and the PCR cycling conditions may greatly influence the outcome.

Changes in the microbial population during traditional acetic acid fermentation of rice vinegar (*Kurosu*) and unpolished rice vinegar (*Kumesu*) were investigated by using Enterobacterial Repetitive Intergenic Consensus-PCR (ERIC-PCR) and RAPD. The analysis of 178 isolates divided them in only two groups thus demonstrating the low diversity at strain level during these processes. The corresponding species was identified as *A. pasteurianus* by sequencing of the 16S rRNA gene (Nanda et al. 2001). AAB spoilers of bottled Australian red wine were also investigated using this technique and the obtained results showed a low diverse group of *A. pasteurianus* strains. Additionally, the strains isolated from the spoiled wine samples were genuinely different from the strains isolated from beer, rice vinegar and cider (Bartowsky et al. 2003).

Amplified Length Fragments Polymorphism (ALFP)

ALFP-PCR fingerprinting is a highly sensitive technique for detection of polymorphisms in DNA and has been developed by Zabeau and Vos (1993) (Vos et al. 1995, Vos and Kuiper 1997). This molecular method is based on PCR amplification of selected restriction fragments from digested whole bacterial genomes. The amplification of the restriction fragments is achieved by using the adapter and restriction site sequence as target sites for primer annealing. The selective amplification is finished by the use of primers that extend into the restriction fragments, amplifying only those fragments in which the primer extensions match the nucleotides flanking the restriction sites.

The method allows specific co-amplification of a high number of restriction fragments. The number of fragments that can be analyzed simultaneously depends on the resolution of the detection system. Typically, 50-100 restriction fragments are amplified and detected on denaturing polyacrylamide gels. This technique provides a very powerful DNA fingerprinting technique for DNAs of any origin or complexity (Vos et al. 1995). Compared with other techniques, AFLP-PCR is more sophisticated and sensitive

This technique was investigated as a tool for fast and accurate identification of AAB to the species level, comparing the *ApaI/TaqI* restriction profiles of 135 well-characterized strains representing 50 AAB species and 15 unknown samples. Based on the results, the authors proposed the reclassification of six strains among the closely related taxa *Ga xylinus/europaeus* (currently *K. xylinus/europaeus*) and *A. orleanensis/cerevisiae*, and concluded that AFLP DNA fingerprinting is suitable for accurate identification and classification of a broad range of AAB, as well as for the determination of intraspecific genetic diversity (Cleenwerck et al. 2009).

Enterobacterial Repetitive Intergenic Consensus-PCR (ERIC-PCR) and Repetitive Extragenic Palindromic-PCR (REP-PCR)

Enterobacterial Repetitive Intergenic Consensus (ERIC) and Repetitive Extragenic Palindromic (REP) elements have been described as consensus sequences derived from highly conserved palindromic inverted repeat regions (dispersed repetitive DNA) found in enteric bacteria (Sechi et al. 1998, Pooler et al. 1996, Versalovic et al. 1994). The amplification of the sequences between these repetitive elements generates DNA fingerprints of multiple Gram-negative and positive species (Pooler et al. 1996, Beyer et al. 1998, Sander et al. 1998, Wieser and Busse 2000). The amplification of Repetitive Element Palindromic by Polymerase Chain Reaction (rep-PCR) (do not confuse with REP-PCR) has also become one of the more powerful molecular tools applicable for identification of bacteria and differentiation of bacterial strains of the same species (Gómez-Gil et al. 2004).

The use of ERIC-PCR and REP-PCR for typing of AAB was tested with 14 reference strains of most of the species of the genera *Acetobacter*, *Gluconobacter* and *Gluconacetobacter*. All the tested strains showed a different ERIC-PCR profile, whilst some of them yielded identical profiles when applying REP-PCR, thus suggesting that the first technique was more adequate for typing of AAB (González et al. 2004). One hundred twenty isolates from traditional wine vinegar fermentations were typed using both the techniques. The results revealed a high degree of strain diversity at the first stage of fermentation, which decreased throughout the process (González et al. 2004). The influence of yeast inoculation and SO_2 addition on AAB population was analyzed on grapes and during wine production processes. The main species during fermentation were identified by 16S rRNA gene as *A. aceti*, and the ERIC and REP fingerprinting results showed a high strain diversity on grapes, which decreased substantially during alcoholic fermentation (González et al. 2005).

So far $(GTG)_5$-rep-PCR fingerprinting was believed to be a promising genotypic tool for rapid and accurate identification of AAB, and to prove it, the method was evaluated on 64 AAB reference strains, including 31 type strains and 132 isolates from Ghanaian fermented cocoa beans, and validated by DNA-DNA hybridization data. Most of the reference strains grouped according to their species designation, indicating the technique is useful for identification at the species level. Moreover, exclusive patterns were obtained for most strains, suggesting it is also adequate for characterization of AAB at the strain level (De Vuyst et al. 2008).

$(GTG)_5$-rep-PCR and AFLP DNA fingerprinting data were used by Cleenwerck et al. (2010) to support the identification of a vinegar isolate as a distinct AAB species and its proposal as the novel taxon *Ga. sucrofermentans*. The population dynamics of AAB in traditional wine vinegar production was determined at strain level by using ERIC-PCR and $(GTG)_5$-rep-PCR. The most widely isolated species was *A. pasteurianus*, and although the two fermentations that were studied showed different degrees of diversity, the number of different strains in both the cases was higher during the early stages of acetification, when acidity reached 2 per cent, in agreement with previous reports (Vegas et al. 2010). Likewise, the study of the AAB community changes during spirit vinegar fermentation revealed a little homogeneous and consistent lowly diverse population during the whole process (Andrés-Barrao 2012, Andrés-Barrao et al. 2016).

Matrix-Assisted Laser Desorption Ionization-Time of Flight Mass Spectrometry (MALDI-TOF MS) for Identification of Acetic Acid Bacteria

A polyphasic approach including molecular techniques, has allowed to improve the taxonomic resolution of the phenotypic analyses that were traditionally used for identification of microorganisms. The application of molecular methods has been a breakthrough for taxonomic studies and for industrial applications as well. They are more accurate and faster than the traditional phenotypic methods, but still have several drawbacks, as the different levels of discrimination, extensive manipulation and the requirement of long incubation times and qualified technicians. Alternative methods, quick and requiring low manipulation, must be developed for high-throughput routine analyses and in this regard MALDI-TOF MS looks very promising.

Principles of MALDI-TOF MS

MALDI-TOF MS provides non-destructive vaporization and ionization of both large and small biomolecules. A typical MALDI-TOF MS instrument is composed of four principal components: (a) a laser beam, (b) a sample ionization chamber, where the laser based vaporization of the sample takes place, (c) a time of flight (TOF) mass analyzer producing a separation of ions in relationship with their mass-to-charge (*m/z*) ratio, and (d) a particle detector that detects ions after its separation (Tonolla et al. 2009, Dekker and Branda 2011,

Clark et al. 2013, Bourassa and Butler-Wu 2015, Trček and Barja 2015). The simplest use of MALDI-TOF MS is in the analysis of intact microbial cells, directly from the culture without any preparation (Andrés-Barrao et al. 2013, Bourassa and Butler-Wu 2015, Trček and Barja 2015). Briefly, a pure colony from a freshly grown culture (whole cell sample) is directly deposited on a conductive metal target plate using a sterile loop or toothpick (Fig. 1). The spotted microorganism is then coated with a solution of a chemical matrix, comprising highly pure aromatic compounds dissolved in organic solvents. While the matrix solution dries in the air, the lysis of the microorganism is produced as a result of the co-crystallization of the matrix and the embedded sample (Clark et al. 2013, Randell 2014, Bourassa and Butler-Wu 2015). The chemical composition of the matrix must be adapted to the nature of the components of the sample to be analyzed (proteins, lipids, etc.) and the specific laser beam used. The commonly used chemical matrix solutions include 2,5-dihydroxybenzoic acid (DBA), α-cyno-4-hydroxyciannamic acid (α-CHCA), sinapinic acid (SA), ferulic acid (FA) and 2-(4-hydroxyphenylazo)benzoic acid (HABA) (Tonolla et al. 2009, Croxatto et al. 2012, Pavlovic et al. 2013).

Although this method to analyze intact cells is highly convenient, it is important to highlight that not all microorganisms can be analyzed in such a way. In certain cases, the whole protein extract needs to be isolated before preparing the mixture with the specific matrix and the subsequent analysis by MALDI-TOF MS. The methods for protein extraction vary widely and must be adapted to the specific microorganism. The most commonly extraction solutions frequently included is ethanol-formic acid (Theel et al. 2012) (Fig. 1).

After sample-matrix crystallization, the target plate is loaded into the specimen ionization chamber of the MALDI-TOF MS instrument, where the sample is bombarded with high-energy photons from a pulsed nitrogen laser beam. The nitrogen laser applies short pulses of energy, producing vibrational excitation of the matrix and desorption of the embedded molecules by clusters of matrix molecules, water and ions (Dekker and Branda 2011, Croxatto et al. 2012, Trček and Barja 2015), leading to the sublimation of the matrix-sample from solid to gas phase; thus occurring the ionization of the sample. The matrix molecules transfer protons to the analyte, resulting in positively-charged analyte cations in the gas phase (Fig. 2). These cations are then accelerated across an electric field within the ionization chamber to a speed that depends on the mass-to-charge (*m/z*) ratio of each specific analyte. The particles leave the ionization chamber to enter a TOF mass analyzer. In this chamber, the particles travel along a field-free flight path and the time required for each ion to reach the detector is precisely measured (Clark et al. 2013, Patel 2013, Trček and Barja 2015, Bourassa and Butler-Wu 2015). Based on the time of flight, the *m/z* ratio of each particle is determined and a mass spectrum is generated, representing both *m/z* and signal intensity of the detected ions. The resulting spectrum is composed of 10-30 peaks, ranging normally from 1000 to 20000 *m/z*. This mass spectrum is unique to each bacterium and corresponds to high-abundance soluble proteins, predominantly ribosomal proteins and other abundant cytosolic proteins (Holland et al. 1996, Fenselau and Demirev 2001, Sun et al. 2006, Hsieh et al. 2008, Dekker and Branda 2011, Welker 2011). A unique spectrum is generated, because protein composition differs between different bacterial genera and species (Clark et al. 2013, Bourassa and Butler-Wu 2015). Then the spectrum must be compared to a database of reference spectra in order to identify the unknown microorganisms. The complete procedure from the time of laser beam pulsation to the comparison of the spectrum generated with the data base interrogation can be completed within a few minutes. The procedure of MALDI-TOF MS-based identification of microorganisms is illustrated schematically in Figs. 1 and 2.

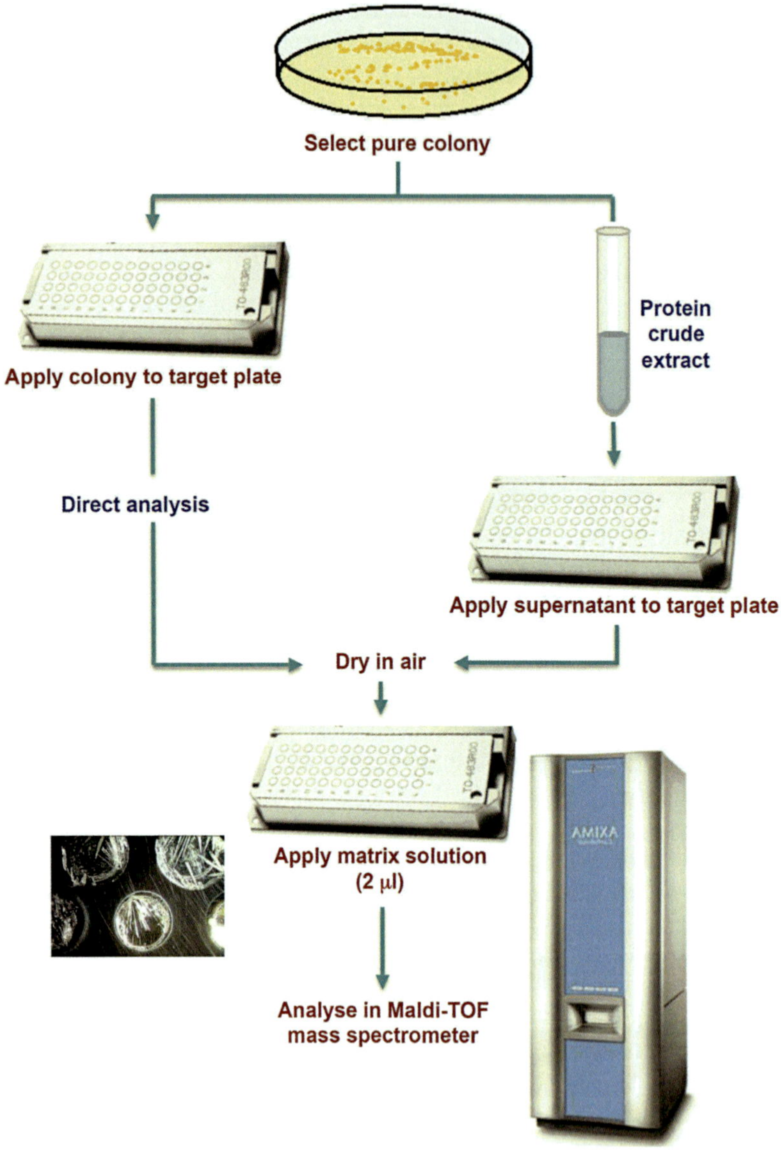

Figure 1: MALDI-TOF MS workflow: Isolation of bacterial culture and direct spotting onto the target plate (alternatively, a small volume of a crude protein extract might be applied to the target plate), drying in the air, layering with the solution matrix, crystallization of the sample-matrix mixture and introduction into the mass spectrometer.

MALDI-TOF MS in Microbiology

Mass spectrometry (MS) is an old and extremely sensitive technique. Historically, it has been widely used in chemistry and in physics as an analytical tool. Matrix-assisted laser desorption/ionization (MALDI) is a soft ionization method used with mass spectrometry for the analysis of biomolecules and large organic molecules (Liyanage and Lay 2005, Hillenkamp and Karas 2007). Although MS has been used in chemistry for decades, its use in the characterization of bacteria was only proposed in 1975 (Anhalt and Fenselau 1975). In this first study, extracted phospholipids and ubiquinones from lyophilized bacteria

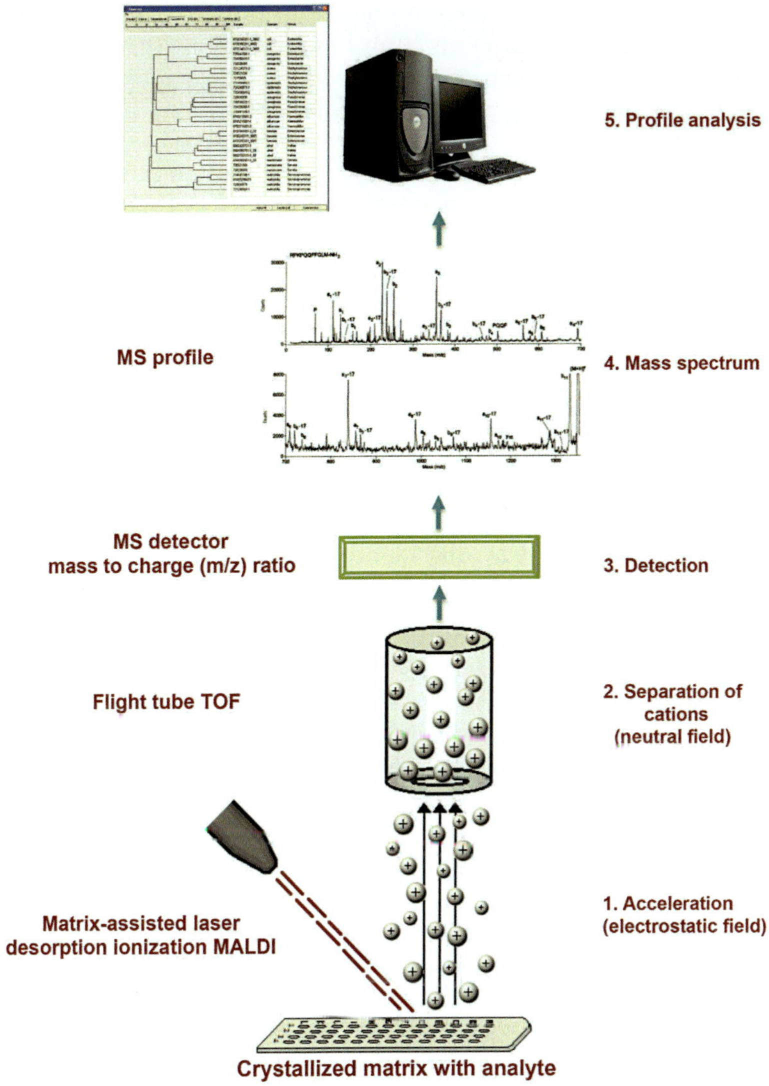

Figure 2: MALDI-TOF MS workflow (*Contd.*): Ionization of sample, detection of accelerated ion, generation of mass spectra and peak-based numeric analysis.

were analyzed. The results demonstrated that the generated spectra are specific to the level of genus or, in some cases, even to the level of species. The first proposal to use MALDI-TOF MS was in 1987 (Karas et al. 1987), but only resulted in 1996 in the spectra of protein biomarkers (Claydon et al. 1996, Despeyroux et al. 1996, Holland et al. 1996, Krishnamurthy and Ross 1996).

MALDI-TOF Mass Spectrometry has been proven a useful tool for characterization and identification of microorganisms within a wide variety of fields, including clinical and veterinarian diagnosis (Benagli et al. 2011, Welker and Moore 2011, Clark et al. 2013), environmental (Ruelle et al. 2004, Siegrist et al. 2007, Giebel et al. 2010) or food- and beverages-related studies (Duskova et al. 2012, Taniwaga et al. 2010, Wieme et al. 2014a). Since the implementation of MALDI-TOF MS as a technique for the characterization of microorganisms, it has been used for identification and typing of bacteria, fungi and viruses. This technique was developed due to the need for a rapid, simple, accurate and cost-effective method for bacterial identification and in particular, for microorganisms related with public safety (Giebel et al. 2010, Li et al. 2015). Although applications of MALDI-TOF MS are not fully exhausted, some limitations are evident, such as the analysis of non-cultivable microorganisms, analysis and quantification of mixed strains and the differentiation of very closely-related species (Welker and Moore 2011). Some authors (Wahl et al. 2002, Warscheid and Fenselau 2004) of studies that evaluated the possibility to identify mixed samples concluded that two or three bacterial species can be identified correctly when the amounts of species are similar, but the identification by MALDI-TOF MS is limited when one species in the mixture is strongly predominant (Anderson et al. 2013, Ford and Burnham 2013). The performance of MALDI-TOF MS can also be influenced by the composition of the culture media (Anderson et al. 2013). Some differences in spectral profiles can be observed using specific culture media (Anderson et al. 2013). As the identification of microorganisms by this technique depends on the comparison to reference databases, future applications in microbiology will depend on the expansion of the existing databases for areas, including food and water safety, plant pathology, pharmacology, veterinary, medicine, industrial microbiology, etc.

Application of MALDI-TOF MS for the Identification of AAB

Although MALDI-TOF MS has been widely used in clinical microbiology to detect the majority of common human pathogens, their use in other biotechnological applications is relatively recent. In the particular case of identification of AAB, the first report demonstrating the suitability of this method to cluster and identify AAB from vinegar samples was published by our group (Andrés-Barrao 2012, Andrés-Barrao et al. 2013).

In these studies, the authors selected 64 strains of 22 representative AAB species, among the genera *Acetobacter* (n = 21), *Gluconacetobacter* (n = 14), *Komagataeibacter* (formerly *Gluconacetobacter*) (n = 17) and *Gluconobacter* (n = 12), and compared their classification by phylogenetic analysis of the 16S rRNA gene with the clustering generated upon analysis by MALDI-TOF MS. In order to establish a reference database of AAB, the authors used an Axima™ Confidence MALDI-TOF mass spectrophotometer (Shimadzu-Biotech, Japan). The instrument was set up with detection in linear positive mode at laser frequency of 50 Hz and to detect a mass range within 2 to 20 kDa. The acceleration voltage was 20 kV and the extraction delay time was 200 ns. For each sample, a minimum of 20 laser shots was used to generate each ion spectrum, where 50 protein mass fingerprints were averaged and processed, using the Launchpad™ v.2.8. software. The average smoothing method was chosen for peak acquisition, with a smoothing filtering width of 50 channels. Peak detection was performed using the threshold-apex peak detection method, with an adaptive voltage threshold that roughly followed the signal noise level, and subtraction of the baseline was set with a filter width of 500 channels. For each sample, a list of the significant spectrum peaks, mass deviations and signal intensities was generated. During the analysis, each target plate was calibrated by using the spectra of *Escherichia coli* K12 (GM40 genotype) as the reference strain. Generated protein mass spectra were imported in the SARAMIS™ software package (Spectral Archive and Microbial Identification System; Anagnostec, Gmbh, Germany) (http://www.mass-spec-capital.com/product/saramis-database-maldi-tof-biomerieux-group-merieux-biological-2001-20066.html) and analyzed using the *m/z* spectra peak between a 3 to 20 kDa mass range and an analytical error for mass accuracy of 800 ppm. Cluster analysis using the single-link agglomerative algorithm was performed to compare the spectra and to produce taxonomic trees. To further evaluate the accuracy of the MS protein fingerprinting analysis, several blind tests were performed: different sets of AAB reference strains were randomly selected and each strain was analyzed in duplicate or quadruplicate (Andrés-Barrao et al. 2013).

transcribed spacer (ITS) region between the 16S-23S rRNA genes (Cleenwerck et al. 2002, Valera et al. 2013).

The erroneous classification of *A. pasteurianus* LMG 1590 and *G. cerinus* LMG 1678 suggested previously upon MALDI-TOF MS analysis was later confirmed by MLSA of housekeeping genes, *dnaK* and *rpoB* (Wieme et al. 2014b, Cleenwerck et al. 2010).

In his later report, Wieme et al. (2014b) used gene sequence analysis of *dnaK* and *rpoB* to reclassify 10 misidentified AAB reference strains as follows: *A. pasteurianus* LMG 1549, LMG 1587, LMG 1597, LMG 1604 and LMG 1698 as *A. cerevisiae/malorum*; *A. pasteurianus* LMG 1701 and *A. lovaniensis* LMG 24630 as *A. fabarum*; *G. cerinus* LMG 1678 and LMG 1743 as *G. japonicus* and *G. thailandicus* respectively; and *K. xylinus* LMG 25 as *K. swingsii*, supporting the results obtained by Andrés-Barrao et al. (2013).

Authors confirmed the identification obtained by peak-base data analysis of 141 AAB isolates by sequence analysis of protein-coding reference genes. 120 isolates (85.1 per cent) were correctly identified at species level. The inconclusive identification of 21 AAB isolates from spoiled beer was corrected by sequence analysis of the genes *dnaK* or *rpoB* (Cleenwerck et al. 2010), and four clusters of AAB species were identified. Two clusters included the species *A. indonesiensis* and *A. persici*, which were not previously known as beer spoilers and hence not included in the reference bacterial set for constructing the mass spectra database. Another cluster represented the recently described novel AAB species *Gluconobacter cerevisiae* (Spitaels et al. 2014). Although further polyphasic analyses were used, only two isolates grouped in the last cluster could not be identified beyond the genus level, and they were tentatively classified as *Gluconacetobacter* sp. (Wieme et al. 2014b).

Conclusion

In the food industry, AAB are associated with the production of fermented food and beverages, either as being directly responsible for the fermentation process or undesired spoilers. A quick and accurate identification of these microorganisms is necessary to ensure the quality of the final products.

The results presented in this chapter demonstrate that MALDI-TOF MS fingerprinting is a suitable tool for a rapid and high-throughput identification of AAB from vinegar and spoiled beer samples. This technique could be appropriate to be implemented as a standard procedure for routine microbiological quality control in vinegar plants as well as in breweries and other fermented food production industries.

Nevertheless, for the unequivocal differentiation of unknown bacteria by MALDI-TOF MS, some further improvements are needed. A larger and accurate mass spectra database must be constructed by including a wider range of members of the AAB group (there are currently described more than 80 species of AAB); as listed in the List of Prokaryotic Names with Standing Nomenclature (LPSN) (Parte 2014), grown on different culture media and/or isolated from different sources. A large number of individual spectra of multiple strains of the same species must be compared and the species-specific peaks must be extracted to construct Super-Spectra. The integration of these Super-Spectra in an extensive database, such as SARAMIS™, will allow a rapid and automated identification of new isolates and the discovery of unknown bacterial species.

AAB are known to be fastidious microorganisms that grow poorly under certain conditions, as for example when inoculated from high-acid (10 per cent) vinegar samples (Trček 2005). Researchers have had relative success in culturing high-acid vinegars by applying different enrichment methods, although the cultivation of these types of vinegars is not yet a routine practice. In this regard, the ultimate challenge for ecological and population studies, as well as quality control analyses during acetic acid fermentation, is the quantification and simultaneous detection of AAB directly harvested from vinegars, as well as from wine, beer and other fermented beverages. The direct analysis of these samples by MALDI-TOF MS will avoid the time-consuming step of *in vitro* cultivation for bacterial isolation while preventing the bias due to the loss of non-cultivable bacterial strains.

Acknowledgments

The research by Dr. François Barja is funded by the Department of Botany and Plant Biology, University of Geneva and the Academic Society of Geneva (SACAD).

References

Alauzet, C., Teyssier, C., Jumas-Bilak, E., Gouby, A., Chiron, R., Rabaud, C., Counil, F., Lozniewski, A. and Marchandin, H. (2010). *Gluconobacter* as well as *Asaia* species, the newly emerging opportunistic human pathogens among acetic acid bacteria. *Journal of Clinical Microbiology* **48:** 3935-3942.

Anderson, K.E., Sheehan, T.H., Mott, B.M., Maes, P., Synder, L., Schwan, M.R., Watson, A., Jones, B.M. and Corby-Haris, V. (2013). Microbial ecology of the hive and pollination landscape: Bacterial associates from floral nectar, the alimentary tract and stored food honey bees (*Apis mellifera*). In *PLoS ONE* **8:** e83125.

Andorrà, I., Landi, S., Mas, A., Guillamón, J.M. and Esteve-Zarzoso, B. (2008). Effect of oenological practices on microbial populations using culture-independent techniques. *Food Microbiology* **25:** 849-856.

Andrés-Barrao, C., Weber, A., Chappuis, M.L., Theiler, G. and Barja, F. (2011). Acetic acid bacteria population dynamics and natural imposition of *Gluconacetobacter europaeus* during submerged vinegar production. *Archives des Sciences* **64:** 99-114.

Andrés-Barrao, C. (2012). Characterization of Acetic Acid Bacteria and Study of the Molecular Strategies Involved in Resistance to Acetic Acid During Oxidative Fermentation. PhD thesis n° 4452 University of Geneva Switzerland.

Andrés-Barrao, C., Benagli, C., Chappuis, M., Ortega Pérez, R., Tonolla, M. and Barja F. (2013). Rapid identification of acetic acid bacteria using MALDI-TOF mass spectrometry fingerprinting. *Systematic and Applied Microbiology*, **36:** 75-81.

Andrés-Barrao, C., Saad, M.M., Cabello Ferrete, E., Bravo, D., Chappuis, M.L., Ortega Pérez, R., Junier, P., Perret, X. and Barja, F. (2016). Metaproteomics and ultrastructure characterization of *Komagataeibacter* spp. involved in high-acid spirit vinegar production. *Food Microbiology* **55:** 112-122.

Anhalt, J.P. and Fenselau, C. (1975). Identification of bacteria using mass spectrometry. *Analytical Chemistry*, **47:** 219-225.

Asai, T. (1934). Taxonomic studies of acetic acid bacteria and allied oxidative bacteria in fruits and a new classification of oxidative bacteria. *Journal of Agricultural Chemical Society of Japan* **10:** 621, 731, 932, 1124.

Asai, T. (1968). *Acetic Acid Bacteria, Classification and Biochemical Activities*. University of Tokyo Press. Tokyo.

Barry, T., Glennon, C.M., Dunican, K. and Gannon, F. (1991). The 16s/23s ribosomal spacer region as a target for DNA probes to identify eubacteria. *Genome* **1:** 51-56.

Bartowsky, E.J., Xia, D., Gibson, R.L., Fleet, G.H. and Henschke, P.A. (2003). Spoilage of bottled red wine by acetic acid bacteria. *Letters in Applied Microbiology* **36:** 307-314.

Bartowsky, E.J. and Henschke, P.A. (2008). Acetic acid bacteria spoilage of bottled red wine—A review. *International Journal of Food Microbiology* **125:** 60-70.

Bartowsky, E.J. (2009). Bacterial spoilage of wine and approaches to minimize it. *Letters in Applied Microbiology* **48:** 149-156.

Beijerinck M.W. (1898). Über die Arten der Essigbakterien. *Zentralblatt für Bakteriologie, Parasitenkunde, Infektionskrankheiten und Hygiene Abteilung* II **4:** 209-216.

Benagli, C., Rossi, V., Dolina, M., Tonolla, M. and Petrin, O. (2011). Matrix-assisted laser desorption ionization-time of flight mass spectrometry for the identification of clinical relevant bacteria. *PLoS ONE* **6:** e16424.

Beyer, W., Mukendi, F.M., Kimming, P. and Böhm, R. (1998). Suitability of reproductive-DNA-sequence-based PCR fingerprinting for characterizing epidemic isolates of *Salmonella enterica serovar* Saintpaul. *Journal of Clinical Microbiology* **36:** 1549-1554.

Bizzini, A., Durussel, C., Bille, J., Greub, G. and Prod'hom, G. (2010). Performance of matrix-assisted laser desorption ionization-time of flight mass spectrometry for identification of bacterial strains routinely isolated in a clinical microbiology laboratory. *Journal of Clinical Microbiology* **48:** 1549-1554.

Boerhaave, H. (1732). Elementa chemiae. *Luduni Batavorum* **2:** 179-207.

Boesch, C., Trček, J., Sievers, M. and Teuber, M. (1998). *Acetobacter intermedius*, sp. nov. *Systematics and Applied Microbiology* **21:** 220-229.

Bourassa, L. and Butler-Wu, S.M. (2015). MALDI-TOF mass spectrometry for microorganisms identification. *Methods in Microbiology* **42:** 37-85.

Bourgeois, J.F. and Barja, F. (2009). The history of vinegar and of its acetification systems. *Archives des Sciences* **62:** 147-160.

Brady, C., Cleenwerck, I., Venter, S., Vancanneyt, M., Swings, J. and Coutinho, T. (2008). Phylogeny and identification of Pantoea species associated with plants, humans and the natural environment based on multilocus sequence analysis (MLSA). *Systematic and Applied Microbiology* **31:** 447-460.

Carr, J.G. and Shimwell J.L. (1961). The acetic acid bacteria 1941-1961: A critical review. *Antoine van Leeuwenhoek* **27:** 386-400.

Castro, C., Cleenwerck, I., Trček, J., Zuluaga, R., De Vos, P., Caro, G., Aguirre, R., Putaux, J.L. and Gañán, P. (2013). *Gluconacetobacter medellinensis* sp. nov., cellulose- and non-cellulose-producing acetic acid bacteria isolated from vinegar. *International Journal of Systematic and Evolutionary Microbiology* **63:** 1119-1125.

Chouaia, B., Gaiarsa, S., Crotti, E., Comandatore, F., Degli Esposti, M., Ricci, I., Alma, A., Favia, G., Bandi, C. and Daffonchio, D. (2014). Acetic acid bacteria genomes reveal functional traits for adaptation to life in insect guts. *Genome Biology and Evolution* **6:** 912-920.

Clark, A.E., Kaleta, E.J., Arora, A. and Wolk, D.M. (2013). Matrix-assisted laser desorption/ ionization time-of-flight mass spectrometry: A fundamental shift in the routine practice of clinical Microbiology. *Clinical Microbiology Reviews* **26:** 547-603.

Claydon, M.A., Davey, S.N., Edwards Jones, V. and Gordon, B.D. (1996). The rapid identification of intact microorganisms using mass spectrometry. *Nature Biotechnology* **14:** 1584-1586.

Cleenwerck, I., Vandemeulebroecke, K., Janssens, D. and Swings, J. (2002). Re-examination of the genus *Acetobacter*, with descriptions of *Acetobacter cerevisiae* sp. nov. and *Acetobacter malorum* sp. nov. *International Journal of Systematic and Evolutionary Microbiology* **52:** 1551-1558.

Cleenwerck, I. and De Vos, P. (2008). Polyphasic taxonomy of acetic acid bacteria: An overview of the currently applied methodology. *International Journal of Food Microbiology* **125:** 2-14.

Cleenwerck, I., De Wachter, M., González, A., De Vuyst, L. and De Vos, P. (2009). Differentiation of species of the family *Acetobacteraceae* by AFLP DNA fingerprinting: *Gluconacetobacter kombuchae* is a later heterotypic synonym of *Gluconacetobacter hansenii*. *International Journal of Systematic and Evolutionary Microbiology* **59:** 1771□1786.

Cleenwerck, I., De Vos, P. and De Vuyst, L. (2010). Phylogeny and differentiation of species of the genus *Gluconacetobacter* and related taxa based on multilocus sequence analyses of housekeeping genes and reclassification of *Acetobacter xylinus* subsp. *sucrofermentans* as *Gluconacetobacter sucrofermentans* (Toyosaki et al. 1996) sp. nov., comb. nov. *International Journal of Systematic and Evolutionary Microbiology* **60:** 2277-2283.

Cohn F. (1872). Untersuchen über Bacterien. *Beitr. z. Biol. d. Pflanz.* **1:** 127-224.

Conner, H.A. and Allgeier, R.J. (1976). Vinegar: Its history and development. *Advances in Applied Microbiology* **20:** 81-133.

Cowan, S.T. (1955). Introduction: The Philosophy of Classification. *Journal of General Microbiology* **12:** 314-321.

Crotti, E., Rizzi, A., Chouaia, B., Ricci, I., Favia, G., Alma, A. Sacchi, L., Bourtzis, K., Mandrioli, M., Cherif, A., Bandi, C. and Daffonchio, D. (2010). Acetic acid bacteria, newly emerging symbionts of insects. *Applied and Environmental Microbiology* **76:** 6963-6970.

Croxatto, A., Prod'hom, G. and Greub, G. (2012). Applications of MALDI-TOF mass spectrometry in clinical diagnostic microbiology. *FEMS Microbiology Reviews* **36:** 380-407.

De Vero, L., Gala, E., Gullo, M., Solieri, L., Landi, S. and Giudici, P. (2006). Application of denaturing gradient gel electrophoresis (DGGE) analysis to evaluate acetic acid bacteria in traditional balsamic vinegar. *Food Microbiology* **23:** 809-813.

De Vero, L. and Giudici, P. (2008). Genus-specific profile of acetic acid bacteria by 16S rDNA PCR-DGGE. *International Journal of Food Microbiology* **125:** 96-101.

De Vuyst, L., Camu, N., De Winter, T., Vandemeulebroecke, K., Van de Perre, V., Vancanneyt, M., De Vos, P. and Cleenwerck, I. (2008). Validation of the (GTG)(5)-rep-PCR fingerprinting technique for rapid classification and identification of acetic acid bacteria, with a focus on isolates from Ghanaian fermented cocoa beans. *International Journal of Food Microbiology* **125:** 79-90.

Dekker, J.P. and Branda, J.A. (2011). MALDI-TOF Mass spectrometry in clinical microbiology laboratory. *Clinical Microbiology Newsletter* **33:** 87-93.

Desmaizières L. (1826). Mycoderma vini. *Annales de Sciences Naturelles* **10:** 42.

Despeyroux, D., Phillpotts, R. and Watts, P. (1996). Electrospray mass spectrometry for detection and characterization of purified cricket paralysis virus (CrPV). *Rapid Communication in Mass Spectrometry* **10:** 937-941.

Döbereiner, J.W. (1823). The origins of heterogenous catalysis by platinium. *Journal of Chemistry (Schweiga)* **38:** 321.

Du Toit, W.J. and Lambrechts, M.G. (2002). The enumeration and identification of acetic acid bacteria from South African red wine fermentations. *International Journal of Food Microbiology* **74:** 57-64.

Du Toit, W.J., Pretorius, I.S. and Lonvaud-Funel, A. (2005). The effect of sulphur dioxide and oxygen on the viability and culturability of a strain of *Acetobacter pasteurianus* and a strain of *Brettanomyces bruxellensis* isolated from wine. *Journal of Applied Microbiology* **98:** 862-871.

Duskova, M., Sedo, O., Ksicova, K., Zdrahal, Z. and Karpiskova, R. (2012). Identification of lactobacilli isolated from food by genotypic methods and MALDI-TOF MS. *International Journal of Food Microbiology* **159:** 107-114.

Dutta, D. and Gachhui, R. (2006). Novel nitrogen-fixing *Acetobacter nitrogenifigens* sp. nov., isolated from Kombucha tea. *International Journal of Systematic and Evolutionary Microbiology* **56:** 1899-1903.

Dutta, D. and Gachhui, R. (2007). Nitrogen-fixing and cellulose-producing *Gluconacetobacter kombuchae* sp. nov. isolated from Kombucha tea. *International Journal of Systematic and Evolutionary Microbiology* **57:** 353-357.

Eisen, J.A. (1995). The RecA protein as a model molecule for molecular systematic studies of bacteria: Comparison of trees of RecAs and 16S rRNAs from the same species. *Journal of Molecular Evolution* **41:** 1105-1123.

Fajardo, V., González I. and Dooley, J. (2009). Application of polymerase chain reaction restriction fragment length polymorphism analysis and lab-on-a-chip capillary electrophoresis for the specific identification of game and domestic meats. *Journal of the Science of Food and Agriculture* **89:** 843-847.

Fenselau, C. and Demirev, A. (2001). Characterization of intact microorganisms by MALDI mass spectrometry. *Mass Spectrometry Reviews* **20:** 157-166.

Ferrer, S., Mañes-Lázaro, R., Benavent-Gil, Y., Yépez, A. and Pardo, I. (2016). *Acetobacter musti* sp. nov. isolated from Bobal grape must. *International Journal of Systematic and Evolutionary Microbiology* **66:** 957-961.

Ford, B.A. and Burnham, C.A. (2013). Optimization of routine identification of clinically relevant Gram-negative bacteria by use of matrix-assisted laser desorption ionization-time of light mass spectrometry and the Briker Biotyper. *Journal of Clinical Microbiology* **51:** 1412-1420.

Franchi, M.A., Tribst, A.A.L. and Cristianini M. (2012). The effect of antimicrobials and bacteriocins on beer spoilage microorganisms. *International Food Research Journal* **19:** 783-786.

Franke-Whittle, I.H., O'Shea, M.G., Leonard, G.J. and Sly, L.I. (2005). Design, development, and use of molecular primers and probes for the detection of *Gluconacetobacter* species in the pink sugarcane mealybug. *Microbial Ecology* **50:** 128-139.

Frateur, J. (1950). Essai sur la systematique de *Acetobacters*. *La Cellule* **53:** 287-392.
Fuentes-Ramírez, L.E., Bustillos-Cristales, R., Tapia-Hernández, A., Jiménez-Salgado, T., Wang, E.T., Martínez-Romero, E. and Caballero-Mellado, J. (2001). Novel nitrogen-fixing acetic acid bacteria, *Gluconacetobacter johannae* sp. nov. and *Gluconacetobacter azotocaptans* sp. nov., associated with coffee plants. *International Journal of Systematic and Evolutionary Microbiology* **51:** 1305-1314.
Gest, H. (2004). The discovery of microorganisms by Robert Hooke and Antoni Van Leeuwenhoek, fellows of the Royal Society. *Notes and Records*. The Royal Society, London **58:** 187-201.
Gevers, D., Cohan, F.M., Lawrence, J.G., Spratt, B.G., Coenye, T., Feil, E.J., Stackebrandt, E., Van der Peer, Y., Vandamme, P., Thompson, F.L. and Swings, J. (2005). Re-evaluating prokaryotic species. *Nature Reviews Microbiology* **3:** 733-739.
Gevers, D. and Coenye, T. (2007). Phylogenetic and genomic analysis. *In:* C. Hurst, R.L. Crawford, J.L. Garland, D.A. Lipson, A.L. Mills and L. Stenzenbach (Eds.). *Manual of Environmental Microbiology*, 3rd edn. ASM Press, Whashington DC, USA, pp. 157-168.
Giebel, R.A., Worden, C., Rust, S.M., Kleinheinz, G.T., Robbins, M., Fredenberg, W. and Sandrin, T.R. (2010). Microbial fingerprint using matrix-assisted laser desorption/ionization time-of-flight mass spectrometry (MALDI-TOF MS) applications and challenges. *Advances in Applied Microbiology* **71:** 149-184.
Gómez-Gil, B., Soto Rodríguez, S., García-Gasca, A., Roque, A., Vázquez-Juarez, R., Thompson, F.L. and Swings J. (2004). Molecular identification of *Vibrio harveyi*-related isolates associated with diseased aquatic organisms. *Microbiology* **150:** 1769-1777.
González, A., Hierro, N., Poblet, M., Rozès, N., Mas, A. and Guillamón, J.M. (2004). Application of molecular methods for the differentiation of acetic acid bacteria in a red wine fermentation. *Journal of Applied Microbiology* **96:** 853-860.
González, A., Hierro, N., Poblet, M., Mas, A. and Guillamón, J.M. (2005). Application of molecular methods to demonstrate species and strain evolution of acetic acid bacteria population during wine production. *International Journal of Food Microbiology* **102:** 295-304.
González, A., Guillamón, J.M., Mas, A. and Poblet, M. (2006a). Application of molecular methods for routine identification of acetic acid bacteria. *International Journal of Food Microbiology* **108:** 141-146.
González, A., Hierro, N., Poblet, M., Mas, A. and Guillamón, J.M. (2006b). Enumeration and detection of acetic acid bacteria by real-time PCR and nested PCR. *FEMS—Microbiology Letters* **251:** 123-128.
González, A. and Mas, A. (2011). Differentiation of acetic acid bacteria based on sequence analysis of 16S-23S rRNA gene internal transcribed spacer sequences. *International Journal of Food Microbiology* **147:** 217-222.
Greenberg, D.E., Porcella, S.F., Stock, F., Wong, A., Conville, P.S., Murray, P.R., Holland, S.M. and Zelazny, A.M. (2006). *Granulibacter bethesdensis* gen. nov., sp. nov., a distinctive pathogenic acetic acid bacterium in the family *Acetobacteraceae*. *International Journal of Systematic and Evolutionary Microbiology* **56:** 2609-2616.
Greenberg, D.E., Porcella, S.F., Zelazny, A.M., Virtaneva, K., Sturdevant, D.E., Kupko, J.J., Barbian, K.D., Babar, A., Dorward, D.W. and Holland, S.M. (2007). Genome sequence analysis of the emerging human pathogenic acetic acid bacterium *Granulibacter bethesdensis*. *Journal of Bacteriology* **189:** 8727-8736.
Gullo, M., Caggia, C., De Vero, L. and Giudici, P. (2006). Characterization of acetic acid bacteria in 'traditional balsamic vinegar'. *International Journal of Food Microbiology* **106:** 209-212.
Hansen, E.C. (1879). Compt. Rend. De Carlsberg. 1q, 49 and 96 A. Klöer: Gesammelte theoretische Abhandlungen über Gärungsorganismen, *Verlag vpon Gustav Fischer, Jena, Jena* (1911). P. 511.
Hansen, H.C. (1894). Recherche sur les bactéries acétifiantes. *Compt. Rend. Lab Carlsberg* **3:** 182-216.
Haruta, S., Ueno, S., Egawa, I., Hashiguchi, K., Fujii, A., Nagano, M., Ishii, M. and Igarashi, Y. (2006). Succession of bacterial and fungal communities during a traditional pot fermentation of rice vinegar assessed by PCR-mediated denaturing gradient gel electrophoresis. *International Journal of Food Microbiology* **109:** 79-87.
Higuchi, R., Dollinger, G., Walsh, P.S. and Griffith, R. (1992). Simultaneous amplification and detection of specific DNA-sequences. *Bio-technology* **10:** 413-417.
Hillenkamp, F. and Karas, M. (2007). The MALDI process and method. *In*: F. Hillenkamp and J. Peter-Katalinić (Eds.). MALDI MS. *A Practical Guide to Instrumentation, Methods and Applications*. Wiley-VCH Verlag GmbH & Co. KGaA, Weinheim, Germany, pp. 1-28.

Holland, R.D., Wilkes, J.G., Raffi, F., Sutherland, J.B., Persons, C.C., Voorhees, K.J. and Lay, J.O. (1996). Rapid identification of intact whole bacteria based on spectral patterns using matrix-assisted laser desorption/ionization with time-of-flight mass spectrometry. *Rapid Communication in Mass Spectrometry* **10:** 1227-1232.

Hooft, F. (1925). Biochemische Ondertoekingen over het geslacht Acetobacter. [dissertation] Journal unknown. Technical University, Delft, The Netherlands.

Hsieh, S.Y., Tsen, C.L., Lee, Y.S., Kuo, A.J., Sun, C.F., Lin, Y.H. and Chen, J.K. (2008). Highly efficient classification and identification or human pathogenic bacteria by MALDI-TOF MS. *Molecular Cell Proteomics* **7:** 448-456.

Huang, C.H., Chang, M.T., Huang, L. and Chu, W.S. (2014). Utilization of elongation factor Tu gene (*tuf*) sequencing and species-specific PCR (SS-PCR) for the molecular identification of *Acetobacter* species complex. *Molecular and Cellular Probes* **28:** 31-33.

Ilabaca, C., Navarrete, P., Mardones, P., Romero, J. and Mas, A. (2008). Application of culture, culture-independent molecular biology-based methods to evaluate acetic acid bacteria diversity during vinegar processing. *International Journal of Food Microbiology* **126:** 245-249.

Janda, J.M. and Abbott, S.L. (2007). 16S rRNA gene sequencing for bacterial identification in the diagnostic laboratory: Pluses, perils and pitfalls. *Journal of Clinical Microbiology* **45:** 2761-2764.

Karas, M., Bachman, D., Bahr, U. and Hillenkamp, F. (1987). Matrix-assisted ultraviolet laser desorption of non-volatile compounds. *International Journal of Mass Spectrometry and Ion Processes* **78:** 53-68.

Kathleen, M.M., Samuel, L., Felecia, C., Ng, K.H., Lesley, M.B. and Kasing, A. (2014). (GTG)5-PCR analysis and 16S rRNA sequencing of bacteria from Sarawak aquaculture environment. *International Food Research Journal* **21:** 915-920.

Kern, C.C., Vogel, R.F. and Behr, J. (2014). Differentiation of *Lactobacillus brevis* strains using matrix-assisted laser desorption/ionization time-of-flight mass spectrometry with respect to their beer spoilage potential. *Food Microbiology* **40:** 18-24.

Konstantinidis, K.T., Ramette, A. and Tiedje, J.M. (2006). The bacterial species definition in the genomic era. Philosophical Transactions of the Royal Society of London. Series B, *Biological Sciences* **361:** 1929-1940.

Kregiel, D. (2015). Health safety of soft drinks: Contents, containers and microorganisms. *Biomedical Research International* **128697:** 1-15.

Krishnamurthy, T. and Ross, P.L. (1996). Rapid identification of bacteria by direct matrix-assited laser desorption/ionization mass spectrometry analysis of whole cells. *Rapid Communication in Mass Spectrometry* **10:** 1992-1996.

Kubista, M., Andrade, J.M., Benftsson, M., Forootan, A., Jonák, J., Lind, K., Sindelka, R., Sjöback, R., Sjögreen, B., Strömbom, L., Ståhlberg, A. and Zoric, N. (2006). The real time polymerase chain reaction. *Molecular Aspects of Medicine* **27:** 95-125.

Kumar, N.S. and Gurusubramanian, G. (2011). Random amplified polymorphic DNA (RAPD) markers and its Applications. *Science Vision* **11:** 116-124.

Kützing, F.T. (1837). Microscopische Untersuchungen über die Hefe und Essigmutter nebst mehreren andern dazu gehörigen vegetabilschen Gebilden. *Journal Fur Praktische Chemie* **11:** 385-391.

Leifson, E. (1954). The flagellation and taxonomy of species of Acetobacter. *Antonie Van Leeuwenhoek* **20:** 102-110.

Li, L., Wieme, A., Spitaels, F., Balzarini, T., Nunes, O.C., Manaia, C.M., Van Landschoot, A., De Vuyst, L., Cleenwerck, I. and Vandamme, P. (2014). *Acetobacter sicerae* sp. nov., isolated from cider and kefir and identification of species of the genus *Acetobacter* by *dnaK, groEL* and *rpoB* sequence analysis. *International Journal of Systematic and Evolutionary Microbiology* **64:** 2407-2415.

Li, L., Praet, J., Borremans, W., Nunes, O.C., Manaia, C.M., Cleenwerck, I., Meeus, I., Smagghe, G., De Vuyst, L. and Vandamme, P. (2015). *Bombella intestini* gen. nov., sp. nov., an acetic acid bacterium isolated from bumble bee crop. *International Journal of Systematic and Evolutionary Microbiology* **65:** 267-273.

Lisdiyanti, P., Kawasaki, H., Seki, T., Yamada, Y., Uchimura, T. and Komagata, K. (2001). Identification of *Acetobacter* strains isolated from Indonesian sources and proposals of *Acetobacter syzygii* sp. nov., *Acetobacter cibinongensis* sp. nov. and *Acetobacter orientalis* sp. nov. *Journal of General Applied Microbiology* **47:** 119-131.

Liu, H., Du, Z., Wang, J. and Yang, R. (2007). Universal sample preparation method for characterization of bacteria by matrix-assisted laser desorption/ionization time-of-flight mass spectrometry. *Applied and Environmental Microbiology* **73:** 1899-1907.

Liyanage, R. and Lay, J.O. (2005). An introduction to MALDI-TOF MS. *In:* C.L. Wilkins and J.O. Lay (Eds.). *Identification of Microorganisms by Mass Spectrometry*, vol. 169. John Wiley & Sons, Inc., Hoboken, NJ, USA, pp. 39-60.

Loganathan, P. and Nair, S. (2004). *Swaminathania salitolerans* gen. nov., sp. nov., a salt-tolerant, nitrogen-fixing and phosphate-solubilizing bacterium from wild rice (*Porteresia coarctata Tateoka*). *International Journal of Systematic and Evolutionary Microbiology* **54:** 1185-1190.

López, I., Ruíz-Larrea, F., Cocolin, L., Orr, E., Phister, T., Marshall, M., Vander Gheynst, J. and Mills, D.A. (2003). Design and evaluation of PCR primers for analysis of bacterial populations in wine by denaturing gradient gel electrophoresis. *Applied and Environmental Microbiology* **69:** 6801-6807.

Mariette, I., Schwarz, E., Vogel, R.F. and Hammes, W.P. (1991). Characterization by plasmid profile analysis of acetic acid bacteria from wine, spirit and cider acetators for industrial vinegar production. *Journal of Applied Microbiology* **71:** 134-138.

Matsutani, M., Hirakawa, H., Yakushi, T. and Matsushita, K. (2011). Genome-wide phylogenetic analysis of *Gluconobacter, Acetobacter* and *Gluconacetobacter. FEMS—Microbiology Letters* **315:** 122-128.

Matsutani, M., Hirakawa, H., Saichana, N., Soemphol, W., Yakushi, T. and Matsushita, K. (2012). Genome-wide phylogenetic analysis of differences in thermotolerance among closely related *Acetobacter pasteurianus* strains. *Microbiology* **158:** 229-239.

Mesa, M.M., Macías, M.D., Cantero and Barja, F. (2003). Use of the direct epifluorescent filter technique for the enumeration of viable and total acetic acid bacteria from vinegar fermentation. *Journal of Fluorescence* **13:** 261-265.

Muyzer. G., de Waal, E.C. and Uitterlinden, A.G. (1993). Profiling of complex microbial populations by denaturing gradient gel electrophoresis analysis of polymerase chain reaction-amplified genes coding for 16S rRNA. *Applied and Environmental Microbiology* **59:** 695-700.

Muyzer G. and Smalla, K. (1998). Application of denaturing gradient gel electrophoresis (DGGE) and temperature gradient gel electrophoresis (TGGE) in microbial ecology. *Antonie Van Leeuwenhoek* **73:** 127-41.

Myers, R.M., Maniatis, T. and Lerman, L.S. (1987). Detection and localization of single base changes by denaturing gradient gel electrophoresis. *Methods in Enzymology* **155:** 501-527.

Nanda, K., Taniguchi, M., Ujike, S., Ishihara, N., Mori, H., Ono, H. and Murooka, Y. (2001). Characterization of acetic acid bacteria in traditional acetic acid fermentation of rice vinegar (komesu) and unpolished rice vinegar (kurosu) produced in Japan. *Applied and Environmental Microbiology* **67:** 986-990.

Naser, S.M., Thompson, F.L., Hoste, B., Gevers, D., Dawyndt, P., Vancanneyt, M. and Swings, J. (2005). Application of multilocus sequence analysis (MLSA) for rapid identification of *Enterococcus* species based on rpoA and *pheS* genes. *Microbiology* **57:** 2141-2150.

Navarro, E., Simonet, P., Normand, P. and Bardin, R. (1992). Characterization of natural populations of *Nitrobacter* spp. using PCR/RFLP analysis of the ribosomal intergenic spacer. *Archives of Microbiology* **157:** 107-115.

Papalexandratou, Z., Cleenwerck, I., De Vos, P. and De Vuyst, L. (2009). (GTG)5-PCR reference framework for acetic acid bacteria. *FEMS Microbiology Letters* **301:** 44-49.

Park, E., Yang, H., Kim, Y. and Kim, J. (2012). Analysis of oligosaccharides in beer using MALDI-TOF-MS. *Food Chemistry* **134:** 1658-1664.

Parte, A.C. (2014). LPSN—list of prokaryotic names with standing in nomenclature. *Nucleic Acids Research* **42**(Database issue): D613-616.

Pasteur, L. (1864) Mémoire sur la fermentation acétique. *Annales scientifiques de l'Ecole Normale Supérieur*, **1:** 113-158

Pasteur, L. (1924). *Etudes sur le vinaigre et sur le vin*. (Ed.) Masson et Cie.

Patel, R. (2013). Matrix-Assisted Laser Desorption Ionization-Time of Flight Mass Spectrometry in clinical microbiology. *Clinical Infectious Diseases* **57:** 564-572.

Pavlovic, M., Huber, I., Konrad, R. and Busch, U. (2013). Application of MALDI-TOF MS for the identification of food-borne bacteria. *Open Microbiology Journal* **7:** 135-141.

Persoon, C.H. (1822). Mycologia europaea, seu Completa omnium fungorum in variis Europaeae regionibus detectorum enumeratio. Sectio prima. *Erlangae impensibus Ioanni Iacobi Palmii.*

Petty, C.A. (2007). Detection and identification of microorganisms by gene amplification and sequencing. *Clinical Infectious Diseases* **44:** 1108-1114.

Poblet, M., Rozès, N., Guillamón, J.M. and Mas, A. (2000). Identification of acetic acid bacteria by restriction fragment length polymorphism analysis of a PCR-amplified fragment of the gene coding for 16S rRNA. *Letters in Applied Microbiology* **31:** 63-67.

Pooler, M.R., Ritchie, D.F. and Hartung, J.S. (1996). Genetic relationships among strains of *Xanthomonas fragariae* based on Random Amplified Polymorphic DNA PCR, Repetitive Extragenic Palindromic PCR and Enterobacterial Repetitive Intergenic Consensus PCR data and generation of Multiplexed PCR primers useful for the identification of this phytopathogen. *Applied and Environmental Microbiology* **62:** 3121-3127.

Praet, J., Aerts, M., De Brandt, E., Meeus, I., Smagghe, G. and Vandamme, P. (2016). *Apibacter mensalis* sp. nov.: A rare member of the bumble bee gut microbiota. *International Journal of Systematic and Evolutionary Microbiology* **66:** 1645-1651.

Randell, P. (2014). It's a MALDI but it's a goodie: MALDI-TOF mass spectrometry for microbial identification. *Thorax* **69:** 776-778.

Rasmussen, H.B. (2012). Restriction fragment length polymorphism analysis of PCR-amplified fragments (PCR-RFLP) and gel electrophoresis—valuable tool for genotyping and genetic fingerprinting. *In:* S. Magdeldin (Ed.). *Gel Electrophoresis—Principles and Basics,* In Tech.

Rezzonico, F., Smits, T.H., Born, Y., Blom, J., Frey, J.E., Goesmann, A., Cleenwerck, I., de Vos, P., Bonaterra, A., Duffy, B. and Montesinos, E. (2016). *Erwinia gerundensis* sp. nov., a cosmopolitan epiphyte originally isolated from pome fruit trees. *International Journal of Systematic and Evolutionary Microbiology* (in press).

Rojas, M., González, I. and Fajardo, V. (2009). Identification of raw and heat-processed meats from game bird species by polymerase chain reaction-restriction fragment length polymorphism of the mitochondrial D-loop region. *Poultry Science* **8:** 669-679.

Ruelle, V., El Mouai, B., Zorzi, W., Ledent, P. and Pauw, E.D. (2004). Rapid identification of environmental bacterial strains by matrix-assisted laser desorption/ionization time-of-flight mass spectrometry. *Rapid Communication in Mass Spectrometry* **18:** 2013-2019.

Ruíz, A., Poblet, M., Mas, M. and Guillamón, J.M. (2000). Identification of acetic acid bacteria by RFLP of PCR-amplified 16S rDNA and 16S-23S intergenic spacer. *International Journal of Systematic and Evolutionary Microbiology* **50:** 1981-1987.

Sakamoto, K. and Konings, W.N. (2003). Beer spoilage bacteria and hop resistance. *International Journal of Food Microbiology* **89:** 105-124.

Sander, A., Ruess, M., Bereswill, S., Schuppler, M. and Steinbrueckner, B. (1998). Comparison of different DNA fingerprinting techniques for molecular typing of Bartonella henselae isolates. *Journal of Clinical Microbiology* **36:** 2973-2981.

Sechi, L.A., Zanetti, S., Dupre, I., Delogu, G. and Fadda, G. (1998). Enterobacterial repetitive intergenic consensus sequences as molecular targets for typing of *Mycobacterium tuberculosis* strains. *Journal of Clinical Microbiology* **36:** 128-132.

Šedo, O., Márová, I. and Zdráhal, Z. (2012). Beer fingerprinting by matrix-assisted laser desorption-ionisation-time of flight mass spectrometry. *Food Chemistry* **135:** 473-478.

Sengun, I.Y. and Karabiyikli, S. (2011). Importance of acetic acid bacteria in food industry. *Food Control* **22:** 647-656.

Siegrist, T.J., Anderson, P.D., Huen, W.H., Kleinheinz, G.T., McDermott, C.M. and Sandrin, T.R. (2007). Discrimination and characterization of environmental strains of *Escherichia coli* by matrix-assisted laser desorption/ionization time-of-flight mass spectrometry (MALDI-TOF-MS). *Journal Microbiological Methods* **68:** 554-562.

Solieri, L. and Giudici, P. (2009). Vinegars of the world. *In:* L. Solieri and P. Giudici (Eds.). *Vinegars of the World.* Springer-Verlag, Italy, pp. 1-16.

Spitaels, F., Li, L., Wieme, A., Balzarini, T., Cleenwerck, I., Van Landschoot, A., De Vuyst, L. and Vandamme, P. (2013). *Acetobacter lambici* sp. nov., isolated from fermenting lambic beer. *International Journal of Systematic and Evolutionary Microbiology* **64:** 1083-1089.

Spitaels, F., Wieme, A., Balzarini, T., Cleenwerck, I., Van Landschoot, A., De Vuyst, L. and Vandamme, P. (2014). *Gluconobacter cerevisiae* sp. nov., isolated from the brewery environment. *International Journal of Systematic and Evolutionary Microbiology* **64:** 1134-1141.

Spitaels, F., Wieme, A.D., Janssens, M., Aerts, M., Van Landschoot, A., De Vuyst, L. and Vandamme, P. (2015). The microbial diversity of an industrially produced lambic beer shares members of a traditionally produced one and reveals a core microbiota for lambic beer fermentation. *Food Microbiology* **49:** 23-32.

Sun, L., Teramoto, K., Sato, H., Torimura, M., Tao, H. and Shintani, T. (2006). Characterization of ribosomal proteins as biomarker for matrix-assisted laser desorption/ionization mass spectral identification of *Lactobacillus plantarum*. *Rapid Communications in Mass Spectrometry* **20:** 3789-3798.

Tanigawa, K., Kawabata, H. and Watanabe, K. (2010). Identification and typing of *Lactococus lactis* by matrix-assisted laser desorption/ionization time-of-flight mass spectrometry. *Applied and Environmental Microbiology* **76:** 4055-4062.

Teuber, M., Sievers, M. and Andresen, A. (1987). Characterization of the microflora of high acid submerged vinegar fermenters by distinct plasmid profiles. *Biotechnology Letters* **4:** 265-268.

Theel, E.S., Schmitt, B.H., Hall, L., Cunningham, S.A., Walchak, R.C. and Patel, R. (2012). Formic acid-based direct, on-plate testing of yeast and *Corynebacterium* species by Bruker Biotyper matrix-assisted laser desorption ionization time-of-flight mass spectrometry. *Journal of Clinical Microbiology* **50:** 3093-3095.

Thompson, R.T. (1852). Acetic acid bacteria: Classification and biochemical activities. *Liebigs Annals.* **83:** 89. Original not seen.

Tonolla, M., Benagli, C., De Respinis, S., Gaia, V. and Petrini, O. (2009). Mass spectrometry in the diagnostic laboratory. *Pipette* **3:** 20-25.

Torija, M.J., Mateo, E., Guillamón, J.M. and Mas, A. (2010). Identification and quantification of acetic acid bacteria in wine and vinegar by TaqMan-MGB probes. *Food Microbiology* **27:** 257-265.

Trček, J. and Teuber, M. (2002). Genetic and restriction analysis of the 16S-23S rDNA internal transcribed spacer regions of the acetic acid bacteria. *FEMS Microbiology Letters* **208:** 69-75.

Trček, J. (2005). Quick identification of acetic acid bacteria based on nucleotide sequences of the 16S-23S rDNA internal transcribed spacer region and of the PQQ-dependent alcohol dehydrogenase gene. *Systematic Applied Microbiology* **28:** 735-745.

Trček, J. (2015). Plasmid analysis of high acetic acid-resistant bacterial strains by two-dimensional agarose gel electrophoresis and insights into the phenotype of plasmid pJK2-1. *Annals of Microbiology* **65:** 1287-1292.

Trček, J. and Barja, F. (2015). Updates on quick identification of acetic acid bacteria with a focus on the 16S-23S rRNA gene internal transcribed spacer and the analysis of cell proteins by MALDI-TOF mass spectrometry. *International Journal of Food Microbiology* **196:** 137 144.

Valera, M.J., Laich, F., González S.S., Torija, M.J., Mateo, E. and Mas, A. (2011). Diversity of acetic acid bacteria present in healthy grapes from the Canary Islands. *International Journal of Food Microbiology* **151:** 105-112.

Valera, M.J., Torija, M.J., Mas, A. and Mateo, E. (2013). *Acetobacter malorum* and *Acetobacter cerevisiae* identification and quantification by Real-Time PCR with TaqMan-MGB probes. *Food Microbiology* **36:** 30-39.

Vargha, M., Takats, Z., Konopka, A. and Nakatsu, C.H. (2006). Optimization of MALDI-TOF MS for strain level differentiation of *Arthrobacter* isolates. *Journal of Microbiological Methods* **66:** 399-409.

Vegas, C., Mateo, E., González, A., Jara, C., Guillamón, J.M., Poblet, M., Torija, M.J. and Mas, A. (2010). Population dynamics of acetic acid bacteria during traditional wine vinegar production. *International Journal of Food Microbiology* **38:** 130-136.

Versalovic, J., Schneider, M., De Bruijn, F.J. and Lupski, J.R. (1994). Genomic fingerprinting of bacteria using repetitive sequence-based polymerase chain reaction. *Methods in Molecular and Cellular Biology* **5:** 25-40.

von Knieriem and Mayer (1873). Landw. Versuchsstation 16: 305 cited by Asai 1968 and Vaughen (1954). Original not seen.

Vos, P., Hogers, R., Bleeker M., Reijans, M., van de Lee, T., Hornes, M., Frifiters, A., Pot, J., Peleman, J., Kuiper, M. and Xabeau, M. (1995). AFLP: A new technique for DNA fingerprinting. *Nucleic Acids Research* **23:** 4407-4414.

Vos, P. and Kuiper, M. (1997). AFLP Analysis. *In:* G. Caetano-Anollés and P.M. Gresshoff (Eds.). *DNA Markers: Protocols, Applications and Overviews*. Wiley, New York, pp. 115-131.

Wahl, K.I., Wunschel, C.S., Jarman, K.H., Valentine, N.B., Petersen, C.E., Kingsley, M.T., Zartolas, K.A. and Saenz, A.J. (2002). Analysis of microbial mixture by matrix-assisted laser desorption/ionization time-of-flight mass spectrometry. *Analytical Chemistry* **74:** 6191-6199.

Warscheid, B. and Fenselau, C. (2004). A target proteomics approach to the rapid identification of bacterial cell mixtures by matrix-assisted laser desorption/ionization time-of-flight mass spectrometry. *Proteomics* **4:** 2877-2892.

Welker, M. (2011). Proteomics for routine identification of microorganism. *Proteomics* **11:** 3143-3153.

Welker, M. and Moore, E.R.B. (2011). Applications of whole-cell matrix-assisted laser desorption/ionization time-of-flight mass spectrometry in systematic microbiology. *Systematic and Applied Microbiology* **34:** 2-11.

Wieme, A.D., Spitaels, F., Aerts, M., De Bruyne, K., Van Landschoot, A. and Vandamme, P. (2014a). Effects of growth medium on matrix-assisted laser desorption-ionization time-of-flight mass spectra: A case study of acetic acid bacteria. *Applied and Environmental Microbiology* **80:** 1528-1538.

Wieme, A.D., Spitaels, F., Aerts, M., De Bruyne, K., Van Landschoot, A. and Vandamme, P. (2014b). Identification of beer-spoilage bacteria using matrix-assisted laser desorption/ionization time-of-flight mass spectrometry. *International Journal of Food Microbiology* **185:** 41-50.

Wieser, M. and Busse, H.J. (2000). Rapid identification of Staphylococcus epidermidis. *International Journal of Systematic Evolutionary* **50:** 1087-1093.

Williams J.G., Kubelik, A.R., Livak, K.J., Rafalski, J.A. and Tingey, S.V. (1990). DNA polymorphisms amplified by arbitrary primers are useful as genetic markers. *Nucleic Acids Research* **8:** 6531-6535.

Yetiman, A.E. and Kesmen, Z. (2015). Identification of acetic acid bacteria in traditionally produced vinegar and mother of vinegar by using different molecular techniques. *International Journal of Food Microbiology* **204:** 9-16.

Yukphan, P., Potacharoen, W., Tanasupawat, S., Tanticharoen, M. and Yamada, Y. (2004). *Asaia krungthepensis* sp. nov., an acetic acid bacterium in the α-Proteobacteria. *International Journal of Systematic Evolutionary Microbiology* **54:** 313-316.

Zabeau, M. and Vos, P. (1993). Selective restriction fragment amplification: A general method for DNA fingerprinting. European Patent Office, publication 0 534 858 A1, bulletin 93/13.

8

Preservation of Acetic Acid Bacteria

Luciana De Vero*, Maria Gullo and Paolo Giudici
Unimore Microbial Culture Collection (UMCC), Department of Life Sciences,
University of Modena and Reggio Emilia, Via Amendola 2,
42122 Reggio Emilia, Italy

Introduction

Increased knowledge on physiological, biochemical and genetic properties of acetic acid bacteria (AAB) promotes advancements in biotechnology and innovative fermentation processes. Accordingly, the microbial collections of well-characterized AAB strains are fundamental sources for selecting cultures with desired properties for biotechnological applications (De Vero et al. 2010). Commonly, culture collections have the fundamental role of safeguarding long-term accessibility of relevant strains by ensuring the maintenance of their 'authenticity'. Nevertheless, they provide more than just strains, but also essential data, expertise and useful services, which are fundamental for supporting industrial researches.

In the case of AAB, the crucial role of a microbial collection is to find and validate appropriate methodologies for culture preservation, especially for strains which are difficult to cultivate.

In this chapter, we discuss conditions for AAB isolation and cultivation as well as how different preservation methods affect their cultivability. Moreover, management of an AAB collection will also be reviewed, analyzing the importance of strains quality controls, the role of databases and open platforms.

Culture Media

AAB can be isolated from different sources, ranging from fermented beverages, like vinegar, wine, cider or beer, to different habitats like plants, flowers and fruits. Additionally, some AAB species occur as symbionts of insects (Mamlouk and Gullo 2013) and human pathogens (e.g., *Granulibacter bethesdensis*, isolated from a chronic granulomatous disease patient) (Greenberg et al. 2006). Many isolation media, which mainly differ in the carbon source that can be single or multiple, have been formulated to recover AAB; a list of these isolation media is reported in Table 1.

Isolation and Enrichment Media

Commonly different culture media need to be tested for isolating AAB from specific sources as the strains can have different requirements. For instance, a common medium is GYC, especially useful to isolate AAB from sources rich in sugar (Gullo et al. 2006).

*Corresponding author: luciana.devero@unimore.it

Table 1: Isolation and cultivation media for acetic acid bacteria

Media	*Components amount*	*References*
Glucose Yeast extract Carbonate (GYC) medium		
Glucose	10 per cent	Sievers and Swings 2005,
Yeast extract	1 per cent	Gullo et al. 2006,
Calcium carbonate	2 per cent	Gullo and Giudici 2008,
Bacteriological agar	1.5 per cent	Sengun et al. 2011
Glucose Yeast extract (GY) medium		
Glucose	2 per cent	Yamada and Yukphan 2008,
Yeast extract	1 per cent	Sengun et al. 2011
Bacteriological agar	2 per cent	
Acetic acid Ethanol (AE) medium		
Glucose	0.5 per cent	Entani et al. 1985, Sievers et al. 1992,
Yeast extract	0.3 per cent	Sokollek and Hammes 1997,
Peptone	0.4 per cent	Yamada et al. 1999,
Bacteriological agar	0.9 per cent	Gullo et al. 2006,
EtOH	3 per cent	Gullo and Giudici 2008,
AcOH	3 per cent	Sengun et al. 2011
Yeast extract Peptone Mannitol (YPM) medium		
Yeast extract	0.5 per cent	Gullo et al. 2006,
Peptone	0.3 per cent	Gullo and Giudici 2008,
Mannitol	2.5 per cent	Sengun et al. 2011
Bacteriological agar	1.2 per cent	
Reinforced Acetic acid Ethanol (RAE) medium		
Glucose	4 per cent	Sokollek and Hammes 1997,
Yeast extract	1 per cent	Zahoor et al. 2006,
Peptone	1 per cent	Trcek et al. 2007,
$Na_2HPO_4 \times 2H_2O$	0.338 per cent	Sengun et al. 2011,
Citric acid x H_2O	0.15 per cent	Mamlouk and Gullo 2013
Ethanol	2 per cent (v/v)	
Acetic acid	1 per cent (v/v)	
Bacteriological agar	2 per cent	
Ethanol	2 per cent	

(*Contd.*)

Frateur Medium		
Yeast Extract	1 per cent	Frateur 1950,
$CaCO_3$	2 per cent	Beheshti and Shafiee 2009
Bacteriological agar	2 per cent	
Ethanol	2 per cent	
Carr medium		
Yeast extract	3 per cent	Carr 1968,
Bacteriological agar	2 per cent	Kittelman et al. 1989,
Bromocresol green	0.002 per cent	Beheshti and Shafiee 2009,
EtOH (v/v)	2 per cent (v/v)	Gullo et al. 2012
Malt Yeast extract Agar (MYA) medium		
Malt extract	1.5 per cent	Gullo et al. 2006,
Yeast extract	0.5 per cent	Mamlouk and Gullo 2013
Bacteriological agar	1.5 per cent	
EtOH	6 ml	

This medium, formulated by Carr and Passmore (1979), includes calcium carbonate to neutralize acids generated by AAB. The colonies of AAB, grown on GYC plates after a period of incubation of 48 hours at 25-30°C in aerobic conditions, are easily recognized by the surrounding area of calcium carbonate clearing (Fig. 1).

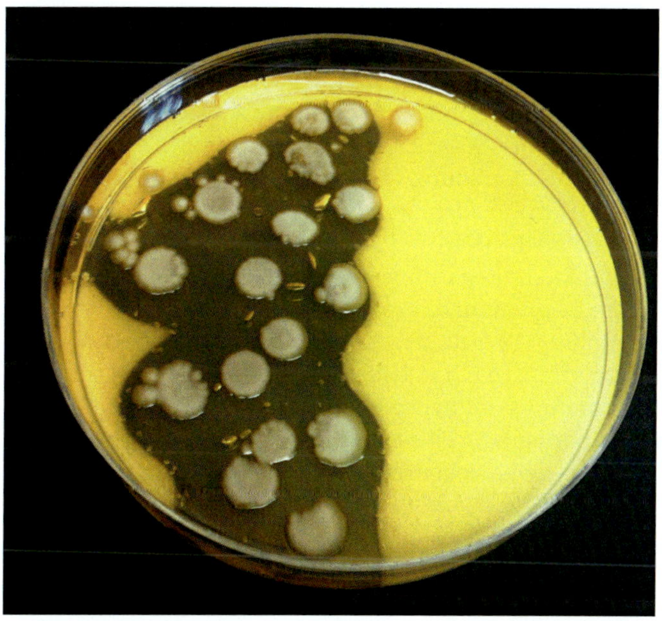

Figure 1: AAB growth on GYC medium plate.

Media containing acetic acid and ethanol, like AE and RAE, are indicated for the high acetic acid-producing strains, as those isolated from spirit vinegars (Sokollek and Hammes 1997, Sengun and Karabiyikli 2011). Other isolation media, as YPM and MYA, which contain mannitol or malt extract, respectively, are also widely used to recover different AAB species (Gullo and Giudici 2008).

Sometimes, enrichment media is required to isolate AAB that occur as minor in mixed cultures or slow growing strains. In these cases, it is possible to favor only the growth of desired microorganisms by making specific culture media with defined chemical composition and/or by changing the physical conditions of the incubation.

For instance, to control the growth of other bacterial species, the media can be acidified to pH 4.4, since AAB are tolerant to low pH (Drysdale and Fleet 1988). Antibiotics, like pimaricin and cycloheximide, can be added to the basal medium to control yeast growth, while penicillin may be used to control the growth of lactic acid bacteria.

Several enrichment media have been tested for AAB isolation from flowers and fruits, which are good sources for recovering different species (Lisdiyanti et al. 2001, Kersters et al. 2006, Sharafi et al. 2010, Arifuzzaman et al. 2014). In particular, three enrichment broths, namely potato medium (0.5 per cent glucose, 1 per cent yeast extract, 1 per cent peptone, 2 per cent glycerol and 1.5 per cent potato extract), seed culture medium (0.5 per cent glucose, 0.5 per cent yeast extract, 0.5 per cent peptone, 2 per cent glycerol) and sterile distilled water, each of them supplemented with four per cent ethanol, have been proved to be effective for selective isolation of AAB from flower and fruit samples (Sudsakda et al. 2007). Moreover, Yamada and co-workers (2000) described an enrichment medium without acetic acid and containing 2.0 per cent D-sorbitol, 0.5 per cent peptone, 0.3 per cent yeast extract, 0.01 per cent cycloheximide (pH 3.5 adjusted with hydrochloric acid), especially useful for the isolation of strains belonging to the *Asaia* genus whose growth is inhibited by 0.35 per cent of acetic acid.

Maintenance Media

The viability of AAB strains as well as the stability of their phenotypic traits, such as resistance to acetic acid and ethanol as well as the ability to synthesize cellulose, are strongly affected by the preservation method. For instance, the maintenance at +4°C needs frequent transplants to restore optimal media conditions reducing physiological stress of cells due to acids production (Sokollek et al. 1998, Schüller et al. 2000). For this reason, media having ethanol as carbon source are not suitable for maintenance at +4°C (Gullo and Giudici 2008). Generally, the availability of carbon sources and energy critically affect the AAB growth in defined media (Rao and Stokes 1953). As the sole source of nitrogen, AAB can use inorganic ammonia. Regarding the effects of essential amino acids, it has been reported that glutamate, glutamine, proline and histidine stimulate the growth of AAB, while others have inhibitory effects, as valine for *Gluconobacter oxydans* and threonine and homoserine for *Acetobacter aceti* (Drysdale and Fleet 1988).

Also high temperatures influenceAAB viability and other cellular features after long time storage. In particular, cell respiration can be affected during storage at a high temperature (35°C) since dehydrogenases are especially heat sensitive (Shafiei et al. 2013).

Short- and Long-term Preservation

The preservation of culture collection as well as the deposit of a microorganism with any of the international depositary authorities or culture collections, under the Budapest Treaty of the United Nations World Intellectual Property Organization (WIPO 1980), requires that the microorganisms resist long-term preservation and revival procedures. Actually, there are several techniques available for the preservation of microorganisms, each of which has particular advantages and disadvantages. Moreover, different microorganisms could require special preservation methods in order to ensure optimal storage without any morphological, physiological or genetic change. Indeed, at least two different techniques should be applied for each microbial culture and the cell stocks should be kept in different places to reduce the risk of accidental loss (OECD 2007).

At least one technique should be able to guarantee long-term maintenance and stability of the culture by reducing the metabolic activity as much as possible. Anyway, the applied preservation method used for microorganisms needs to be described in culture collections database or reported in publications.

In the following paragraphs, the main techniques for short- and long-term maintenance of AAB are reported.

Sub-culturing

The sub-culturing methods comprise periodical transplant of the cultures in a fresh medium in order to ensure growth and viability of cells. The transplant is performed aseptically to avoid contamination and, successively, the culture grown is stored at low temperature (0-4°C) to reduce its metabolic activity. Serial sub-cultures are widely used especially if the cultures are required frequently and quickly. However, sub-culturing involves, usually, short periods between transplants, variable from one strain to another (at least every three months). This is because even at 0-4°C the microorganisms continue to grow and utilize nutrients as well as to produce wastes in the medium that can cause cell death. Consequently, the sub-culturing method is time consuming and not suitable for maintaining a large number of cultures. Furthermore, the main disadvantages of this method are the risk of contamination and genetic changes due to the development of variants and mutants. For these reasons, continuous growth techniques should be used just for short-term maintenance.

Storage Under Mineral Oil

The preservation of cultures under mineral oil is a simple and economical method for storage of microorganisms in laboratories with limited resources and facilities, as it does not require expensive equipment and consumables. By this method cultures are covered with sterile mineral oil to prevent dehydration and to slow down the metabolic activity (Smith et al. 2001). In this dormant state, the culture can be preserved for a time ranging from months to years, depending on the microorganism. Usually, the cultures under mineral oil are stored in glass tubes located at controlled room temperatures (range from 15°C to 18°C). The revitalization of the culture is done by streaking a colony on to a suitable agar medium. The preservation with mineral oil can be particularly advantageous for a long viability of species that do not survive other preservation methods; however, strains instability and the risk of contamination are main disadvantages.

Freeze-drying

Freeze-drying is one of the commonest preferred methods for long-term maintenance of microorganisms, mainly because the cultures show high viability and stability after recovering and are easily stored in small ampoules for several years. By freeze-drying, the metabolic activity of cells is reduced to a minimum that preserves the culture in an inactive but viable state. This is achieved by a complex process that involves rapid freezing of the culture and the removal of water by sublimation (Greaves 1964). However, the freeze-dried cells need to maintain a residual moisture content of 1-3 per cent (w/v) for their long-term shelf-life (Gherna 2010). Moreover, it is fundamental to suspend the cells to be lyophilized in a medium containing cryoprotective agents against freezing and drying injuries (Greaves 1964). Indeed, these agents can prevent both ice crystal formation and total desiccation, which damages DNA and kills the cells, as well as to neutralize harmful effects, for instance from toxic electrolytes and carbonyl radicals (Malik and Claus 1987).

In particular, the fluid-mosaic structure of the cell membranes can be affected by the phase changes that occur during the cooling and drying processes. In fact, the liquid crystalline structure of the membranes may change to the gel phase, leading to consequent leakage of the membranes (Smith 2009, De Vero and Giudici 2013). Moreover, variation in cell water content can induce unwanted chemical reactions or cause disorder in specific enzymes. For instance, the antioxidant defense systems could be inactive under water stress and, consequently, free radicals, normally removed by these systems, can promote peroxidation of lipids during storage of freeze-dried cells (Shafiei et al. 2013).

Suitable cryoprotectants, such as polyvinylpyrrolidone (PVP), dimethyl sulfoxide (DMSO), glycerol, dextran, skimmed milk, mannitol, inositol, sucrose, malt extract and trehalose can be used in order to preserve strain viability. Regarding the cell concentration for freeze-drying, it is generally related to the protective medium used usually, a cell concentration of 1×10^8 cells/mL can be suitable, even if a higher initial concentration allows increase in the probability of a greater survival of viable cells (Morgan et al. 2006). The effect of cryoprotectants on the survival rate of freeze-dried AAB used as starter cultures for vinegar making has been investigated by several researchers (Sokollek et al. 1998, Ndoye et al. 2007, Shafiei et al. 2013). For instance, Sokollek and Hammes (1997) used 20 per cent malt extract during the lyophilization process of *Acetobacter* strains and observed no loss of their original properties as a starter after one year (Sokollek and Hammes 1997). Nomura and collaborators (1998) prepared a cell suspension of *A. aceti* with 30 per cent sucrose in McIlvaine buffer (pH 6) for freeze-drying and this proved efficient in avoiding the decrease of aldehyde oxidase activity during the process and consequent storage.

Furthermore, mannitol (20 per cent) was also successfully used during the lyophilization process of two thermoresistant *A. senegalensis* and *A. pasteurianus* strains that showed no loss of viability after storage for six months at +4°C (Ndoye et al. 2007).

Nevertheless, it is important to underline that there is no unique preservation method for all the strains and, consequently, protocols regarding suitable suspension media and cell concentration, kind of cryoprotectant and cooling rates have to be tested in a specific way.

Moreover, even if no loss in viability is shown, quality controls are essential after lyophilization because genetic stability of strains may be affected, causing loss of important properties.

Cryopreservation

Cryopreservation is one of the most valuable methods for maintaining genetic stability and viability of bacterial cultures for a long time. The frozen samples can be stored in ultra-low temperature freezers at around –70°C and –150°C or in tanks containing vapor phase of nitrogen (about –150°C) or liquid nitrogen (–196°C). Several factors can influence the efficiency of the cryopreservation method, including the strain, growth and recovery medium, pH of the medium, amount of cell water, temperature and period of storage, cooling and warming rate (Hubálek 2003). Regarding cooling, the optimal rate should be around –1°C/min. over the critical phase, in order to prevent any mechanical injury to the cell as well as the physical and chemical changes of the cell solutions (Moussa et al. 2008,

Smith 2009). To achieve this rate of cooling it is possible to use special equipments with programmable temperature control or simple commercial systems, like the 'Mr. Frosty' System (Nalgene, Rochester, NY) (De Vero and Giudici 2013). Other procedures can be followed to obtain slow cooling according to the descriptions reported by Gherna (2010). As in the case of freeze-drying, also for cryopreservation, it is necessary to use a suitable cryoprotectant to prevent cell damage during the process. Glycerol (10-25 per cent) and DMSO (5 per cent) are widely used since they can diffuse across cell membranes by protecting the cells externally and internally, unlike other cryoprotectants (Gherna 2010). A final concentration of 25 per cent glycerol is suitable for cryopreservation of different AAB species (Cleenwerck et al. 2007). In particular, it has been reported that the maintenance of phenotypic and genotypic stability of *Acetobacter pasteurianus* strain AB0220 over nine years of preservation at –80°C with glycerol (Gullo et al. 2012). However, glycerol is not suitable for preservation of cellulose-producing AAB, such as *Komagataeibacter xylinus*, because glycerol can affect the cellulose structure. In these cases, DSMO is more feasible since it is able to ensure stability and high survival rate without compromising the production and structure of cellulose (Wiegand and Klemm 2006).

Recovery of the Preserved Cultures

The revitalization of preserved biomass allows the cells to become active again and capable to reproduce themselves. To avoid cell damage due to the ice recrystallization, it is fundamental to thaw rapidly the cryopreserved microorganisms. Practically, a fast warming rate can be obtained by using a water bath at +37°C in which the cryovials are placed until the melting of the ice (Smith 2009). When the cryoprotective agent used is potentially toxic or growth-inhibiting, it is necessary to quickly remove it after thawing through several steps of short centrifugation and washing with the growth medium (Heylen et al. 2012). Sometimes, cells can show a prolonged lag phase after recovering because of chemical and temperature stress subjected during the time of storage. Therefore, prolonged incubation period is suggested to allow viable and active cultures. Moreover, the addiction of supplementing elements to standard growth medium, like cell-free extracts, short peptides, alternative energy, carbon sources or generally cyclic adenosine monophosphate (cAMP) and homolactones, have proved efficient in increasing the cultivability of 'unculturable' bacteria (Bruns et al. 2002, Nichols et al. 2008, Vartoukian et al. 2010).

Quality Control

The success rate of different methodologies used for strain maintenance is proven through quality control of the recovered microbial cultures, including the evaluation of viability, purity, authenticity and phenotypic proprieties (Janssens et al. 2010).

Viability after preservation can be checked through confirmation of growth in a specific medium. Moreover, evidence on purity is obtained through microscopic observation of the recovered cell suspension and, when possible, through streaking on the appropriate solid growth medium in order to check cells and colony morphology, respectively. Other methods like epifluorescence staining method or FISH (Fluorescence in Situ Hybridization) techniques allow a quick identity and purity check for those AAB cultures not able to grow on solid media (Schmid et al. 2005, Zwirglmaier 2005, Baena-Ruano et al. 2006, Heylen et al. 2012). Basic phenotypic characterization of AAB include Gram-staining, KOH test, catalase production, formation of water-soluble brown pigments and cellulose production, according to the protocols described in literature (Swings et al. 1992). Moreover, the evaluation of ethanol oxidation and acetate assimilation by chalk-ethanol tests on Frateur and modified Carr and Passmore media should be performed to confirm the stability of these important physiological traits (Sievers and Swing 2005, Gullo et al. 2012). A description of other common tools useful for AAB characterization is reported by Gullo and Giudici (2009). Nevertheless, in order to assess the genomic or physiological integrity of the strain cultures, after preservation, it is crucial to set and record their distinguished characters with fixed and reproducible tests. Data recording is an essential prerequisite; therefore, it must be accurate and presented in a standardized format (Smith and Ryan 2012).

Genetic Stability

Complete understanding of the AAB genetic background is fundamental for assessing the stability of industrially useful strains. Generally, the genetic stability of AAB, similar to other organisms, may be affected by casual events like mutations or genomic rearrangements that can lead to evolution and cause deficiency in important physiological properties – ethanol oxidation, acetic acid resistance or cellulose production (Ohmori et al. 1982, Azuma et al. 2009, Wang et al. 2015). In particular, genetic instability has been investigated in different *A. pasteurianus* strains, for which specific insertion sequence elements (namely IS1380 and IS1452) were identified as responsible for the loss of ethanol oxidation (Takemura et al. 1991, Kondo and Horinouchi 1997). Other findings were reported by Azuma et al. (2009), who investigated the mutability of the *A. pasteurianus* strain IFO 3283 (NBRC 3283), which formed a multi-phenotype cell complex after a series of slant passages for 21 years. The genome analysis of isolates from the cell complex revealed the presence of three single nucleotide mutations and five transposon insertions.

The existence of hypermutable AAB highlights the drawbacks related to the culture maintenance of certain strains in collections. Indeed, it is highly unlikely that all the strain copies are genetically identical after a long time, especially if they have been subjected to several recoverings. DNA fingerprinting methods, able to molecular characterize the microorganisms through the generation of strain-specific banding patterns, should be performed in different laboratories with standardized protocols in order to compare the results. The fingerprinting methods applied to AAB are extensively described in literature, and include $(GTG)_5$ REP (Repetitive Extragenic Palindromic)-PCR, Ribotyping, ERIC (Enterobacterial Repetitive Intergenic Consensus sequences)-PCR and RAPD (Random Amplified Polymorphic DNA)-PCR (De Vuyst et al. 2008, Cleenwerck et al. 2009, Gullo et al. 2012).

All the tests performed on the strain cultures preserved in different collections as well as the molecular technique used for their characterization need to be recorded and retained in public and integrated databases.

AAB Culture Collections Worldwide

The majority of AAB strains are maintained in a rather small number of well-known Biological Research Centers (BRC) affiliated to the World Federation for Culture Collection (WFCC) and distributed worldwide (Table 2).

Indeed, several AAB strains are kept in laboratory collections 'non-affiliated' in official organizations (De Vero and Giudici 2013) and commonly used for different purposes, such as teaching, researches and biotechnological applications. However, it is important to underline that the use of uncharacterized or contaminated microbial cultures is extremely disadvantageous for both research and industrial processes. Only qualified collections can guarantee high quality standards and services with respect to the regulations established

Table 2: Biological Research Centers which hold acetic acid bacteria strains

Acronym	*Name*	*Country*	*Website URL*
ATCC	American Type Culture Collection	USA	http://www.atcc.org/
AWRI MCC	Australian Wine Research Institute Microorganisms Culture Collection	Australia	http://www.awri.com.au
BCC	BIOTEC Culture Collection	Thailand	http://www.biotec.or.th/bcc/
BCCM/LMG	Belgian Coordinated Collections of Microorganisms/ LMG Bacteria Collection	Belgium	http://bccm.belspo.be/index.php
BCRC	Bioresources Collection and Research Center	Taiwan	http://www.bcrc.firdi.org.tw
CCBAU	Culture Collection of Beijing Agricultural University	China	http://grbio.org/institution/culture-collection-beijing-agricultural-university
CCTCC	China Center for Type Collection	China	http://www.cctcc.org/
CECT	ColecciónEspañola De Cultivos Tipo	Spain	http://www.cect.org
CIP	Collection de l'Institut Pasteur	France	http://www.crbip.pasteur.fr
DSMZ	Deutsche Sammlung von. Mikroorganismen und Zellkulturen GmbH	Germany	http://www.dsmz.de/
JCM	Japan Collection of Microorganisms	Japan	http://jcm.brc.riken.jp/en/
KCTC	Korean Collection for Type Cultures	Korea	http://www.abrcn.net/jundb_ko/
MCC-MNH	Microbial Culture Collection, Museum of Natural History	The Philippines	http://mnh.uplb.edu.ph/index.php/access-collections

(Contd.)

Table 2: (*Contd.*)

Acronym	*Name*	*Country*	*Website URL*
NBRC	NITE Biological Resource Center	Japan	http://www.nbrc.nite.go.jp/e/index.htm
NCCB	Netherlands Culture Collection of Bacteria	Netherlands	http://www.cbs.knaw.nl/nccb
NCIM	National Collection of Industrial Microorganisms	India	http://www.ncl-india.org/files/NCIM/Catalogue.aspx
NCIMB	National Collection of Industrial, Food and Marine Bacteria	United Kingdom	http://www.ncimb.co.uk
NRIC	NODAI Research Institute Culture Collection	Japan	http://nodaiweb.university.jp/nric/index_e.html
UMCC	Unimore Microbial Culture Collection	Italy	http://www.umcc.unimore.it
VTTCC	VTT Culture Collection	Finland	http://culturecollection.vtt.fi/

Sources: StrainsInfo (www.straininfo.net) and World Data Centre for Microorganisms (WDCM, www.wdcm.org).

by the Nagoya Protocol on Access and Benefit Sharing and the Convention on Biological Diversity (CBD 2011, Sette et al. 2013).

Rules and Services

The main role of microbial collections is to preserve 'authenticated biological materials' including not only microorganisms, but also genomic DNA, cloned genes, plasmids and so on. Furthermore, they are repositories for the strains cited in scientific publications or applied in researches and industrial processes, for which they can provide different kinds of services (Smith 2003, De Vero et al. 2013). Indeed, microbial culture collections can contribute effectively in research and technological development because they make the strains readily available to users after carrying out their screening, preservation, identification and certification (Sette et al. 2013). They can meet the high standards of quality and expertise demanded by the international community of scientists and industry for delivery of biological information and materials. Moreover, culture collectors have the expertise to select and implement starter cultures suitable for industrial processes (Fig. 2).

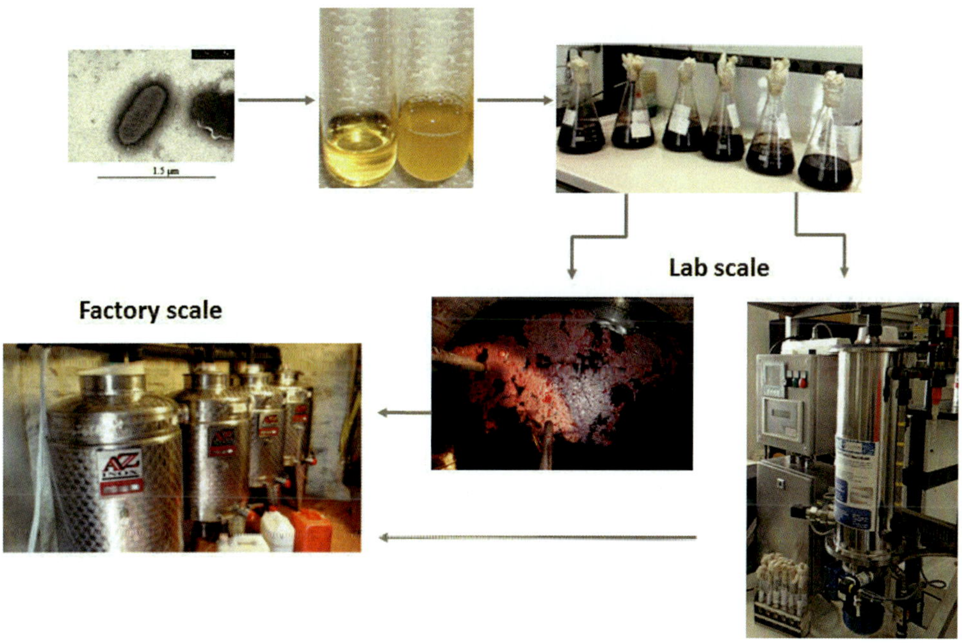

Figure 2: Workflow scheme of AAB starter culture screening for vinegar production performed by Unimore Microbial Culture Collection.

Database and Catalogues

According to the *OECD Best Practice Guidelines for Biological Resource Centers*, collections database should include the Minimum Data Set (MDS), which provides information on the strains, such as identity and taxonomy, other collection numbers, hazard information, biological origin, date of isolation, country of origin, growth requirements and methods of preservation and bibliographic references. An example of a strain's record reported according to these criteria is from the UMCC Database (Table 3). Other information can be also reported in a strain record, such as annotations, pictures, electrophoretic gel images as well as connection links to other databases. Indeed, even the most complete database in

Table 3: Database record of the strain *Acetobacter pasteurianus* UMCC 1754 deposit at the Unimore Microbial Culture Collection

UMCC 1754	
General information	
Taxonomy	
Species name:	*Acetobacter pasteurianus*
Type of organism:	Bacteria
Collections	
Other collection number:	AB0220=DSM 25273
Biosafety and restrictions	
Biological safety level:	BSL-1
Origin	
Substrate of isolation:	Traditional Balsamic Vinegar
Isolated by:	Gullo Maria
Isolation year:	2002
Latitude, longitude coordinates where collected:	44.6167°; 10.6167°
Locality:	Provinces of Reggio Emilia and Modena
Country:	Italy
Identified by:	Gullo Maria
Molecular method used:	• PCR/RFLP analysis of 16S-23S-5S rDNA (RsaI); • Eric-PCR; • 16S rRNA gene sequencing (Accession n. HE650905.1)
Miscellaneous info	
Application:	Selected starter culture for vinegar production
Maintenance conditions:	Cryopreservation at –80°C
Annotations:	No cellulose production
Media and growth conditions	
Growth medium:	GYC
Growth temperature:	+28°C
Bibliography	
References:	Gullo et al. 2009. *Appl. Environ. Microbiol.* 75: 2585-2589 Gullo et al. 2012. *Curr. Microbiol.* 64: 576-580
DNA	
Genbank:	Accession numberof AB0220 partial 16S rRNA gene, available from: http://www.ncbi.nlm.nih.gov/nuccore/HE650905.1

Source: Unimore Microbial Culture Collection (UMCC) Database (available on line at: http://biolomics.umcc.unimore.it/. Accessed: June 2016).

terms of records and associated characteristics cannot contain all the needed information that a researcher may want. For instance, detailed taxonomic descriptions or sequence data of the strains may be reported in other repositories and it may be useful to link all the information (Robert et al. 2015). According to this consideration, it is essential that data stored in different databases are a part of a network of interoperable repositories.

Open Platform Integrating Microbial Information

Actually, several platforms have been tried in order to standardize and integrate the information among different culture collection databases as well as to allow easy retrieval of biological material. Among them, **StrainInfo.net** reports the data on the strain deposits in different BRCs as 'strain passports' which also permit to follow the exchange history of the strains from one BRC to one other. In this open platform, strain designations, historic information on deposit, growth characteristics, genomics and other data as well as relevant publications are shown to the user in an integrated way by using a Microbial Common Language (MCL). Moreover, StrainInfo aggregates taxonomic information for its taxon passports from several external resources, such as the List of Prokaryotic names with Standing in Nomenclature (LPSN, http://www.bacterio.net/), the Prokaryotic Nomenclature Up-to-Date (PNU, https://www.dsmz.de/bacterial-diversity/prokaryotic-nomenclature-up-to-date.html) and Mycobank (http://www.mycobank.org/) (Robert et al. 2005, Verslyppe et al. 2014).

Like StrainInfo, the Global Catalogue of Microorganisms **(GCM)** (http://gcm.wfcc.info/datastandards) acts as integration of catalogue information from each of the collections reported in the WDCM constructing a global catalogue database. In addition to the MDS information, GCM database contains extensive citation, patent and gene or genome information related to each strain. Moreover, species name and strain numbers can be further used to identify and extract information from public database, such as:

- Genbank (http://www.ncbi.nlm.nih.gov/genbank/)
- NCBI Genome (http://www.ncbi.nlm.nih.gov/genome/)
- UniProt (http://www.uniprot.org/uniprot/)
- PDB (Protein Data Bank; http://www.wwpdb.org/)

Users can access the data via a web interface, as the GCM database runs on a platform with both Java and MySQL server, and results may be displayed in different formats allowing refining of results by using filters (Wu et al. 2013).

Conclusion

Microbial collections are important bioresourses that encourage innovation in food, industrial, environmental and medical fields. Microbial collections of AAB strains can provide critical insights into their genome, transcriptome, proteome and metabolome while combining phenotypic and molecular traits with industrial strain performance to develop potential new microbial processes.

Apart from the applications of accurate preservation methods to guarantee the strain authenticity and stability for a long time, microbial collections are responsible for cataloging and sharing of detailed information on the collected strains. For this purpose, the databases available online allow the management of a considerable amount of biological data with easier accessibility thanks to the increasingly wide databases interoperability. Moreover, microbial collections can offer innovative research and scientific services to customers and stakeholders in order to exploit the AAB diversity for novel and functional starter cultures.

References

Arifuzzaman, M., Hasan M.Z., Rahman, S.M.B. and Pramanik, M.K. (2014). Isolation and characterization of *Acetobacter* and *Gluconobacter* spp. from sugarcane and rotten fruits. *Research and Reviews in Biosciences* **8:** 359-365.

Azuma, Y., Hosoyama, A., Matsutani, M., Furuya, N., Horikawa, H., Harada, T., Hirakawa, H., Kuhara, S., Matsushita, K., Fujita, N. and Shirai, M. (2009). Whole-genome analyses reveal genetic instability of *Acetobacterpasteurianus*. *Nucleic Acids Research* **37:** 5768-5783.

Baena-Ruano, S., Jimenez-Ot, C., Santos-Duenas, I.M., Cantero-Moreno, D., Barja, F. and Garcia-Garcia, I. (2006). Rapid method for total, viable and non-viable acetic acid bacteria determination during acetification process. *Process Biochemistry* **41:** 1160-1164.

Beheshti Maal, K. and Shafiee, R. (2009). Isolation and Identification of an *Acetobacter* strain from Iranian white-red cherry with high acetic acid productivity as a potential strain for cherry vinegar production in food and agriculture biotechnology. *World Academy of Science, Engineering and Technology* **54:** 201-204.

Bruns, A., Cypionka, H. andOvermann, J. (2002). Cyclic AMP and acyl homoserine lactones increase the cultivation efficiency of hetero-trophic bacteria from the Central Baltic Sea. *Applied and Environmental Microbiology* **68:** 3978-3987.

Carr, J.G. (1968). Identification of acetic acid bacteria. *In:* B.M. Gibbs and D.A. Shapton (Eds.). *Identification Methods for Microbiologists*, Part B. Academic Press, London, pp. 1-8.

Carr, J.G. and Passmore, S.M. (1979). Methods for identifying acetic acid bacteria. *In:* F.A. Skinner and D.W. Lovelock (Eds.). *Identification Methods for Microbiologists*. Academic Press, UK, pp. 33-47.

CBD (The Convention on Biological Diversity). (2011). *Year in Review*. Available from: https://www.cbd.int/doc/reports/cbd-report-2011-en.pdf (Accessed: June 2016).

Cleenwerck, I., Camu, C., Engelbeen, K., De Winter, T., Vandemeulebroecke, K., De Vos, P. and De Vuyst, L. (2007). *Acetobacter ghanensis* sp. a novel acetic acid bacterium isolated from traditional heap fermentations of Ghanaian cocoa beans. *International Journal of Systematic and Evolutionary Microbiology* **57:** 1647-1652.

Cleenwerck, I., de Wachter, M., Gonzalez, A., de Vuyst, L. and de Vos, P. (2009). Differentiation of species of the family Acetobacteraceae by AFLP DNA fingerprinting: *Gluconacetobacter kombuchae* is a later heterotypic synonym of *Gluconacetobacter hansenii*. *International Journal of Systematic and Evolutionary Microbiology* **59:** 1771-1786.

De Vero, L., Gullo, M. and Giudici, P. (2010). Acetic Acid Bacteria, Biotechnological Applications. *In:* M.C. Flickinger (Ed.). *Encyclopedia of Industrial Biotechnology: Bioprocess, Bioseparation and Cell Technology*. John Wiley & Sons Inc., New York, pp.1-17.

De Vero, L. and Giudici, P. (2013). Significance and management of acetic acid bacteria culture collections. *Acetic Acid Bacteria* **2:** e9.

De Vuyst, L., Camu, N., De Winter, T., Vandemeulebroecke, K., Van de Perre, V., Vancanneyt, M., De Vos, P. and Cleenwerck, I. (2008). Validation of the $(GTG)_5$-rep-PCR fingerprinting technique for rapid classification and identification of acetic acid bacteria, with a focus on isolates from Ghanaian fermented cocoa beans. *International Journal of Food Microbiology* **125:** 79-90.

Drysdale, G.S. and Fleet, G.H. (1988). Acetic acid bacteria in winemaking: A review. *American Journal of Enology and Viticulture* **39:** 143-154.

Entani, E., Ohmori, S., Masai, H. and Suzuki, K.I. (1985). *Acetobacter polyoxogenes* sp. nov., a new species of an acetic acid bacterium useful for producing vinegar with high acidity. *The Journal of General and Applied Microbiology* **31:** 475-490.

Frateur, J. (1950). Essaisur la systématique des Acetobacters. *Cellule* **53:** 285-392.

GCM (Global Catalogue of Microorganisms). Available from: http://gcm.wfcc.info/datastandards (Accessed: June 2016).

Genbank (Genetic Sequence Database). Available from: http://www.ncbi.nlm.nih.gov/genbank/ (Accessed: June 2016).

Gherna, R.L. (2010). Culture preservation. *In:* M.C. Flickinger (Ed.). *Encyclopedia of Industrial Biotechnology: Bioprocess, Bioseparation and Cell Technology*. John Wiley & Sons Inc., New York, pp. 1-18.

Greaves, R.I.N. (1964). Fundamental aspects of freeze-drying bacteria and living cells. *In:* L. Rey (Ed.). *Aspects Theoretiqueset Industrials de la Lyophilisation*. Hermann, Paris, pp. 407-410.

Greenberg, D.E., Porcella, S.F., Stock, F., Wong, A., Conville, P.S., Murray, P.R., Holland, S.M. and Zelazny, A.M. (2006). *Granulibacter bethesdensis* gen. nov., sp. nov., a distinctive pathogenic acetic acid bacterium in the family Acetobacteraceae. *International Journal of Systematic and Evolutionary Microbiology* **56:** 2609-2616.

Gullo, M., Caggia, C., De Vero, L. and Giudici, P. (2006). Characterization of acetic acid bacteria in 'traditional balsamic vinegar'. *International Journal of Food Microbiology* **106:** 209-212.

Gullo, M. and Giudici, P. (2008). Acetic acid bacteria in traditional balsamic vinegar: Phenotypic traits relevant for starter cultures selection. *International Journal of Food Microbiology* **125:** 46-53.

Gullo, M. and Giudici, P. (2009). Acetic acid bacteria taxonomy from early descriptions to molecular techniques. *In:* L. Solieri and P. Giudici (Eds.). *Vinegars of the World.* Springer-Verlag, Italia, pp. 41-60.

Gullo, M., De Vero, L. and Giudici, P. (2009). Succession of selected strains of *Acetobacter pasteurianus* and other acetic acid bacteria in traditional balsamic vinegar. *Applied and Environmental Microbiology* **75:** 2585-2589.

Gullo, M., Mamlouk, D., De Vero, L. and Giudici, P. (2012). *Acetobacter pasteurianus* strain AB0220: Cultivability and phenotypic stability over 9 years of preservation. *Current Microbiology* **64:** 576-580.

Heylen, K., Hoefman, S., Vekeman, B., Peiren, J. and De Vos, P. (2012). Safeguarding bacterial resources promotes biotechnological innovation. *Applied Microbiology and Biotechnology* **94:** 565-574.

Hubálek, Z. (2003). Protectants used in the cryopreservation of microorganisms. *Cryobiology* **46:** 205-229.

Janssens, D., Arahal, D.R., Bizet, C. and Garay, E. (2010). The role of public biological resource centers in providing a basic infrastructure for microbial research. *Research in Microbiology* **161:** 422-429.

Kersters, K., Lisdiyanti, P., Komagata, K. and Swings, J. (2006). The family *Acetobacteraceae*: The genera *Acetobacter, Acidomonas, Asaia, Gluconacetobacter, Gluconobacter* and*Kozakia. In:* M. DworkinS. Falkow, E. Rosenberg, K-H. Schleifer and E. Stackebrandt (Eds.). The Prokaryotes—A Handbook on the Biology of Bacteria, 3rd. edn, vol. 5. Springer-Verlag, New York, pp. 163-200.

Kittelman, M., Stamm, W.W, Follmann, H. and Truper, H.G. (1989). Isolation and classification of acetic acid bacteria from high percentage vinegar fermentations. *Applied Microbiology and Biotechnology* **30:** 47-52.

Kondo, K. and Horinouchi, S. (1997). Characterization of the genes encoding the three-component membrane-bound alcohol dehydrogenase from *Gluconobacter suboxydans*and their expression in *Acetobacter pasteurianus. Applied and Environmental Microbiology* **63:** 1131-1138.

Lisdiyanti, P., Katsura, K., Seki, T., Yamada, Y., Uchimura, T. and Komagata, K. (2001). Identification of *Acetobacter* strains isolated from Indonesian sources, and proposals of *Acetobacter syzygii* sp. nov.,*Acetobacter cibinongensis* sp. nov. and *Acetobacter orientalis* sp. nov. *Journal of General and Applied Microbiology* **47:** 119-131.

LPSN (List of Prokaryotic Names with Standing in Nomenclature). Available from: http://www.bacterio.net/ (Accessed: June 2016).

Malik, K.A. and Claus, D. (1987). Bacterial culture collections: Their importance to biotechnology and microbiology. *Biotechnology and Genetic Engineering Reviews* **5:** 137-198.

Mamlouk, D. and Gullo, M. (2013). Acetic acid bacteria: Physiology and carbon sources oxidation. *Indian Journal of Microbiology* **53:** 377-384.

Morgan, C.A., Herman, N., White, P.A. and Vesey, G. (2006). Preservation of micro-organisms by drying: A review. *Journal of Microbiological Methods* **66:** 183-193.

Moussa, M., Dumont, F., Perrier-Cornet, J.M. and Gervais, P. (2008). Cell inactivation and membrane damage after long-term treatments at sub-zero temperature in the supercooled and frozen states. *Biotechnology and Bioengineering* **101:** 1245-1255.

Mycobank. Available from: http://www.mycobank.org/ (Accessed: June 2016).

NCBI Genome (National Center for Biotechnology Information). Available from: http://www.ncbi.nlm.nih.gov/genome/ (Accessed: June 2016).

Ndoye, B., Weekers, F., Diawara, B., Guiro, A.T. and Thonart, P. (2007). Survival and preservation after freeze-drying process of thermoresistant acetic acid bacteria isolated from tropical products of Subsaharan Africa. *Journal of Food Engineering* **79:** 1374-1382.

Nichols, D., Lewis, K., Orjala, J., Mo, S., Ortenberg, R., O'Connor, P., Zhao, C., Vouros, P., Kaeberlein, T. and Epstein, S.S. (2008). Short peptide induces an 'uncultivable' microorganism to grow *in vitro. Applied and Environmental Microbiology* **74:** 4889-4897.

Nomura, Y. and Matsushita, K. (1998). Preparation and preservation of freeze-dried cells of acetic acid bacteria with aldehyde oxidase activity. *Bioscience, Biotechnology and Biochemistry* **62:** 1134-1137.

OECD (Organization for Economic Co-operation and Development). (2007). Best-practice guidelines. Available from: https://www.oecd.org/sti/biotech/38777417.pdf (Accessed: June 2016).

Ohmori, S., Uozumi, T. and Beppu, T. (1982). Loss of acetic acid resistance and ethanol oxidizing ability in an *Acetobacter strain. Agricultural and Biological Chemistry* **46:** 381-389.

PDB (Protein Data Bank). Available from: http://www.wwpdb.org/ (Accessed: June 2016).

PNU (Prokaryotic Nomenclature Up-to-Date). Available from: https://www.dsmz.de/bacterial-diversity/prokaryotic-nomenclature-up-to-date.html (Accessed: June 2016).

Rao, M.R. and Stokes, J.L. (1953). Nutrition of the acetic acid bacteria. *Journal of Bacteriology* **65:** 405-412.

Robert, V., Stegehuis, G. and Stalpers, J. (2005). The MycoBank Engine and Related Databases. Available from: http://www.mycobank.or (Accessed: June 2016).

Robert, V., Cardinali, G. and Duong, V.U. (2015). *Biodiversity Bioinformatics, Challenges and Opportunities*. Proceedings of Microbial Diversity Perugia – The challenge of complexity – MD2015, SIMTREA, Firenze, Italy, pp. 145-149.

Schmid, M.W., Lehner, A., Stephan, R., Schleifer, K.H. and Meier, H. (2005). Development and application of oligonucleotide probes for *in situ* detection of thermotolerant *Campylobacter* in chicken faecal and liver samples. *International Journal of Food Microbiology* **105:** 245-255.

Schüller, G., Hertel, C. and Hammes, W.P. (2000). *Gluconacetobacter entanii* sp. nov., isolated from submerged high-acid industrial vinegar fermentations. *International Journal of Systematic and Evolutionary Microbiology* **50:** 2013-2020.

Sengun, I.Y. and Karabiyikli, S. (2011). Importance of acetic acid bacteria in food industry. *Food Control* **22:** 647-656.

Sette, L.D., Pagnocca, F.C. and Rodrigues, A. (2013). Microbial culture collections as pillars for promoting fungal diversity, conservation and exploitation. *Fungal Genetics and Biology* **60:** 2-8.

Shafiei, R., Delvigne, F. and Thonart, P. (2013). Flow-cytometric assessment of damages to *Acetobacter senegalensis* during freeze-drying process and storage. *Acetic Acid Bacteria* **2:** e10.

Sharafi, S.M., Rasooli, I. and Beheshti-Maal, K. (2010). Isolation, characterization and optimization of indigenous acetic acid bacteria and evaluation of their preservation methods. *Iranian Journal of Microbiology* **2:** 38-45.

Sievers, M., Sellmer, S. and Teuber, M. (1992). *Acetobacter europaeus* sp. nov., a main component of industrial vinegar fermenters in Central Europe. *Systematic and Applied Microbiology* **15:** 386-392.

Sievers, M. and Swings, J. (2005). Family Acetobacteraceae. *In:* G.M. Garrity (Ed.). *Bergey's Manual of Systematic Bacteriology*, 2nd edn., Springer, New York, pp. 41-95.

Smith, D., Ryan, M.J. and Day, J.G. (2001). *The UK National Culture Collection (UKNCC) Biological Resource: Properties, Maintenance and Management*. London, UK National Culture Collection.

Smith, D. (2003). Culture collections over the world. *International Microbiology: The Official Journal of the Spanish Society for Microbiology* **6:** 95-100.

Smith, D. (2009). Culture collections and biological resource centers (BRCs). *In:* M.C. Flickinger (Ed.). *Encyclopedia of Industrial Biotechnology: Bioprocess, Bioseparation and Cell Technology*. John Wiley & Sons Inc., New York, pp. 1-19.

Smith, D. and Ryan, M. (2012). Implementing best practices and validation of cryopreservation techniques for microorganisms. *The Scientific World Journal* 1-9.

Sokollek, S.J. and Hammes, W.P. (1997). Description of a starter culture preparation for vinegar fermentation. *Systematic and Applied Microbiology* **20:** 481-491.

Sokollek, S.J., Hertel, C. and Hammes, W.P. (1998). Cultivation and preservation of vinegar bacteria. *Journal of Biotechnology* **60:** 195-206.

StrainsInfo. Available from: www.straininfo.net (Accessed: June 2016).

Sudsakda, S., Srichareon, W. and Pathom-Aree, W. (2007). Comparison of three enrichment broths for the isolation of thermotolerant acetic acid bacteria from flowers and fruits. *Research Journal of Microbiology* **2:** 792-795.

Swings, J., Gillis, M. and Kersters, K. (1992). Phenotypic identification of acetic acid bacteria. *In:* R.G. Board, D. Jones and F.A. Skinner (Eds.). *Identification Methods in Applied and Environmental Microbiology*. Blackwell Scientific, Oxford, pp. 103-110.

Takemura, H., Horinouchi, S. and Beppu, T. (1991). Novel insertion-sequence IS1380 from *Acetobacter pasteurianus* is involved in loss of ethanol-oxidizing ability. *Journal of Bacteriology* **173:** 7070-7076.

Trcek, J., Jernejc, K. and Matsushita, K. (2007). The highly tolerant acetic acid bacterium *Gluconacetobacter europaeus* adapts to the presence of acetic acid by changes in lipid composition, morphological properties and PQQ-dependent ADH expression. *Extremophiles* **11:** 627-635.

UMCC Database (Unimore Microbial Culture Collection). Available from: http://biolomics.umcc.unimore.it/ (Accessed: June 2016).

UniProt (The Universal Protein Resource). Available from: http://www.uniprot.org/uniprot/ (Accessed: June 2016).

Vartoukian, S.R., Palmer, R.M. and Wade, W.G. (2010). Strategies for culture of 'unculturable' bacteria. *FEMS Microbiology Letters* **309:** 1-7

Verslyppe, B., De Smet, W., De Baets, B., De Vos, P. and Dawyndt, P. (2014). Strain info introduces electronic passports for microorganisms. *Systematic and Applied Microbiology* **37:** 42-50.

Yamada, Y., Hosono, R., Lisdyanti, P., Widyastuti, Y., Saono, S., Uchimura, T. and Komagata, K. (1999). Identification of acetic acid bacteria isolated from Indonesian sources, especially of isolates classified in the genus *Gluconobacter*. *Journal of General and Applied Microbiology* **45:** 23-28.

Yamada, Y., Katsura, K., Kawasaki, H., Widyastuti, Y., Saono, S., Seki, T., Uchimura, T. and Komagata, K. (2000). *Asaia bogorensis* gen. nov., sp. nov., an unusual acetic acid bacterium in the α-Proteobacteria. *International Journal of Systematic and Evolutionary Microbiology* **50:** 823-829.

Yamada, Y. and Yukphan, P. (2008). Genera and species in acetic acid bacteria. *International Journal of Food Microbiology* **125:** 15-24.

Wang, B., Shao, Y., Chen, T., Chen, W. and Chen, F. (2015). Global insights into acetic acid resistance mechanisms and genetic stability of *Acetobacter pasteurianus* strains by comparative genomics. *Scientific Reports* **5:** (18330)e.

WDCM (World Data Center for Microorganisms). Available from: www.wdcm.org (Accessed: June 2016).

Wiegand, C. and Klemm, D. (2006). Influence of protective agents for preservation of *Gluconacetobacter xylinus* on its cellulose production. *Cellulose* **13:** 485-492.

WIPO (Monthly Review of the World Intellectual Property Organization) (1980). Tables of Contents, 16th year, Geneva 34, Chemin des Colombettes.

Wu, L., Sun, Q., Sugawara, H., Yang, S., Zhou, Y., McCluskey, K., Vasilenko, A., Suzuki, K., Ohkuma, M., Lee, Y., Robert, V., Ingsriswang, S., Guissart, F., Philippe, D. and Ma, J. (2013). Global catalogue of microorganisms (GCM): A comprehensive database and information retrieval, analysis and visualization system for microbial resources. *BMC Genomics* **14:** 1-10.

Zahoor, T., Siddique, F. and Farooq, U. (2006). Isolation and characterization of vinegar culture (*Acetobacter aceti*) from indigenous sources. *British Food Journal* **108:** 429-439.

Zwirglmaier, K. (2005). Fluorescence in situ hybridisation (FISH)—The next generation. *FEMS Microbiology Letters* **246:** 151-158.

PART II

IMPORTANCE OF ACETIC ACID BACTERIA IN FOOD INDUSTRY

9

Microbiology of Fermented Foods

Ilkin Yucel Sengun[1] and Michael P. Doyle[2*]

[1] Ege University, Faculty of Engineering, Department of Food Engineering, 35100, Bornova, Izmir, Turkey

[2] University of Georgia, Center for Food Safety, Department of Food Science & Technology, 1109 Experiment Street, Griffin, GA 30223 USA

Introduction

Fermentation is one of the oldest methods of food preservation. Many types of fermented foods, known throughout the world, have been prepared and consumed for centuries for their nutritional value, flavor and stability. Many are not only consumed as food sources, but are also used for therapeutic purposes. The major benefits of food fermentation include antimicrobial metabolites such as organic acids, ethanol and bacteriocins which are produced by microorganisms responsible for fermentation, providing safe foods, reducing the volume of the raw material and the amount of energy required for cooking and improving the nutritional value and organoleptic quality of food.

Certain microorganisms, including bacteria, yeast and molds, can be used to produce a variety of fermented foods. The microbes involved in fermented foods include at least 195 species of bacteria and 69 species of molds and yeasts (Bourdichon et al. 2012). Lactic acid bacteria (LAB) play an important role in fermentation of many commonly consumed foods, such as yogurt, cheese, bread, sausage, pickles and olives. They improve the sensory characteristics of food and inhibit the growth of undesirable microorganisms by producing a variety of inhibitory substances, such as lactic acid and bacteriocins. Acetic acid bacteria (AAB) are another important group of bacteria which are essential in vinegar and cocoa fermentations. Yeasts play an important role in the production of bread and many alcoholic beverages, such as beer and wine. Many Asian foods, such as soy sauce and tempeh and several kinds of cheeses, such as Brie and Camembert, are produced by a variety of molds. Many of these processes of fermentations can be performed spontaneously with natural microflora, or by backslopping, or by addition of starter cultures.

Fermented foods commonly known around the world as well as many that are locally produced will be addressed in this chapter, with the microbiological diversity of these foods being highlighted. The types of foods that will be addressed include dairy products (yogurt, cultured buttermilk, cultured sour cream, acidophilus milk, kefir, koumiss and cheese), cereal-based products (bread, boza and soya sauce), fruit and vegetable products (sauerkraut, kimchi, olives, pickles and wine), meat products (dry fermented sausages and semi-dry fermented sausages), locally produced traditional fermented foods and innovative functional foods. Vinegar and cocoa fermentations will be described in detail in Chapters 10 and 11.

*Corresponding author: mdoyle@uga.edu

Fermented Dairy Products

Many fermented dairy products are consumed as food and beverages throughout the world. In the early years, fermented dairy products were produced by spontaneous fermentation, which was subsequently optimized with backslopping, i.e. addition of previously fermented food to raw ingredients, resulting in the dominance of the best-adapted strains (Leroy and De Vuyst 2004). The predominant microbiota vary depending on the source of milk, treatments applied to milk (e.g., pasteurization), addition of starter cultures or the previous product batch if backslopping is used, fermentation conditions and the microbiota of the environment and containers used (Marsh et al. 2014). Today, the most popular fermented dairy products are commonly produced through the use of starter cultures to obtain high-quality standardized end-products. In general, homofermentative LAB are the dominant microflora of fermented dairy products, whereas heterofermentative LAB are used as secondary cultures for flavor development. Starter cultures used in the manufacture of fermented dairy products include both mesophilic and thermophilic bacteria. *Streptococcus thermophilus*, *Lactobacillus delbrueckii* subsp. *bulgaricus*, *Lactococcus lactis* subsp. *lactis* and *L. lactis* subsp. *cremoris* are widely used for milk fermentation as conventional starter cultures and have Generally Recognized as Safe (GRAS) status. They metabolize lactose to produce acid and give new functionalities to the products (Doyle et al. 2013, Wu et al. 2015).

Starter cultures used in dairy products should be evaluated for taxonomy, history of use, their ecology, industrial applications, pathogenic potential, undesirable metabolite production, end use and presence of acquired antibiotic resistance factors (Chamba and Jamet 2008). The main function of starter cultures used in dairy products is to ferment lactose to lactic acid, which is important in stimulating the growth of specific groups of microbes, while inhibiting the development of undesirable microorganisms. The milk protein, casein, precipitates at a pH value of 4.6, which is the isoelectric point of casein and a coagulum is formed. Moreover, small organic molecules, which are important to impart a characteristic flavor to the product, such as acetaldehyde, diacetyl, acetic acid and ethanol are also produced (Hutkins 2008). Hence, microorganisms used in the manufacture of fermented dairy products have a critical importance to obtain specific products that have defined acidity, viscosity, flavor, body and mouth feel. Although thermophiles, such as *Str. thermophilus* and starter lactobacilli, are paired to obtain the maximum rates of acid development for specific products, mesophiles can also be paired with thermophiles (Johnson and Steele 2013). Moreover, it is very important to choose appropriate strain combinations for obtaining high quality products, including bioactive peptides that have specific health claims. LAB also provide special health benefits to fermented dairy products that include antitumor activity, reduction of serum cholesterol, prevention of gastrointestinal infections and antimutagenic activity (Shiby and Mishra 2013).

The major dairy products throughout the world include yogurt, acidophilus milk, cultured buttermilk and sour cream, kefir, koumiss and cheese. Some of the fermented milk products, such as yogurt and cultured cream, are mainly produced by lactic fermentation, whereas others, such as kefir and koumiss, are produced by the combined growth of LAB and yeasts. The fermented milk products produced mainly by commercial starter cultures include yogurt, acidophilus milk, cultured buttermilk, cultured sour cream and kefir.

Fermented Milk Products

Yogurt

Yogurt is one of the most important cultured dairy products consumed throughout the world. The popularity of this product is related to its positive nutritional benefits and its function in maintaining health. There are various types of yogurt which are classified according to the type of milk used, fat content, production method, flavor and physical state. Concentrated yogurt, which is the most common type, has different local names in many countries, such as *Labneh* (in the eastern Mediterranean), *Laban zeer* (in Egypt, and Sudan), *Torba* (in Turkey), *Stragisto* (in Greece), *Syuzma* (in Russia), *Mastou* and *Mast* (in Iraq and Iran), *Ititu* (in Ethiopia), Greek-style (in the United Kingdom), *Chakka* (in India), and *Ymer* (in Denmark) (Tamime et al. 2011).

Yogurt production involves three main steps: (1) the mix preparation, (2) fermentation, and (3) packaging (Corrieu and Beal 2016). In the first step, milk used for yogurt production is heated to 85-88°C for 30 min. to destroy heat-resistant bacteria and denature the whey proteins. Milk at 43°C is inoculated with yogurt starter cultures, i.e. a mixture of *Str. thermophilus* and *Lb. delbrueckii* subsp. *bulgaricus* in the ratio of 1:1, at a rate of 2.5-3.0 per cent. Then the mixture is incubated at 43°C for four to six hours until the pH decreases to approximately 4.4 to 4.6 (Hutkins 2008). The incubation temperature has a critical role in influencing the growth rates of yogurt cultures. Initially, *Str. thermophilus* hydrolyzes lactose via a β-galactosidase and produces acids and CO_2, which provides conditions suitable for the growth of *Lb. delbrueckii* subsp. *bulgaricus*. Then small peptides and amino acids are produced by the lactobacilli, which enhance the growth of *Str. thermophilus*. As a result, yogurt cultures, through synergistic growth during the fermentation period, produce desired levels of acid and flavor compounds. *Str. thermophilus* is the main acid producer, whereas *Lb. delbrueckii* subsp. *bulgaricus* is mostly responsible for the production of flavor compounds, such as acetaldehyde (the main flavor contributor), carbonyl compounds, acetone, acetoin and diacetyl. If the temperature is increased above 43°C, *Str. thermophilus* becomes the dominant microbe in yogurt and produces a product containing more acid and fewer flavors. However, if the temperature is decreased below 43°C, *Lb. delbrueckii* subsp. *bulgaricus* is favored, subsequently producing less acid and more flavors. Hence, balanced growth of these two bacteria is very important to obtain yogurt with optimum flavor, in which the acetaldehyde content should range between 23 and 41 ppm, and the pH between 4.0 and 4.4 (Erkmen and Bozoglu 2008). Kinetic models are useful to predict the behavior of yogurt cultures at different pH values and temperatures (Aghababaie et al. 2015).

During yogurt fermentation, the milk protein is partly degraded by bacterial proteolytic activity, which enhances the digestibility of the product. Additionally, hydrolyzation of lactose makes the product more suitable for consumers suffering from lactose intolerance (Wouters et al. 2002). Health claims for yogurt to improve lactose digestion have been scientifically confirmed by the European Food Safety Authority (EFSA 2010). Although *Str. thermophilus* and *Lb. delbrueckii* subsp. *bulgaricus* are the dominant bacteria in yogurt production and are used as commercial starter cultures in industrial production, other LAB strains, such as *Enterococcus faecium, Pediococcus acidilactici, P. pentosaceus, Lb. citreum, Lb. fermentum* and *Lb. pseudomesenteroides*, are present in home-made yogurts (Sengun et al. 2009). Furthermore, probiotic lactobacilli (*Lb. acidophilus, Lb. casei, Lb. paracasei* subsp. *paracasei, Lb. rhamnosus, Lb. helveticus, Lb. johnsonii, Lb. gasseri* and *Lb. plantarum*) and bifidobacteria (*Bifidobacterium adolescentis, B. breve, B. bifidum, B. infantis, B. animalis* subsp. *lactis* and *B. longum*) can also be used as a co-culture of yogurt to generate bioactive peptides possessing antioxidant, antimutagenic and anticancer components from milk proteins (Ng et al. 2011, Tamime et al. 2011, Sah et al. 2016). When 100 grams of yogurt containing at least 10^7 CFU g^{-1} of live cells are consumed daily, there may be health benefits associated with the product (Granato et al. 2010).

A variety of yogurt products have been reported to have antihypertensive, antithrombotic, antiamnesic, antioxidant, antimicrobial and antidiabetic activities, angiotensin I-converting enzyme (ACE)-inhibitor, hypocholesterolemic properties, immunomodulation activity, and/or the ability to inhibit demineralization and promote remineralization of enamel caries (Hafeez et al. 2014).

Acidophilus Milk

Acidophilus milk is produced by fermenting milk with *Lb. acidophilus*. It is produced by inoculating heat-treated milk with *Lb. acidophilus* at the ratio of 5 per cent, and then incubating the mix at 37-38°C for 18-24 hours until the acidity reaches 1 per cent. Sweet acidophilus milk is also produced by *Lb. acidophilus*. The culture is added to heat-treated milk, not for fermentation, but for the milk to serve as a carrier (Kongo and Malcata 2016). Some yogurts are also co-fermented with *Lb. acidophilus*. However, acid and microbial inhibitory metabolites, such as H_2O_2 produced by yogurt cultures, reduce *Lb. acidophilus* populations during refrigerated storage (Ng et al. 2011). Hence, microencapsulation of *Lb. acidophilus* is a strategy used to improve the survival of lactobacilli in yogurt. Milk products containing *Lb. acidophilus* at sufficiently high levels can have therapeutic effects, such as preventing or controlling intestinal disorders, improving lactose digestion, exerting anticarcinogenic activity, controlling serum cholesterol levels and modulating the immune system (Desrouillères et al. 2015, Kongo and Malcata 2016).

Cultured Buttermilk

Cultured buttermilk is a dairy drink produced by inoculating skim milk or low-fat milk with 0.5 per cent of LAB and incubating the mix at 20-22°C for 12-14 hours until the pH decreases to ca. 4.5. *L. lactis* subsp. *cremoris* and *L. lactis* subsp. *lactis* are commonly used as acid producers in the production of buttermilk, whereas *Leuconostoc lactis, Leu. mesenteroides* subsp. *cremoris* and *L. lactis* subsp. *lactis* are used for flavor production (Hutkins 2008, Johnson and Steele 2013). These strains can ferment citric acid to produce diacetyl, which is the main flavor contributor in buttermilk. If the temperature is too high (>24°C), acid producers become dominant and the citrate-fermenting species will not grow well, thereby producing a product containing more acid and less flavors. However, if the temperature is too low (<20°C), the growth of acid producers can be inhibited, resulting in insufficient acid production and negatively affecting the formation of diacetyl (Hutkins 2008). Therefore, incubation temperature is of critical importance in obtaining a proper balance between the acid and flavor producers. Cultured buttermilk is also an ingredient used in bakery products (Shiby and Mishra 2013).

Cultured Sour Cream

Cultured sour cream is produced in a similar manner as buttermilk. It is produced from heat-treated (at 75-80°C) and homogenized (at >13 MPa) cream, which has a fat content between 12-30 per cent. *L. lactis* subsp. *lactis, L. lactis* subsp. *cremoris, Leu. mesenteroides* subsp. *dextranicum* and *Leu. lactis* could be used as starter cultures in the manufacture of sour cream (Hutkins 2008). Starter cultures used in the production of sour cream are best grown under processing conditions at 21-22°C for 12-14 hours to produce a balanced acid and flavor profile. Hence, the fermentation is stopped at an acidity of 0.6 per cent (Shiby and Mishra 2013).

Kefir

Kefir is an acidic-alcoholic dairy product with high probiotic activity. It is produced by fermenting milk by addition of commercial kefir starter cultures or kefir grains, which are aggregates of a mixture of microorganisms, mainly LAB, acetic acid bacteria, yeasts and molds (Wszolek et al. 2006). A symbiotic relationship between bacteria and yeast produces an aggregation of granules, which is enhanced by a decrease of pH, subsequently form kefir grains. During fermentation, a variety of metabolites are formed in kefir, such as lactic acid, acetic acid, alcohol, aromatic compounds and CO_2 (Magalhaes et al. 2011). A variety of factors affect the quality of kefir, such as the microbiological quality of kefir grains, the amount of grain, fermentation conditions, sanitation and storage conditions (Guzel-Seydim et al. 2010). Hence, starter cultures containing suitable LAB and yeast may be used in kefir production not only to control the fermentation, but also for the post-fermentation, which occurs during cold storage of kefir, during which undesirable changes can occur in the product.

Lb. kefiri represents ca. 80 per cent of the *Lactobacillus* spp. in the final fermented beverage and the remainder largely includes *Lb. acidophilus, Lb. delbrueckii* subsp. *bulgaricus, Lb. paracasei* subsp. *paracasei, Lb. plantarum* and *Lb. kefiranofaciens.* Additionally, *Saccharomyces cerevisiae, Sac. unisporus, Candida kefyr* and *Kluyveromyces marxianus* subsp. *marxianus* are the predominant yeasts in kefir samples of different origins. Acetic acid bacteria, such as *Acetobacter aceti, A. fabarum, A. lovaniensis, A. orientalis* and *A. syzygii* have also been isolated from various kefir samples (Prado et al. 2015). Some species of LAB isolated from kefir, such as *Lb. acidophilus* and *Lb. kefiranofaciens*, can have probiotic properties (Santos et al. 2003). *Bifidobacterium* spp. and probiotic yeasts, such as *Sac. boulardii*, may also be used as adjunct cultures in kefir production (Wszolek et al. 2006).

There is a growing commercial interest in kefir consumption due to its beneficial health effects. Biological activities of kefirs that have been reported include antiallergic, antiinflammatory, anticarcinogenic, antimutagenic, hypocholesterolemic, improving lactose digestion and wound healing (Hafeez et al. 2014). These health benefits have been linked to the microflora and their metabolic products produced in kefir, such as organic acids and polysaccharides. Kefiran is the main polysaccharide produced by *Lb. kefiranofaciens*, which improves the viscosity of the product and has biological activities that include antimicrobial, antioxidant, antitumor, antiinflammatory, immunomodulator and cholesterol-lowering effects (Serafini et al. 2014, Prado et al. 2015).

There are also a variety of traditional fermented milk products, which are known by different local names such as *dahi, koumiss, matsoni, mutandabota, skyr* and *villi* (Table 1).

The products listed in Table 1 are produced by traditional methods and contain unique cultures that lend to the regional character of the product.

Koumiss

One of the most popular traditional fermented milk products is koumiss. It is a fermented drink with low alcohol content (approximately 2 per cent), originally produced from unpasteurized mare's milk in Asia and Russia. Hard curd formation occurs only when cow's milk is used instead of mare's milk. A similar beverage known as kumis in Columbia is produced from cow milk (Chaves-López et al. 2014). In the first step of koumiss production, milk is inoculated with 30 per cent of a specially prepared bulk starter (yeast culture and LAB, activated in milk) or mixed with a previously fermented (backslop) koumiss. Fermentation is carried out at 20-30°C with agitation for flavor formation. After reaching the desired acidity (0.5-1.5 per cent lactic acid), it is stored at 4°C and consumed within three days (Chaves-López et al. 2011). Koumiss fermentation includes *Lb. acidophilus, Lb. fermentum, Lb. paracasei* and *Lb. kefiranofaciens* as the dominant LAB and a variety of yeasts (Table 1). Lactobacilli play a major role in koumiss fermentation by affecting acidity, aroma and texture of the product, whereas yeasts are responsible for the formation of ethanol and carbon dioxide (Wouters et al. 2002). Koumiss is also an important source of enterococci which can produce in fermented milks Angiotensin-I Converting Enzyme (ACE)-inhibitory activity

Table 1: Overview of the microbial diversity of fermented milks from different geographical locations

Product name	*Product definition*	*Microorganisms*	*Country*	*References*
Ayran	Yogurt drink	*Lb. delbrueckii* subsp. *bulgaricus*, *Str. thermophilus*	Turkey	Kocak and Avsar 2009
Dahi	Similar to yogurt	*Lb. delbrueckii* subsp. *bulgaricus*, *Lactobacillus* spp., *Str. thermophilus*	India	Hutkins 2008
Gariss	Fermented camel milk	*Lb. alimentarium*, *Lb. animalis*, *Lb. brevis*, *Lb. divergens*, *Lb. fermentum*, *Lb. gasseri*, *Lb. paracasei* subsp. *paracasei*, *Lb. plantarum*, *Lb. rhamnosus*, *Lactobacillus* sp.	Sudan, Somalia	Ashmaig et al. 2009
Gioddu	Fermented sheep or goat milk	*L. lactis* subsp. *lactis*, *Lb. casei* subsp. *casei*, *Lb. delbrueckii* subsp. *bulgaricus*, *Leu. mesenteroides*, *Str. thermophilus*, *K. marxianus*, *Sac. cerevisiae*	Sardinia	Ortu et al. 2007
Koumiss	Alcoholic fermented mare's milk	*E. faecalis*, *L. lactis*, *Lb. acidophilus*, *Lb. buchneri*, *Lb. fermentum*, *Lb. jensenii*, *Lb. kefiranofaciens*, *Lb. kefiri*, *Lb. kitasatonis*, *Lb. paracasei*, *Leu. mesenteroides*, *Str. thermophilus*	Central Asia, Russia	Hao et al. 2010
		Issatchenkia orientalis, *Kazachstania unispora*, *K. marxianus*, *Pi. mandshurica*, *Sac. cerevisiae*		Watanabe et al. 2008
Matsoni	Similar to yogurt prepared from milk of cow, sheep, goat, buffalo	*L. lactis* subsp. *cremoris*, *L. lactis* subsp. *lactis*, *Lb. acidophilus*, *Lb. delbrueckii* subsp. *bulgaricus*, *Lb. delbrueckii* subsp. *lactis*, *Str. thermophilus*, *Ca. famata* *G. candidum*, *K. lactis*, *K. marxianus*, *Sac. cerevisiae*	Georgia Armenia	Bokulich et al. 2015

(*Contd.*)

Nunu	Similar to yogurt	*E. faecium, E. italicus, Lactococcus* spp., *Lb. fermentum, Lb. helveticus, Lb. plantarum, Leu. mesenteroides, W. confuse, Ca. parapsilosis, Ca. rugosa, Ca. tropicalis, Ga. geotrichum, Pi. kudriavzevii, Sac. cerevisiae*	Ghana	Akabanda et al. 2013
Skyr	Fermented milk	*Lb. delbrueckii* subsp. *bulgaricus, Str. thermophilus*	Iceland	Hutkins 2008
Suusac	Fermented camel milk	*Lb. curvatus, Lb. raffinolactis, Lb. salivarius, Leu. mesenteroides*	Kenya, Somalia	Lore et al. 2005
Villi	Fermented milk with ropy texture	*Lactococcus* spp., *Leuconostoc* spp., *G. candidum*	Finland	Tamime et al. 2011
Yoba mutandabota	Yoghurt-like consistency	*L. rhamnosus* yoba	Southern Africa	Mpofu et al. 2016

Ca.: Candida, E.: Enterococcus, G.: Geotrichum, Ga.: Galactomyces, K.: Kluyveromyces, L.: Lactococcus, Lb.: Lactobacillus, Leu.: Leuconostoc, Pi.: Pichia, Str.: Streptococcus, Sac.: Saccharomyces, W.: Weissella

(Chaves-López et al. 2011). The complete genome of a novel strain of *Lb. paracasei* isolated from koumiss in China, suggests a probiotic potential in humans (Wang et al. 2015). Koumiss is considered not only to be a food, but some also consider it to be a medicinal remedy with positive effects for a variety of diseases, such as tuberculosis, diabetes and cardiovascular and neurological diseases (Li et al. 2006).

Cheese

There are over a thousand varieties of cheese throughout the world, which are often classified on the basis of texture, i.e. soft, semi-hard, or hard, or simply on the origin and processing method, i.e. industrial or traditional cheeses. Production steps for most cheese varieties include gel formation, whey removal, acid production, salt addition and a ripening period (Beresford et al. 2001). The milk could be either raw or pasteurized to create significant perceived quality differences between them. Although pasteurization is believed by some cheese purists to adversely affect cheese quality, it is mostly preferred for the safety of the product (Yoon et al. 2016). The milk can be coagulated by proteolytic enzymes or organic acids produced by LAB, or alternatively by the combination of moderate acid addition (pH 6.0) and high heat application. However, the most common method to coagulate milk proteins is the addition of the enzyme chymosin or rennet, which affects a specific peptide bond in certain types of casein molecules (Hutkins 2006). Amino acids, which are produced by conversion of casein, are further metabolized to aldehydes, alcohols, ketones, amines, acids, esters and sulphur-containing compounds by amino acid-converting enzymes and contribute to cheese flavor (Wouters et al. 2002). After the curd is formed, it is cut and pressed to remove the remaining liquid, which is whey. Cold storage is necessary to stop or limit additional acid production by starter bacteria. Appropriate packaging, such as oxygen limiting, is important to prevent mold growth on cheese.

Table 2 lists many of the most popular cheeses produced in certain regions around the world. The microflora of cheese is very complex and affects the overall characteristics of cheese. The microorganisms present in cheese are influenced by continuous changes occurring in environmental conditions. The industrial manufacture of cheese is accomplished by starter cultures, which are selected on the basis of the type of cheese to be produced and provide product safety with reproducible organoleptic and structural properties (Wouters et al. 2002, Montel et al. 2014). Other selection criteria for starter cultures include performance and handling and stability properties (Wouters et al. 2002).

Table 2: Microorganisms involved in the manufacture of different cheese varieties

Product name	*Microorganisms*	*References*
Blue	*L. lactis* subsp. *cremoris*, *L. lactis* subsp. *lactis*, *Penicillium roqueforti*	Johnson and Steele 2013
Brick, Limburger	*L. lactis* subsp. *cremoris*, *L. lactis* subsp. *lactis*, *Br. linens*, *Micrococcus* spp., *G. candidum*	Boldyreva et al. 2016
Camembert	*L. lactis* subsp. *cremoris*, *L. lactis* subsp. *lactis*, *Leu. mesenteroides*, *Str. thermophilus Lb. fermentum*, *Lb. paracasei*,	Firmesse et al. 2008
	Lb. plantarum, *Penicillium camemberti*	Beresford et al. 2001
Cotija Mexican cheese	*Lb. plantarum*, *Lb. mesenteroides*, *W. paramesenteroides*	Escobar-Zepeda et al. 2016

(*Contd.*)

Danish Danbo cheeses	*Bavariicoccus seileri, C. casei, C. variabile, L. lactis* subsp. *lactis, Marinilactibacillus psychrotolerans, Psychrobacter* spp., *S. equorum, D. hansenii*	Gori et al. 2013
Dutch-type cheeses (Guoda, Edam)	*L. lactis* subsp. *lactis, L. lactis* subsp. *cremoris, Lb. casei/paracasei, Lb. plantarum, Lb. rhamnosus*	Porcellato et al. 2013
Graviera Kritis Greek cheese	*E. gallinarum, E. faecalis, E. faecium, Lb. paracasei, Leu. lactis, Leu. mesenteroides*	Bozoudi et al. 2016
Graviera Naxou Greek cheese	*E. faecium, L. lactis* subsp. *lactis, L. garvieae, Lb. brevis, Leu. mesenteroides, Leu. pseudomesenteroides*	Bozoudi et al. 2016
Kaşar cheese	*Lb. casei, Lb. plantarum, P. acidilactici*	Aydemir et al. 2015
Morocco soft white cheeses	*E. durans, E. faecalis, E. faecium, E. saccharominimus, L. lactis, L. raffinolactis, Lb. brevis, Lb. buchneri, L. garvieae, Lb. paracasei, Lb. plantarum, Lb. rhamnosus, Leu. citreum, Leu. mesenteroides, Leu. pseudomesenteroides*	Ouadghiri et al. 2005
Mozzarella	*E. faecalis, L. garvieae, L. lactis, Lb. casei* subsp. *casei, Lb. delbrueckii* subsp. *bulgaricus, Lb. fermentum, Lb. helveticus, Lb. plantarum, Str. thermophilus*	de Candia et al. 2007
Shanklish Syrian cheese	*Lb. casei, Lb. fermentum, Lb. helveticus*	Albesharat et al. 2011
Slovakian bryndza cheese	*L. lactis* subsp. *lactis, L. raffinolactis, Lb. brevis, Lb. delbrueckii, Leu. pseudomesenteroides, Str. thermophilus, Str. macedonicus, D. hansenii, Ga. geotrichum, Mu. fragilis, Y. lipolytica*	Chebeňová-Turcovská et al. 2011
Suero costeño Colombian cream cheese	*Lb. acidophilus, Lb. brevis, Lb. delbrueckii* subsp. *delbrueckii, Lb. paracasei* subsp. *paracasei, Lb. pentosus, Lb. rhamnosus*	Cueto et al. 2007
Swiss-type cheeses	*Lb. helveticus, Pr. freudenreichii* subsp. *shermanii, Str. thermophilus*	O'Sullivan et al. 2016
Traditional Mountain cheese	*E. faecalis, L. lactis, Str. thermophilus*	Carafa et al. 2016

Br.: Brevibacterium, C.: Corynebacterium, D.: Debaryomyces, E.: Enterococcus, G.: Geotrichum, Ga.: Galactomyces, L.: Lactococcus, Lb.: Lactobacillus, Leu.: Leuconostoc, Mu.: Mucor, P.: Pediococcus, Pr.: Propionibacterium, Str.: Streptococcus, S.: Staphylococcus, W.: Weissella, Y.: Yarrowia

Mesophilic cultures could be used for Cheddar, Gouda, Edam, Blue and Camembert productions, whereas thermophilic cultures are used for Emmental, Gruyère, Parmesan and Grana, which are exposed to high temperatures during the production period (Beresford et al. 2001). Mesophilic lactococci are usually used as starter cultures in a variety of cheese types. The most commonly used starter cultures for cheesemaking are *L. lactis* subsp. *lactis, L. lactis* subsp. *cremoris* and *L. lactis* subsp. *lactis* var. *diacetylactis*.

L. lactis subsp. *lactis* grows quickly and rapidly produces lactic acid, whereas *L. lactis* subsp. *cremoris* and *L. lactis* subsp. *lactis* var. *diacetylactis* mainly produce flavor compounds such as diacetyl (Beresford et al. 2001). Thermophilic starter LAB, such as *Lb. delbrueckii* subsp. *bulgaricus*, *Lb. casei*, *Lb. fermentum*, *Lb. helveticus* and *Str. thermophilus*, are the dominant viable microflora from the beginning of cheesemaking to at least six months of ripening (Montel et al. 2014). Mesophilic lactobacilli, which are the main components of the nonstarter LAB population, are essential determinants for developing satisfactory and diverse sensory characteristics of ripened cheese varieties (Gobbetti et al. 2015). *Br. linens* is used as a secondary culture for red color formation in the manufacture of smear surface-ripened cheeses, such as Appenzeller, Brick, Limburger, Munster and Tilsiter (Rattray and Fox 1999). Swiss-type cheeses, such as Gruyère and Emmental, contain *Pr. freudenreichii* which is responsible for their characteristic flavor and eye formation due to the production of CO_2 (Thierry et al. 2011). During Emmental cheese manufacture, starter LAB provide lactate and peptides for dairy propionibacteria, which cannot utilize nitrogen and are largely not able to ferment lactose from milk (Cousin et al. 2012).

Certain molds, mainly *Penicillium* species that are not toxigenic, are used to produce mold-ripened cheeses with different organoleptic characteristics. White or blue molds could be used in cheese production as secondary cultures and have different color and growth characteristics. *Penicillium camemberti* grows on the outside of cheeses, such as Camembert and Brie, whereas *Penicillium roqueforti* grows inside of cheeses such as Roquefort, Blue Stilton, Danish Blue and Gorgonzola. *G. candidum* also plays a key role in the ripening of Camembert cheese and contributes to typical cheese flavor (Wouters et al. 2002, Sengun et al. 2008). *Mu. mucedo* and *Mu. racemosus* grow as the ripening agent on the surface and throughout the interior of the Norwegian cheese, Gamalos (Rage 1993). Most cheeses made from unpasteurized milk harbor a wide diversity of yeast species, mainly *Candida, Pichia, Saccharomyces, Rhodotorula* and *Trichosporon* (Montel et al. 2014).

Enterococcus species metabolizes citrate and proteins and therefore has an important role in developing the sensory profile of fermented dairy products (Chaves-López et al. 2011). Although they could be used as starter cultures for the production of fermented milk products with a number of nutritional and therapeutic attributes, several species, such as *E. faecalis* and *E. faecium*, have been associated with human infections (Carlos et al. 2009). These species are also able to form in the product biogenic amines which are considered toxic when their concentration exceeds 100 mg/kg (Buňková et al. 2013). In dairy products, the biogenic amine concentration can range from a few milligrams to tens of milligrams per kilogram of food, and cheese is one of the most important dairy products in which biogenic amines occur, thereby leading to serious health risks (Buňková et al. 2013). However, cheese is an important food, rich in protein, fat, minerals, vitamins, biologically active amino acids, peptides and fatty acids, having several health benefits such as antihypertensive, anticarcinogenic and antiosteoporotic activities (Jerónimo and Malcata 2016).

The microbiota of fermented dairy products are a very diverse group which originate from a variety of sources and include microbes that can produce bacteriocins and/or have probiotic properties. In recent years, there has been an increased interest in producing functional foods from nutritive by-products obtained after milk processing. In these types of products, LAB perform an essential role in preserving a highly nutritious product.

Fermented Meat Products

Fresh meat is a highly perishable food that can be preserved by salting, fermenting and/or drying. For fermented meat, ingredients such as minced meat, fat, salts, curing agents and spices are combined and stuffed into casings to provide selective conditions for the growth of certain Gram-positive bacteria such as LAB, which produce organic acids, carbon dioxide and alcohols in the product (Leroy et al. 2013, Ricke et al. 2013). The combination of fermentation with salting and drying produces shelf-stable meat products. It is very common to apply these techniques together for meat processing in the sequence of salting, stuffing, fermenting and drying to create a hurdle effect for spoilage and pathogenic microbes (Leroy et al. 2013).

Fermented sausages represent the largest category of fermented meat products throughout the world. Sausage fermentation is commonly carried out by metabolic activities of LAB and/or coagulase-negative staphylococci. In general, LAB enhance the quality of fermented sausages, especially by restricting the growth of some undesirable microorganisms, while coagulase-negative staphylococci prevent the formation of off-flavors and rancidity (Casaburi et al. 2005). The diversity of sausage microbiota is influenced by the type of raw materials and their origins, ingredient formulations, condiments, additives, meat grinding size, casing diameter, smoking intensity, fermentation and drying conditions (Vignolo et al. 2010, Pecanac et al. 2015). Hence, there is a wide range of fermented sausages with unique sensorial characteristics mainly originating in the Mediterranean region (Table 3).

Fermented sausage production involves reducing the particle size of the raw meat, mixing of all the ingredients, stuffing into casings and ripening (fermentation and aging). Fermented sausages are divided into two groups, i.e. dry (slowly-fermented) and semi-dry (quickly-fermented) sausages. Although the steps in sausage manufacture appear to be similar, there are some differences. These products rely on both reduced pH and reduced water activity for microbial stability. The water activity for dry and semi-dry sausages range from <0.85-0.91 and 0.90-0.94, respectively. The final pH value of dry sausages (pH 5.0-5.3) is usually higher than that of semi-dry sausages (pH 4.7-5.1). However, the percentage of lactic acid of dry sausages (0.5-1.0 per cent) is less than that of semi-dry sausages (0.5-1.3 per cent). A large concentration of lactic acid with many other fermentative breakdown products produce a tangy flavor, especially in semi-dry sausages. Dry fermented sausages are dried or aged after fermentation to remove 25-50 per cent of the moisture and are held without refrigeration. Semi-dry sausages lose up to 15 per cent moisture, are generally heat processed and dried with smoke, packed and chilled after processing, and are then refrigerated (Vignolo et al. 2010, Ricke et al. 2013). A variety of fermentation temperatures are used in dry sausage manufacture in Europe. For example, German and Italian dry sausage fermentations are performed at relatively high temperatures (18-24°C) for one to two days, whereas Greek, Argentinean and certain Italian dry sausages are fermented for seven days at these temperatures. For traditional French, Spanish and Portuguese dry sausages, lower fermentation temperatures (10-17°C) for one week are commonly used. For semi-dry sausages, especially those produced in the United States, the fermentation time is shortened to one day by using higher temperatures, which is usually raised slowly to over 35°C (Vignolo et al. 2010).

A key step in sausage production is fermentation, which is performed by indigenous microbiota or by starter cultures. Complex biochemical and physical reactions occur during sausage fermentation, resulting in decreased pH, dehydration, nitrate reduction, nitrosomyoglobin formation, protein solubilization and gelification in the product (Casaburi et al. 2007). Moreover, the initial microflora also change during the fermentation period

Table 3: Overview of microbial diversity of fermented sausages from different geographical locations

Country	*Product name/ definition*	*Major ingredients*	*Microorganisms*	*Reference*
Argentina	llama (Fermented sausage)	llama meat, pork fat, salt, sugar, sodium nitrite, black pepper, garlic, red wine	*Ca. deformans, Ca. parapsilosis, Ca. pararugosa, Ca. zeylanoides, Cry. curvatus, Cry. kuetzingii, D. hansenii, Metschnikowia pulcherrima, R. diobovatum, R. mucilaginosa, R. slooffiae, To. delbrueckii, Tr. japonicum, Tr. montevideense, Y. lipolytic*	Mendoza et al. 2014
Greece	Fermented sausage	Pork meat, beef meat, pork back fat, salt, sugars, sodium nitrate/nitrite/ascorbate, skim milk powder, spices	*Lb. plantarum, Lb. rhamnosus, Lb. sakei, S. cohnii, S. gallinarum, S. saprophyticus S. simulans, S. xylosus*	Drosinos et al. 2007
	Fermented sausage	Pork meat, beef meat, pork back fat, salt, sugars, sodium nitrite/nitrate/ascorbate, skim milk powder, pepper, garlic, clove, white wine	*E. faecium/durans, Lb. alimentarius, Lb. casei/paracasei, Lb. curvatus, Lb. paraplantarum, Lb. paraplantarum/plantarum, Lb. plantarum, Lb. sakei, Leu. citreum*	Rantsiou et al. 2005
Hungary	Fermented sausage	Pork meat, pork back fat, salt, sugars, clove, red/black pepper, garlic, paprika	*Lb. curvatus, Lb. paraplantarum/plantarum, Lb. plantarum, Lb. sakei, Leu. citreum, Leu. mesenteroides, W. paramesenteroides/hellenica, W. viridescens*	Rantsiou et al. 2005
Italy	Ventricina Vastese (Fermented sausage)	Pork meat, salt, sweet and hot chili pepper	*Lb. sakei, S. equorum, S. xylosus, Te. koreensis*	Amadoro et al. 2015
	Ciauscolo salami (Fermented sausage)	Meat, salt, potassium nitrate, pepper, garlic, white wine	*Carnobacterium* spp., *E. faecalis, Lb. brevis, Lb. johnsonii, Lb. lactis, Lb. paracasei, Lb. paraplantarum, Lb. plantarum, Lb. sakei, Lactococcus* spp.,	Federici et al. 2014

(*Contd.*)

			Leu. mesenteroides, *P. pentosaceus*, *W. hellenica*	
	Fermented sausage	Pork meat, lard, salt, nitrite, nitrate, black pepper	*L. garvieae*, *L. lactis*, *Lb. brevis*, *Lb. casei*, *Lb. curvatus*, *Lb. paraplantarum*, *Lb. plantarum*, *Lb. sakei*, *Leu. carnosum*, *Leu. mesenteroides*, *W. hellenica*, *W. paramesenteroides*	Urso et al. 2006
	Fermented sausage	Pork meat, lard, salt, sugar, nitrate, pepper, nutmeg, cinnamon, coriander	*E. gilvus*, *Lb. pentosus*, *Lb. sakei*, *Leu. carnosum*, *S. cohnii*, *S. equorum*, *S. saprophyticus*, *S. succinus*, *S. xylosus*	Greppi et al. 2015
	Fermented sausage	Pork meat, pork back fat, salt, sugars, sodium nitrite/ascorbate, black pepper	*E. pseudoavium*, *Leu. citreum*, *L. lactis* subsp. *lactis*, *Lb. curvatus*, *Lb. paraplantarum*, *Lb. paraplantarum/pentosus*, *Lb. plantarum*, *Lb. sakei*, *W. paramesenteroides/hellenica*	Rantsiou et al. 2005
Portugal	Alheira (Fermented sausage)	Various meat, olive oil and/or fat, bread, spices,	*E. faecalis*, *E. faecium*, *Lb. brevis*, *Lb. paracasei*, *Lb. paraplantarum*, *Lb. plantarum*, *Lb. rhamnosus*, *Lb. sakei*, *Lb. zeae*, *Leu. mesenteroides*, *P. acidilactici*, *P. pentosaceus*, *W. cibaria*, *W. viridescens*	Albano et al. 2009
Serbia	Sremski/Lemeski kulen (Fermented sausages)	Meat of mature pigs, ground red paprika	*Lb. brevis*, *Lb. curvatus* subsp. *curvatus*, *Lb. fermentum*, *Lb. paracasei*, *Lb. plantarum*, *Lb. pentosus*, *Lb. salivarius*	Vasilev et al. 2015
Spain	Galician chorizo (Dry fermented sausage)	Lean pork, pork back fat, salt, garlic, paprika	*Lb. sakei*, *S. equorum*	Fonseca et al. 2013a
	Dry fermented sausage	Lean pork meat, pork back fat	*E. faecium*, *Lb. coryniformis*, *Lb. paracasei*, *Lb. plantarum*, *Lb. sakei*	Landeta et al. 2013
	Salchichón (Dry fermented sausage)	Iberian pork, NaCl, sodium nitrate/ascorbate, pepper, dextrin, dextrose, lactose	*P. acidilactici*, *S. vitulus*	Casquete et al. 2012

(*Contd.*)

Table 3: (*Contd.*)

Country	*Product name/ definition*	*Major ingredients*	*Microorganisms*	*Reference*
Switzerland	Dry fermented sausages	Wildlife meat (deer, wild boar, chamois) or meat from pork, cattle, sheep	*E. faecalis, E. faecium, Lb. curvatus, Lb. sakei, P. pentosaceus, S. equorum, S. xylosus, Streptococcus* spp., *Staphylococcus* spp.	Marty et al. 2012
Thai	Nham (Fermented pork)	Pork meat	*Lb. plantarum, W. cibaria*	Pringsulaka et al. 2012
Thailand	Mum (Fermented sausage)	Beef, bovine liver spleen, salt, rice powder, garlic, spices	*L. lactis, Lb. brevis, Lb. fermentum, Lb. plantarum, Lb. sakei, Leu. mesenteroides, P. pentosaceus*	Wanangkarn et al. 2014
Turkey	Sucuk (Dry fermented sausage)	Beef, beef back fat, sheep tail fat, salt, sugar, nitrite, garlic, pepper, cumin, pimento	*Lb. curvatus, Lb. sakei, W. viridescens*	Kesmen et al. 2012
	Sucuk (Dry fermented sausage)	Beef meat, beef fat/sheep tail fat, salt, glucose, nitrate/nitrite, cumin, garlic, pepper, allspice	*L. curvatus, L. fermentum, Lb. plantarum, P. acidilactici, P. pentosaceus, Ko. rosea, S. carnosus, S. equorum, S. saprophyticus, S. xylosus*	Kaban and Kaya 2008
Vietnam	Nem chua	Ground pork, pig skin, garlic, leaf 'Oi' (*Psidium guajava*), banana leaves (for wrapping)	*L. garvieae, L. lactis, Lb. acidipiscis, Lb. brevis, Lb. crispatus, Lb. farciminis, Lb. fermentum, Lb. helveticus, Lb. fuchuensis, Lb. pentosus, Lb. paracasei, Lb. plantarum, Lb. namurensis, Lb. rossiae, Leu. citreum, Leu. fallax, P. acidilactici, P. pentosaceus, P. stilesii, Vagococcus* spp., *W. cibaria, W. paramesenteroides*	Nguyen et al. 2013

Ca.: Candida, Cry.: Cryptococcus, D.: Debaryomyces, E.: Enterococcus, L.: Lactococcus, Lb.: Lactobacillus, Leu.: Leuconostoc, Ko.: Kocuria, P.: Pediococcus, R.: Rhodosporidium, S.: Staphylococcus, Te.: Tetragenococcus, To.: Torulaspora, Tr.: Trichosporon, W.: Weissella, Y.: Yarrowia and LAB constitute the dominant microflora (Kaban 2013). *Lb. curvatus, Lb. plantarum, Lb. sakei* and coagulase-negative cocci (CNC) represented by the *Staphylococcus* and *Kocuria* genera are commonly the predominant microflora in the dry fermented sausage, whereas other LAB, such as *Lb. alimentarius, Lb. brevis, Lb. casei, Lb. farciminis, Lb. paracasei, Lb. pentosus, Lb. versmoldensis, Te. koreensis, Weissella* spp., leuconostocs and pediococci are usually present in significantly lower numbers (Hammes and Hertel 2009, Talon and Leroy 2011). The selection of microorganisms depends on temperature, humidity and ingredients of each production plant (Cocolin et al. 2011). Although some manufacturers still rely on spontaneous fermentation, most prefer to use starter cultures, which play an important role in the production of high-quality fermented sausage. The advantages of using starter cultures over spontaneous meat fermentation include ensuring a predictable rate of pH reduction, reducing/inhibiting the growth of undesirable microorganisms, providing stability and safety, shortening the fermentation time, reducing product defects and producing high quality product with desired flavor, color and textural properties (Leroy et al. 2006, Hutkins 2008, Casquete et al. 2012).

The functionality of starter cultures depends on the raw materials, ingredients, sausage type, the ripening conditions and the technology applied. An appropriate choice of starter culture combinations is fundamental for obtaining high quality products with expected sensory quality (Tabanelli et al. 2012). LAB starters, to be used in fermented sausage production, should have certain criteria, such as rapid acid production, homofermentative metabolism, robust growth at different temperatures, survival and growth during the entire fermentation and ripening process, nitrate reduction, ability to express catalase, no lactose fermentation, no production of peroxide, biogenic amines and ropy slime, antagonistic activity toward pathogens or undesirable microorganisms, and improvement of the nutritional, microbiological and sensory quality of the product (Buckenhüskes 1993). The natural microflora in traditional fermented sausages, which are well-adapted to the harsh environment of sausage making, could be used as starter culture in fermented sausage production. However, fermented sausage production is mostly carried out with defined starter cultures comprising LAB or a combination of LAB, coagulase-negative staphylococci (CNS) and yeasts or molds. In the USA, pediococci are the predominant starter cultures, whereas lactobacilli are the most common starter cultures used in fermented sausages production in Europe (Rantsiou and Cocolin 2006, Landeta et al. 2013).

Fermentation temperature is an important factor that should be considered when choosing a suitable starter culture in sausage production. *P. acidilactici* is used for high temperature fermentation (32-40°C) for fast acidification and lower pH values in semi-dry sausages, whereas *Lb. sake* and *Lb. curvatus* are most suitable for dry sausage fermentations performed at low temperature (<25°C) (Vignolo et al. 2010). Different starter culture combinations can be used successfully for the manufacture of a variety of dry fermented sausages, such as *Lb. sakei* and *Bacillus pumilus* in dry fermented sausage (Herranz et al. 2006), *Lb. curvatus* and *S. xylosus* in the traditional fermented sausages of Vallo di Diano (Southern Italy) (Casaburi et al. 2007), *D. hansenii, Y. lipolytica* and *Lb. plantarum* in dry fermented sausages (Patrignani et al. 2007), *Lb. curvatus, Lb. sakei, Streptomyces griseus* and *P. acidilactici* in rapid-fermented sausages of the USA (Candogan et al. 2009), *P. acidilactici* and *S. vitulus* in traditional Spanish Iberian dry fermented sausages (Casquete et al. 2012), and *Lb. plantarum* and *S. xylosus* in traditional Tunisian dry fermented sausages (Essid and Hassouna 2013).

LAB used in the manufacture of sausages can produce a variety of metabolites that possess antimicrobial activity, such as organic acids (lactic and acetic), bacteriocins, hydrogen peroxide, antimicrobial enzymes and reuterin. Lactic acid production and pH reduction contribute to product safety by creating unfavorable conditions for the survival and growth of pathogens. Some LAB also used in fermented meat products as a starter culture produce bacteriocins, which can provide product safety and increased shelf life with improved quality by inhibiting closely related LAB and other Gram-positive bacteria, such as *Listeria monocytogenes, Clostridium botulinum, Clostridium perfringens, Bacillus cereus, Brochothrix thermosphacta, E. faecalis/faecium* and *S. aureus* (Tichaczek et al. 1994, Castellano et al. 2008, Vignolo et al. 2010, Gao et al. 2014, Casaburi et al. 2016). Bacteriocin-producing LAB originating from a variety of fermented sausages include *L. lactis* (nisin producer), *Lb. sakei* (lactocin and sakacin producer), *Lb. brevis* (brevicin producer), *Lb. curvatus* (lactocin and curvaticin producer), *Lb. plantarum* (plantaricin producer), *Leu. mesenteroides* (leucin producer), *P. acidilactici* and *P. pentosaceus* (pediocin producers), *E. faecium* and *E. casseliflavus* (enterocin producers) and *W. cibaria* (bacteriocin N23 producer) (Castellano et al. 2008, Pringsulaka et al. 2012). Hydrogen peroxide formed by LAB is bacteriostatic to Gram-positive bacteria and bactericidal to Gram-negative bacteria (Albano et al. 2009). Although many commercial starter cultures can produce hydrogen peroxide, bacteriocin-producing bacteria are usually weak hydrogen peroxide producers (Kröckel 2013).

Some starter cultures used for sausage fermentation present the risk of biogenic amine production. During sausage fermentation, tyramine and histamine can be produced by some strains of *E. faecalis, E. faecium, Lb. brevis, Lb. casei/paracasei* and *Lb. plantarum* (Komprda et al. 2010). However, production of biogenic amines during sausage fermentation can be mitigated by using cultures such as *Lb. plantarum, Lb. sakei* or *Micrococcus varians* (Leuschner and Hammes 1998, Tosukhowong et al. 2011).

Natural sausage fermentation involves a variety of CSN bacteria, which include *S. xylosus, S. capitis, S. carnosus, S. cohnii, S. equorum, S. epidermidis, S. haemolyticus, S. lentus, S. pasteuri, S. saprophyticus, S. simulans, S. sciuri, S. succinus, S. vitulinus, S. warneri, Ko. kristinae, Ko. varians* and *Mic. caseolyticus* (Iacumin et al. 2012, Marty et al. 2012). The survival and growth of CNS bacteria depend on the degree of acidity produced during fermentation (Kaban 2013). CNS bacteria are beneficial during sausage fermentation as they contribute to flavor development, typical cured-red color formation and nitrate reduction (Vignolo et al. 2010). The proteolytic and lipolytic activities of CSN bacteria contribute to the sensory profile of fermented sausages by producing amino acids, amines, free fatty acids, aldehydes, ketones and alcohols (Leroy et al. 2006, Toldrá and Wai-Kit 2008, Fonseca et al. 2013b). Volatile compound production in fermented sausages depends on the activity of specific starter cultures, such as *S. carnosus* + *P. acidilactici, S. carnosus* + *Lb. sakei* or *S. carnosus* + *P. pentosaceus*, which produce the cured meat odor from 2-pentnose, 2-hexanone and 2-heptanone. *S. saprophyticus* and *S. warneri* produce acetoin, diacetyl, 1,3-butanediol and 2,3-butanediol which contribute a butter-like odor. *S. saprophyticus* and *Lb. plantarum,* reduce rancidity and produce large amounts of acid, respectively (Leroy et al. 2006, Toldrá and Wai-Kit 2008).

The yeast population in dry fermented sausages produces amino acids, free fatty acids and flavor compounds, delivers antioxidant, proteolytic, and lipolytic activity, and improves sausage appearance (Flores et al. 2015). *D. hansenii* and *Ca. zeylanoides* are the dominant yeasts in fermented sausages (Cocolin et al. 2006, Andrade et al. 2010, Cano-García et al. 2013, Mendoza et al. 2014, Flores et al. 2015). Dry fermented sausages produced with reduced fat and salt content as healthful alternatives often have less desirable sensory characteristics (Yıldız-Turp and Serdaroğlu 2008, Flores et al. 2015). Hence, yeasts, such as *D. hansenii* that produce desirable flavor and aroma characteristics, could be useful for improving the sensory characteristics of reduced-fat/low-salt dry fermented sausages (Flores et al. 2015). Mold cultures, such as *Mucor* spp., *Penicillium nalgiovense* and *Penicillium aurantiogriseum* may also be used to enhance the sensory quality of fermented sausages (Bruna et al. 2001, Garcia et al. 2001).

Fermented Cereal-based Foods

Fermentation is an important practice for improving the nutritional properties of cereals that lack certain essential amino acids (e.g. lysine), have low starch availability, are low in protein content or have antinutrients, such as phytic acid, polyphenols and tannins (Blandino et al. 2003). Hence, the fermentation of cereals can:

- Decrease the carbohydrate content
- Lower the level of some nondigestible poly/oligosaccharides
- Synthesize certain amino acids
- Improve the availability of B group vitamins
- Provide optimum pH conditions for enzymatic degradation of phytate and, therefore, increase the levels of soluble iron, calcium and zinc manifold
- Improve the protein quality and amino acid content
- Increase the dietary fiber content
- Improve the shelf-life, taste, aroma and texture of the final product

Cereals such as rice, wheat, corn and sorghum are used in many parts of the world to produce fermented foods. Hence, the major fermented cereal-based products can be classified as rice-based fermented foods (e.g. *idli*, *dosa* and *dhokla*), wheat-based fermented foods (e.g. soy sauce, *tarhana* and *kishk*), corn-based fermented foods (e.g. *ogi*, *pozol* and *kenkey*), sorghum-based fermented foods (e.g. *injera* and *kisra*) and cereal-based fermented beverages (e.g. *boza, sake, chichi* and *mahewu*) (Blandino et al. 2003). The same product is often known by different names in different regions. For example, *tarhana* and *tarhana*-like products are known as *kishk* in Syria, Palestine, Jordan, Lebanon and Egypt; *kushuk* in Iraq and Iran; *talkuna* in Finland; *tarhana* in Turkey; *thanu* in Hungary and *trahanas* in Greece (Sengun et al. 2009). Similarly, a variety of fermented soybean products consumed in Asian countries are known by different names, such as *chungkookjang* and *doenjang* in Korea and as *natto* and *miso* in Japan (Kwon et al. 2010). These fermentations are characterized by a complex microbial ecosystem. They are mainly produced by LAB and yeasts. Many of the most popular cereal-based fermented foods produced in certain regions of the world are listed in Table 4.

Cereal fermentations are mainly carried out by spontaneous fermentation or backslopping under local climatic conditions. The microbiota of these fermentations are influenced by the process parameters and the type of cereals used, and therefore, their sensory characteristics and quality are variable.

Bread

Commercial starter cultures are used for sourdough fermentation and include LAB and yeasts which may be frozen, freeze-dried or spray-dried preparations (including *Lb. brevis, Lb. casei, Lb. delbrueckii, Lb. fermentum, Lb. plantarum, Lb. sanfranciscensis, P. pentosaceus, P. acidilactici, Sac. cerevisiae, Sac. cerevisiae* var. *chevalieri* and *To. delbrueckii*) or cereal-based preparations (including *Lb. brevis, Lb. casei, Lb. crispatus, Lb. fermentum, Lb. helveticus, Lb. paracasei, Lb. paralimentarius, Lb. plantarum, Lb. pontis, Lb. sanfranciscensis, Leu. lactis, Ca. milleri, Sac. cerevisiae* and *Sac. pastorianus*) (Brandt 2014). In addition, *E. faecium, Lb. delbrueckii* subsp. *bulgaricus, Lb. brevis, Lb. helveticus, Lb. paralimentarius, Lb. sakei, Lb. sanfranciscensis, P. pentosaceus, W. cibaria* and *K. marxianus* could also be used as starter cultures in sourdough fermentation (Plessas et al. 2011). Hence, selected strains are used as starter cultures in industrial bakeries to improve the quality of bakery products. The main criterion for selecting the most suitable combination of cultures depends on the intended fermentation process and the type of cereal used. For example, wheat bran, which is an

Table 4: Examples of cereal-based fermented foods produced in different regions of the world

Product name	*Major ingredients*	*Product definition*	*Fermentative microorganisms*	*Country*	*Reference*
Boza	Millet, rice, wheat, maize, cracked wheat, water	Beverage	*L. garvieae, L. lactis* subsp. *lactis, Lb. brevis, Lb. coryniformis, Lb. fermentum, Lb. graminis, Lb. paracasei* subsp. *paracasei, Lb. paraplantarum, Lb. pentosus, Lb. plantarum, Lb. rhamnosus, Leu. citreum, Leu. lactis, Pediococcus* spp., *P. parvulus, Str. macedonicus, W. confuse*	Turkey, Balkan countries	Heperkan et al. 2014
			L. lactis subsp. *casei, L. lactis* subsp. *lactis, Lb. buchneri, Lb. coryniformis, Lb. fermentum, Lb. parabuchneri, Leu. citreum, P. ethanolidurans, P. parvulus, W. confusa, W. oryzae, G. fragrans, Ga. geotrichum, Meyerozyma guilliermondii, Pi. fermentans, Pi. norvegensis, Pichia* spp., *Torulaspora* spp.		Osimani et al. 2015
Cassava beer (Chicha)	Cassava tuber (*Manihot esculenta*)	Alcoholic beverage	*A. pasteurianus, Lb. acidophilus, Lb. delbrueckii, Lb. fermentum, Lb. reuteri*	Amazon	Colehour et al. 2014
Champùs	Maize, pineapple or "lulo" (*Solanum quitoense*), panela syrup, spices	Low alcoholic beverage	*Ga. geotrichum, Hanseniaspora* spp., *Pi. kudriavzevii, Pi. fermentans, Pi. kluyveri* var. *kluyveri, Sac. cerevisiae, To. delbrueckii, Zy. fermentati*, LAB	Colombia	Osorio-Cadavid et al. 2008
Chhang, Jau chhang, Sura	Rice (in Chhang), barley (in Jau chhang), dheli (in Sura)	Alcoholic beverages	*E. faecium, Lb. casei, Lb. plantarum, P. pentosaceus, Sac. cerevisiae, Ca. tropicalis, Pi. kudriavzevii, Sac. fibuligera*	India	Thakur et al. 2015

(*Contd.*)

Chicha	Rice, water, sugar	Beverage	*B. cereus, B. subtilis, E. durans, E. faecium, E. hirae, E. lactis, E. ludwigii, Lb. casei, Leu. lactis, Aci. baumannii, Cronobacter sakazakii, Escherichia coli, Klebsiella pneumonia, Bifidobacterium* sp., *Enterobacter* sp., *Streptomyces* sp.	Brazil	Puerari et al. 2015
Dolo and Pito	Sorghum malt flour, dried yeast, yolga extract (*Grewia bicolor*)	Alcoholic beverage (sorghum beer)	*Lb. delbrueckii* subsp. *bulgaricus, Lb. delbrueckii* subsp. *delbrueckii, Lb. fermentum, P. acidilactici*	West Africa	Sawadogo-Lingani et al. 2007
Hawaijar	Soybean, fig plant (for wrapping)	Alkaline, sticky fermented food	*B. amyloliquefaciens, B. cereus, B. licheniformis, B. subtilis, S. aureus, S. sciuri, Alkaligenes* sp., *Proteus mirabilis, Providencia rettgers*	India	Singh et al. 2014
Idli	Rice, dehusked blackgram	Steamed cake for breakfast	*E. faecalis, L. lactis, Lb. delbrueckii, Lb. fermentum, Leu. mesenteroides, P. cerevisiae, G. candidum, T. candida, T. holmii, Tr. pullulans*	India, Sri Lanka	Blandino et al. 2003
Ikivunde	Cassava	Fermented powder mix	*Lb. brevis, Lb. fermentum, Lb. plantarum, Leu. mesenteroides, G. candidum*	Burundi	Aloys and Angeline 2009
Kinema	Soybean, fresh fern, firewood ash, water	Sticky, curry, alkaline	*E. faecium, B. cereus, B. circulans, B. licheniformis, B. subtilis, B. thuringiensis, B. sphaericus, Ca. parapsilosis, G. candidum*	India	Tamang 2015
Kishk	Yoghurt, bulgur (cracked wheat)	Dried solid powder	*Lb. brevis, Lb. delbrueckii* subsp. *bulgaricus, Lb. casei, Lb. fermentum, Lb. plantarum, Str. thermophilus*	Lebanon, Syria, Egypt	Albesharat et al. 2011

(*Contd.*)

Table 4: (*Contd.*)

Product name	*Major ingredients*	*Product definition*	*Fermentative microorganisms*	*Country*	*Reference*
Kivunde	Cassava	Fermented powder mix	*Lb. plantarum*	Tanzania	Kimaryo et al. 2000
Koozh	Millet, rice	Paste as staple	*W. koreensis*	India	Anandharaj et al. 2015
Kunu-zaki	Millet	Paste as staple	*L. lactis* subsp. *lactis*, *Lb. amylolyticus*, *Lb. delbrueckii* subsp. *bulgaricus*, *Lb. fermentum*, *Str. gallolyticus* subsp. *macedonicus*, *Str. lutetiensis*, *W. confusa*, *Bacillus* spp.	Nigeria	Oguntoyinbo et al. 2011
Natto	Soybean	Fermented food	*B. amyloliquefaciens*, *B. subtillis*, *B. tequilensis*	Japan	Murata et al. 2013
Nukadoko	Rice bran mash	Fermented rice bran bed used for pickling vegetables	*E. faecalis*, *Lb. curvatus*, *Lb. plantarum*, *P. pentosaceus*, *W. cibaria*, *W. paramesenteroides*	Japan	Ono et al. 2014
Ogi	Maize or sorghum	Paste as staple	*L. lactis* subsp. *lactis*, *Lb. amylolyticus*, *Lb. delbrueckii* subsp. *bulgaricus*, *Lb. fermentum*, *Lb. pantheris*, *Lb. plantarum*, *Lb. vaccinostercus*, *W. confusa*, *Bacillus* spp.	Nigeria	Oguntoyinbo et al. 2011
Omegisool	Millet	Alcoholic beverage	*Lb. pentosus*, *Lb. plantarum* subsp. *plantarum*, *Lb. sakei* subsp. *sakei*, *P. acidilactici*, *P. pentosaceus*	Korea	Oh and Jung 2015

(*Contd.*)

Poto poto	Maize, sugar, water	Weaning food	*Bacillus* sp., *Enterococcus* sp., *Escherichia coli*, *Lb. acidophilus*, *Lb. delbrueckii*, *Lb. casei*, *Lb. gasseri*, *Lb.plantarum/paraplantarum*, *Lb. reuteri*	Congo	Abriouel et al. 2006
Sake	Rice	Alcoholic beverage	*As. oryzae*, *Sac. sake*, *Hansenula anomala*	Japan	Aidoo et al. 2006
Soy sauces	Soybeans and wheat	Condiment	*As. oryzae*, *As. sojae*, *Candida* spp., *Zy. rouxii*	Japan, China	Aidoo et al. 2006
Tarhana	Wheat flour, yogurt, vegetable mix, bakers yeast, salt, Tarhana herb (*Echinophora sibthorpiana*), spices	Solid powder, dried seasoning for soups	*E. faecium*, *Lb. casei*, *Lb. delbrueckii* subsp. *bulgaricus*, *Lb. fermentum*, *Lb. paraplantarum*, *Lb. plantarum*, *Leu. citreum*, *Leu. pseudomesenteroides*, *P. acidilactici*, *P. pentosaceus*, *Str. thermophilus*, *W. cibaria*	Turkey	Sengun et al. 2009
Thua nao	Soybean seeds, banana leaves (for wrapping)	Alkaline fermented foods	*B. licheniformis*, *B. megaterium*, *B. pumilus*, *B. subtilis*	Thailand	Chukeatirote 2015
Tungrymbai	Soybean, *Clinogyne dichotoma* leaves (for wrapping)	Alkaline, sticky, curry, soup	*B. licheniformis*, *B. pumilus*, *B. subtilis*	India	Chettri and Tamang 2015
Yanyanku and Ikpiru	Malcavene bean (*Hibiscus sabdariffa*)	Condiment	*B. altitudinis*, *B. amyloliquefaciens*, *B. aryabhattai*, *B. cereus*, *B. circulans*, *B. flexus*, *B. licheniformis*, *B. safensis*, *B. subtilis*, *Aneurinibacillus* spp., *Brevibacillus* spp., *Lysinibacillus* spp., *Paenibacillus* spp.	Benin	Agbobatinkpo et al. 2013

A.: *Acetobacter*, *Aci.*: *Acinetobacter*, *As.*: *Aspergillus*, *B.*: *Bacillus*, *Ca.*: *Candida*, *E.*: *Enterococcus*, *G.*: *Geothrichum*, *Ga.*: *Galactomyces*, *L.*: *Lactococcus*, *Lb.*: *Lactobacillus*, *Leu.*: *Leuconostoc*, *P.*: *Pediococcus*, *Pi.*: *Pichia*, *Sac.*: *Saccharomyces*, *S.*: *Staphylococcus*, *Str.*: *Streptococcus*, *T.*: *Torulopsis*, *To.*: *Torulaspora*, *Tr.*: *Trichosporon*, *W.*:*Weissella*, *Zy.*: *Zygosaccharomyces* excellent source of dietary fiber, has undesirable technological and sensory properties, such as bitterness and grittiness. Hence, to mitigate the negative effects of wheat bran, *Lb. plantarum* and *Lb. pentosus* could be used for fermentation of bread enriched with wheat bran, instead of the more traditional sourdough cultures, such as *Lb. sanfranciscensis* and *Lb. brevis* (Prückler et al. 2015). Kefir grains could also be used in bread fermentation to enhance the microbiological quality, volatile composition, reduced staling rate and preservation time of the product (Plessas et al. 2011).

Soy Sauce

Soy sauce is a popular liquid condiment prepared mainly from soybean and wheat. In the first stage of its production, a cooked soybean-wheat mixture is inoculated with *Aspergillus oryzae* and then fermented at 25-30°C for three days. At the end of the fermentation period, the mixture known as koji is immersed in a 22-25 per cent brine solution and left to ferment for six to eight months. During this stage, known as moromi fermentation, halotolerant LAB, such as *P. halophilus*, produce acid and then yeasts, such as *Zy. rouxii* and *Candida* species undergo an alcoholic fermentation and produce flavor compounds. Finally, the aged-mash is pressed and the liquid portion is pasteurized and bottled (Blandino et al. 2003, Rhee et al. 2011). A variety of biological activities of soy sauce has been attributed to soy sauce, including antihypertensive, antiplatelet, antioxidant, antimicrobial, anticataract and anticarcinogenic, as well as to promoting digestion (Kataoka 2005).

Boza

Boza is one of the most well-known cereal-based fermented beverages in Turkey, Balkan countries and Russia. It is produced from wheat, millet, maize, rye and other cereals. In the first step of boza production, one volume of the cereals is mixed with five volumes of water. The mixture is then boiled, cooled, distributed into containers and fermented at 20-22°C for one to two days (Heperkan et al. 2014). A wide variety of microorganisms has been recovered from boza fermentation (Table 4), including LAB, such as *Lactobacillus* spp., *Lactococcus* spp., *Leuconostoc* spp., *Pediococcus* spp., *Enterococcus* spp. and *Weissella* spp., and yeasts. *Sac. cerevisiae, Leu. mesenteroides* subsp. *mesenteroides* and *W. confusa* have been identified as ideal starter culture mixtures to achieve desirable sensory characteristics (Zorba et al. 2003). However, starter cultures are not commonly used in boza fermentation. At the end of fermentation, the pH value of boza is ca. 3.0-4.0, which provides optimum conditions for enzymatic degradation of phytates and enhanced safety of the product (Kabak and Dobson 2011). Boza is a good source of probiotic bacteria, such as *L. lactis* subsp. *lactis, Lb. paracasei, Lb. plantarum, Lb. pentosus, Lb. rhamnosus, Leu. lactis* and *P. pentosaceus*, which produce bacteriocins and antimicrobial compounds (Altay et al. 2013). It is both nutritious and provides beneficial health effects. However, biogenic amines, such as putrescine, spermidine and tyramine can be formed in boza, mainly by LAB due to the decarboxylation of free amino acids when the fermentation is not properly controlled (Yegin and Uren 2008, Elsanhoty and Ramadan 2016).

Fermented Fruits and Vegetable Products

Fermentation is an ancient preservation method that was applied to fruits and vegetables. Most fermented foods and beverages are from household production for daily domestic consumption. In general, vegetables are dry salted or brined for fermentation process, which promote the production of flavor, controls undesirable microorganisms, extracts nutrients and water, and strengthens soft tissue (Liu et al. 2011). The combination of salt with organic acids and biologically active antimicrobial compounds produced during the fermentation period results in products with enhanced safety. Fresh fruits and vegetables naturally contain a variety of microorganisms, mainly dominated by yeasts and fungi. Although LAB are a small part of the microbiota of raw fruits and vegetables, they are the principal microorganisms responsible for the fermentation of these products. The most common LAB isolated from fruits and vegetables include *Leuconostoc, Lactobacillus, Weissella, Enterococcus* and *Pediococcus* genera, while *W. cibaria*/*W. confusa* and, especially, *Lb. plantarum* are the most frequent species (Di Cagno et al. 2013).

There are many types of fermented fruit and vegetable products throughout the world (Table 5). Spontaneous fermentation of these products by the indigenous LAB flora results in variations in the sensory properties of the products; therefore, use of selected starter cultures helps to make the fermentation more uniform. Fermentation of vegetables is usually initiated by *Leuconostoc* spp. These heterofermentative bacteria can grow rapidly in the presence of large concentrations of brine and produce acid and carbon dioxide. The production of acids lowers the pH while carbon dioxide provides the anaerobic conditions, which are important factors in inhibiting the growth of undesirable microorganisms. In the absence of brine, spoilage microbiota can dominate and cause deterioration of the product (Breidt et al. 2013). *Lactobacillus* spp., primarily *Lb. plantarum,* dominate in most vegetable fermentations because of their superior acid tolerance (Table 5).

The primary fermented fruit and vegetable products that are commercially produced include pickles, olives, sauerkraut, wine and vinegar. The most frequently used commercial starter culture for plant-based fermented foods is *Lb. plantarum* (Rodríguez et al. 2009). There are also a variety of fruit- and vegetable-based products that are traditionally fermented and known by different local names, such as *gari*, *shalgam* juice, *mekhalal*, and *hardaliye*. Most of these are produced according to a traditional fermentation process involving locally unique microbial cultures (Table 5). Although LAB is the main group responsible for the fermentation of fruits and vegetables, acetic acid bacteria constitute an important group for products such as vinegar and cocoa. These products will be addressed in detail in separate chapters within this book.

Sauerkraut

Sauerkraut is a kind of vegetable product with a characteristic flavor that is obtained by fermentation of shredded cabbage containing 2-3 per cent NaCl (w/w). It is traditionally produced by backslopping to favor the dominance of the best-adapted strains (Leroy and De Vuyst 2004). For high quality sauerkraut, the temperature is maintained at ca. 21 to 24°C during the fermentation period (Machve 2009). During the initial days of fermentation, the heterofermentative LAB grow rapidly by producing acid and CO_2, which provide suitable conditions for more acid-tolerant homofermentative LAB. Early studies revealed that the primary fermentative microorganisms in sauerkraut fermentation were *Leu. mesenteroides, Lb. brevis, Lb. plantarum* and *P. cerevisiae*. However, in recent years, studies based on molecular identification techniques revealed that the predominant LAB include *Leuconostoc* spp., *Lactobacillus* spp. and *Weissella* spp. (Table 5). Sauerkraut fermentation is essentially complete within two weeks and *Lb. plantarum*, which is the most acid-tolerant species, becomes dominant at the end of fermentation (Plengvidhya et al. 2007). *Lb. brevis* can also be found in fermented cabbage (Olszewska et al. 2015). A combination of commercial starter cultures, such as *Leu. mesenteroides, P. pentosaceus, Lb. plantarum* and *Lb. brevis* can also be used in cabbage fermentation for improving the sensory properties of sauerkraut (Wiander and Ryhänan 2005, Johanningsmeier et al. 2007). The acidity of the final product ranges between 1.7-2.5 per cent, and the pH value is ca. 3.5. After reaching the desired acidity, a heat treatment (at 73.9°C during canning) or refrigeration is applied to stop the fermentation (Machve 2009).

Table 5: Microorganisms involved in fermented fruit and vegetable products in different regions of the world

Product name	*Major ingredients*	*Product definition*	*Microorganisms*	*Country*	*Reference*
Bandji (Palm wine)	Sap from palm tree (*Borassus akeassii*)	Fermented alcoholic beverages	*A. aceti, A. cerevisiae, A. estunensis, A. ghanensis, A. indonesiensis, A. lovaniensis, A. orientalis, A. pasteurianus, A. tropicalis, Komagataeibacter saccharivorans, Gluconobacter oxydans, Lb. fermentum, Lb. paracasei, Lb. plantarum, Lb. nagelii, Leu. mesenteroides, Arthroascus fermentans, Fructobacillus durionis, Ca. tropicalis, Sac. cerevisiae, Str. mitis, Tr. asahii*	Burkina Faso	Ouoba et al. 2012
Cocoa	Cocoa beans (fruits of *Theobroma cacao*)	Fermented powder used in foods	1. Box fermentation: *A. pasteurianus, A. syzygii, Lb. fermentum, Lb. plantarum, Ca. ethanolica, Sac. cerevisiae* 2. Heap fermentation: *A. pasteurianus, A. lovaniensis, Lb. fermentum, Lb. plantarum, Ca. ethanolica, Ha.guilliermondii, Ha.uvarum, Pi. manshurica, Sac. cerevisiae, Schizosaccharomyces pombe*	Ivory Coast	Visintin et al. 2015
Fermented caper berry	Caper berry (*Capparis* sp.), water	Fermented food used as pickle	*Lb. brevis, Lb. fermentum, Lb. paraplantarum, Lb. pentosus, Lb. plantarum*	Spain, Turkey	Pérez-Pulido et al. 2007
Fermented olives	Fermented green and black olives	Fermented food	*Lb. paracasei, Lb. paraplantarum, Lb. pentosus, Lb. plantarum, Leu. mesenteroides, Leu. pseudomesenteroides*	Greece	Doulgeraki et al. 2013

(*Contd.*)

	Bella di Cerignola treated green olives	Fermented food	*E. casseliflavus, E. italicu, L. lactis, Lb. brevis, Lb. casei, Lb. coryniformis, Lb. mali, Lb. paracasei, Lb. plantarum, Lb. pentosus, Lb. rhamnosus, Lb. vaccinostercus, Leu. mesenteroides, Leu. pseudomesenteroides, W. cibaria, W. paramesenteroides*	Italy	De Bellis et al. 2010
Fermented Yacon	Yacon (*Smallanthus sonchifolius*), brine	Fermented food	*Leu. citreum, Leu. mesenteroides, Leu. pseudomesenteroides*	USA	Reina et al. 2015
Fu-tsai	Mustard, salt	Fermented food used as soup or fried with meat	*Lb. farciminis, Leu. mesenteroides, Leu. pseudomesenteroides, W. cibaria, W. paramesenteroides*	Taiwan	Chao et al. 2009
Gari	Breadfruit, cassava	Fermented semi solid meal	*B. subtilis, B. coagulans, Bacillus* sp., *Corynebacterium* sp., *C. manihot, Lb. plantarum, Sac. cerevisae. Sac. fragilis, Sac. rouxii, G. candidum, As. niger, Rhizopus stolonifer*	Nigeria and other West African countries	Adeniran and Ajifolokun 2015
Gherkins	Cucumber, salt	Similar to pickle	*Lb. crispatus*	India	Anandharaj et al. 2015
Hardaliye	Red grape, black mustard seed, cherry leaf	Grape based non-alcoholic beverage	*Lb. acetotolerans, Lb. brevis, Lb. paracasei, Lb. pontis, Lb. pseudoplantarum, Lb. sanfranciscensis, Lb. vaccinostercus*	Turkey	Arici and Coskun 2001
Kimchi	Cabbage, radish, salt, spices, ginger,	Similar to sauerkraut	*Lb. brevis, Lb. plantarum, Lb. sakei, Leu. citreum, Leu gasicomitatum,*	Korea	Park et al. 2014

(Contd.)

Table 5: (*Contd.*)

Product name	*Major ingredients*	*Product definition*	*Microorganisms*	*Country*	*Reference*
	pepper, onion, garlic		*Leu. mesenteroides*		
Maari	Seeds from the Baobab tree (*Adansonia digitata* L.)	Fermented food condiment	*B. subtilis*, *Corynebacterium* sp., *E. faecium*, *P. acidilactici*, *S. sciuri*	Burkina Faso	Parkouda et al. 2010
Mekhalal	Cucumber, cabbage, turnip, carrot capsicum, eggplant etc.	Fermented vegetable mix	*Lb. brevis*, *Lb. plantarum*, *S. epidermidis*	Syria	Albesharat et al. 2011
Pickles	Cucumber, vinegar, salt	Fermented food as a side dish	*Lb. brevis*, *Lb. pentosus*, *Lb. plantarum*, *Leuconostoc* spp.	Asia, Europe, USA	Tamminen et al. 2004
Sauerkraut	Cabbage, salt	Fermented food as a side dish	*Lb. coryniformis*, *Lb. paraplantarum*, *Leu. argentinum*, *Leu. citreum*, *Leu. fallax*, *Weissella* spp.	Europe, USA	Plengvidhya et al. 2007
Shalgam juice	Turnips, black carrot, bulgur (broken wheat) flour, salt, water	Fermented non-alcoholic beverage	*Lb. brevis*, *Lb. buchneri*, *Lb. delbrueckii* subsp. *delbrueckii*, *Lb. paracasei* subsp. *paracasei*, *Lb. plantarum*, *P. pentosaceus*	Turkey	Tanguler and Erten 2012
Wine	Grape	Fermented alcoholic beverages	*Oenococcus oeni*, *Sac. cerevisiae*	Southern Italy	Garofalo et al. 2015
Yan-dong-gua (fermented wax gourd)	Wax gourd, salt, sugar, soybeans Mijiu (rice wine)	Fermented food used as seasoning	*W. cibaria*, *W. paramesenteroides*	Taiwan	Lan et al. 2009

A.: Acetobacter, As.: Aspergillus, B.: Bacillus, C.: Corynebacterium, Ca.: Candida, E.: Enterococcus, G.: Geotrichum, Ha.: Hanseniaspora, L.: Lactococcus, Lb.: Lactobacillus, Leu.: Leuconostoc, P.: Pediococcus, Pi.: Pichia, Str.: Streptococcus, Sac.: Saccharomyces, S.: Staphylococcus, Tr.: Trichosporon, W.: Weissella

Kimchi

Kimchi is a traditional Korean fermented vegetable product, which mainly comprises Chinese cabbage and Asian radishes, and is gaining popularity in many countries because of its vitamin, mineral, fiber and phytochemical content, as well as its proported beneficial properties, including antiaging, antimutagenic, antioxidative, antitumor, antimicrobial, immune stimulator, weight-controlling, digestion-enhancing, lipid-lowering and antiatherogenic activities (Khan and Kang 2016, Lee et al. 2011). Although the main ingredients are cabbage and radishes, the recipe for kimchi may also include a variety of vegetables, such as carrots, mustard, green onions, parsley, ginger, red peppers, and garlic as well as fermented seafood (*jeotgal*) (Jung et al. 2011). Shredded cabbage is placed in a 5-7 per cent salt solution for 12 hours and rinsed with water several times and drained. Then the other ingredients are mixed with the cabbage. Although ingredients and production methods can differ, kimchi fermentation is microbiologically similar to that of sauerkraut. The types of microorganisms that have been identified during kimchi fermentation are mainly *Leuconostoc* spp. and *Lactobacillus* spp. (Table 5). During the first stage of fermentation, *Leuconostoc* spp. constitute the dominant flora. In the later stage, *Leuconostoc* spp. are inhibited by increased levels of lactic acid and are replaced by more acid-tolerant LAB, such as *Lb. plantarum* (Di Cagno et al. 2013). When the acidity reaches 0.5-0.6 per cent, optimum taste is obtained and the fermentation is stopped by refrigeration. Kimchi has a unique characteristic taste and is usually served with cooked rice and soup (Rhee et al. 2011). As an alternative, it could be powdered to use in meat products as an antimicrobial agent as well as a flavoring (Kim et al. 2015). A *Lb. plantarum* isolated from kimchi purportedly had antibacterial activity and was protective against bacterial invasion, as well as could modulate the host immune response to promote a healthy gastrointestinal tract (Khan and Kang 2016). Moreover, several strains of bacteriocin-producing LAB, such as *E. faecium*, *Lb. citreum* and *Lb. sakei*, have been isolated from kimchi (Lee et al. 2011, Rhee et al. 2011). Overall, kimchi's antagonistic effect on microbes is due to the organic acids and antimicrobial compounds produced during fermentation and also due to the antimicrobial activity of certain ingredients, such as garlic (Rhee et al. 2011). Hence, kimchi may be a useful aid to maintaining good health, having strong probiotic potential.

Pickles

Pickles are fermented cucumbers that are produced by submerging washed cucumbers in a 5-7 per cent NaCl brine and then undergoing a spontaneous fermentation by LAB. Carbon dioxide, which is synthesized during respiration and the malolactic fermentation of cucumbers, is removed by replacement with air (Di Cagno et al. 2013). During the early hours of fermentation, *Leuconostoc* spp. and *Lactobacillus* spp. predominate. Subsequently the process is followed by *Lactobacillus* spp. and *Pediococcus* spp. (Singh and Ramesh 2008). *Lb. plantarum* primarily carries out the fermentation, producing lactic acid and lowering the pH (Johanningsmeier and McFeeters 2013). The acidity reaches 1.5 per cent and a pH value of 3.1 to 3.5 at the end of fermentation. After the fermentation process, which takes two to three weeks to complete depending on the temperature, the cucumbers are washed to remove excess salt and then packed in containers and covered with liquor (Breidt et al. 2013). Some bacteriocin-producing LAB, such as *P. acidilactici*, *Lb. sakei* and *Lb. garvieae* have been obtained from fermented cucumbers (Singh and Ramesh 2008, Chen et al. 2012, Gao et al. 2014). Pickled cucumbers produced in India (Gherkin) contain *Lb. crispatus,* which has probiotic properties, is resistant to acidic conditions, bile and antibiotics, is inhibitory to pathogens and has aggregation properties and assimilates cholesterol (Anandharaj et al. 2015).

Fermented Olives

Fermented olives are one of the world's major fermented plant-based products. Olives can be processed (green table olives) or natural (black olives) (Rodríguez et al. 2009). Green olives are treated with lye (1-3 per cent NaOH) before they are placed in brine to start the fermentation, which reduces the bitterness of the olives by degrading oleuropein. Natural black olives in brine are prepared by slow fermentation and lack the NaOH treatment. Hence, natural black olives have greater LAB diversity than processed green olives (Breidt et al. 2013). LAB are primarily responsible for the fermentation of green olives, whereas LAB and yeasts compete in natural black olive fermentations. *Lb. plantarum* and *Lb. pentosus* are the predominant species in most olive fermentations (Hurtado et al. 2012). Other LAB isolated from olives include *Lb. brevis, Lb. casei, Lb. coryniformis, Leu. mesenteroides, Leu. pseudomesenteroides, P. parvulus, P. pentosaceus, E. faecium* and *E. hirae* (Randazzo et al. 2004, Chamkha et al. 2008, Albesharat et al. 2011, Ghabbour et al. 2011, Abriouel et al. 2012, Aponte et al. 2012). Yeasts, such as *Pi. anomala* (recently classified as *Wickerhamomyces anomalus*) *Pi. guillermondii, Pi. kluyveri, Pi. membranifaciens, Ca. boidinii, Ca. parapsilosis, D. hansenii, Rhodotorula glutinis* and *Sac. cerevisiae* (Arroyo-López et al. 2006, Coton et al. 2006, Aponte et al. 2010, Tofalo et al. 2012) also have been isolated from fermented olives.

The microbiota of olive fermentations differ, depending on olive properties (cultivar, geographical origin, etc.) and the processing method (Table 5). For example, *Lb. plantarum* is mainly recovered from green olive fermentation, whereas *Lb. pentosus* and *Leu. mesenteroides* are the primary LAB in black olives (Doulgeraki et al. 2013). During the processing of green olives, the olive's oleuropein, which is inhibitory to LAB growth in olive brines, is eliminated, thereby allowing LAB to grow (Medina et al. 2009, Hurtado et al. 2012). Although table olives are mainly produced by spontaneous fermentation processes, the use of a lactic starter culture aids in maintaining a controlled and more standardized fermentation, thereby reducing the amount of lye solution used and preventing microbial spoilage of olives during the storage period (Benincasa et al. 2015). *Lb. brevis, Lb. casei, Lb. pentosus, Lb. paracasei, Lb. plantarum, Leu. cremoris* and *Leu. mesenteroides* could be used as starter cultures in table olive fermentations (Hurtado et al. 2012). LAB strains, originating from olives, such as *Lb. pentosus, Lb. paraplantarum* and *Lb. plantarum,* may also be useful starter cultures in olive fermentations (Bautista-Gallego et al. 2013, Argyri et al. 2014, 2015, Blana et al. 2016).

Wine

Various types of fruits can be used as substrates to produce fermented beverages or fruit wines that have different chemical and volatile properties of which yeasts are the major contributor (Berenguer et al. 2016). Wine is one of the world's most common alcoholic fermented beverages. It is mainly produced by the fermentation of grapes or grape juice. Wine production involves crushing the grapes to extract the juice, alcoholic fermentation (essential) and malolactic fermentation (optional), bulk storage and maturation of the wine, clarification and packaging steps. Alcoholic fermentation is an important step in the conversion of grape musts into wine, which is usually carried out by indigenous or added starter yeasts, mainly *Sac. cerevisiae* strains (Sun et al. 2016). The malolactic fermentation, which involves decarboxylation of malic acid into lactic acid and CO_2 during or at the end of alcoholic fermentation, is carried out by LAB, such as *Lactobacillus, Leuconostoc, Oenococcus* and *Pediococcus* (Garofalo et al. 2015). The fermentation greatly influences the volatile characteristics of wines (Sun et al. 2016). Wines produced with *Sac. cerevisiae* as starter cultures often have better oenological characteristics (Aponte and Blaiotta 2016). *Oenococcus oeni*, which is the main bacterium involved in malolactic fermentation, can be used as a commercial starter culture for controlling wine fermentation, but it is not always beneficial because of its inability to tolerate the harsh environment of wine (Ruiz et al. 2010). *Lb. plantarum* could be an alternative for controlling malolactic fermentation, as it can survive the harsh alcoholic conditions of wine and enhances the wine's global aromatic intensity (Sun et al. 2016). A low temperature and slow fermentation are used in white wine production (10-21°C/1-2 weeks) for the formation of desirable volatile compounds. A higher temperature is needed for red wines (24-27°C/3-5 days) in order to enable the extraction of color compounds from the grape skins. Depending on the type of wine, the alcohol concentration typically ranges between 8 to 13 per cent by volume, with an acetic acid content of 12-14 per cent (w/v) (Machve 2009).

Although using starter cultures helps to standardize the fermentation, not all starters perform equally well during processing. The success of an inoculum depends on the fermentable carbohydrate level, pH, buffering capacity and presence of inhibitory compounds in the substrates. Hence, the main criteria to be considered for selection of starter cultures in vegetable and fruit fermentations include tolerance to low pH and low temperature, growth rate, acidification rate, phenol and salt tolerance, completion of fermentation, malolactic fermentation, pectinolytic activity, antimicrobial and antioxidant activity, and synthesis of biogenic compounds, antimicrobial compounds, hydrogen peroxide, exopolysaccharides and aromatic compounds (Di Cagno et al. 2013). In recent years, the popularity of functional probiotic foods has increased with greater consumer awareness and researcher knowledge of their health benefits (Peres et al. 2012). Although dairy products are the main vehicles of probiotic bacteria, fermented fruits and vegetables can also serve as vehicles of probiotics because of the microstructure of the product and prebiotic compounds (e.g. oligosaccharides) that are naturally occurring in plants (Peres et al. 2012). Probiotic strains, that are mainly used in vegetable and fruit fermentations, include *Lb. acidophilus, Lb. casei, Lb. delbrueckii, Lb. plantarum, Lb. rhamnosus, Leu. mesenteroides* and species of the genus *Bifidobacterium*, which survive in the gastrointestinal tract and arrive in the intestines alive in sufficient numbers; hence they have been promoted for the prevention and treatment of various types of diseases (Peres et al. 2012, Di Cagno et al. 2013).

Innovative Functional Foods Involving Probiotics

Production of functional foods, which includes a variety of methods such as modified processing techniques or fortification with different substances, has become a major focus of the food industry. According to the EU concerted action initiative, "a food can be considered as functional if it has been satisfactorily demonstrated to affect beneficially one or more target functions in the body beyond adequate nutritional effects in a way that is relevant to either an improved state of health and well-being and/or a reduction of risk of disease" (Shiby and Mishra 2013). The functional characteristics of fermented foods are mostly associated with probiotic bacteria or their metabolites that are produced during food fermentation. Probiotics are defined as "live microorganisms that, when administered in adequate amounts, confer a health benefit on the consumer" (FAO/WHO 2002). They should be stable and viable for long storage periods, survive in the stomach at low pH levels, colonize the epithelium of the gastrointestinal tract, be antagonist to infectious disease microbes and not be pathogenic (Kalantzopoulos 1997). For a probiotic to be effective, it should be consumed regularly at adequate levels (Peres et al. 2012). Three major vehicles of probiotics include fermented milk products, foods and drinks supplemented with live probiotic cells and pharmaceutical products in the form of tablets, capsules and granules (Doyle et al. 2013). Probiotic microbes, such as *B. infantis, B. lactis,*

Clostridium butyricum, E. faecium, Lb. acidophilus, Lb. casei, Lb. helveticus, Lb. plantarum, Lb. rhamnosus and *Sac. boulardii*, are usually used as adjunct starter cultures in fermented foods with novel functions (Wu et al. 2015). *Pr. freudenreichii* subsp. *shermanii* and *Pr. freudenreichii* subsp. *freudenreichii* also have probiotic properties, and are well-known strains used in cheese ripening (Cousin et al. 2012).

Many health benefits have been purported from the consumption of foods containing viable probiotic microbes. Some probiotic strains help to modulate the host immune system and influence the microbial composition of the intestine, which is very important, especially for lactose-intolerant individuals. Some probiotics have a direct antagonistic effect on certain groups of microorganisms, by affecting their metabolism in the gut or by stimulating the systemic or mucosal immunity of the host. Other beneficial effects of probiotics may include lowering of serum cholesterol and suppressing cancer (Ceapa et al. 2013). A specific fermented milk product containing *Lb. acidophilus, Lb. casei* and *Lb. rhamnosus* purportedly has the potential chemo-preventive activity against colon carcinogenesis (Desrouillères et al. 2015).

Today's consumers are more interested in functional food products, which are mainly probiotic, prebiotic and symbiotic products, for weight loss and management, heart health, eye health, preventing/suppressing cancer, increasing energy and stamina, and improved memory. Hence, the use of selected starter cultures for the development of novel food formulations with enhanced nutritional and therapeutic characteristics are an expanding new product opportunity for the food industry (Table 6).

Milk-based Products

Fermented milk products are an important source of bioactive peptides to exert positive effects on body function and health. Hence, different strategies can be used to enhance the production of bioactive peptides from milk proteins for the development of novel functional foods. Such strategies include using the proteolytic system of LAB, food-grade enzymes or recombinant DNA technology for the production of bioactive peptides of interest (Hafeez et al. 2014).

Fermented milk products have the ability to protect bacteria against digestive enzymatic and bile stresses and therefore, probiotic bacteria remain viable and metabolically active in dairy products for a long period of time. Thus these live bacteria can be transported to the gut in high levels (Cousin et al. 2012). The most common probiotic dairy products contain LAB that have a number of nutritional and therapeutic attributes, such as improved digestion and bioavailability of milk constituents, protection of the gastrointestinal tract from harmful bacteria, alleviation of lactose intolerance and anticancer and hypocholesterolemic effects (Shiby and Mishra 2013). Certain types of LAB can also be used to bio-enrich fermented milk products, such as to increase folate content (Laiño et al. 2014, 2015).

Meat-based Products

Fermented meat products play a critical role in the food industry, not only as a nutrient source, but also as a rich source of bacterial biodiversity, which can provide beneficial health effects. Raw, fermented sausages can be a suitable vehicle for consuming probiotic microorganisms; however, most of the known probiotic bacteria are unable to survive in the harsh environment of sausages that includes sodium chloride, nitrite, acid and a low water activity. Processing steps and long-term storage of fermented sausage also adversely affect the survival and growth of probiotic bacteria (Kröckel 2013). Therefore, it is important to select hardy, stress-tolerant microorganisms for use as probiotic strains in fermented sausages. Some bacterial strains isolated from fermented sausages, such

Table 6: Innovative products involving probiotics and their beneficial effects

Product name	*Ingredients*	*Probiotic cultures used*	*Beneficial effects*	*Reference*
Milk-based products				
Bifighurt	Milk	*B. bifidum*, *Str. thermophilus*	Reduction of serum cholesterol	Shiby and Mishra 2013
Fermented beverage	Whey	Kefir grain	New functional beverage as vehicle of viable probiotic cells	Balabanova and Panayotov 2011
Fermented milk	Milk, IMO[a] (prebiotic)	*B. lactis*, *Lb. acidophilus*	Improvement of intestinal health and positive effect on the humoral and cell-mediated immunity of host	Wang et al. 2012
Fermented milk	Whole milk, sweet whey, sucrose, mango pulp, iron, amino acid chelate	*Lb. acidophilus*, *Lb. delbrueckii* subsp. *bulgaricus*, *Str. thermophilus*	Improvement of nutrient intake and the nutritional status of preschool children with increased iron bioavailability	Silva et al. 2008
Fermented milk beverage	Defatted milk, sugar, fructose, glucose-fructose solution, soybean, hemicellulose, flavoring agents, sucralose	*Lb. casei*	Intestine conditioning effect	Matsumoto et al. 2010
Fiber-rich probiotic yogurt	Milk base (from skim milk powder), pineapple peel powder	*Lb. acidophilus*, *Lb. delbrueckii* subsp. *bulgaricus*, *Lb. casei*, *Lb. paracasei* subsp. *paracasei*, *Str. thermophilus*	Fiber-enriched product for decreased risk of various chronic diseases and probiotic starters generating bioactive peptides possessing potential anticancer, antioxidant, and antimutagenic activities	Sah et al. 2016
Folate bio-enriched fermented milk	Milk	Folate-producing strains of *Lb. amylovorus*, *Lb. delbrueckii* subsp. *bulgaricus*, *Str. thermophilus*	Prevention of folate deficiency, which mitigates neural tube defects, and cardiovascular and colon cancer risks	Laiño et al. 2014

(*Contd.*)

Table 6: (*Contd.*)

Product name	*Ingredients*	*Probiotic cultures used*	*Beneficial effects*	*Reference*
Kombucha fermented milk	Milk, Kombucha tea (*Camellia sinensis*)	Kombucha inoculum (combination of yeasts, AAB, LAB)	Significant ACE inhibitory activity, antihypertensive, antioxidative and free radical scavenging activity	Hrnjez et al. 2014
Probiotic butter	Cream	*Lb. acidophilus*, *B. bifidum*	Recommended as a source of viable probiotic cells	Erkaya et al. 2015
Progurt	Milk	*L. lactis* subsp. *diacetylactis*, *L. lactis* subsp. *cremoris*, *Lb. acidophilus*, *B. bifidum*	Alleviates lactose maldigestion	Shiby and Mishra 2013
Meat-based products				
Fermented sausage	Lean pork, back fat pork, NaCl, $NaNO_2$, KNO_3, sodium ascorbate, glucose, black pepper, water	*Lb. casei/paracasei*, *Lb. rhamnosus*	Antagonistic activity against pathogens	Rubio et al. 2014
Fermented lamb sausage	Lamb, salt, sucrose, ginger, pepper	*Lb. plantarum*	Improvement of safety and quality of the product, and potential functional properties	Arief et al. 2014
Cereal-based products				
Fermented cereal beverage	Oat flour, barley flour or malt flour, water	*Lb. acidophilus*, *Lb. plantarum* or *Lb. reuteri*	New functional beverage as vehicle of viable probiotic cells	Salmerón et al. 2015
Fermented pigeon pea	Pigeon pea, water	*Bacillus subtilis*	Beneficial to cardiovascular health with elevated antioxidant levels, and increased nattokinase and ACEI activities	Lee et al. 2015
Fermented protein- fortified cassava flour	Cassava flour fortified with protein (zeolin, sporazein, sporazein plus pro-vitamin A) or pro-vitamin A	*Lb. plantarum*	Protein- and vitamin- enriched product with reduced unacceptable aromatic compounds	Rosales-Soto et al. 2016

(*Contd.*)

Fermented red beans	Red bean	*Bacillus subtilis*, *Lb. delbrueckii* subsp. *bulgaricus*	New product with high antioxidant properties and fibrinolytic activity	Jhan et al. 2015
Fermented soy milk	Soy milk	*Lb. lactis* subsp. *lactis*	Prevention of inflammatory bowel disease and other observed phenomena (shortening of colon length, breaking of epithelial cells, lowered liver and thymus weights, enlargement of spleen)	Kawahara et al. 2015
Fermented soy milk	Soy milk	*Lb. plantarum*	Inhibition of melanogenesis by suppressing tyrosinase activity	Chen et al. 2013
Fermented tofu	Soy milk	*Lb. acidophilus*, *Lb. casei*	Enhanced keeping quality and sensory properties of the product	Serrazanetti et al. 2013
Probiotic cereal, pseudocereal and cereal-leguminous products	Buckwheat, barley, oat, chickpea + oat soya	*Lb. rhamnosus*	New products as vehicles of viable probiotic cells	Kocková and Valík 2014
Probiotic-fermented genetically modified soy milk	Genetically modified soy milk	*L. lactis* subsp. *lactis*	Improve hypercholesterolemia and reduce atherosclerotic risks	Tsai et al. 2014
Rice-based fermented beverage	Rice	*Lb. fermentum*	Improved functional composition (nutrients, minerals), digestibility and therapeutic activities	Ghosh et al. 2015
Nilamadana (fermented cereal-based food)	Kodo millet	*Penicillium roqueforti*	Increased levels of soluble fat, reduced sugar, increased potassium, calcium and flavonoids and reduction in cholesterol level	Dwivedi et al. 2015
Fruit/vegetable-based products				
Fermented carrot juice	Carrot juice	*B. bifidum*, *B. lactis*	New product as probiotic source with improved nutritional quality	Kun et al. 2008

(*Contd.*)

Table 6: (*Contd.*)

Product name	*Ingredients*	*Probiotic cultures used*	*Beneficial effects*	*Reference*
Fermented chestnut purees	Chestnuts, water	*Lb. casei*, *Lb. rhamnosus*	New healthy food as vehicle of viable probiotic cells	Blaiotta et al. 2012
Fermented coconut water	Coconut water, sucrose, yeast extract, soy protein hydrolysate	*B. animalis*, *Lb. plantarum*	New functional beverages as vehicle of viable cells	Camargo Prado et al. 2015
Fermented pomegranate juice	Pomegranate juice	*Lb. acidophilus*, *Lb. delbrueckii*, *Lb. paracasei*, *Lb. plantarum*	New functional beverages as vehicle of viable probiotic cells	Mousavi et al. 2011
Fermented tomato juice	Tomato juice, FOS[b]	*B. breve*, *B. longum*	New functional beverage as vehicle of viable probiotic cells with improved taste	Koh et al. 2010

[a]IMO: isomaltooligosaccharide, [b]FOS: fructooligosaccharides,
B.: Bifidobacterium, L.: Lactococcus, Lb.: Lactobacillus, Str.: Streptococcus

as *Lb. casei/paracasei, Lb. fermentum, Lb. plantarum, Lb. rhamnosus* and *P. acidilactici*, have been characterized as probiotic bacteria (Ammor and Mayo 2007, Vuyst et al. 2008, Ruiz-Moyano et al. 2011, Kröckel 2013, Rubio et al. 2014, Trząskowska et al. 2014, Jofré et al. 2015). Microencapsulation of probiotic strains, which do not otherwise survive in the sausage environment, could be an alternative (Khan et al. 2011). Immobilization supports, such as wheat grains, may also be used to protect probiotic bacteria from the harsh environment of dry fermented sausages (Sidira et al. 2014).

Cereal-based Products

Although the primary vehicles of probiotic microbes are milk-based products, there has been, in recent years, an increased interest in developing cereal-based probiotic-containing fermented foods. Probiotic strains sensitive in many fermented foods can survive in cereal matrices, which contain native prebiotic substances that protect the cells against stress in the gut (Peres et al. 2012). As revealed in Table 6, a variety of microorganisms (natural or from starter cultures) can grow in cereal-based foods and enrich the product by producing vitamins, available minerals, amino acids, modified dietary fibers, phytochemicals, digestive enzymes, and probiotic and bioactive components, which can provide significant beneficial effects and add value for the consumers (Ghosh et al. 2015). Cereal-based fermented products, especially fermented soybean foods, are very popular in Asian countries where the incidence of type 2 diabetes is less than in Western countries (Kwon et al. 2010). After approval by the US FDA of soy foods for heart disease risk reduction, their consumption and popularity increased rapidly in the West countries (Serrazanetti et al. 2013).

Fruit/Vegetable-based Products

The use of probiotic LAB provides additional health benefits for fruits and vegetables, which are naturally rich sources of important nutrients, such as dietary fibers, antioxidants, vitamins and minerals (Kohajdova et al. 2006). However, successive applications depend on the fruit or vegetable matrix and the ingredients used. Artichokes, carrots, cabbage, red beet juices and tomatoes provide suitable environments for probiotic bacteria fermentation (Di Cagno et al. 2013), whereas orange and apple juices are unsuitable substrates for functional foods because of their low pH and lack of suitable nutrients for probiotic bacteria growth (Ding and Shah 2008). *Lb. acidophilus, Lb. delbrueckii, Lb. paracasei, Lb. plantarum, B. animalis, B. lactis, B. bifidum, B. breve* and *B. longum* are probiotic strains that can be used for the fortification of fermented fruit and vegetable products (Table 6). The shelf-life of probiotic strains in fermented fruits and vegetables is improved by applying encapsulation technology. However, further studies are needed to determine the efficacy and long-term safety of these probiotic applications (Prakash et al. 2011).

Conclusions

Currently, a variety of fermented foods which are commercially or traditionally produced from different commodities, such as milk, meat, cereals, fruits and vegetables, are consumed by people worldwide. Fermented foods constitute a significant part of many diets, not only as nutritive foods, but also as a rich source of beneficial microorganisms, including lactic acid bacteria of which many possess probiotic properties. Fermenting microorganisms can transform raw materials into final products that have improved shelf-life sensory properties and health benefits. Although these foods can have a variety of beneficial effects, any potential safety issues should be evaluated by appropriate studies. In the last decade, more detailed information on the biodiversity of fermented foods has been revealed by the introduction of molecular biology-based methods. Studies elucidating the dynamics and biodiversity of these microorganisms in food fermentations will provide the knowledge needed for processing and preserving these foods for the benefit of international communities. Furthermore, the use of novel starter culture strains with industrially important functionality can serve in developing microbiologically safe new products with enhanced nutritional, sensory and health properties.

References

Abriouel, H., Ben Omar, N., López, R.L., Martínez-Cañamero, M., Keleke, S. and Gálvez, A. (2006). Culture-independent analysis of the microbial composition of the African traditional fermented foods *poto poto* and *dégué* by using three different DNA extraction methods. *International Journal of Food Microbiology* **111:** 228-233.

Abriouel, H., Ben Omar, N., Cobo, A., Caballero, N., Fernández Fuentes, M.Á., Pérez-Pulido, R. and Gálvez, A. (2012). Characterization of lactic acid bacteria from naturally-fermented Manzanilla Aloreña green table olives. *Food Microbiology* **32:** 308-316.

Adeniran, H.A. and Ajifolokun, O.M. (2015). Microbiological studies and sensory evaluation of breadfruit and cassava co-fermented into *gari* analogue. *Nigerian Food Journal* **33:** 39-47.

Agbobatinkpo, P.B., Thorsen, L., Nielsen, D.S., Azokpota, P., Akissoe, N., Hounhouigan, J.D. and Jakobsen, M. (2013). Biodiversity of aerobic endospore-forming bacterial species occurring in Yanyanku and Ikpiru, fermented seeds of *Hibiscus sabdariffa* used to produce food condiments in Benin. *International Journal of Food Microbiology* **163:** 231-238.

Aghababaie, M., Khanahmadi, M. and Beheshti, M. (2015). Developing a kinetic model for co-culture of yogurt starter bacteria growth in pH-controlled batch fermentation. *Journal of Food Engineering* **166:** 72-79.

Aidoo, K. E., Rob Nout, M. J. and Sarkar, P. K. (2006). Occurrence and function of yeasts in Asian indigenous fermented foods. *FEMS Yeast Research* **6:** 30-39.

Akabanda, F., Owusu-Kwarteng, J., Tano-Debrah, K., Glover, R.L.K., Nielsen, D.S. and Jespersen, L. (2013). Taxonomic and molecular characterization of lactic acid bacteria and yeasts in *nunu*, a Ghanaian fermented milk product. *Food Microbiology* **34:** 277-283.

Albano, H., van Reenen, C.A., Todorov, S.D., Cruz, D., Fraga, L., Hogg, T., Dicks, L.M.T. and Teixeira, P. (2009). Phenotypic and genetic heterogeneity of LAB Isolated from 'Alheira', a traditional fermented sausage produced in Portugal. *Meat Science* **82:** 389-398.

Albesharat, R., Ehrmann, M.A., Korakli, M., Yazaji, S. and Vogel, R.F. (2011). Phenotypic and genotypic analyses of lactic acid bacteria in local fermented food, breast milk and faeces of mothers and their babies. *Systematic and Applied Microbiology* **34:** 148-155.

Aloys, N. and Angeline, N. (2009). Traditional fermented foods and beverages in Burundi. *Food Research International* **42:** 588-594.

Altay, F., Karbancıoglu-Güler, F., Daskaya-Dikmen, C. and Heperkan, D. (2013). A review of traditional Turkish fermented non-alcoholic beverages: Microbiota, fermentation process and quality characteristics. *International Journal of Food Microbiology* **167:** 44-56.

Amadoro, C., Rossi, F., Piccirilli, M. and Colavita, G. (2015). *Tetragenococcus koreensis* is part of the microbiota in a traditional Italian raw fermented sausage. *Food Microbiology* **50:** 78-82.

Ammor, M.S. and Mayo, B. (2007). Selection criteria for lactic acid bacteria to be used as functional starter cultures in dry sausage production: An update. *Meat Science* **76:** 138-146.

Anandharaj, M., Sivasankari, B., Santhanakaruppu, R., Manimaran, M., Rani, R. P. and Sivakumar, S. (2015). Determining the probiotic potential of cholesterol-reducing Lactobacillus and Weissella strains isolated from gherkins (fermented cucumber) and South Indian fermented *koozh*. *Research in Microbiology* **166:** 428-439.

Andrade, M.J., Rodriguez, M., Casado, E.M., Bermudez, E. and Cordoba, J.J. (2010). Effect of selected strains of *Debaryomyces hansenii* on the volatile compound production of dry fermented sausage 'salchichón'. *Meat Science* **85:** 256-264.

Aponte, M. and Blaiotta, G. (2016). Selection of an autochthonous *Saccharomyces cerevisiae* strain for the vinification of 'Moscato di Saracena', a southern Italy (Calabria Region) passito wine. *Food Microbiology* **54:** 30-39.

Aponte, M., Blaiotta, G., La Croce, F., Mazzaglia, A., Farina, V., Settanni, L. and Moschetti, G. (2012). Use of selected autochthonous lactic acid bacteria for Spanish-style table olive fermentation. *Food Microbiology* **30:** 8-16.

Aponte, M., Ventorino, V., Blaiotta, G., Volpe, G., Farina, V., Avellone, G., Lanza, C.C. and Moschetti, G. (2010). Study of green Sicilian table olive fermentations through microbiological, chemical and sensory analyses. *Food Microbiology* **27:** 162-170.

Argyri, A.A., Nisiotou, A.A., Pramateftaki, P., Doulgeraki, A.I., Panagou, E.Z. and Tassou, C.C. (2015). Preservation of green table olives fermented with lactic acid bacteria with probiotic potential under modified atmosphere packaging. *LWT—Food Science and Technology* **62:** 783-790.

Argyri, A.A., Nisiotou, A.A., Malouchos, A., Panagou, E.Z. and Tassou, C.C. (2014). Performance of two potential probiotic *Lactobacillus* strains from the olive microbiota as starters in the fermentation of heat-shocked green olives. *International Journal of Food Microbiology* **171:** 68-76.

Arici, M. and Coskun, F. (2001). Hardaliye: Fermented grape juice as a traditional Turkish beverage. *Food Microbiology* **18:** 417-421.

Arief, I.I., Wulandari, Z., Aditia, E.L. and Baihaqi, M. (2014). Physicochemical and microbiological properties of fermented lamb sausages using probiotic *Lactobacillus plantarum* IIA-2C12 as starter culture. *Procedia Environmental Sciences* **20:** 352-356.

Arroyo-López, F.N., Durán Quintana, M.C., Ruiz Barba, J.L., Querol, A. and Garrido- Fernández, A. (2006). Use of molecular methods for the identification of yeast associated with table olives. *Food Microbiology* **23:** 791-796.

Ashmaig, A., Hasan, A. and EL-Gaali, E. (2009). Identification of lactic acid bacteria isolated from traditional Sudanese fermented camel's milk (*gariss*). *African Journal of Microbiology Research* **3:** 451-457.

Aydemir, O., Harth, H., Weckx, S., Dervişoğlu, M. and De Vuyst, L. (2015). Microbial communities involved in Kaşar cheese ripening. *Food Microbiology* **46:** 587-595.

Balabanova, T. and Panayotov, P. (2011). Obtaining functional fermented beverages by using the kefir grains. *Procedia Food Science* **1:** 1653-1659.

Bautista-Gallego, J., Arroyo-López, F. N., Rantsiou, K., Jiménez-Díaz, R., Garrido-Fernández, A. and Cocolin, L. (2013). Screening of lactic acid bacteria isolated from fermented table olives with probiotic potential. *Food Research International* **50:** 135-142.

Benincasa, C., Muccilli, S., Amenta, M., Perri, E. and Romeo, F.V. (2015). Phenolic trend and hygienic quality of green table olives fermented with *Lactobacillus plantarum* starter culture. *Food Chemistry* **186:** 271-276.

Berenguer, M., Vegara, S., Barrajón, E., Saura, D., Valero, M. and Martí, N. (2016). Physicochemical characterization of pomegranate wines fermented with three different *Saccharomyces cerevisiae* yeast strains. *Food Chemistry* **190:** 848-855.

Beresford, T.P., Fitzsimons, N.A., Brennan, N.L. and Cogan, T.M. (2001). Recent advances in cheese microbiology. *International Dairy Journal* **11:** 259-274.

Blaiotta, G., Di Capua, M., Coppola, R. and Aponte, M. (2012). Production of fermented chestnut purees by lactic acid bacteria. *International Journal of Food Microbiology* **158:** 195-202.

Blana, V.A., Polymeneas, N., Tassou, C.C. and Panagou, E.Z. (2016). Survival of potential probiotic lactic acid bacteria on fermented green table olives during packaging in polyethylene pouches at 4 and 20°C. *Food Microbiology* **53:** 71-75.

Blandino, A., Al-Aseeri, M.E., Pandiella, S.S., Cantero, D. and Webb, C. (2003). Cereal-based fermented foods and beverages. *Food Research International* **36:** 527-543.

Bokulich, N.A., Amiranashvili, L., Chitchyan, K., Ghazanchyan, N., Darbinyan, K., Gagelidze, N., Sadunishvili, T., Goginyan, V., Kvesitadze, G., Torok, T. and Mills, D.A. (2015). Microbial biogeography of the transnational fermented milk matsoni. *Food Microbiology* **50:** 12-19.

Boldyreva, E.M., Hill, A., Griffiths, M. and Marcone, M. (2016). The quality and safety of washed-rind cheeses with a focus on antilisterial protection. *International Dairy Journal* **55:** 26-37.

Bourdichon, F., Casaregola, S., Farrokh, C., Frisvad, J.C., Gerds, M.L., Hammes, W.P., Harnett, J., Huys, G., Laulund, S., Ouwehand, A., Powell, I.B., Prajapati, J.B., Seto, Y., Ter Schure, E., Van Boven, A., Vankerckhoven, V., Zgoda, A., Tuijtelaars, S. and Hansen, E.B. (2012). Food fermentations: Microorganisms with technological beneficial use. *International Journal of Food Microbiology* **154:** 87-97.

Bozoudi, D., Pavlidou, S., Kotzamanidis, C., Georgakopoulos, P., Torriani, S., Kondyli, E., Claps, S., Belibasaki, S. and Litopoulou-Tzanetaki, E. (2016). Graviera Naxou and Graviera Kritis Greek PDO cheeses: Discrimination based on microbiological and physicochemical criteria and volatile organic compounds profile. *Small Ruminant Research* **136:** 161-172.

Brandt, M.J. (2014). Starter cultures for cereal based foods. *Food Microbiology* **37:** 41-43.

Breidt, F., McFeeters, R.F., Perez-Diaz, I. and Lee, C.H. (2013). Fermented vegetables. *In:* M.P. Doyle and R.L. Buchanan (Eds.). *Food Microbiology: Fundamentals and Frontiers*, 4th edn. ASM Press, Washington, D.C., pp. 841-855.

Bruna, J.M., Ordóñez, J.A., Fernández, M., Herranz, B. and de la Hoz, L. (2001). Microbial and physico-chemical changes during the ripening of dry fermented sausages superficially inoculated with or having added an intracellular cell-free extract of *Penicillium aurantiogriseum*. *Meat Science* **59:** 87-96.

Buckenhüskes, H.J. (1993). Selection criteria for lactic acid bacteria to be used as starter cultures for various food commodities. *Federation of European Microbiological Societies Microbiology Review* **12:** 253-271.

Buňková, L., Adamcová, G., Hudcová, K., Velichová, H., Pachlová, V., Lorencová, E. and Buňka, F. (2013). Monitoring of biogenic amines in cheeses manufactured at small-scale farms and in fermented dairy products in the Czech Republic. *Food Chemistry* **141:** 548-551.

Camargo Prado, F., De Dea Lindner, J., Inaba, J., Thomaz-Soccol, V., Kaur Brar, S. and Soccol, C.R. (2015). Development and evaluation of a fermented coconut water beverage with potential health benefits. *Journal of Functional Foods* **12:** 489-497.

Candogan, K., Wardlaw, F.B. and Acton, J.C. (2009). Effect of starter culture on proteolytic changes during processing of fermented beef sausages. *Food Chemistry* **116:** 731-737.

Cano-García, L., Flores, M. and Belloch, C. (2013). Molecular characterization and aromatic potential of *Debaryomyces hansenii* strains isolated from naturally fermented sausages. *Food Research International* **52:** 42-49.

Carafa, I., Clementi, F., Tuohy, K. and Franciosi, E. (2016). Microbial evolution of traditional mountain cheese and characterization of early fermentation cocci for selection of autochtonous dairy starter strains. *Food Microbiology* **53:** 94-103.

Carlos, A.R., Santos, J., Semedo-Lemsaddek, T., Barreto-Crespo, M.T. and Tenreiro, R. (2009). Enterococci from artisanal dairy products show high levels of adaptability. *International Journal of Food Microbiology* **129:** 194-199.

Casaburi, A., Blaiotta, G., Mauriello, G., Pepe, I. and Villani, F. (2005). Technological activities of *Staphylococcus carnosus* and *Staphylococcus simulans* strains isolated from fermented sausages. *Meat Science* **71:** 643-650.

Casaburi, A., Aristoy, M.C., Cavella, S., Monaco, R.D., Ercolini, D., Toldrá, F. and Villani, F. (2007). Biochemical and sensory characteristics of traditional fermented sausages of Vallo di Diano (Southern Italy) as affected by the use of starter cultures. *Meat Science* **76:** 295-307.

Casaburi, A., Di Martino, V., Ferranti, P., Picariello, L. and Villani, F. (2016). Technological properties and bacteriocins production by *Lactobacillus curvatus* 54M16 and its use as starter culture for fermented sausage manufacture. *Food Control* **59:** 31-45.

Casquete, R., Benito, M.J., Martín, A., Ruiz-Moyano, S., Aranda, E. and Córdoba, M.G. (2012). Microbiological quality of salchichón and chorizo, traditional Iberian dry-fermented sausages from two different industries, inoculated with autochthonous starter cultures. *Food Control* **24:** 191-198.

Castellano, P., Belfiore, C., Fadda, S. and Vignolo, G. (2008). A review of bacteriocinogenic lactic acid bacteria used as bioprotective cultures in fresh meat produced in Argentina. *Meat Science* **79:** 483-499.

Ceapa, C., Wopereis, H., Rezaïki, L., Kleerebezem, M., Knol, J. and Oozeer, R. (2013). Influence of fermented milk products, prebiotics and probiotics on microbiota composition and health. Best Practice & Research. *Clinical Gastroenterology* **27:** 139-155.

Chamba, J.F. and Jamet, E. (2008). Contribution to the safety assessment of technological microflora found in fermented dairy products. *International Journal of Food Microbiology* **126:** 263-266.

Chamkha, M., Sayadi, S., Bru, V. and Godon, J.J. (2008). Microbial diversity in Tunisian olive fermentation brine as evaluated by small subunit rRNA—single strand conformation polymorphism analysis. *International Journal of Food Microbiology* **122:** 211-215.

Chao, S.-H., Wu, R.-J., Watanabe, K. and Tsai, Y.-C. (2009). Diversity of lactic acid bacteria in suan-tsai and fu-tsai, traditional fermented mustard products of Taiwan. *International Journal of Food Microbiology* **135:** 203-210.

Chaves-López, C., Serio, A., Grande-Tovar, C.D., Cuervo-Mulet, R., Delgado-Ospina, J. and Paparella, A. (2014). Traditional fermented foods and beverages from a microbiological and nutritional perspective: The Colombian heritage. *Comprehensive Reviews in Food Science and Food Safety* **13:** 1031-1048.

Chaves-López, C., Serio, A., Martuscelli, M., Paparella, A., Osorio-Cadavid, E. and Suzzi, G. (2011). Microbiological characteristics of *kumis*, a traditional fermented Colombian milk, with particular emphasis on enterococci population. *Food Microbiology* **28:** 1041-1047.

Chebeňová-Turcovská, V., Ženišová, K., Kuchta, T., Pangallo, D. and Brežná, B. (2011). Culture-independent detection of microorganisms in traditional Slovakian bryndza cheese. *International Journal of Food Microbiology* **150:** 73-78.

Chen, Y.-M., Shih, T.-W., Chiu, C. P., Pan, T.-M. and Tsai, T.-Y. (2013). Effects of lactic acid bacteria-fermented soy milk on melanogenesis in B16F0 melanocytes. *Journal of Functional Foods* **5:** 395-405.

Chen, Y.S., Wu, H.C., Lo, H.Y., Lin, W.C., Hsu, W.H., Lin, C.W., Lin, P.Y. and Yanagida, F. (2012). Isolation and characterisation of lactic acid bacteria from *jiang-gua* (fermented cucumbers), a traditional fermented food in Taiwan. *Journal of the Science of Food and Agriculture* **92:** 2069-2075.

Chettri, R. and Tamang, J.P. (2015). *Bacillus* species isolated from tungrymbai and bekang, naturally fermented soybean foods of India. *International Journal of Food Microbiology* **197:** 72-76.

Chukeatirote, E. (2015). Thua nao: Thai fermented soybean. *Journal of Ethnic Foods* **2:** 115-118.

Cocolin, L., Dolci, P. and Rantsiou, K. (2011). Biodiversity and dynamics of meat fermentations: The contribution of molecular methods for a better comprehension of a complex ecosystem. *Meat Science* **89:** 296-302.

Cocolin, L., Urso, R., Rantsiou, K., Cantoni, C. and Comi, G. (2006). Dynamics and characterization of yeasts during natural fermentation of Italian sausages. *Federation of European Microbiological Societies Yeast Research* **6:** 692-701.

Colehour, A.M., Meadow, J.F., Liebert, M.A., Cepon-Robins, T.J., Gildner, T.E., Urlacher, S.S., Bohannan, B.J.M., Snodgrass, J.J. and Sugiyama, L.S. (2014). Local domestication of lactic acid bacteria via cassava beer fermentation. *PeerJ* **2:** e479.

Corrieu, G. and Béal, C. (2016). Yogurt: The product and its manufacture. *In:* B. Caballero, P. M. Finglas and F. Toldrá (Eds.). *Encyclopedia of Food and Health*. Academic Press, Oxford, pp. 617-624.

Coton, E., Coton, M., Levert, D., Casaregola, S. and Sohier, D. (2006). Yeast ecology in French cider and black olive natural fermentations. *International Journal of Food Microbiology* **108:** 130-135.

Cousin, F.J., Louesdon, S., Maillard, M.-B., Parayre, S., Falentin, H., Deutsch, S.-M., Boudry, G. and Jan, G. (2012). The first dairy product exclusively fermented by *Propionibacterium freudenreichii*: A new vector to study probiotic potentialities *in vivo*. *Food Microbiology* **32:** 135-146.

Cueto, C., García, D., Garcés, F. and Cruz, J. (2007). Preliminary studies on the microbiological characterization of lactic acid bacteria in suero costeño, a Colombian traditional fermented milk product. *Revista Latinoamericana de Microbiologia* **49:** 12-18.

De Bellis, P., Valerio, F., Sisto, A., Lonigro, S.L. and Lavermicocca, P. (2010). Probiotic table olives: Microbial populations adhering on olive surface in fermentation sets inoculated with the probiotic strain *Lactobacillus paracasei* IMPC2.1 in an industrial plant. *International Journal of Food Microbiology* **140:** 6-13.

de Candia, S., De Angelis, M., Dunlea, E., Minervini, F., McSweeney, P.L.H., Faccia, M. and Gobbetti, M. (2007). Molecular identification and typing of natural whey starter cultures and microbiological and compositional properties of related traditional Mozzarella cheeses. *International Journal of Food Microbiology* **119:** 182-191.

Desrouillères, K., Millette, M., Vu, K.D., Touja, R. and Lacroix, M. (2015). Cancer preventive effects of a specific probiotic fermented milk containing *Lactobacillus acidophilus* CL1285, *L. casei* LBC80R and *L. rhamnosus* CLR2 on male F344 rats treated with 1,2-dimethylhydrazine. *Journal of Functional Foods* **17:** 816-827.

Di Cagno, R., Coda, R., De Angelis, M. and Gobbetti, M. (2013). Exploitation of vegetables and fruits through lactic acid fermentation. *Food Microbiology* **33:** 1-10.

Ding, W.K. and Shah, N.P. (2008). Survival of free and microencapsulated probiotic bacteria in orange and apple juices. *International Food Research Journal* **15:** 219-232.

Doulgeraki, A.I., Pramateftaki, P., Argyri, A.A., Nychas, G.-J.E., Tassou, C.C. and Panagou, E.Z. (2013). Molecular characterization of lactic acid bacteria isolated from industrially-fermented Greek table olives. *LWT—Food Science and Technology* **50:** 353-356.

Doyle, M.P., Steenson, L.R. and Meng, J. (2013). Bacteria in food and beverage production. *In:* E. Rosenberg (Ed.). *The Prokaryotes, Bacteriology and Biotechnology*, 4th edn. Springer-Verlag, New York, pp. 241-256.

Drosinos, E.H., Paramithiotis, S., Kolovos, G., Tsikouras, I. and Metaxopoulos, I. (2007). Phenotypic and technological diversity of LAB and staphylococci isolated from traditionally fermented sausages in Southern Greece. *Food Microbiology* **24:** 260-270.

Dwivedi, M., Vasantha, K.Y., Sreerama, Y. N., Haware, D.J., Singh, R.P. and Sattur, A.P. (2015). Nilamadana, new fungal fermented cereal based food. *Journal of Functional Foods* **15:** 217-224.

EFSA (European Food Safety Authority). (2010). Available from: http://www.efsa.europa.eu/fr/efsajournal/doc/1763.pdf. (Accessed: June 2016).

Elsanhoty, R.M. and Ramadan, M.F. (2016). Genetic screening of biogenic amines production capacity from some lactic acid bacteria strains. *Food Control* **68:** 220-228.

Erkaya, T., Ürkek, B., Doğru, Ü., Çetin, B. and Şengül, M. (2015). Probiotic butter: Stability, free fatty acid composition and some quality parameters during refrigerated storage. *International Dairy Journal* **49:** 102-110.

Erkmen, O. and Bozoglu, F. (2008). *Food Microbiology 4. Beneficial Uses of Microorganisms for Food Preservation and Health*. İlke Publishing Company, Ankara.

Escobar-Zepeda, A., Sanchez-Flores, A. and Quirasco Baruch, M. (2016). Metagenomic analysis of a Mexican ripened cheese reveals a unique complex microbiota. *Food Microbiology* **57:** 116-127.

Essid, I. and Hassouna, M. (2013). Effect of inoculation of selected *Staphylococcus xylosus* and *Lactobacillus plantarum* strains on biochemical, microbiological and textural characteristics of a Tunisian dry fermented sausage. *Food Control* **32:** 707-714.

FAO/WHO (Food and Agriculture Organization/World Health Organization). (2002). *Guidelines for the Evaluation of Probiotics in Food*. London, Ontario.

Federici, S., Ciarrocchi, F., Campana, R., Ciandrini, E., Blasi, G. and Baffone, W. (2014). Identification and functional traits of lactic acid bacteria isolated from Ciauscolo salami produced in Central Italy. *Meat Science* **98:** 575-584.

Firmesse, O., Alvaro, E., Mogenet, A., Bresson, J.-L., Lemée, R., Le Ruyet, P., Bonhomme, C., Lambert, D., Andrieux, C., Doré, J., Corthier, G., Furet, J.-P. and Rigottier-Gois, L. (2008). Fate and effects of Camembert cheese micro-organisms in the human colonic microbiota of healthy volunteers after regular Camembert consumption. *International Journal of Food Microbiology* **125:** 176-181.

Flores, M., Corral, S., Cano-García, L., Salvador, A. and Belloch, C. (2015). Yeast strains as potential aroma enhancers in dry fermented sausages. *International Journal of Food Microbiology* **212:** 16-24.

Fonseca, S., Cachaldora, A., Gómez, M., Franco, I. and Carballo, J. (2013a). Monitoring the bacterial population dynamics during the ripening of Galician chorizo, a traditional dry fermented Spanish sausage. *Food Microbiology* **33:** 77-84.

Fonseca, S., Cachaldora, Gómez, A.M., Franco, I. and Carballo, J. (2013b). Effect of different autochthonous starter cultures on the volatile compounds profile and sensory properties of Galician chorizo, a traditional Spanish dry fermented sausage. *Food Control* **33:** 6-14.

Gao, Y., Li, D. and Liu, X. (2014). Bacteriocin-producing *Lactobacillus sakei* C2 as starter culture in fermented sausages. *Food Control* **35:** 1-6.

Garcia, M.L., Casas, C., Toledo, V.M. and Selgas, M.D. (2001). Effect of selected mould strains on the sensory properties of dry fermented sausages. *European Food Research and Technology* **212:** 287-291.

Garofalo, C., El-Khoury, M., Lucas, P., Bely, M., Russo, P., Spano, G. and Capozzi, V. (2015). Autochthonous starter cultures and indigenous grape variety for regional wine production. *Journal of Applied Microbiology* **118:** 1395-1408.

Ghabbour, N., Lamzira, Z., Thonart, P., Cidalia, P., Markaouid, M. and Asehraou, A. (2011). Selection of oleuropein-degrading lactic acid bacteria strains isolated from fermenting Moroccan green olives. *Grasas Y Aceites* **62:** 84-89.

Ghosh, K., Ray, M., Adak, A., Halder, S.K., Das, A., Jana, A., Parua Mondal, S., Vágvölgyi, C., Das Mohapatra, P.K., Pati, B.R. and Mondal, K.C. (2015). Role of probiotic *Lactobacillus fermentum* KKL1 in the preparation of a rice-based fermented beverage. *Bioresource Technology* **188:** 161-168.
Gobbetti, M., De Angelis, M., Di Cagno, R., Mancini, L. and Fox, P.F. (2015). Pros and cons for using non-starter lactic acid bacteria (NSLAB) as secondary/adjunct starters for cheese ripening. *Trends in Food Science & Technology* **45:** 167-178.
Gori, K., Ryssel, M., Arneborg, N. and Jespersen, L. (2013). Isolation and identification of the microbiota of Danish farmhouse and industrially produced surface-ripened cheeses. *Microbial Ecology* **65:** 602-615.
Granato, D., Branco, G.F., Cruz, A.G., Faria, J. de A.F. and Shah, N.P. (2010). Probiotic dairy products as functional foods. *Comprehensive Reviews in Food Science and Food Safety* **9:** 455-470.
Greppi, A., Ferrocino, I., La Storia, A., Rantsiou, K., Ercolini, D. and Cocolin, L. (2015). Monitoring of the microbiota of fermented sausages by culture independent rRNA-based approaches. *International Journal of Food Microbiology* **212:** 67-75.
Guzel-Seydim, Z., Kök-Tas, T. and Greene, A.K. (2010). Kefir and koumiss: Microbiology and technology. *In:* F. Yildiz (Ed.). *Development and Manufacture of Yogurt and Functional Dairy Products*. CRC Press, Boca Raton, U.S., pp. 143-163.
Hafeez, Z., Cakir-Kiefer, C., Roux, E., Perrin, C., Miclo, L. and Dary-Mourot, A. (2014). Strategies of producing bioactive peptides from milk proteins to functionalize fermented milk products. *Food Research International* **63:** 71-80.
Hammes, W.P. and Hertel, C. (2009). Genus I. *Lactobacillus Beijerinck* 1901, 212AL. *In:* P. De Vos, G.M. Garrity, D. Jones, N.R. Krieg, W. Ludwig, F.A. Rainey, K.H. Schleifer and W.B. Whitman (Eds.). *The Firmicutes. Bergey's Manual of Systematic Bacteriology*, 2nd edn, vol. 3. Springer, Dordrecht, Heidelberg London, New York, pp. 465-511.
Hao, Y., Zhao, L., Zhang, H., Zhai, Z., Huang, Y., Liu, X. and Zhang, L. (2010). Identification of the bacterial biodiversity in koumiss by denaturing gradient gel electrophoresis and species-specific polymerase chain reaction. *Journal Dairy Science* **93:** 1926-1933.
Heperkan, D., Daskaya-Dikmen, C. and Bayram, B. (2014). Evaluation of lactic acid bacterial strains of boza for their exopolysaccharide and enzyme production as a potential adjunct culture. *Process Biochemistry* **49:** 1587-1594.
Herranz, B., Fernández, M., de la Hoz, L. and Ordóñez, J.A. (2006). Use of bacterial extracts to enhance amino acid breakdown in dry fermented sausages. *Meat Science* **72:** 318-325.
Hrnjez, D., Vaštag, Ž., Milanović, S., Vukić, V., Iličić, M., Popović, L. and Kanurić, K. (2014). The biological activity of fermented dairy products obtained by kombucha and conventional starter cultures during storage. *Journal of Functional Foods* **10:** 336-345.
Hurtado, A., Reguant, C., Bordons, A. and Rozès, N. (2012). Lactic acid bacteria from fermented table olives. *Food Microbiology* **31:** 1-8.
Hutkins, R.W. (2006). *Microbiology and Technology of Fermented Foods*. Blackwell Publishing Ltd., U.S.
Hutkins, R.W. (2008). Cultured dairy products. *In: Microbiology and Technology of Fermented Foods*, Chapter 4. Wiley-Blackwell, Hoboken, New Jersey, USA.
Iacumin, L., Manzano, M. and Comi, G. (2012). Catalase-positive cocci in fermented sausage: Variability due to different pork breeds, breeding systems and sausage production technology. *Food Microbiology* **29:** 178-186.
Jerónimo, E. and Malcata, F.X. (2016). Cheese: Composition and health effects. *In:* B. Caballero, P. M. Finglas and F. Toldrá (Eds.). *Encyclopedia of Food and Health*. Academic Press, Oxford, pp. 741-747.
Jhan, J.-K., Chang, W.-F., Wang, P.-M., Chou, S.-T. and Chung, Y.-C. (2015). Production of fermented red beans with multiple bioactivities using co-cultures of *Bacillus subtilis* and *Lactobacillus delbrueckii* subsp. *bulgaricus*. *LWT— Food Science and Technology* **63:** 1281-1287.
Jofré, A., Aymerich, T. and Garriga, M. (2015). Probiotic Fermented Sausages: Myth or Reality? *Procedia Food Science* **5:** 133-136.
Johanningsmeier, S.D. and McFeeters, R.F. (2013). Metabolism of lactic acid in fermented cucumbers by *Lactobacillus buchneri* and related species, potential spoilage organisms in reduced salt fermentations. *Food Microbiology* **35:** 129-135.
Johanningsmeier, S., McFeeters, R.F., Fleming, H.P. and Thompson, R.L. (2007). Effect of *Leuconostoc mesenteroides* starter culture on fermentation of cabbage with reduced salt concentrations. *Journal of Food Science* **72:** 166-172.

Johnson, M.E. and Steele, J.L. (2013). Fermented Dairy Products. *In:* M.P. Doyle and R.L. Buchanan (Eds.) *Food Microbiology: Fundamentals and Frontiers*, 4th edn. ASM Press, Washington, D.C., pp. 825-839.

Jung, J.Y., Lee, S.H., Kim, J.M., Park, M.S., Bae, J.W., Hahn, Y., Madsen, E.L. and Jeon, C.O. (2011). Metagenomic analysis of *kimchi*, a traditional Korean fermented food. *Applied and Environmental Microbiology* **77:** 2264-2274.

Kabak, B. and Dobson, A.D.W. (2011). An introduction to the traditional fermented foods and beverages of Turkey. *Critical Reviews in Food Science and Nutrition* **51:** 248-260.

Kaban, G. (2013). Sucuk and pastırma: Microbiological changes and formation of volatile compounds. *Meat Science* **95:** 912-918.

Kaban, G. and Kaya, M. (2008). Identification of lactic acid bacteria and Gram-positive catalase-positive cocci isolated from naturally fermented sausage (*sucuk*). *Journal of Food Science* **73:** M385-M388.

Kalantzopoulos, G. (1997). Fermented products with probiotic quality. *Anaerobe* **3:** 185-190.

Kataoka, S. (2005). Functional effects of Japanese style fermented soy sauce (*shoyu*) and its components. *Journal of Bioscience and Bioengineering* **100:** 227-234.

Kawahara, M., Nemoto, M., Nakata, T., Kondo, S., Takahashi, H., Kimura, B. and Kuda, T. (2015). Anti-inflammatory properties of fermented soy milk with *Lactococcus lactis* subsp. lactis S-SU2 in murine macrophage RAW264.7 cells and DSS-induced IBD model mice. *International Immunopharmacology* **26:** 295-303.

Kesmen, Z., Yetiman, A.E., Gulluce, A., Kacmaz, N., Sagdıc, O., Cetin, B., Adiguzel, A., Sahin, F. and Yetim, H. (2012). Combination of culture-dependent and culture-independent molecular methods for the determination of lactic microbiota in sucuk. *International Journal of Food Microbiology* **153:** 428-435.

Khan, I. and Kang, S.C. (2016). Probiotic potential of nutritionally improved *Lactobacillus plantarum* DGK-17 isolated from Kimchi—A traditional Korean fermented food. *Food Control* **60:** 88-94.

Khan, M.I., Arshad, M.S., Anjum, F.M., Sameen, A., Rehman, A. and Gill, W.T. (2011). Meat as a functional food with special reference to probiotic sausages. *Food Research International* **44:** 3125-3133.

Kim, H.W., Heo, C., Han, D.-J., Kim, C.-J., Kim, K.-T., Park, B.-Y., Ahn, D. U. and Paik, H.-D. (2015). Predicting the Growth Kinetics of Total Microflora in *Kimchi* Powder—Treated Pork Snack Sticks. *Journal of Food Safety* **35:** 172-178.

Kimaryo, V.M., Massawe, G.A., Olasupo, N.A. and Holzapfel, W.H. (2000). The use of a starter culture in the fermentation of cassava for the production of 'kivunde', a traditional Tanzanian food product. *International Journal of Food Microbiology* **56:** 179-190.

Kocak, C. and Avsar, Y.K. (2009). Ayran: Microbiology and technology. *In:* F. Yildiz (Ed.). *Development and Manufacture of Yogurt and Functional Dairy Products*. CRC Press, Boca Raton, U.S., pp. 123-141.

Kocková, M. and Valík, Ľ. (2014). Development of new cereal-, pseudocereal- and cereal-leguminous-based probiotic foods. *Czech Journal of Food Sciences* **32:** 391-397.

Koh, J.H., Kim, Y. and Oh, H. (2010). Chemical characterization of tomato juice fermented with *Bifidobacteria*. *Journal of Food Science* **75:** C428-C432.

Kohajdova, Z., Karovicova, J. and Greifova, M. (2006). Lactic acid fermentation of some vegetable juices. *Journal of Food and Nutrition Research* **45:** 115-119.

Komprda, T., Sládková, P., Petirová, E., Dohnal, V. and Burdychová, R. (2010). Tyrosine- and histidine-decarboxylase positive lactic acid bacteria and enterococci in dry fermented sausages. *Meat Science* **86:** 870-877.

Kongo, J.M. and Malcata, F.X. (2016). Acidophilus Milk. *In:* B. Caballero, P. M. Finglas and F. Toldrá (Eds.). *Encyclopedia of Food and Health*. Academic Press, Oxford, pp. 6-14.

Kröckel, L. (2013). The role of lactic acid bacteria in safety and flavour development of meat and meat products. *In:* M. Kongo (Ed.). *Lactic Acid Bacteria—R & D for Food, Health and Livestock Purposes*. InTech, Croatia, pp. 129-152.

Kun, S., Rezessy-Szabo, J.M., Nguyen, Q.D. and Goston-Hoschke, A. (2008). Changes of microbial population and some components in carrot juice during fermentation with selected *Bifidobacterium* strains. *Process Biochemistry* **43:** 816-821.

Kwon, D.Y., Daily, J.W., Kim, H.J. and Park, S. (2010). Antidiabetic effects of fermented soybean products on type 2 diabetes. *Nutrition Research* **30:** 1-13.

Laiño, J.E., Juarez del Valle, M., Savoy de Giori, G. and LeBlanc, J.G.J. (2014). Applicability of a Lactobacillus amylovorus strain as co-culture for natural folate bio-enrichment of fermented milk. *International Journal of Food Microbiology* **191:** 10-16.

Laiño, J.E., Zelaya, H., Juárez del Valle, M., Savoy de Giori, G. and LeBlanc, J.G. (2015). Milk fermented with selected strains of lactic acid bacteria is able to improve folate status of deficient rodents and also prevent folate deficiency. *Journal of Functional Foods* **17:** 22-32.

Lan, W.-T., Chen, Y.-S. and Yanagida, F. (2009). Isolation and characterization of lactic acid bacteria from Yan-dong-gua (fermented wax gourd), a traditional fermented food in Taiwan. *Journal of Bioscience and Bioengineering* **108:** 484-487.

Landeta, G., Curiel, J.A., Carrascosa, A. V, Muñoz, R. and de las Rivas, B. (2013). Technological and safety properties of lactic acid bacteria isolated from Spanish dry-cured sausages. *Meat Science* **95:** 272-280.

Lee, B.-H., Lai, Y.-S. and Wu, S.-C. (2015). Antioxidation, angiotensin converting enzyme inhibition activity, nattokinase, and antihypertension of *Bacillus subtilis* (natto)-fermented pigeon pea. *Journal of Food and Drug Analysis* **23:** 750-757.

Lee, H., Yoon, H., Ji, Y., Kim, H., Park, H., Lee, J., Shin, H. and Holzapfel, W. (2011). Functional properties of *Lactobacillus* strains isolated from *kimchi*. *International Journal of Food Microbiology* **145:** 155-161.

Leroy, F. and De Vuyst, L. (2004). Lactic acid bacteria as functional starter cultures for the food fermentation industry. *Trends in Food Science & Technology* **15:** 67-78.

Leroy, F., Geyzen, A., Janssens, M., De Vuyst, L. and Scholliers, P. (2013). Meat fermentation at the crossroads of innovation and tradition: A historical outlook. *Trends in Food Science & Technology* **31:** 130-137.

Leroy, F., Verluyten, J. and De Vuyst, L. (2006). Functional meat starter cultures for improved sausage fermentation. *International Journal of Food Microbiology* **106:** 270-285.

Leuschner, R.G.K. and Hammes, W.P. (1998). Tyramine degradation by micrococci during ripening of fermented sausages. *Meat Science* **49:** 289-296.

Li, X.K., Li, K.X. and Zou, S.D. (2006). Koumiss. *Dairy Industry* **7:** 58-60.

Liu, S.N., Han, Y. and Zhou, Z.J. (2011). Lactic acid bacteria in traditional fermented Chinese foods. *Food Research International* **44:** 643-651.

Lore, T.A., Mbugua, S.K. and Wangoh, J. (2005). Enumeration and identification of microflora in suusac, a Kenyan traditional fermented camel milk product. *LWT—Food Science and Technology* **38:** 125-130.

Machve, K.K. (2009). Fermentation by microorganisms. *In: Fermentation Technology*. Mangalam Publishers, Delhi, India, pp. 65-117.

Magalhaes, K.T., Pereira, G.V.M., Campos, C.R., Dragone, G. and Schwan, R.F. (2011). Brazilian kefir: Structure, microbial communities and chemical composition. *Brazilian Journal of Microbiology* **42:** 693-702.

Marsh, A.J., Hill, C., Ross, R.P. and Cotter, P.D. (2014). Fermented beverages with health-promoting potential: Past and future perspectives. *Trends in Food Science & Technology* **38:** 113-124.

Marty, E., Buchs, J., Eugster-Meier, E., Lacroix, C. and Meile, L. (2012). Identification of staphylococci and dominant lactic acid bacteria in spontaneously fermented Swiss meat products using PCR-RFLP. *Food Microbiology* **29:** 157-166.

Matsumoto, K., Takada, T., Shimizu, K., Moriyama, K., Kawakami, K., Hirano, K., Kajimoto, O. and Nomoto, K. (2010). Effects of a probiotic fermented milk beverage containing *Lactobacillus casei* strain Shirota on defecation frequency, intestinal microbiota, and the intestinal environment of healthy individuals with soft stools. *Journal of Bioscience and Bioengineering* **110:** 547-552.

Medina, E., García, A., Romero, C., de Castro, A. and Brenes, M. (2009). Study of the anti-lactic acid bacteria compounds in table olives. *International Journal of Food Science and Technology* **44:** 1286-1291.

Mendoza, L.M., Padilla, B., Belloch, C. and Vignolo, G. (2014). Diversity and enzymatic profile of yeasts isolated from traditional llama meat sausages from north-western Andean region of Argentina. *Food Research International* **62:** 572-579.

Montel, M.-C., Buchin, S., Mallet, A., Delbes-Paus, C., Vuitton, D.A., Desmasures, N. and Berthier, F. (2014). Traditional cheeses: Rich and diverse microbiota with associated benefits. *International Journal of Food Microbiology* **177:** 136-154.

Mousavi, Z.E., Mousavi, S.M., Razavi, S.H., Eman-Djomeh, Z. and Kiani, H. (2011). Fermentation of pomegranate juice by probiotic lactic acid bacteria. *World Journal of Microbiology and Biotechnology* **27:** 123-128.

Mpofu, A., Linnemann, A.R., Nout, M.J.R., Zwietering, M.H., Smid, E.J. and den Besten, H.M.W. (2016). Inactivation of bacterial pathogens in yoba mutandabota, a dairy product fermented with the probiotic *Lactobacillus rhamnosus* yoba. *International Journal of Food Microbiology* **217:** 42-48.

Murata, D., Sawano, S., Ohike, T., Okanami, M. and Ano, T. (2013). Isolation of antifungal bacteria from Japanese fermented soybeans, *natto*. *Journal of Environmental Sciences* 25 Suppl **1:** S127-31.

Ng, E.W., Yeung, M. and Tong, P.S. (2011). Effects of yogurt starter cultures on the survival of *Lactobacillus acidophilus*. *International Journal of Food Microbiology* **145:** 169-175.

Nguyen, D.T.L., Van Hoorde, K., Cnockaert, M., De Brandt, E., De Bruyne, K., Le, B.T. and Vandamme, P. (2013). A culture-dependent and -independent approach for the identification of lactic acid bacteria associated with the production of nem chua, a Vietnamese fermented meat product. *Food Research International* **50:** 232-240.

O'Sullivan, D.J., McSweeney, P.L.H., Cotter, P.D., Giblin, L. and Sheehan, J.J. (2016). Compromised *Lactobacillus helveticus* starter activity in the presence of facultative heterofermentative *Lactobacillus casei* DPC6987 results in atypical eye formation in Swiss-type cheese. *Journal of Dairy Science* **99:** 2625-2640.

Oguntoyinbo, F.A., Tourlomousis, P., Gasson, M.J. and Narbad, A. (2011). Analysis of bacterial communities of traditional fermented West African cereal foods using culture independent methods. *International Journal of Food Microbiology* **145:** 205-210.

Oh, Y.J. and Jung, D.S. (2015). Evaluation of probiotic properties of *Lactobacillus* and *Pediococcus* strains isolated from Omegisool, a traditionally fermented millet alcoholic beverage in Korea. *LWT—Food Science and Technology* **63:** 437-444.

Olszewska, M.A., Kocot, A.M. and Łaniewska-Trokenheim, Ł. (2015). Physiological functions at single-cell level of *Lactobacillus* spp. isolated from traditionally fermented cabbage in response to different pH conditions. *Journal of Biotechnology* **200:** 19-26.

Ono, H., Nishio, S., Tsurii, J., Kawamoto, T., Sonomoto, K. and Nakayama, J. (2014). Monitoring of the microbiota profile in *nukadoko*, a naturally fermented rice bran bed for pickling vegetables. *Journal of Bioscience and Bioengineering* **118:** 520-525.

Ortu, S., Felis, G.E., Marzotto, M., Deriu, A., Molicotti, P., Sechi, L.A., Dellaglio, F. and Zanetti, S. (2007). Identification and functional characterization of *Lactobacillus* strains isolated from milk and Gioddu, a traditional Sardinian fermented milk. *International Dairy Journal* **17:** 1312-1320.

Osimani, A., Garofalo, C., Aquilanti, L., Milanović, V. and Clementi, F. (2015). Unpasteurised commercial boza as a source of microbial diversity. *International Journal of Food Microbiology* **194:** 62-70.

Osorio-Cadavid, E., Chaves-Lopez, C., Tofalo, R., Paparella, A. and Suzzi G. (2008). Detection and identification of wild yeasts in Champùs, a fermented Colombian maize beverage. *Food Microbiology* **25:** 771-777.

Ouadghiri, M., Amar, M., Vancanneyt, M. and Swings, J. (2005). Biodiversity of lactic acid bacteria in Moroccan soft white cheese (Jben). *FEMS Microbiology Letters* **251:** 267-271.

Ouoba, L.I.I., Kando, C., Parkouda, C., Sawadogo-Lingani, H., Diawara, B. and Sutherland, J.P. (2012). The microbiology of Bandji, palm wine of Borassus akeassii from Burkina Faso: Identification and genotypic diversity of yeasts, lactic acid and acetic acid bacteria. *Journal of Applied Microbiology* **113:** 1428-1441.

Park, K.Y., Jeong, J.K., Lee, Y.E. and Daily, J.W. (2014). Health benefits of kimchi (Korean fermented vegetables) as a probiotic food. *Journal of Medicinal Food* **17:** 6-20.

Parkouda, C., Thorsen, L., Compaoré, C.S., Nielsen, D.S., Tano-Debrah, K., Jensen, J.S., Diawara, B. and Jakobsen, M. (2010). Microorganisms associated with Maari, a Baobab seed fermented product. *International Journal of Food Microbiology* **142:** 292-301.

Patrignani, F., Iucci, L., Vallicelli, M., Guerzoni, M. E., Gardini, F. and Lanciotti, R. (2007). Role of surface-inoculated *Debaryomyces hansenii* and *Yarrowia lipolytica* strains in dried fermented sausage manufacture. Part 1: Evaluation of their effects on microbial evolution, lipolytic and proteolytic patterns. *Meat Science* **75:** 676-686.

Pecanac, B., Djordjevic, J., Baltic, M.Z., Djordjevic, V., Nedic, D.N., Starcevic, M., Dojcinovic, S. and Baltic, T. (2015). Comparison of bacteriological status during ripening of traditional fermented sausages filled into different diameter artificial casings. *Procedia Food Science* **5:** 223-226.

Peres, C.M., Peres, C., Hernández-Mendoza, A. and Malcata, F.X. (2012). Review on fermented plant materials as carriers and sources of potentially probiotic lactic acid bacteria—With an emphasis on table olives. *Trends in Food Science & Technology* **26:** 31-42.

Pérez-Pulido, R., Ben Omar, N., Abriouel, H., Lucas López, R., Martínez Cañamero, M., Guyot, J.P. and Gálvez, A. (2007). Characterization of lactobacilli isolated from caper berry fermentations. *Journal of Applied Microbiology* **102:** 583-590.

Plengvidhya, V., Breidt, F., Lu, Z. and Fleming, H. P. (2007). DNA fingerprinting of lactic acid bacteria in sauerkraut fermentations. *Applied and Environmental Microbiology* **73:** 7697-7702.

Plessas, S., Alexopoulos, A., Voidarou, C., Stavropoulou, E. and Bezirtzoglou, E. (2011). Microbial ecology and quality assurance in food fermentation systems. The case of kefir grains application. *Anaerobe* **17:** 483-485.

Porcellato, D., Østlie, H. M., Brede, M. E., Martinovic, A. and Skeie, S. B. (2013). Dynamics of starter, adjunct non-starter lactic acid bacteria and propionic acid bacteria in low-fat and full-fat Dutch-type cheese. *International Dairy Journal* **33:** 104-111.

Prado, M.R., Blandón, L.M., Vandenberghe, L.P.S., Rodrigues, C., Castro G.R., Thomaz-Soccol, V. and Soccol, C.R. (2015). Milk kefir: Composition, microbial cultures, biological activities and related products. *Frontiers in Microbiology* **6:** 1177.

Prakash, S., Tomaro-Duchesneau, C., Saha, S. and Cantor, A. (2011). The gut microbiota and human health with an emphasis on the use of microencapsulated bacterial cells. *Journal of Biomedical Biotechnology* **98:** 12-14.

Pringsulaka, O., Thongngam, N., Suwannasai, N., Atthakor, W., Pothivejkul, K. and Rangsiruji, A. (2012). Partial characterisation of bacteriocins produced by lactic acid bacteria isolated from Thai fermented meat and fish products. *Food Control* **23:** 547-551.

Prückler, M., Lorenz, C., Endo, A., Kraler, M., Dürrschmid, K., Hendriks, K., Soares da Silva, F., Auterith, E., Kneifel, W. and Michlmayr, H. (2015). Comparison of homo- and heterofermentative lactic acid bacteria for implementation of fermented wheat bran in bread. *Food Microbiology* **49:** 211-219.

Puerari, C., Magalhães-Guedes, K.T. and Schwan, R.F. (2015). Physicochemical and microbiological characterization of *chicha*, a rice-based fermented beverage produced by Umutina Brazilian Amerindians. *Food Microbiology* **46:** 210-217.

Rage, A. (1993). North European varieties of cheese. IV. Norwegian cheese varieties. *In:* P.F. Fox (Ed.). *Cheese: Chemistry, Physics and Microbiology. Major Cheese Groups*, 2nd edn, vol. 2. London: Chapman and Hall, pp. 257-260.

Randazzo, C., Restuccia, C., Romano, A.D. and Caggia, C. (2004). *Lactobacillus casei*, dominant species in naturally fermented Sicilian green olives. *International Journal of Food Microbiology* **90:** 9-14.

Rantsiou, K. and Cocolin, L. (2006). New developments in the study of the microbiota of naturally fermented sausages as determined by molecular methods: A review. *International Journal of Food Microbiology* **108:** 255-267.

Rantsiou, K., Drosinos, E.H., Gialitaki, M., Urso, R., Krommer, J., Gasparik-Reichardt, J., Tóth, S., Metaxopoulos, I., Comi, G. and Cocolin, L. (2005). Molecular characterization of *Lactobacillus* species isolated from naturally fermented sausages produced in Greece, Hungary and Italy. *Food Microbiology* **22:** 19-28.

Rattray, F.P. and Fox, P.F. (1999). Aspects of enzymology and biochemical properties of *Brevibacterium linens* relevant to cheese ripening: A review. *Journal of Dairy Science* **82:** 891-909.

Reina, L.D., Pérez-Díaz, I.M., Breidt, F., Azcarate-Peril, M.A., Medina, E. and Butz, N. (2015). Characterization of the microbial diversity in yacon spontaneous fermentation at 20°C. *International Journal of Food Microbiology* **203:** 35-40.

Rhee, S., Lee, J.-E. and Lee, C.-H. (2011). Importance of lactic acid bacteria in Asian fermented foods. *Microbial Cell Factories* **10:** S5.

Ricke, S.C., Koo, O.K. and Keeton, J.T. (2013). Fermented meat, poultry and fish products. *In*: M.P. Doyle and R.L. Buchanan (Eds.). *Food Microbiology: Fundamentals and Frontiers*, 4th edn. ASM Press, Washington, D.C., pp. 857-880.

Rodríguez, H., Curiel, J.A., Landete, J.M., de las Rivas, B., López de Felipe, F., Gómez-Cordovés, C., Mancheño, J.M. and Muñoz, R. (2009). Food phenolics and lactic acid bacteria. *International Journal of Food Microbiology* **132:** 79-90.

Rosales-Soto, M.U., Gray, P.M., Fellman, J.K., Mattinson, D.S., Ünlü, G., Huber, K. and Powers, J.R. (2016). Microbiological and physico-chemical analysis of fermented protein-fortified cassava (*Manihot esculenta* Crantz) flour. *LWT—Food Science and Technology* **66:** 355-360.

Rubio, R., Jofré, A., Martín, B., Aymerich, T. and Garriga, M. (2014). Characterization of lactic acid bacteria isolated from infant faeces as potential probiotic starter cultures for fermented sausages. *Food Microbiology* **38:** 303-311.

Ruiz-Moyano, S., Martín, A., Benito, M.J., Hernández, A., Casquete, R. and de Guia Córdoba, M. (2011). Application of *Lactobacillus fermentum* HL57 and *Pediococcus acidilactici* SP979 as potential probiotics in the manufacture of traditional Iberian dry-fermented sausages. *Food Microbiology* **28:** 839-847.

Ruiz, P., Izquierdo, P.M., Sesena, S. and Palop, M.L. (2010). Selection of autochthonous *Oenococcus oeni* strains according to their oenological properties and vinification results. *International Journal of Food Microbiology* **137:** 230-235.

Sah, B.N.P., Vasiljevic, T., McKechnie, S. and Donkor, O.N. (2016). Physicochemical, textural and rheological properties of probiotic yogurt fortified with fiber-rich pineapple peel powder during refrigerated storage. *LWT—Food Science and Technology* **65:** 978-986.

Salmerón, I., Thomas, K. and Pandiella, S.S. (2015). Effect of potentially probiotic lactic acid bacteria on the physicochemical composition and acceptance of fermented cereal beverages. *Journal of Functional Foods* **15:** 106-115.

Santos, A., San Mauro, M., Sanchez, A., Torres, J.M. and Marquina, D. (2003). The antimicrobial properties of different strains of *Lactobacillus* spp. isolated from kefir. *Systematic and Applied Microbiology* **26:** 434-437.

Sawadogo-Lingani, H., Lei, V., Diawara, B., Nielsen, D.S., Møller, P.L., Traoré, A.S. and Jakobsen, M. (2007). The biodiversity of predominant lactic acid bacteria in dolo and pito wort for the production of sorghum beer. *Journal of Applied Microbiology* **103:** 765-777.

Sengun, I.Y., Nielsen, D.S., Karapinar, M. and Jakobsen, M. (2009). Identification of lactic acid bacteria isolated from *Tarhana*, a traditional Turkish fermented food. *International Journal of Food Microbiology* **135:** 105-111.

Sengun, I.Y., Yaman, D.B. and Gonul, S.A. (2008). Mycotoxins and mould contamination in cheese. *World Mycotoxin Journal* **1:** 291-298.

Serafini, F., Turroni, F., Ruas-Madiedo, P., Lugli, G.A., Milani, C., Duranti, S., Zamboni, N., Bottacini, F., van Sinderen, D., Margolles, A. and Ventura, M. (2014). Kefir fermented milk and *kefiran* promote growth of Bifidobacterium bifidum PRL2010 and modulate its gene expression. *International Journal of Food Microbiology* **178:** 50-59.

Serrazanetti, D.I., Ndagijimana, M., Miserocchi, C., Perillo, L. and Guerzoni, M.E. (2013). Fermented tofu: Enhancement of keeping quality and sensorial properties. *Food Control* **34:** 336-346.

Shiby, V.K. and Mishra, H.N. (2013). Fermented milks and milk products as functional foods—A review. *Critical Reviews in Food Science and Nutrition* **53:** 482-496.

Sidira, M., Karapetsas, A., Galanis, A., Kanellaki, M. and Kourkoutas, Y. (2014). Effective survival of immobilized *Lactobacillus casei* during ripening and heat treatment of probiotic dry-fermented sausages and investigation of the microbial dynamics. *Meat Science* **96:** 948-955.

Silva, M.R., Dias, G., Ferreira, C.L.L.F., Franceschini, S.C.C. and Costa, N.M.B. (2008). Growth of preschool children was improved when fed an iron-fortified fermented milk beverage supplemented with *Lactobacillus acidophilus*. *Nutrition Research* **28:** 226-232.

Singh, A.K. and Ramesh, A. (2008). Succession of dominant and antagonistic lactic acid bacteria in fermented cucumber: Insights from a PCR-based approach. *Food Microbiology* **25:** 278-287.

Singh, T.A., Devi, K.R., Ahmed, G. and Jeyaram, K. (2014). Microbial and endogenous origin of fibrinolytic activity in traditional fermented foods of Northeast India. *Food Research International* **55:** 356-362.

Sun, S.Y., Gong, H. S., Liu, W.L. and Jin, C.W. (2016). Application and validation of autochthonous Lactobacillus plantarum starter cultures for controlled malolactic fermentation and its influence on the aromatic profile of cherry wines. *Food Microbiology* **55:** 16-24.

Tabanelli, G., Coloretti, F., Chiavari, C., Grazia, L., Lanciotti, R. and Gardini, F. (2012). Effects of starter cultures and fermentation climate on the properties of two types of typical Italian dry fermented sausages produced under industrial conditions. *Food Control* **26:** 416-426.

Talon, R. and Leroy, S. (2011). Diversity and safety hazards of bacteria involved in meat fermentations. *Meat Science* **89:** 303-309.

Tamang, J.P. (2015). Naturally fermented ethnic soybean foods of India. *Journal of Ethnic Foods* **2:** 8-17.

Tamime, A. Y., Wszolek, M., Božanić, R. and Özer, B. (2011). Popular ovine and caprine fermented milks. *Small Ruminant Research* **101:** 2-16.

Tamminen, M., Joutsjoki, T., Sjöblom, M., Joutsen, M., Palva, A. and Ryhänen, E.L. (2004). Screening of lactic acid bacteria from fermented vegetables by carbohydrate profiling and PCR-ELISA. *Applied Microbiology* **39:** 439-444.

Tanguler, H. and Erten, H. (2012). Occurrence and growth of lactic acid bacteria species during the fermentation of *shalgam* (salgam), a traditional Turkish fermented beverage. *LWT—Food Science and Technology* **46:** 36-41.

Thakur, N., Saris, P.E.J. and Bhalla, T.C. (2015). Microorganisms associated with amylolytic starters and traditional fermented alcoholic beverages of north-western Himalayas in India. *Food Bioscience* **11:** 92-96.

Thierry, A., Deutsch, S.-M., Falentin, H., Dalmasso, M., Cousin, F.J. and Jan, G. (2011). New insights into physiology and metabolism of *Propionibacterium freudenreichii. International Journal of Food Microbiology* **149:** 19-27.

Tichaczek, P.S., Vogel R.F. and Hammes, W.P. (1994). Cloning and sequencing of sakP encoding sakacin P, the bacteriocin produced by *Lactobacillus sake* LTH 673. *Microbiology* **140:** 361-367.

Tofalo, R., Schirone, M., Perpetuini, G., Suzzi, G. and Corsetti, A. (2012). Development and application of a real-time PCR-based assay to enumerate total yeasts and *Pichia anomala, Pichia guillermondii* and *Pichia kluyveri* in fermented table olives. *Food Control* **23:** 356-362.

Toldrá, F. and Wai-Kit, N. (2008). Flavor development. *In: Dry-Cured Meat Products*, Chapter 8. Wiley-Blackwell, United States.

Tosukhowong, A., Visessanguan, W., Pumpuang, L., Tepkasikul, P., Panya, A. and Valyasevi, R. (2011). Biogenic amine formation in Nham, a Thai fermented sausage, and the reduction by commercial starter culture, *Lactobacillus plantarum* BCC 9546. *Food Chemistry* **129:** 846-853.

Trząskowska, M., Kołożyn-Krajewska, D., Wójciak, K. and Dolatowski, Z. (2014). Microbiological quality of raw-fermented sausages with *Lactobacillus casei* LOCK 0900 probiotic strain. *Food Control* **35:** 184-191.

Tsai, T.-Y., Chen, L.-Y. and Pan, T.-M. (2014). Effect of probiotic-fermented, genetically modified soy milk on hypercholesterolemia in hamsters. *Journal of Microbiology, Immunology, and Infection = Wei Mian Yu Gan Ran Za Zhi* **47:** 1-8.

Urso, R., Comi, G. and Cocolin, L. (2006). Ecology of lactic acid bacteria in Italian fermented sausages: Isolation, identification and molecular characterization. *Systematic and Applied Microbiology* **29:** 671-680.

Vasilev, D., Aleksic, B., Tarbuk, A., Dimitrijevic, M., Karabasil, N., Cobanovic, N. and Vasiljevic, N. (2015). Identification of lactic acid bacteria isolated from Serbian traditional fermented sausages Sremski and Lemeski Kulen. *Procedia Food Science* **5:** 300-303.

Vignolo, G., Fontana, C. and Fadda, S. (2010). Semidry and dry fermented sausages. *In*: F. Toldrá (Ed.). *Handbook of Meat Processing*. Blackwell Publishing, USA, pp. 379-398.

Visintin, S., Alessandria, V., Valente, A., Dolci, P. and Cocolin, L. (2015). Molecular identification and physiological characterization of yeasts, lactic acid bacteria and acetic acid bacteria isolated from heap and box cocoa bean fermentations in west Africa. *International Journal of Food Microbiology* **216:** 69-78.

Vuyst, L. De, Falony, G. and Leroy, F. (2008). Probiotics in fermented sausages. *Meat Science* **80:** 75-78.

Wanangkarn, A., Liu, D.-C., Swetwiwathana, A., Jindaprasert, A., Phraephaisarn, C., Chumnqoen, W. and Tan, F.-J. (2014). Lactic acid bacterial population dynamics during fermentation and storage of Thai fermented sausage according to restriction fragment length polymorphism analysis. *International Journal of Food Microbiology* **186:** 61-67.

Wang, G., Xiong, Y., Xu, Q., Yin, J. and Hao, Y. (2015). Complete genome sequence of *Lactobacillus paracasei* CAUH35, a new strain isolated from traditional fermented dairy product koumiss in China. *Journal of Biotechnology* **214:** 75-76.

Wang, S., Zhu, H., Lu, C., Kang, Z., Luo, Y., Feng, L. and Lu, X. (2012). Fermented milk supplemented with probiotics and prebiotics can effectively alter the intestinal microbiota and immunity of host animals. *Journal of Dairy Science* **95:** 4813-4822.

Watanabe, K., Fujimoto, J., Sasamoto, M., Dugersuren, J., Tumursuh, T. and Demberel S. (2008). Diversity of lactic acid bacteria and yeasts in *airag* and *tarag*, traditional fermented milk products of Mongolia. *World Journal of Microbiology and Biotechnology* **24:** 1313-1325.

Wiander, B. and Ryhänan, E.L. (2005). Laboratory and large-scale fermentation of white cabbage into sauerkraut and sauerkraut juice by using starters in combination with mineral salt with low NaCl content. *European Food Research and Technology* **220:** 191-195.

Wouters, J.T., Ayad, E.H., Hugenholtz, J. and Smit, G. (2002). Microbes from raw milk for fermented dairy products. *International Dairy Journal* **12:** 91-109.

Wszolek, M., Kupiec-Teahan, B., Skov Gulard, H. and Tamime, A.Y. (2006). Production of *kefir, koumiss* and other related products. *In:* A.Y. Tamime (Ed.). *Fermented Milks.* Blackwell Publishing, Oxford, United Kingdom, pp. 174-216.

Wu, Q., Cheung, C.K.W. and Shah, N.P. (2015). Towards galactose accumulation in dairy foods fermented by conventional starter cultures: Challenges and strategies. *Trends in Food Science & Technology* **41:** 24-36.

Yegin, S. and Üren, A. (2008). Biogenic amine content of boza: A traditional cereal-based, fermented Turkish beverage. *Food Chemistry* **111:** 983-987.

Yıldız-Turp, G. and Serdaroğlu, M. (2008). Effect of replacing beef fat with hazelnut oil on quality characteristics of *sucuk*—A Turkish fermented sausage. *Meat Science* **78:** 447-454.

Yoon, Y., Lee, S. and Choi, K.H. (2016). Microbial benefits and risks of raw milk cheese. *Food Control* **63:** 201-215.

Zorba, M., Hancioglu, O., Genc, M., Karapınar, M. and Ova, G. (2003). The use of starter cultures in the fermentation of boza, a traditional Turkish beverage. *Process Biochemistry* **38:** 1405-1411.

10

Vinegars

Paolo Giudici*, Luciana De Vero and Maria Gullo
Unimore Microbial Culture Collection (UMCC), Department of Life Sciences,
University of Modena and Reggio Emilia, Via Amendola 2,
42122, Reggio Emilia, Italy

Introduction

The history of fermentation is strictly linked to vinegar production. Among fermented foods, however, vinegar is recognized as a poor product; in many cases, the raw materials have higher nutritional values than the final product. Vinegar is generally used as a food flavoring, a preservative and a beverage with healthy properties. Acetic acid has several uses in the pharmaceutical, cosmetic, chemical and textile industries, where it is produced via chemical synthesis.

Vinegar is the product of double scalar fermentation, which is performed by yeasts and acetic acid bacteria (AAB) from sources of fermentable carbohydrates, such as apples, pears, grape, honey, sap of plants, cereals and hydrolyzed starch. In addition, distilled ethanol from fermented raw materials can be used to produce vinegar. In the latter case, vinegar is referred to as 'spirit vinegar' or 'white vinegar'; its use is widespread in northern Europe and the Americas. In some countries, vinegar is also produced from acetic acid via wood pyrolysis; however, this practice is less common, as the fermentation of agricultural raw material is a priority for international organizations, such as the Food and Agriculture Organization (FAO), the World Health Organization (WHO), the European Union (EU) and countries that engage in sustainable production.

A separate category is flavored vinegars, which differ from fruit vinegars. Fruit vinegars are obtained through double fermentation of fruit juices and referred to as the 'vinegar of…' followed by the name of the initial raw material. Flavored vinegars are produced from spirit vinegar (in countries where its production is permitted) or wine that has been flavored with herbs and/or sweeteners and fruit juices to obtain a sweet and sour taste. Balsamic vinegar of Modena (protect geographic indication, PGI) is flavored and sweetened with concentrated grape must and caramel. Traditional balsamic vinegar of Modena (protected designation of origin, PDO) and traditional balsamic vinegar of Reggio Emilia (PDO) are produced via a complex procedure—the addition of any ingredient is forbidden.

The critical steps in vinegar production include raw material preparation and fermentation. Fruits generally require fewer treatments than cereals. Fruits are perishable and abundant in water; their transportation and preservation are difficult. As a result, industries should be located near a raw material production site. Conversely, cereals are easily stored and preserved, which is beneficial for industries that are located far from the production site of raw materials.

*Corresponding author: paolo.giudici@unimore.it

In developing countries, vinegar production is an acceptable method for preserving vegetables and highly perishable fruit via fermentation, especially in countries with hot and humid weather, when decomposition processes are fast. Vinegar production from local agricultural sources has economic benefits, such as the promotion of local human resources and raw materials. For these reasons, projects that involve the development of fermentation technologies and vinegar production are an objective of the FAO.

Vinegar Definition

Vinegar production is regulated by an extensive set of statutes and product definitions, which vary among countries. FAO/WHO defines vinegar as any liquid that is fit for human consumption and is exclusively produced from suitable products that contain starch and/or sugars through the process of double fermentation—alcoholic fermentation followed by acetous fermentation. The maximum residual ethanol content in wine vinegar and other vinegars must be 1.5 per cent and 1 per cent, respectively.

In the US, the Food and Drug Administration (FDA) requires that vinegar products contain a minimum acidity of 4 per cent. This qualification ensures the minimum strength of vinegars that are sold at the retail level. No standards of identity have been established for vinegar; however, the FDA has established *Compliance Policy Guides* regarding the labelling of vinegars, such as cider, wine, malt, sugar and spirit and vinegar blends (FDA 1995).

European countries have regional standards for vinegar that is produced or sold in these countries. In contrast to US law, the EU established a threshold for acidity and ethanol content. 'Vinegar of X' is a general definition for products with a minimum acidity of 5 per cent (w/v) and a maximum ethanol content of 0.5 per cent (v/v). Wine vinegar is exclusively obtained by acetous fermentation of wine and has a minimum acidity of 6 per cent (w/v) and a minimum ethanol content of 1.5 per cent (v/v) (EC 1999).

The Chinese National Standard definitions (CNS 2005) use the word 'vinegar' to denote both fermented and artificial vinegars. In the former National Industrial Standard of vinegar, vinegar was classified into three grades based on the concentration of acetic acid (3.5-4.5 per cent, 4.5-6 per cent and >6 per cent). Recently, a New National Standard Code of Condiments was issued by the Chinese State Administration Bureau for Quality and Technology, in which the definition of vinegar has been introduced and vinegars are classified as either brewed or artificial (acetic acid blended with other ingredients, such as flavors). Each major vinegar has a separate local quality criterion and grading system.

Vinegar Grouping

Many different kinds of vinegars are produced worldwide that differ for their raw material, production system and use. In Table 1 a list of vinegars is provided; this list is not considered to be exhaustive due to the numerous types of emerging vinegars.

Table 1: Vinegars produced worldwide

Raw material	*Intermediate*	*Vinegar name*	*Geographical diffusion*
[a]Cereal (rice, millet, bran, wheat, sorghum, barley), peas and soy	Pei/Moromi	Komesu, Kurosu (in Japanese) Heicu (in Chinese)	East and Southeast Asia
		Cereal vinegar	China, East Asia
	Malt/Beer	Malt/beer vinegar	North and Central Europe, United States

(*Contd.*)

Palm sap	Palm wine (toddy, tari, tuack, tuba, etc.)	Palm vinegar, toddy vinegar	Southeast Asian, Africa
Bamboo sap	Fermented bamboo sap	Bamboo vinegar	Japan, Korea
Tea and sugar	Kombucha	Kombucha vinegar	Russia, Asia (China, Japan, Indonesia)
Onion	Onion alcohol	Onion vinegar	East and Southeast Asia
Sugarcane	Fermented sugar cane juice	Cane vinegar	France, United States
	Basi	Sukang iloko	Philippines
	Fermented sugar cane juice	Kibizu	Japan
Apple	Cider	Cider/Apple vinegar	United States, Canada
Grape	Raisin wine	Raisin vinegar	Turkey, Middle East
	Red or white wine	Vinegar; wine vinegar	Widespread
	Sherry wine	Sherry (Jerez) vinegar	Spain
	Wine (partially fermented cooked must)	Traditional balsamic vinegar	Italy
	Blend of wine vinegar and concentrated/cooked must	Balsamic vinegar	Widespread
Coconut	Coconut water wine	Coconut water vinegar	Philippines, Sri Lanka
Date	Dates wine	Date vinegar	Middle East
Red Date	Jujube wine	Jujube vinegar	China
Berry (raspberry, blackcurrant, blackberry, mulberry and cranberry)	'Berry' wine	Vinegar of followed by 'berry name'	East and Southeast Asia
Plum	[b]Umeboshi/ plum wine	Ume-su Plum vinegar	Japan Widespread
Kaki	Persimmon wine	Persimmon vinegar	South Korea
		Kakisu	Japan
Whey	Fermented whey	Whey vinegar	Europe
Honey	Hydromel/ mead		Europe, America, Africa

[a]Cereals are processed individually or mixed.
[b]Umeboshi are pickled ume fruits. Ume is a species of fruit-bearing tree of the genus Prunus, which is often called a plum, but is actually more closely related to the apricot. Adapted from Giudici et al. 2006, Solieri and Giudici 2009.

The majority of vinegars have vegetal origin with two exceptions—whey and honey. Whey is the milk serum residual of cheesemaking; it is rich in lactose and/or in the corresponding hydrolyzed sugars, galactose and glucose, depending on the cheese-making technology that is employed. Sour whey is substantially contaminated by lactic acid bacteria (LAB) and needs to be pasteurized prior to alcoholic fermentation. Honey is extremely abundant in sugars (70-80 per cent w/w)—predominantly sucrose, fructose and glucose. The ratio is influenced by the botanic origin of the nectar that is collected by the bees. Honey is always diluted prior to alcoholic fermentation and the maximum ethanol content in honey wine is 17 per cent (v/v) (Steinkraus 1996). This alcoholic beverage is well known all over the world with different names, such as mead, ambrosia, metheglin, hydromel, aguamiel, medovukha and ogol; it is a possible base for the production of vinegar.

The use of vinegar in Far East regions is more common than in Western countries. In the Far Eastern countries, a multitude of vinegars differing in raw material and manufacturing are produced, especially in China, where due to its vast population, a multitude of unique local traditions exist (Table 2).

Table 2: Production area and names of main Chinese vinegars

Chinese Name	*English name*	*Production area*
山西陈醋	Shanxi mature vinegar	Shanxi Province
镇江香醋	Zhenjiang aromatic vinegar	Zhenjiang, Jiangsu Province
保宁药醋	Baoning herbal vinegar	Langzhong, Sichuan Province
永春红曲老醋	Yongchun monascus aged vinegar	Yongchun, Fujian Province
浙江玫瑰醋	Zhejiang rose vinegar	Hangjia Lake area, Zhejiang Province
凉州熏醋	Liangzhou fumigated vinegar	Wuwei, Gansu Province
高粱熏醋	Sorghum fumigated vinegar	Shanxi Province
太平米醋	Taiping millet vinegar	Taiping country, Shanxi Province
苦荞麦保健醋	Bitter Buckwheat healthy vinegar	Shanxi Province
运城柿子醋	Yuncheng persimmon vinegar	Yuncheng, Shanxi Province
台湾凤梨醋	Taiwan pineapple vinegar	Taiwan Province
葡萄果醋	Grape vinegar	Hebei Province
南六堡大曲醋	Nanliubao Daqu vinegar	Nanliubao, Shanxi Province
白醋	White vinegar	"

Rice vinegars are traditionally divided into two categories—the first category is clear (amber) vinegar, which is consumed for daily culinary uses; the second category is black vinegar, which is primarily used as a health drink or medicament. Many other types of vinegar belong to these two main categories. Recently, the classification of rice vinegars has been complicated by the proliferation of many new commercial names for vinegars.

In the Chinese market, more than 20 types of homemade vinegars are sold—the majority of these vinegars are brewed with starchy materials such as rice, sorghum, corn, barley and wheat; however, some fruit vinegars and fruit vinegar drinks have been brewed since the early 1990s (Liu et al. 2004, Anonymous 2016a). Chinese vinegars are also referred to as cereal vinegars. The most famous Chinese vinegars are Shanxi aged vinegar, Zhenjiang aromatic vinegar, Sichuan bran vinegar and Fujian *Monascus* vinegar.

These vinegars use sorghum, sticky rice, wheat bran and red yeast rice as the main raw material (Table 2) (Chen et al. 2009). In addition to the raw material, Chinese vinegars can be classified into different groups according to the colors, the special flavors and the production processes, such as black (brown) vinegar, red vinegar, white vinegar, smoky vinegar and herbal vinegar (Anonymous 2016b).

Geographical Origin

Wine and vinegar can be easily differentiated in European countries that traditionally produce both products, as the names are well consolidated with a precise meaning. Wine has a minimum content of acetic acid of 1.2 g/L and a maximum ethanol content of 9-15 per cent. Vinegar has a minimum titratable acidity of 6 per cent and a maximum residual ethanol content of 1.5 per cent. However, national regulations can establish different limits. For instance, in Italy, the residual ethanol content of wine vinegar that is produced via artisanal methods increased from 1.5 to 4 per cent (Anonymous 2014). Fermented alcoholic beverages and vinegars share many steps and some products do not match either the definition of vinegar or the definition of wine. To classify vinegars, a minimum of three parameters should be established: lowest acceptable threshold of acetic acid, upper limit for ethanol and lowest acceptable value of the acetic acid/ethanol ratio.

The acidity and residual ethanol are two main parameters that are useful for establishing an all-inclusive worldwide definition of vinegar. The concentrations of acetic acid and ethanol content are dependent on the raw material, the microorganisms that are involved in the fermentation process, the technology employed and the culture and 'vinegar lore'. For instance, a common practice in China is to mix vinegar and wine to improve the flavor and safety of a product. In Europe, vinegar is considered to be a flavoring or a preservative food ingredient; with few exceptions, it is generally sharp and sour. In Asia and Africa, vinegar is a drink with a less sour taste. Many sweetened fruit vinegars in China and Southeast Asia are characterized by their low acidity and aromatic flavors. In Africa, several fermented beverages can spontaneously acidify to produce alcoholic-acetous products, which are not easily classified as alcoholic beverages or vinegars. In Japan, rice vinegars are referred to as Kurosu and Komesu, which differ based on the type of rice. Komesu is obtained from polished rice, whereas Kurosu is made from unpolished rice. Komesu vinegar is clear, has low acidity and is used as a condiment of fish and vegetables, whereas Kurosu is black and abundant in amino acids and vitamins and is appreciated as a beverage when diluted with water. It covers 20 per cent of the Japanese vinegar market, which corresponded to 21.46 billion yen in 2004.

In some western countries, primarily the US and Canada, apple/cider vinegar is considered to be an old folk remedy for treating diseases; it is mixed with fruit juice.

Botanical Species

Several botanical species can be employed in vinegar production as they have two main basic attributes: (1) they are safe for human and animal consumption, and (2) they comprise a direct or indirect fermentable sugar source. General considerations can be made based on the chemical composition of the edible parts and their fermentation attitude. Proposed groupings are as follows:

- Acidic and easily fermentable: pH less than 3.5; glucose, fructose and sucrose are the main constituents, e.g. berries, grape, apple and plums
- Moderate acidity and easily fermentable: pH between 3.5 and 4.5; tomato
- Low acidity and easily fermentable: pH greater than 4.5; figs, dates and palm sap
- Non-fermentable: hydrolysis is required prior to fermentation (rice, sorghum and barley)

Fermentation Technologies

Commercial vinegar is produced by fast or slow fermentation processes. Slow methods are generally employed to produce traditional vinegars; fermentation slowly proceeds over the course of weeks or months. Traditional methods are commonly referred to as 'surface culture fermentations' (SCFs) as AAB abundantly grow on the media surface, where oxygen concentrations are high. Fast methods primarily refer to submerged fermentation (SF) that is employed in industrial vinegar production (Adams 1998). Solid static fermentation (SSF) refers to the growth of microorganisms in the absence of a flux of water, which is applied to produce cereal vinegars in the Eastern countries.

Surface Culture Fermentation

SCF includes several variations of methods that are no longer practiced (e.g. Boerhaave method). Other methods, such as the 'half full' barrel, are typical of homemade productions and are not suitable for producing a high quantity of vinegar.

AAB develop a biofilm with variable thickness and texture that is dependent on the species that conduct the fermentation. For instance, strains of *Komagataeibacter xylinus* form large cellulose layers (Fig. 1) that collapse to the bottom of the container; new are formed on the surface until suitable carbon sources (primarily glucose and ethanol) occur.

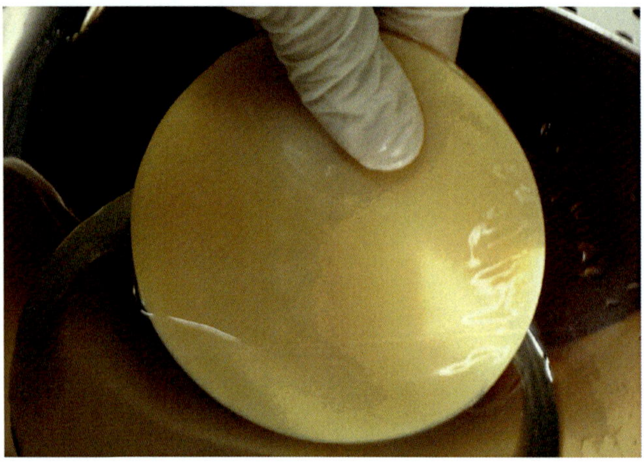

Figure 1: Cellulose layer formed by AAB in static cultivation.

In a vinegar environment, other species develop thin and fragile biofilms that are preferred to thick biofilms. In an SCF system, acetification occurs without strict parameter control; as a result, the final products are often not consistent with the expected results. However, this system does not require substantial investments.

Among the vinegars that are produced by SCF, the Orléans system remains current although it is an ancient method. In this case, acetic acid fermentation occurs in wooden barrels with a volume of approximately 230 litres and two holes that are horizontally arranged on the anterior bottom. A large hole (diameter of 6 cm) is used to withdraw vinegar and fill wine, whereas a small hole enables air entry. Once the fermentation commences, approximately 40 litres of vinegar are withdrawn and 10 litres of wine are added weekly until the original volume (40 litres) is restored. After a week from the last refilling, another 40 litres of vinegar are withdrawn. Similar procedures are adopted by small vinegar producers for special productions or in the absence of a suitable applied technology (Fig. 2).

Figure 2: Production of wine vinegar in wood barrel by surface culture fermentation.

The acetification process is controlled by measuring the ethanol content and titratable acidity that enable withdrawal of the final product when ethanol is not completely exhausted; the new wine is subsequently added. By this procedure, the viability of the AAB culture is maintained and acetate oxidation is avoided.

A similar system was developed by our group to produce a base wine for traditional balsamic vinegar production (Gullo et al. 2009). The main steps that affect these processes are fermentation start-up and scaling-up. The first phase is the AAB culture preparation and involves strains that are capable of producing acetified liquids with a suitable amount of acetic acid in the planned time (start-up). The volume is increased (scaling-up) by adding aliquots of base wine (cooked and fermented), avoiding the disruption of the previously formed biofilm.

The sequential addition of wine needs to follow precise rules to maintain the AAB culture viability. In particular, wine is added before the ethanol content and titratable acidity fall below 2.5 per cent. After reaching suitable volumes, the processes are transferred into final containers and wine is added as previously described.

Submerged Fermentation

SF is generally operated in a semi-continuous mode, which enables the production of vinegars with a minimum acetic acid content in the range of 8 to 14 per cent. SF is performed in steel bioreactors that are equipped with a self-aspirating aeration system that enables the formation of very fine air bubbles and homogenous air-liquid dispersion.

The important process parameters are forced air, temperature and the time and modality of the withdrawal and filling. At the end of one cycle, a third of the final product is maintained in the bioreactor as inoculum for the following cycle. Efficient processes in the bioreactors are necessary. In the first process, the fermentation is conducted until the range of ethanol content is approximately 2-3 per cent. Then, vinegar is transferred to the second bioreactor, where fermentation proceeds until the ethanol is depleted. Using a single bioreactor, the ethanol oxidation is completed in one step and the bioreactor is empty when the ethanol content ranges from 0.5-0.3 per cent. The start-up of the next cycle is slow with respect to the two bioreactor systems as AAB culture is less active. If the discharge is late, the AAB culture is substantially affected and the new cycle requires more time. The duration of a cycle generally varies between 18 and 30 hours (Garcia et al. 2009).

Solid State Fermentation

SSF refers to the acetic fermentation that occurs on solid substrates. The substrates can be inert and act as the surface of microbial growth or the source of nutrients, as in the case of vinegars. In countries that produce wine, a method similar to SSF is used for vinegar production from marc during the fermentation of must in vats. During the tumultuous fermentation, marc is pushed upwards to the surface of the liquid; due to the high surface that makes contact with air, they become an optimum solid substrate for AAB growth. Rational oenological practices prevent these process conditions, but comprise a rapid and guaranteed method for producing vinegar.

The increase in surface contact between wine and air has been exploited to produce vinegar. The most well-known example is the Schutzenbach method, which is performed by a barrel with wood shavings inside it, on which both wine and air are counter currently circulated. The SSF system is prevalent for acetification of cereal vinegar, which occurs in large tanks (urns) that contain cereal bran and wine (Pei or Moromi) and are frequently shuffled. The shuffling is conducted at regular intervals, manually or by an automated system inside the urns (Fig. 3).

Figure 3: Manual shuffling during Chinese cereal vinegar production.

This operation has a triple function: (1) to aerate, (2) to maintain an evenly moist mass, and (3) to dissipate heat. AAB adhere to the surfaces of glumes and husks and form a highly active surface per volume. In industrial systems for the production of citric acid, the bed substrate is employed; it comprises a middle method between SSS and SCF, in which the substrate is continuously sprayed with culturing medium. A similar system was developed by our research group to produce cellulose (Fig. 4).

Mass Balance and Yield

The theoretical conversion yield of ethanol to acetic acid is 100 per cent as one mole of acetic acid is formed from one mole of ethanol according to the following reaction:

$$1\ C_2H_5OH + 1\ O_2 \longrightarrow 1\ CH_3COOH + 1\ H_2O\ (\Delta G° = -455\ kJ/mol)$$

(a)

(b)

Figure 4 (a and b): Prototype bioreactor for cellulose production.

The yield can exceed 100 per cent if calculated by weight or volume of ethanol and weight of acetic acid. This inaccurate expression of yield is prevalent among vinegar producers as they buy wine based on ethanol content (in volume) and sell vinegar based on the acetic acid content, which is expressed in g/L. According to this previous reaction, 1 molar ethanol solution (4.6 g/100 mL) yields rise to 6 g of acetic acid in a volume of 102 mL, 4.6 g/100 mL corresponds to 5.8 mL ethanol/100 mL and 6 g/102 mL yield 5.9 g of acetic acid per 100 mL. In this case, the maximum theoretical yield (calculated based on the volume of ethanol and the weight of acetic acetic) is 102 per cent.

To calculate the maximal concentration of acetic acid, which can be obtained by complete fermentation, the sum of the vol per cent ethanol and the g acetic acid per 100 mL is referred to as the 'total concentration'. For instance, a wine with 8.5 per cent ethanol and 2 per cent acetic acid content has a total concentration of 10.5 per cent. Consequently, the quotient of the total concentration of the mash produces the concentration yield (Ebner et al. 1996). The difference between the theoretical yield and the real yield (95-98 per cent) is attributed to the notion that ethanol is also a carbon source for cell components and a part of ethanol can be depleted via acetate oxidation and by evaporation.

Two important parameters, especially in SF, are acetification rate and the productivity, such as the quantity of acetic acid during time (g of acetic acid × 100 mL × hours) and the amount of acetic acid produced per hour, which is referred to as the bioreactor volume.

The variables that significantly affect yield, productivity and rate are the final ethanol content and the speed of charge and discharge. An optimized process is a compromise of the final parameters.

Vinegar Microbial Community

Many microorganisms serve a positive role in vinegar production and belong to the phylogenetic *Phila*: *Proteobacteria* class *Alphaproteobacteria* (acetic acid bacteria); *Firmicutes* (lactic acid bacteria); and *Ascomycota* (*Saccharomyces, Monascus, Rhizopus, Aspergillus* and *Penicillium*).

Final fermentation is always AAB fermentation, which occurs after alcoholic fermentation; in some cases, fermentations are separate phases, such as in wine vinegar where alcoholic fermentation always precedes acetic acid fermentation. In other vinegars, the process is more complex and a greater amount of microorganisms are observed in all the phases. The most representative example is cereal vinegar. During the preparation of Koji or Qu, molds, yeasts and AAB are simultaneously present. Although molds (*Aspergillus, Mucor, Rhizopus* and *Monascus* spp.) are dominant, yeasts became dominant in the following step to obtain 'Pei', and AAB complete the fermentation.

The factors that determine the dominance of some microorganisms are dependent on intrinsic conditions (media composition) and extrinsic parameters, such as temperature and humidity. As an example, molds are favored with respect to yeasts due their lower values of water activity; for this reason, Koji/Qu has a lower relative humidity than Pei.

Acetic Acid Bacteria

AAB are obligate aerobic bacteria that are capable of partially oxidizing a variety of carbohydrates, which release corresponding metabolites (organic acids, aldehydes and ketones) in the surrounding media. They are Gram-negative or Gram-variable, non-spore forming, ellipsoidal to rod-shaped cells that can occur in single, pairs or short chains. They can be motile due to the presence of peritrichous or polar flagella and are catalase positive and oxidase negative. The optimum pH for growth is in the range 5-6.5, but they can grow at lower pH values (approximately 3), as in vinegars. The optimum growth temperature varies between 28°C and 30°C, and some species are thermotolerant. They are capable of producing pigments and different types of exopolysaccharides, such as cellulose (Sievers and Swings 2005). AAB are well known for their role in vinegar production; of the 18 genera that are currently recognized, only *Acetobacter, Gluconobacter, Gluconacetobacter* and *Komagataeibacter* include species that have relevance in fermented foods and only *Acetobacter, Gluconacetobacter* and *Komagataeibacter* are employed in vinegar production (Table 3).

After their initial description as vinegar bacteria more than 150 years ago, they have undergone many taxonomic changes, which cause a significant amount of trouble for people who are not taxonomists and have to work with AAB. Technologists continue to use old and invalid names of AAB and may not recognize new species. We briefly discuss the relevant reclassifications of genera that impact vinegars.

Acetobacter is the oldest genus; its description dates from early studies of vinegar microbiology (1898); the name defines 'vinegar bacteria'. In 1968, the genus *Gluconobacter* was introduced, based on the capacity to oxidize ethanol and glucose. The *Gluconobacter* species, which strongly oxidizes glucose and weakly oxidizes ethanol, did not oxidize acetate and lactate to carbonate and only exhibits polar flagellation if motile, were included in the new genus.

The genus *Acetobacter* included organisms that only exhibited peritrichous flagellation if motile, strongly oxidized ethanol, weakly oxidized glucose and completely oxidized acetate and lactate.

The genus *Gluconacetobacter* was established based on the differences in the ubiquinone (UQ) system. The species of *Acetobacter*, which were equipped with UQ-10, were transferred to this genus and the species with UQ-9 remained in the genus *Acetobacter*. In 2013, the genus *Gluconacetobacter* was separated into two phylogenetically and phenotypic groups, named as *Gluconacetobacter liquefaciens* and *Gluconacetobacter xylinus*. The latter group included species that were transferred to the new established genus *Komagataeibacter* (Yamada et al. 2012). Species of the genus *Komagataeibacter* have relevance to vinegar; for example, the species *Komagataeibacter europaeus*, that is highly tolerant to acetic acid and ethanol, is responsible for the production of high-acidity vinegars (Gullo et al. 2014).

From the industrial point of view, taxonomic issues do not have substantial relevance. Instead, the following phenotypic traits of bacterial cultures are very important in industry:

- no production of a thick 'mother' of vinegar, which affects the yield and quality of the product. This trait is widespread among strains of the species *K. xylinus* and other closely related species;
- high tolerance to ethanol and the ability to grow in wines with more than 12 per cent ethanol.

Table 3: Characteristics of acetic acid bacteria of fermented beverages and vinegars

Species	*Overoxidation*	*Polysaccharides production*	*Acetic acid requirement*	*Growth on ethanol as the only carbon source*	*Resistance (w/v) (per cent)*		*Isolation source*
					Ethanol	Acetic acid	
A. pasteurianus	+	v	-	v	10	6	Cereal vinegar, traditional balsamic vinegar, wine, beer, cider, wine vinegar, fruit vinegar
A. aceti	+	-	-	v	12	5	Wine vinegar, traditional balsamic vinegar, fruit vinegar, wine
A. oeni	+	/	/	+	10	/	Wine
A. cerevisiae	+	/	-	/	/	/	Beer
A. pomorum	+	-	-	v	<10	/	Cider vinegar
A. nitrogenifigens	+	/	/	+	10	/	Kombucha tea, wine
K. europaeus	+	v	+	+	5-10	14	Wine, wine vinegar, traditional balsamic vinegar
K. oboediens	+	v	-	+	<3	/	Wine vinegar
K. hansenii	+	+	-	v	<3	/	Wine, wine vinegar, traditional balsamic vinegar, kombucha tea
K. intermedius	+	+	-	+	/	7	Cider vinegar
K. xylinus	+	+	-	v	/	/	Wine vinegar, fruits
G. oxydans	-	-	-		<5	/	Fruit, fruit juice, wine, beer

+: positive; -: negative; v: variable; /: not determined.
Adapted from Gullo and Giudici 2014.

Vinegar production in small and large industries is performed without the use of a selected starter culture of AAB. A common practice is to use mixed AAB cultures from previous fermentation batches (backslopping practice). Selected strains are not applied in the industry for two reasons: the first reason is technological and the second reason is economical. First, the strong variation in the selective pressure between the beginning of the process and the end of the process requires the ongoing adaption of the culture. The adaptation of the culture substantially affects the processes in terms of a slowdown or stop to fermentation (Gullo et al. 2012). Second, the backslopping practice is effective and inexpensive.

In submerged fermentation, the practice of backslopping is described as follows: approximately one-third of the previous culture volume is maintained as inoculum of the subsequent batch, and new wine is slowly added to prevent physiological stress of the culture. In these conditions, acetic acid fermentation is conducted by strains of different species; their abundance is dependent on the selective pressure of the medium and the process parameters.

In any fermentation process, the use of selected microorganisms is practiced only if the selected strain has specific traits that offer advantages compared with the backslopping practice. Although selected AAB are scarcely applied in the vinegar industry, selected strains may be introduced in the future, especially if they have exploitable phenotypic traits. Table 3 lists the traits that are frequently associated with vinegar production; some of these traits are species-specific and some, strain-specific.

In industrial vinegar, high acetic acid resistance is the most important technological trait of AAB. For this reason, knowledge of the resistance mechanisms is a priority of the applied research in the field. Other relevant traits include the ability to oxidize ethanol as a preferred carbon source and rapidly produce acetic acid. These traits assure a high yield and prevent the growth of undesirable microorganisms that affect the product quality.

K. europaeus is the dominant species in SF as it resists a high concentration of acetic acid with respect to other species and requires this acid to grow. Acetic acid resistance in AAB is the result of many mechanisms that explain their activity in the presence of acetic acid and ethanol during high-acidity vinegar production. In *Acetobacter* strains, that grow on ethanol as a single carbon source, the overexpression of gene coding for Krebs cycle enzymes (e.g. acetyl-CoA synthase and phosphoenolpyruvate carboxylase) was observed. In addition, a proton force-dependent motive and ABC-transporter-like efflux pump systems are involved in pumping acetic acid from cytoplasm to the outside of cells, which permits the growth of cells at high acetic acid concentrations. In these conditions, changes in the membrane lipid composition have been reported as an additional mechanism for acetic acid resistance. These mechanisms are particularly active in *K. europaeus*, which is the most important species that performs submerged fermentation (Gullo et al. 2014).

Yeasts

Yeasts are unicellular fungi; their main characteristic is alcoholic fermentation of sugars. Although yeasts are a homogeneous group of organisms, they include a high number of species (currently more than 700 species). The most important and known species in industrial transformation is *Saccharomyces cerevisiae*.

Genera and species that are involved in vinegar production belong to the oenological field with some exceptions; for instance, the genus *Kluyveromyces* is involved in whey vinegar and the osmophilic species are involved in honey and concentrated grape must in traditional balsamic vinegar production. Strain selection, which is very important in production of high-quality wine, is not considered for vinegar production. The importance of specific strains is not demonstrated and detailed studies about the potential effect of a specific strain on the sensorial quality of vinegar are not available. A single exception is the fermentation of cooked must to produce traditional balsamic vinegar, in which the importance of some species of the genus *Zygosaccharomyces* was observed (Solieri and Giudici 2008).

Zygosaccharomyces genus belongs to the *Saccharomycetaceae* family and exhibits the following general characteristics: multiplication by multilateral budding and spheroidal, ellipsoidal or elongated cells that can form pseudohyphae; the cells are prevalently haploid; sporification is generally preceded by cell conjugation and rarely between cells and buds; asci are persistent and contain one to four spherical or ellipsoidal spores. Currently, nine species exhibit alcohol tolerance, osmotolerance and fermentation vigour. This genus contains *Z. bailii, Z. bisporus, Z. cidri, Z. fermentati, Z. florentinus, Z. mellis, Z. microellipsoideus, Z. mrakii* and *Z. rouxii*. Other genera and species that are detected in fermented and cooked must include *Hanseniaspora* with the species *Hansenispora valbyensis* and *Hanseniaspora hosmophila, Candida stellata, Saccharomycodes ludwigii* and *Saccharomyces cerevisiae*, which always appear in fermented and cooked must (Solieri et al. 2006).

Similar to wine fermentation, we can reasonably hypothesize that the yeast strains influence the composition and sensorial proprieties of wine for vinegar production. Few data justify a specific selection of yeast strains for vinegar and special vinegars. However, variations in secondary metabolites of alcoholic fermentation, such as succinic and malic acids, glycerol and acetoin and higher alcohols, have been observed (Gobbi et al. 2014).

Some oxidative yeasts are very common and dangerous for wine and vinegars as they produce CO_2 and H_2O from ethanol respiration, which causes a decrease in yield. These yeasts form a characteristic whitish and fragile film that is easily distinguished from AAB biofilm.

The species *Pichia membranaefaciens* is the most common cause of film formation and off-flavor; however, other species can appear. Wines that are strongly contaminated by these yeasts are not suitable for vinegar production. In whey vinegar, yeasts ferment lactose and species such as *Kluyveromyces marxianus* var. *marxianus* can develop. These yeasts have relevance and are industrially cultivated to produce biomasses or ethanol from milk whey. They thrive in residual whey from unpasteurized milk, grow three to four days after lactic fermentation and appear in association with other yeasts and LAB in kefir preparation.

Molds

Molds are an arbitrary, but useful taxonomic group, including multicellular fungi that belong to different divisions. They have a mycelium that is formed by *hyphae* with septa (diameter ranges from 5-10 µm) with a thick wall of chitin, are strictly aerobic and reproduce by sporification. The term spore defines the dissemination structures—from hyphae fragmentation to the formation of an ad hoc structure, such as sporangia, conidiophores and sexual spores.

Two families of molds include genera and species that are involved in cereal vinegars production: *Aspergillaceae* (genera of *Aspergillus* and *Monascus*) and *Mucoraceae* (genera of *Mucor* and *Rhyzopus*). Other genera can develop in cereal vinegar, but their importance is not easily detectable. With the exception of sake and a few other alcoholic beverages, the fermentation of Eastern wine and cereal vinegars is performed without the use of selected starter cultures. Fermentations are conducted by 'backslopping' of mixed cultures, including different species whose relative abundance is not defined.

The genus *Aspergillus* includes common molds that develop on the surfaces of several substrates and foods at low values of relative humidity. The most relevant characteristic of the genus *Aspergillus* is the form of the carpophore that consists of a globular vesicle (sporangia) from which the chain of spherical conidia arise.

Many species (more than 100) are included in the *Aspergillus* genus. They differ by the color of the vegetative and external mycelium, the pigmentation of the basal mycelium and the underlying substrate, colony size, texture of the basal mycelium and other characteristics.

Generally, aspergilli are not pathogenic; however, they are dangerous due to their ability to form toxic compounds (mycotoxins). The most well-known mycotoxins are aflatoxins and ochratoxins that are mainly produced by *Aspergillus flavus, Aspergillus parasiticus* and *Aspergillus ochraceus*. Other mycotoxins at different degrees of toxicity are produced by other species. For these reasons, the growth of molds in foods is always undesired. In general, aspergilli use a high number of natural compounds, including starch and cellulose. Several species are industrially exploited to produce commodities (for instance, citric acid) and extract enzymes. *Aspergillus oryzae* is used to ferment rice flour for the preparation of many Eastern alcoholic beverages and vinegars.

Within the *Monascus* genus, the species *M. purpureus,* whose color is reddish purple, is known by the name 'ang-khak rice mold' as it imparts the characteristic color to rice products, as in vinegars.

Red rice is also part of the products of traditional Chinese pharmacopoeia. *M. purpureus* and other related species (e.g. *M. ruber* and *M. pilosus*) produce statins, such as monacolin K or lovastatin, which is known for cholesterol synthesis inhibition. However, these species also produce other toxic compounds, such as citrinin; their safety is not absolute and should be evaluated on a case-by-case basis (Li et al. 2004).

The genus *Mucor* includes common mold species of soil, spoiling vegetables and bread. They are utilized in cheese manufacturing during microbial rennet production and in Easter fermented foods such as 'tempe'. In the genus *Mucor,* more than 50 species, which are generally nonpathogenic for humans, and homoeothermic animals are included. These species are not able to grow at temperatures more than 37°C, with the exception of the species *Mucor indicus* and related species. Distinguishing features of the *Mucor* species are rapid growth, colonies with a cottony appearance and colors that range from yellow, white and grey according to the age of the culture and the development of globular or spherical sporangia put on high columelle without apophyses. A collar is evident and they do not form stolons or rhizoids, which makes them distinguishable from the genus *Rhizopus* and other genera of the *Mucoraceae* family. Sporangia have sexual spores in a thin wall and are generally globular or ellipsoidal with a smooth wall or are finely ornamented. Sexual spores (*zygospores*) are formed upon conjugation of special *hyphae* (gametangia) that form the *zygosporangium,* which contains a single *zygospora* in the fusion point.

The sexual spores are typically genetic recombinants and can germinate by producing a new asexual cycle. Self-compatible species (homothallic), which can combine between the hyphae of the same thallus or the heterothallic species (with sexual differentiation), belong to the *Mucor* genus. Similar to species of the *Aspergillus* genus, species of *Mucor* are often associated with 'Qu', which is the starter for the preparation of cereal beverages.

The genus *Rhizopus* includes common mold species that grow on the surface of organic material and spoil fruits, vegetables, bread and soil. Macroscopically, they appear as a white mat that tends to be grey or yellow; hyphae propagate with the formation of stolons and rhizoids with globular-vesicular terminal sporangia. Sexual spores are formed after the conjugation of two differentiate hyphae and they produce genetic recombination.

Some species are pathogenic for humans, whereas other species are pathogenic for plants. The species *R. nigricans* and *R. arrhizus* are saprophytes; however, they may be pathogens of plants that produce soft rot (general term that indicates the degradation of vegetable tissues by eukaryotic and prokaryotic organisms).

Species of the genus *Rhizopus* are employed in industrial fermentation to produce organic acids (fumaric and lactic acids) and synthesize carotene and cortisone. In food, the species *R. oligosporus* is employed in the production of tempe and *R. orizae* in alcoholic beverages, such as sake and vinegar.

Lactic Acid Bacteria and Spore-forming Bacilli

Other bacteria that can grow in vinegar environments are LAB and spore-forming bacilli. LAB can develop in the production of fermented beverages, both before and after the fermentation process. In some productions, they are beneficial and used as a starter; in others, they are undesired due to their spoiling activity. In beer production, their controlled growth before fermentation is beneficial as they contribute to product stability and produce lactic acid.

In wine, they can be advantageous by reducing fixed acidity via malic acid fermentation, which contributes to ageing of the product. In contrast, their growth is undesired in the following cases: (i) after alcoholic fermentation of beer as they cause product cloudiness; (ii) in wines and similar beverages where they grow at the expense of residual sugars, which cause spoilage; the spoiled product can only be employed in a distillery. For vinegar production, the growth of LAB is not necessary and generally does not affect the activity of AAB. Different LAB are involved in the production of fermented milks and belong to all known genera and many species. Although the dominance of a specific strain is dependent on many factors, the most important factor is temperature.

In milk whey for the production of vinegar, the alcoholic fermentation is possible due to lactose hydrolysis that provides glucose and galactose. Yeasts do not ferment lactose, with the exception of some species of the *Kluyveromyces* genus. Then LAB should not deplete lactose. For this reason whey is pasteurized prior to alcoholic fermentation and acetic acid fermentation.

Spore-forming bacilli are generally spoiling and toxigenic organisms in foods. However, some species have a positive role in African and Eastern fermented foods that are made from cereals and legumes. In vinegars, the role of spore-forming bacilli is limited to Qu preparation, in which the microbial association is primarily formed by molds and *Bacillus* spp., whose role is not completely defined (Chen et al. 2009).

Nematodes

Nematodes are organisms that can thrive during vinegar production. They are roundworms that are able to grow and survive in high-acidity environments, such as vinegars, where they eat bacteria. They belong to the Phylum *Nematoda,* which comprises unsegmented worms (ametameric), have a thin cylindrical body and are usually microscopic in size, but can have a length of one meter in some cases.

The recognition of genera and species is particularly complex in nematodes due to the limited number of diagnostic characters.

Nematodes include more than 20,000 species, primarily parasitic species. They are ubiquitous and widespread in Nature—in freshwater, marine water and several terrestrial environments. They are simple organisms with a complete digestive system with two different orifices to uptake nutrients and excrete faeces. In addition to residual vegetables, many species eat bacteria, yeasts, fungi, protozoa and algae. Some nematodes grow and reproduce on damaged fruits and exist in environments where alcoholic beverages (wine, cider) are produced. These nematodes are spread by midges and find optimum growth conditions in vinegars. The nematode body wall has longitudinal muscle fiber, but is not transverse and only moves with wave movements; this feature renders nematodes recognizable even by a superficial visual analysis of liquids (Petroni and Gullo 2004).

The optimum values of temperature for growth is in the range between 20 and 30°C; they are not able to grow at temperatures above 50°C. The growth of nematodes in vinegars is well known and is primarily ascribed to the species *Turbatrix aceti*, which was formerly described as *Anguillula aceti*. *Turbatrix aceti* has a cylindrical body that is transparent and thin at both ends; it is long with a maximum length of a couple of millimeters; however, its width is not greater than one tenth of a millimeter. The effect of *Turbatrix aceti* in vinegar production is not well studied and the previous studies report inconsistent opinions. It is described as a dangerous organism as it destroys the AAB biofilm in static fermentations; other authors attribute a beneficial role to *Turbatrix aceti* as eating bacteria increases the amount of nutrients obtained via the extraction of metabolites that are consumed by AAB. However, current regulations prohibit the marketing of vinegars that are colonized by nematodes.

In industrial vinegar production, nematodes are removed from the final product by filtration and pasteurization. Artisanal vinegars that are produced in a static system and pasteurization and/or filtration cannot be applicable; a corrective action is to immerse electric heaters in vinegars until a minimum temperature of 55°C is attained.

Vinegar Types

In this section, we report the description of some widespread vinegars; they appear according to typology instead of real world diffusion. Vinegars from sugared and acid raw materials (fruit vinegars and TBV), vinegar from amylaceous raw materials (beer vinegar and cereal vinegar), whey vinegar as an example of vinegars from animal sources and finally vinegars from roots and tubers.

Fruit Vinegars

Fruit for vinegar production should be healthy without residual pesticide, spoiling organisms and mycotoxin or other dangerous compounds. Juices must have a chemical composition that is suitable for alcoholic fermentation, especially acidity and sugar content. Low pH values (approximately 3.5) are necessary to prevent fermentations by LAB and/or anaerobic bacilli, which form undesirable sensorial compounds. For instance, suitable apple juices have a sugar concentration of 10-13 per cent, from which the potential acidity is approximately 6.7-7.4 per cent, respectively. Juices with lower sugar contents (less than 7-8 per cent of the theoretical ethanol) should be supplemented with concentrated apple juice (60-80 per cent of sugar content). The transformation processes begin with the inspection and selection of the raw material, followed by washing and mashing, which differs according to the type of fruit. The best described process is the process of apple vinegar, which is often produced from concentrated juices instead of direct juice because concentrated juices can produce vinegar independent from the season and have a starting material in constant composition.

Apple juice is fermented with the same yeasts (*S. cerevisiae*) that are used to produce cider or wine; however, a specific strain selection can be advantageous for more rational processes and the sensorial quality of vinegar. After alcoholic fermentation, the must is centrifuged to remove scums and the clear juice is supplemented with at a minimum of 1.5 per cent acetic acid to stabilize the product until acetic acid fermentation. This practice is also employed to stabilize the alcoholic liquid in wine vinegar production. The majority of acetic acid fermentation of fruit and wine vinegar is performed in submerged conditions.

After fermentation, vinegar requires other treatments, such as clarification by filtration or gelatin addition and static separation. In some cases, the color must be corrected by active carbon or adsorbents of polymeric phenolic compounds. Vinegar can be pasteurized to prevent additional growth of AAB and bottled with or without antioxidants like ascorbic acid or sulphite.

Traditional Balsamic Vinegar

Traditional balsamic vinegar (TBV) is a condiment that is obtained from alcoholic and acetous fermentation of cooked grape must and aged in a set of barrels that are referred to as 'batteria'. Due to its historical and manufacturing attributes, TBV received a PDO (protected designation of origin). Note that a double PDO was assigned: the first PDO was assigned to Traditional Balsamic Vinegar of Modena and the second to Traditional Balsamic Vinegar of Reggio Emilia (Disciplinare di produzione 2000a, b). The differences between the two designations are less than the differences detected for TBVs of the same province; thus, using two PDOs for the same product is questionable.

Moreover, a PGI (protected geographical indication) is also attributed to Balsamic Vinegar of Modena. These designations are easily confused, since they share the terminology: Traditional Balsamic Vinegar of Modena, Traditional balsamic vinegar of Reggio Emilia and Balsamic Vinegar of Modena.

In addition, a high number of vinegars, condiments and other products from grapes, which are similar to the PDO products for their use, color, sweet and sour taste, pungent odor, density and viscosity, are available on the market. However, the different commercial categories imply raw materials and technologies, which render TBV easily distinguishable from other categories, both for their composition and sensorial proprieties. The significant differences between TBV and other balsamic vinegars arise from the complex production technology and the long ageing period of TBV. However, TBV is employed as the archetype of balsamic vinegars, which give to other products, homonyms in part, the quality of precious seasoning. Among the multitude of commercial balsamic products, consumers are easily confused by products with very different prices, which complicate their selection.

With the exception of TBV, all other products including Balsamic Vinegar of Modena (PGI), are produced by rapid and modern fermentation processes by blending ingredients and with limited or without ageing.

To clarify, a categorization of main balsamic products is as follows:

- Generic balsamic vinegar: produced all over the world; with no reference to any recognized geographic origin; it is simply referred to as balsamic vinegar; can be produced by submerged and surface culture fermentations
- Balsamic vinegar with PGI, Balsamic Vinegar of Modena: acetic oxidation of wine vinegar generally performed in submerged conditions
- Traditional balsamic vinegar: two PDOs for Traditional Balsamic Vinegar of Modena and Traditional Balsamic Vinegar of Reggio Emilia, respectively; only the surface culture fermentation is permitted
- Balsamic condiments: heterogeneous group, including glazes, sauces, jellies and other food dressings without any legal protection; can be produced by submerged and surface culture fermentations

From an analytical point of view, raw materials, technology and a long ageing period render TBV easily recognizable. As demonstrated in our recent paper, simple determinations of Brix, acidity, pH, reducing sugars and their ratios enable us to correctly differentiate TBV from other balsamic vinegars (Giudici et al. 2015).

Peculiarities of TBV are mainly attributed to four factors:

- must with appropriate sugar/acidity ratio

- scalar and direct fermentation of cooked must
- absence of ingredients that differ from cooked must
- long ageing period in wood barrels

Traditional Balsamic Vinegarmaking

The rational procedure to make TBV differs from the procedure that is described in the production disciplinary and can be divided into four phases:

- Must preparation and cooking
- Transformation of cooked must in base wine (alcoholic fermentation)
- Transformation of base wine in base vinegar (acetic acid fermentation)
- Ageing in a set of barrels (batteria).

Codification of TBV production in four distinct phases is based on our studies, which began prior to 1990 and demonstrated that AAB growth follows alcoholic fermentation and determined that it is not reasonable to hypothesize a concomitant growth of yeasts and AAB. Prior to these studies, a common conviction that alcoholic and acetic fermentations were associated in a sort of commensalisms between yeasts and AAB prevailed. Currently, the notions that the alcoholic and acetic acid fermentations are scalar and that they must be performed in separated containers is almost universally accepted. The management of 'batteria' is reminiscent of sherry vinegar preparation, in which barrels are never empty but yield product to a downstream barrel and receive product from an upstream barrel.

The pouring operation is referred to as 'rincalzo' and is made once a year. 'Rincalzo' maintains a constant volume in a single barrel (replacing the volume that is lost by evaporation). At the beginning of the process, barrels are filled with base vinegar (acetified cooked must). In the subsequent years 'rincalzo' is made to restore the lost volume, using the product of the preceding barrel of the set. The larger barrel receives new base vinegar, whereas the other barrels receive partially aged vinegar, whose age increases in the direction of the final (smallest) barrel.

After a suitable number of years, the production disciplinary indicates 12 years from the build-up of 'batteria'; an aliquot of product is withdrawn from the last barrel (designated number 1).

The storage of TBV in barrels is a dynamic process that is operated by partially transferring vinegar from one barrel to another with the consequence that the product of a different production year with a different concentration of solutes is mixed. By this procedure, each barrel contains a blend of vinegars. The residence time is a weighted average value; the calculation requires the development of algorithms that are based on the paradox of Achilles and the Tortoise (Giudici and Rinaldi 2007).

During the long ageing period, vinegar undergoes a concentration of solutes and significant chemical and physical changes. The following macroscopic variables underwent these changes: water content, specific weight, viscosity, color, Brix degree, titratable acidity and pH. An indefinite number of chemical compounds, already occurring and newly formed, are accumulated, especially polymers with high molecular weights as melanoidines. These factors contribute to making TBV a peculiar product that differs from other vinegars that are designated with similar names.

More detailed literature on TBV can be found in Giudici et al. 2015, a book that summarizes the state of the art on balsamic vinegars.

Vinegars from Amylaceous Raw Materials

Rice, barley, millet and other cereals can be used to produce vinegar (Fig. 5a). The production of alcoholic beverages from amylaceous raw materials is not easy, as polysaccharides are not directly fermented by yeasts and bacteria. These raw materials need to be treated with enzymes to release simple sugars that enable fermentation. This operation is currently performed by fermentation industries (distilleries) that produce ethanol for human consumption or fuel beginning with cereals or potato containing starch and inulin (in tropical and subtropical countries, such as Central and South America).

Polysaccharides are formed by chains of different lengths; they differ by the type of sugar or allocation on the chain. The most well-known polysaccharides are cellulose, starch, dextrin, inulin and glycogen. They are constituents of tissues and cells, as in the case of cellulose, or they are accumulated as reserve substances in seeds, fruits, tubers or roots (starch and inulin). After complete hydrolysis by chain breaking, the constituent sugars are released—glucose in the case of starch and fructose in the case of inulin.

Starch is a polysaccharide that consists of molecules arranged in semi-crystalline form in concentric layers that form granules. As the granules differ in aspect and size in plant species, their microscopic examination simplifies the determination of their origin. The typical structure of the granules may be destroyed enzymatically or destroyed by gelatinization in hot water. The presence of starch in any form is easily detectable, as it turns blue when it reacts with iodine.

From a chemical point of view, starch is not a simple compound as its chain is composed of two fractions (amylose and amylopectin). Amylase, which is the enzyme that hydrolyses starch, is formed by two fractions that are active at different temperatures: α-amylase, from which complex compounds are formed (dextrins), and ß-amylase, which ends the hydrolysis and releases maltose. Maltose is a disaccharide that consists of two glucose molecules that can be hydrolyzed by another enzyme (maltase) in glucose.

Beer Vinegar

Beer vinegar, which is also referred to as malt vinegar, is widespread in Anglo-Saxon countries, where barley is prepared according to a process that is comparable to beer production in its initial steps with different subsequent phases.

Barley

Barley (*Hordeum vulgare*) is a cereal that is cultivated worldwide, especially in countries that produce beer. Barley has a spike that is formed from six rows of caryopses; however, several cultivars have only two or four rows of caryopses. Continental European beers are made with cultivars with two rows (distichous), whereas the cultivar employed in the United Kingdom has six rows of caryopses. Barley sowing occurs in autumn and ripening commences in the beginning of summer. A spike always has a long bristle at its apex; after threshing, a caryopsis retains its glumes. Similar to other cereals, caryopses are primarily formed by starch that is not fermented by yeasts. For this reason, fermentation is preceded by complex preparation of the raw material, of which the most important process is malting, which comprises caryopsis germination (Grierson 2009).

Malt Production

Malt production is performed in dedicated industries that are referred to as 'malt houses'. The process of malt production includes moistening, germination and drying of caryopses. Barley caryopses are immersed in water until the humidity increases from approximately 12 to 45 per cent. Then, moist and swollen caryopses are transferred to aerated and conditioned plants, where they start to germinate. When a radicle is approximately three-fourth of the length of a caryopsis, the process is interrupted by drying to stop the vitality of fresh barley and preserve the product. After drying, malt is stored until milling. Milling is performed in a similar manner to methods that are applied during the production of wheat from cereals. During germination, the vegetative activity of seeds is awakened and the enzymes are re-activated. The enzyme activity is very intense, but limited in time as it is interrupted by drying; it causes partial hydrolysis of starch and proteins with the release of simple compounds that are used by seeds to grow. Drying stops the hydrolysis process, but does not deactivate the enzymes that will act in the following step—must preparation.

Must malt preparation by infusion: Must preparation is performed by different methods that primarily differ according to the temperature of the infusion step. The traditional method is the simplest method and is performed in a single stage at approximately 65.5°C. During this process, starch and proteins are hydrolyzed. From this stage, the processes of beer and vinegar production differ. In beer production, the hydrolysis is stopped when starch is not detectable (by the iodine method). In this manner, the released products are glucose, maltose, maltotriose and polymers of different molecular sizes (dextrin). Dextrins are gummy compounds that are not fermentable; they remain in the product by conferring persistence to the foam. In contrast to beer, the occurrence of dextrin in vinegar production is undesirable. For this reason, the process is not interrupted until hydrolysis is complete to obtain only glucose at the end of the process. Therefore, all initial starch is converted to fermentable sugar. Boiling is always performed to stabilize must in the case of beer and if required, in vinegar production. Filtered and diluted must has a high pH value and the acidity needs to be corrected prior to alcoholic fermentation.

Must malt composition: To prepare must, malt flour is mixed with water (approximately 32-64 per cent). The flour consists of approximately 65 per cent carbohydrates (starch, saccharose, other sugars and dextrins); they constitute 20-40 per cent of the initial mass. After hydrolysis and boiling, the must is diluted to obtain 10-15 per cent of extract that is composed of fermentable carbohydrates. The other difference between beer production and vinegar production is the addition of hop for beer production, which provides the typical bitter taste.

Fermentations: In beer production, *Saccharomyces carlsbergensis* strains are primarily employed for alcoholic fermentation; they are psychrophiles that are capable of fermenting at temperatures below 10°C. Strains of this species are also employed to produce 'lager' beer. In the United Kingdom, 'ale' beer is produced by *Saccharomyces cerevisiae*, which is also exploited to produce vinegar. After alcoholic fermentation, acetic acid fermentation primarily occurs with a mixed culture of *Acetobacter* and *Komagataeibacter*. The last step is ageing.

Cereal Vinegars

Cereal vinegars making can be divided into eight steps according to the methods that are applied in China and Japan: (1) Qu preparation; (2) cereal preparation; (3) saccharification; (4) alcohol fermentation; (5) acetic acid fermentation; (6) vinegar extraction; (7) ripening and (8) packaging.

Qu preparation: Qu (in Chinese) or Koji (in Japanese) is the starter that is used in SSF (Fig. 5b). It is made from grains such as rice, sorghum, wheat, barley and beans such as peas and soy. The microbial compositions of these grains are heterogeneous and dependent on different factors, such as raw material, humidity and temperature.

In China, several types of Qu, namely, Mai Qu, Xiao Qu, Da Qu, Hong Qu and Fu Qu, which differ in aspect, raw material and geographical origin but always contain molds, yeasts and bacteria, are employed (Table 4).

The most frequently detected molds are *Aspergillus*, *Rhizopus*, *Mucor* and *Monascus*. When *Monascus* spp. prevail, vinegar is red.

Molds secrete a high amount of hydrolytic enzymes, such as amylases, glucoamylases, proteases, lipases and tannases, which release simple molecules from complex vegetable constituents.

The growth of yeasts and bacteria is strongly affected by hydrolyzed compounds. Among bacteria, the species of the genera *Bacillus* and *Lactobacillus* are frequently reported; however, their role is not completely understood. Qu is desiccated; a portion is added to cereals prior to alcoholic fermentation and the remaining fraction is used as a starter to produce new Qu.

Cereal preparation, saccharification and alcohol fermentation: Cereals are crushed, placed in water for a period of 12 or 24 hours according to the season and steamed in typical baskets. Steaming causes starch gelification, which favors enzymatic hydrolysis. The alcoholic fermentation of cooked cereals is performed in special ceramic urns and mixed with crushed Qu and water.

(a)

(b)

(c)

Figure 5: (a) Urns for solid state fermentation of cereal vinegars. (b) Bricks of Qu (units in centimeters). (c) Pei/Moromi in fermentation (alcoholic) urns.

Table 4: Chinese vinegars and their characteristics

	Ingredient	*Geographic distribution*	*Main microorganisms*	*Color*
SHANXI MATURE VINEGAR	Sorghum, barley, bran, chaff, pea, salt, water	Nearly all over China	*Mucor* spp., *A. oryzae*, *Monascus* spp., *S. cerevisiae*, *Hansenula* spp. and *Acetobacter pasteurianus*	Browed-red color
ZHENJIANG AROMATIC VINEGAR	Sticky rice, rice hull, salt water	Eastern of China	*Rhizopus* spp., *A. oryzae*, *S. cerevisiae*, *Acetobacter pasteurianus* and *Gluconobacter* spp.	Dark brown
BAONING MEDICAL VINEGAR	Sticky rice, wheat bran, rice hull, salt, herbal medicines, water	Central and western regions of China	*Aspergillus niger*, *A. oryzae, S. cerevisiae*, *Acetobacter* spp. and *Lactobacillus* spp.	Black
FUJIAN MONASCUS VINEGA	Sticky rice, sesame, salt, water	Fujian province	*A. oryzae, Monascus* spp., *S. cerevisiae* and *Acetobacter* spp.	Brown
ZHEJIANG ROSE RICE VINEGAR	Sticky rice, salt	Zhejiang province	*A. oryzae, Monascus* spp., *S. cerevisiae*, *Acetobacter* spp. and *Lactobacillus* spp.	Rose-colored
SHANGHAI RICE VINEGAR	Sticky rice, wheat bran, rice hull, soda, salt	Shanghai Region	*Rhizopus* spp., *S. cerevisiae* and *Acetobacter* spp.	Brown
KAZUO MATURE VINEGAR	Grain, sorghum, wheat bran, salt, water	North East of China	*Rhizopus* spp., *Saccharomyces cerevisiae* and *Acetobacter* spp.	Black
FRUIT VINEGAR	Fruit juices, sugar, salt	Nearly all over China	*S. cerevisiae* and *Acetobacter* spp.	Light brown

At this stage, a yeast culture (*S. cerevisiae*) is added to support the growth of naturally occurring yeasts, such as *S. cerevisiae*, *Hansenula anomala* and *Candida* spp.

During alcoholic fermentation, the temperature increases. Therefore, the mixture is transferred to other empty urns once a day to decrease the temperature to approximately 28°C.

The final product is rice wine (Pei and Moromi in Chinese and Japanese, respectively) (Fig. 5c) that is acetified in similar urns where rice bran is added.

Acetic acid fermentation: Acetic acid fermentation is performed in a submerged system or by SSF. The solid substrate is composed of rice bran poured with Pei. The mass is converted to aerate on a daily basis to cool and to evenly re-distribute the Pei.

In the presence of oxygen, AAB have strong metabolic activity and the temperature in the urn should be stabilized at temperatures varying between 38 and 40°C. The process is rapid (approximately 12 days); afterwards, the urns are closed.

Vinegar extraction: The rice bran, which is soaked in vinegar, is placed in percolation tanks to collect the vinegar that is released by gravity. Three washing steps with defined amounts of water or washing vinegar derived from previous fermentations are performed to produce different types of vinegar (Fig. 6). In practice, the objective of the second and third steps is to wash the material of the subsequent cycle. Vinegars are subsequently mixed in defined portions, salted and aged in urns that are placed outside. The addition of salt to vinegar inhibits additional AAB growth.

Figure 6: Chinese cereal vinegar washing.

Ageing: The vinegar remains in closed ceramic urns for months or years. During this period, it undergoes many chemical-physical changes that contribute to sensorial modification of the product. Colored pigments, such as melanoidines (as in traditional balsamic vinegar) or other polymers, are formed.

Simultaneously solid fraction deposits and vinegar become shiny and clear. Prior to packaging, aromatic substances, spices, rice sugar, soy sauce and other ingredients can be added to vinegar.

An Old Story About Chinese Vinegar

Two old Chinese characters, which are no longer employed, but indicate vinegar are 醯 (xi)′ and 酢 (cu)′.

At the time of the Zhou dynasty, the character 醯 (xi) was used; currently, however, only the ′"酢" (cu)′ is used. It is interesting to note that these types indicate both wine and vinegar. The type "醋" is composed of two parts: "酉" and '昔', which mean from 7 pm to 9 pm and 21 days, respectively. The meaning of these two symbols is linked to Chinese mythology. It is said that a Mister Dukang (杜康) erroneously put a sort of marc in a large vessel. After 21 days from 7 pm to 9 pm, he opened the vessel and was fascinated by the aroma of the liquid. He drank and discovered that the liquid was sweet and slightly sour. He thought for a while and then decided to indicate it with symbols 酉 and 昔 that together is '醋'.

For clarity, the Chinese characters are reported below in enlarged form.

Milk Whey Vinegar

Vinegars from milk are the products of three scalar fermentations: (1) lactic acid fermentation, (2) alcoholic fermentation and (3) acetic acid oxidation. Although they can be obtained from any milk, but milk that yields the best vinegars originate from equine milk (horse or donkey) and whey cow milk. Whey is the residual liquid of cheese preparation; it remains after curdling and casein extraction. Although it is abundant in useful compounds, its use is difficult. In Italy, whey primarily originates from cow with fewer contributions from buffalo, sheep and goat. Ricotta cheese is produced from whey; after hot-acid treatment, the derived liquid is referred to as cottage cheese whey or second cheese whey. With the exception of the absence of fats and proteins in whey and albumin in second cheese whey, the composition of the originating milk is similar. Both whey and second cheese whey contain traces of nitrogen compounds, vitamins, minerals and lactose—which is the main compound (approximately 5 per cent). They are yellowish opalescent liquids due to a high amount of riboflavin (B2 vitamin is also known as lactofavine). Similar to cheese, their composition facilitates the growth of many microorganisms.

Whey contains a rich bacterial microflora that is primarily composed of homofermentative LAB, which also occurs during cheese ripening. Different species of LAB are present according to the adopted cheesemaking technology. Whey is a very unstable product as it undergoes lactic fermentation from the boiling step. In a period of 24 hours, the pH decreases to less than 4, lactic fermentation ceases and a high amount of lactose remains in the liquid.

Alcoholic fermentation only begins if cheese manufacturing is performed with raw milk and an indigenous starter. If whey is derived from pasteurized milk, a yeast starter that ferments lactose by producing ethanol must be added. At the end of alcoholic fermentation, acetic acid fermentation occurs, which is performed by adding an inoculum of AAB.

Whey is used in the production of vinegar, which is appreciated in many European countries. It is considered and recognized by the European legislation that defines it: "Vinegar from whey is the product obtained from concentrated whey without intermediate distillation." Whey vinegar and honey vinegar are the only two vinegars that are made from raw material of animal origin.

It is appreciated as it is sweeter than other vinegars, is non-irritating and does not attack the intestinal mucosa. Therefore, it can be consumed by people whose stomachs do not tolerate vinegars. These characteristics are attributed to a relatively high pH (3.4) with respect to other vinegars having lower pH, and presence of phosphates and amino acids.

Vinegars from Roots and Tubers

Many plants accumulate carbohydrates in underground organs, such as roots (beet, carrot, chicory, turnip and radish) or tubers (potato, topinambur and manioc). Some of these raw materials have been employed in the industrial production of ethanol, in which fermentation is performed without hydrolysis, that is, if the original carbohydrate is a simple sugar, or after hydrolysis if the carbohydrate is a polysaccharide. The following carbohydrates are produced by the raw material:

- saccharose from beet
- starch from potatoes
- inulin from chicory, topinambur and manioc
- direct fermentable sugars from carrot, turnip and radish

Must for vinegar preparation can be obtained according to the raw material characteristics by pressure or enzymatic activity on polysaccharides after cooking and homogenization. As roots and tubers contain few acids, must should be corrected with citric or tartaric acid (or lemon juice) to decrease the pH to less than 5.5. The addition of ammonium phosphate and yeast extract may be necessary to increase the nitrogen compounds, which enable regular fermentation.

To adjust the sugar and polysaccharides concentrations, appropriate dilutions or additions are performed to obtain approximately 10 per cent of fermentable sugar suitable for alcoholic fermentation.

Vinegar Spoilage and Safety

Microbiological defects of vinegar can occur during the process steps and compromises the suitability of intermediary products by affecting the product during packaging and marketing.

The partial or total acetic acid fermentation breakdown is the main source of vinegar defects, which cause a lack of acidity and the presence of undesirable microorganisms (molds and yeasts).

Fermentation interruption may be attributed to phage infections, especially in high-acidity vinegars, where fermentation is performed by a single strain or a few strains of the same species. In these processes, phage sensitivity has serious consequences with respect to fermentation processes at lower selective pressure, where a greater number of strains at different phage sensitivity occur.

In the industry, the risk of phage infection is minimized by processes such as substrate pasteurization, the use of sterile air and the separation of bioreactors for high-acidity vinegar from other acetifying liquids.

Other microbiological defects include overoxidation, which is a resistance mechanism of AAB at high acetic acid concentration. Factors that determine overoxidation include a high oxygen concentration in the absence of ethanol. To reduce the risk of overoxidation in industrial vinegar production, these two parameters (oxygen and ethanol) are monitored in real time. Conversely, the overoxidation risk is higher in less controlled fermentation systems.

Other vinegar defects arise from the production of exopolysaccharides (cellulose, levans and dextrans) by AAB, which affects both the technological steps and the sensorial quality of vinegar (Gullo and Giudici 2008).

Although the main vinegar defects are caused by AAB metabolism, a common defect (crystallization) is attributed to yeast metabolism in the production of some vinegars, for instance TBV. Crystallization can occur during production or after bottling and is correlated to fructosophilic yeasts (Giudici et al. 2004), which do not consume glucose and favor crystallization.

Conclusion

Through the years, vinegar was evolved from a poor condiment to an important industrial food preservative and, recently, to a valuable product. It is not a rare event to find special vinegar with extremely high prices, some time unjustified. The perception of vinegar as a product of value is one of the main reasons that prompted the food industry to bring to market new products based on vinegar. In particular, products with a high service content: ready to use and easy to use. These last two qualities have acquired more and more importance for the consumer in general, but also in specific of food condiments.

Today, the success of vinegar is no longer the common bottle and of little value, but the functional package, so as to simplify its use. Even more important is the type of vinegar, which can no longer be considered a trivial commodity, but a series of specific products for respective uses. Vinegars with high viscosity, sweet, not aggressive sensory; vinegars with claimed health benefits; vinegars for special food; vinegars; vinegars for sweet and savory dishes; vinegars that are declaimed like the most prestigious wines. In short, it is no longer correct to speak of vinegar, but it becomes more respectful of reality to speak of condiments in the plural: no vinegar, but many, many vinegars and, of course, your preferred vinegar.

Acknowledgment

We would like to thank to Dr. Jiajia Wu (College of Life Science, China Jiliang University) for her contribution on Chinese vinegar names.

References

Adams, M.R. (1998). Vinegar. *In:* J.B.B. Wood (Ed.). *Microbiology of Fermented Foods*, vol.1. Blackie Academic and Professional, London, pp. 1-44.

Anonymous (2014). Decreto-Legge convertito con modificazioni dalla L. 11 agosto 2014, n. 116 (in S.O. n. 72, G.U. 20/8/2014, n. 192).

Anonymous (2016a). Available from: http://www.tech-food.com (Accessed: April 2016).

Anonymous (2016b). Available from: http://www.cntwp.com (Accessed: April 2016)

Chen, F., Li, L., Qu, J. and Chen, C. (2009). Cereal Vinegars Made by Solid-State Fermentation in China. *In:* L. Solieri and P. Giudici (Eds.). *Vinegars of the World*. Springer, Milan, pp. 243-259.

CNS (Chinese National Standard) (2005). Edible Vinegar. No.14834, N5239. Ministry of Economic Affairs, Taiwan, Republic of China.

Disciplinare di produzione, Aceto Balsamico Tradizionale di Modena. MiPAF 15/05/2000. *Gazzetta Ufficiale della Repubblica Italiana*, 124, May 30, 2000 (a).

Disciplinare di produzione. Aceto Balsamico Tradizionale di Reggio Emilia. MiPAF 15/05/2000. *Gazzetta Ufficiale della Repubblica Italiana*, 124, May 30, 2000 (b).

EC (Council Regulation) (1999). No 1493/1999 of 17 May 1999□9 on the common organisation of the market in wine. *Official Journal of the European Communities* **L179:** 1-84.

Ebner, H., Sellmer, S. and Follmann, H. (1996). Acetic acid. *In:* H.J. Rehm and G. Reed (Eds.). *Biotechnology*, 2nd edn. vol. 6. VCH Weinheim, pp. 381-401.

FDA (Food and Drug Administration) (1995). Compliance Policy Guides (CPG) Sec. 525.825. *Vinegar, Definitions—Adulteration with Vinegar Eels.* Available from: http://www.fda.gov/ICECI/ComplianceManuals/CompliancePolicyGuidance Manual/ucm074471.htm (Accessed: April 2016).

García-García, I., Santos-Dueñas, I.M., Jiménez-Ot, C., Jiménez-Hornero, J.E. and Bonilla-Venceslada, J.L. (2009). Vinegar engineering. *In:* L. Solieri and P. Giudici (Eds.). *Vinegars of the World*. Springer, Milan, pp. 97-120.

Giudici, P., Pulvirenti, A., De Vero, L., Landi, S. and Gullo, M. (2004). Vinegar crystallization of traditional balsamic vinegar. *Industrie delle bevande* **33:** 426-429.

Giudici, P., Gullo, M., Solieri, L., De Vero, L., Landi, S., Pulvirenti, A. and Rainieri, S. (2006). *Le fermentazioni dell'aceto balsamico tradizionale*. Edizioni Diabasis. ISBN: 9788881034215.

Giudici, P. and Rinaldi, G. (2007). A theoretical model to predict the age of traditional balsamic vinegar. *Journal of Food Engineering* **81:** 121-127.

Giudici, P., Lemmetti, F. and Mazza, S. (2015). *Balsamic Vinegars: Tradition, Technology, Trade*. Springer, Switzerland, International Publishing.

Gobbi, M., De Vero, L., Solieri, L., Comitini, F., Oro, L., Giudici, P. and Ciani, M. (2014). Fermentative aptitude of non-*Saccharomyces* wine yeast for reduction in the ethanol content in wine. *European Food Research and Technology* **239:** 41-48.

Grierson, B. (2009). Malt and Distilled Malt Vinegar. *In:* L. Solieri and P. Giudici (Eds.). *Vinegars of the World*. Springer, Milan, Italy, pp. 136-143.
Gullo, M. and Giudici, P. (2008). Acetic acid bacteria in traditional balsamic vinegar: Phenotypic traits relevant for starter cultures selection. *International Journal of Food Microbiology* **125:** 46-53.
Gullo, M., De Vero, L. and Giudici, P. (2009). Succession of selected strains of *Acetobacter pasteurianus* and other acetic acid bacteria in traditional balsamic vinegar. *Applied and Environmental Microbiology* **75:** 2585-2589.
Gullo, M., Mamlouk, D., De Vero, L. and Giudici, P. (2012). *Acetobacter pasteurianus* strain AB0220: Cultivability and phenotypic stability over 9 years of preservation. *Current Microbiology* **64:** 576-580.
Gullo, M. and Giudici, P. (2014). I batteri acetici. *In:* G. Suzzi and R. Tofalo (Eds.). *Microbiologia Enologica*. Edagricole, Bologna, Italy, pp. 133-148.
Gullo, M., Verzelloni, E. and Canonico, M. (2014). Aerobic submerged fermentation by acetic acid bacteria for vinegar production: Process and biotechnological aspects. *Process Biochemistry* **49:** 1571-1579.
Li, Y.G., Zhang, F., Wang, Z.T. and Hu, Z.B. (2004). Identification and chemical profiling of monacolins in red yeast rice using high-performance liquid chromatography with photodiode array detector and mass spectrometry. *Journal of Pharmaceutical and Biomedical Analysis* **35:** 1101-1112.
Liu, D., Zhu, Y., Beeftink, R., Ooijkaas, L. and Rinzema, A. (2004). Chinese vinegar and its solid-state fermentation process. *Food Reviews International* **20:** 407-424.
Petroni, G. and Gullo, M. (2004). Le anguillule dell'aceto: Stato dell'arte e prospettive. Proceedings of *Anche la tradizione va studiata. Ricerche preliminari per l'individuazione di starter per l'aceto balsamico tradizionale*. Italy, pp. 71-78.
Sievers, M. and Swings, J. (2005). Family *Acetobacteraceae*. *In:* G.M. Garrity (Ed.). *Bergey's Manual of Systematic Bacteriology*, 2nd edn, vol. 2. Springer, New York, pp. 41-95.
Solieri, L., Landi, S., De Vero, L. and Giudici, P. (2006). Molecular assessment of indigenous yeast population from traditional balsamic vinegar. *Journal of Applied Microbiology* **101:** 63-71.
Solieri, L. and Giudici, P. (2008). Yeasts associated to Traditional Balsamic Vinegar: Ecological and technological features. *International Journal of Food Microbiology* **125:** 36-45.
Solieri, L. and Giudici, P. (2009). Vinegars of the World. *In:* L. Solieri and P. Giudici (Eds.). *Vinegars of the World*. Springer-Verlag Italia, pp. 1-16.
Steinkraus, K.H. (1996). *Handbook of Indigenous Fermented Foods*, 2nd edn. Marcel Dekker, Inc. New York.
Yamada, Y., Yukpan, P., Vu, H.T.L., Muramatsu, Y., Ochaikul, D. and Nakagawa, Y. (2012). Subdivision of the genus *Gluconacetobacter* Yamada, Hoshino and Ishikawa 1998: The proposal of *Komagatabacter* gen. nov., for strains accommodated to the *Gluconacetobacter xylinus* group in the *α-Proteobacteria*. *Annals of Microbiology* **62:** 849-859.

11

Impact of Acetic Acid Bacteria on Cocoa Fermentation

Yasmine Hamdouche[1], Corinne Teyssier[2] and Didier Montet[1*]

[1] Cirad, UMR 95 Qualisud TA B-95/16 73, rue Jean-François Breton 34398 Montpellier cedex 5, France

[2] University of Montpellier, UMR 95 Qualisud TA B-95/16 73, rue Jean-François Breton 34398 Montpellier cedex 5, France

Introduction

Cocoa is one of the most traded commodities in the world. The total production was estimated at 4.2 million tons in 2014-2015 (ICCO 2015). Fifty tropical countries cultivate cocoa, but the world production is dominated by Ivory Coast, Ghana and Indonesia. While cocoa is largely produced in developing countries, the derived products are mainly consumed in the industrialized world. The buyers of imported cocoa beans in these countries are essentially the chocolate-processing and manufacturing industry.

Commercial cocoa corresponds to cocoa beans that have undergone different post-harvest processing steps. Cocoa fermentation is a crucial step because it participates and prints the marketable and organoleptic quality of cocoa beans in order to enhance the chocolate taste and flavor. At present, cocoa is fermented spontaneously by contamination of naturally-occurring yeasts, lactic acid bacteria (LAB) and acetic acid bacteria (AAB). The action of these microorganisms causing biochemical reactions contributes to the formation of aroma precursors. If appropriately carried out, the fermentation process is necessary for the production of high quality chocolates, since incorrectly fermented cocoa will produce chocolate that will not contain enough specific aroma compounds.

This chapter describes cocoa fermentation, the microbiota associated with this process and the fermentation methods used. A special focus is on the acetic acid bacteria involved in the fermentation process.

Cocoa Post-harvest Processing

The fruit of the cocoa tree, called 'pod', consists of a fleshy and hard pericarp partially lignified, which contains mucilage and 30 to 60 seeds, called beans. The cocoa postharvest processing represents all the technological operations essential to obtain a merchant cocoa from fresh beans with pulp contained in the harvested pods. Cocoa pods' harvest is done throughout the year, but the great harvest season takes place from October to December, mainly in West Africa (Barel 2013). The mature pods are harvested by hand or with a machete. The harvested pods are stored for some days or immediately opened, depending on the agricultural practice. The pods' opening is either done by cutting with a machete

*Corresponding author: didier.montet@cirad.fr

or by hitting the pod against the ground or with a wooden club, causing the bursting of the fruit and the subsequent release of beans surrounded by pulp. Fresh cocoa beans are fermented under the action of microorganisms using different fermentation methods (Schwan and Wheals 2004). The fermented beans are sundried in order to reduce the water content of the fermented beans and avoid contamination during the storage step. Fermented and dried beans represent merchant cocoa.

Traditional Cocoa Fermentation

Definitions

The cocoa beans are sterile inside the pod, but during the opening process, the environment contaminates the beans coated with pulp and then the fermentation takes place. The duration of this step varies according to the variety of cocoa and agricultural practice (Saltini et al. 2013). Fermentation-induced biochemical reactions occur in the pulp and the cotyledons under the action of microorganisms. The objectives of the fermentation are: (i) removing the mucilage by the microbial action, (ii) causing death of the embryo to prevent sprouting, and (iii) triggering biochemical reactions in the cotyledons to produce aroma precursors (Schwan and Wheals 2004).

The fermentation process is biphasic in time. The first phase, which occurs in the pulp, corresponds to alcoholic fermentation where sugars are converted into alcohols. During the second phase, 'acetic fermentation', ethanol is oxidized into acetic acid, which diffuses into the beans, causing death of the embryo and activating endogenous enzymes necessary for the production of aroma precursors.

Fermentation Methods

The most common cocoa bean fermentation methods are carried out in heaps and box (Camu et al. 2007, 2008, Daniel et al. 2009, Guehi et al. 2010, Papalexandratou et al. 2011a, 2011b, Bankoff et al. 2014, Visintin et al. 2016). For the heap fermentation, cocoa beans are placed on plantain leaves and covered by leaves. Sixty per cent of the produced cocoa is processed by this practice. The advantage of this method is to reduce the number of brewing, as it ensures good gas exchange between the beans and the external environment, but, it does not protect the beans against temperature variations (Barel 2013). The cocoa beans, fermented in wooden boxes, are generally of good quality because the boxes have good thermal insulation and are naturally contaminated by microorganisms that are essential to the success of the fermentation (Barel 2013).

The microbiota associated with box and heap cocoa fermentation were widely assessed (Camu et al. 2007, 2008, Lagunes Gálvez et al. 2007, Nielsen et al. 2007, Daniel et al. 2009, Papalexandratou et al. 2011a, 2011b, 2013, De Melo Pereira et al. 2013, Hamdouche et al. 2015), but only few studies compared the microbial ecology of the two methods. Global microbial analyses of fermented cocoa beans using PCR-DGGE did not show a significant difference for the two traditional methods. The number of strains identified by 16S DNA sequencing is not much different (Hamdouche 2015). Further study based on molecular characterization (rep-PCR) highlighted that the fermentation method affected the biodiversity of dominating strains in cocoa fermentation as *Acetobacter pasteurianus* and *Lactobacillus fermentum* (Visintin et al. 2016). The same study showed that the temperature was higher in box fermentation than in heap fermentation. Although, cocoa fermentation methods influence the microbiota, the same cocoa quality was obtained for wooden box, plastic box and heap fermentations (Bankoff et al. 2014).

Microbial Flora Succession during Fermentation

Several studies have described the succession of microbial groups during cocoa beans fermentation. Yeasts are the first group that start the fermentation followed by lactic acid bacteria (LAB), and then acetic acid bacteria (AAB). Some aerobic spore-forming bacteria, like *Bacillus* spp., appear at the end of fermentation. Indeed, the microbial flora relay takes place only if the environmental conditions are adequate for the development of different microbial groups (Ardhana and Fleet 2003, Nielsen et al. 2007).

Yeasts Involved in Cocoa Fermentation

Yeasts are the dominant microorganisms formed during the first two days of cocoa fermentation. They ensure alcoholic fermentation by transforming sugars (sucrose, glucose and/or fructose) into ethyl alcohol and carbon dioxide. These reactions occur under anaerobic conditions due to the density of the pulp, which prevents air to seep in between the beans. Some yeasts have a good pectinolytic activity, which permit hydrolysis of the cocoa pulp, causing its liquefaction (Schwan et al. 1997, Crafack et al. 2013). During this phase, the maximum load of yeasts varies from 6 to 8 log CFU g of cocoa (Lagunes Gálvez et al. 2007, Nielsen et al. 2007).

The alcoholic fermentation is slightly exothermic (40°C) and is accompanied by a slight increase in pH, which can be caused by the consumption of citric acid by yeasts. Several studies on fermented beans showed that *Hanseniaspora* was the most isolated yeast genus active during the initial fermentation stage (Ardhana and Fleet 2003, Nielsen et al. 2005, Lagunes Gálvez et al. 2007, Nielsen et al. 2007, 2008, De Melo Pereira et al. 2013). *Hanseniaspora guilliermondii* was frequently isolated from Ghanaian fermented cocoa (Nielsen et al. 2005, 2007). *Hanseniaspora opuntiae* was also isolated (Papalexandratou et al. 2013) as well as *Saccharomyces cerevisiae* in fermented cocoa beans from Indonesia (Ardhana and Fleet 2003), Brazil (De Melo Pereira et al. 2013) and the Dominican Republic (Lagunes Gálvez et al. 2007). Others yeast species were also detected in cocoa fermentation, like *Kloeckera apis, Candida tropicalis, Candida krusei, Candida zemplinina, Candida diversa, Issatchenkia orientalis* and *Pichia membranifaciens,* which played an important role in alcoholic fermentation (Ardhana and Fleet 2003, Nielsen et al. 2005, Lagunes Gálvez et al. 2007, Nielsen et al. 2007, 2008, De Melo Pereira et al. 2013).

Bacteria Involved in Cocoa Fermentation

Lactic Acid Bacteria (LAB)

Cocoa pulp acidity and availability of substrates promote the development of LAB that reach a growth ranging from 7 to 7.8 log CFU/g of cocoa beans (Lagunes Gálvez et al. 2007, Nielsen et al. 2007). LAB convert citric acid and sugars of cocoa pulp into lactic acid and sometimes into acetic acid and ethanol as seen in the case of heterolactic fermentation. Previous studies showed that *Lactobacillus fermentum*, *Lactobacillus plantarum* followed by *Leuconostoc pseudomesenteroides* were the most frequently found LAB in fermented cocoa (Camu et al. 2007, Lefeber et al. 2011, Papalexandratou et al. 2011a, 2011b, 2013, Hamdouche et al. 2015). Other LAB genera, such as *Fructobacillus*, *Enterococcus* and *Weissella* (Camu et al. 2007, Lefeber et al. 2011, Papalexandratou et al. 2011b) were also detected in cocoa fermentation.

Acetic Acid Bacteria (AAB)

Many acetic acid bacteria (AAB) are fastidious, meaning that they can be difficult to culture in laboratory situations, because they need a high quantity of oxygen to perform the oxidation of ethanol in acetic acid following the reaction:

$$\text{Ethanol} + O_2 \longrightarrow \text{Acetic acid} + \text{Water}$$

This process could be improved by agitation. Assimilation of citric acid by LAB generates a slight rise in pH to 4.5 and temperature to 37°C of cocoa pulp (Camu et al. 2007). This environment, together with the high level of ethanol and oxygen, allow the start of the second phase of fermentation, which is called acetic fermentation. AAB convert ethanol into acetic acid by oxidation. This reaction is highly exothermic (45-50°C). The maximum growth of AAB varies from 7.8 to 8 log CFU/g during cocoa beans fermentation (Nielsen et al. 2007, Lagunes Gálvez et al. 2007). Usually AAB found in fermented beans belong to *Acetobacter* genus while *A. pasteurianus* is the most dominant species. *A. aceti, A. tropicalis* and *Gluconobacter oxydans* are the species isolated in most cocoa fermentation while *A. cerevisiae, A. fabarum, A. ghanensis, A. lovaniensis*-like species, *A. indonesiensis, A. malorum, A. nitrogenifigens, A. lovaniensis, A. orientalis, A. peroxydans, A. pomorum, A. senegalensis, A. syzygii, Asaia sp., Gluconobacter oxydans* and *Komagataeibacter xylinus* are the less frequently isolated species from cocoa fermentation (Sengun 2015). Some works show dynamic and succession of AAB communities during fermentation (Neilsen et al. 2007, Papalexandratou et al. 2011a) and drying step (Hamdouche et al. 2015).

Others

When fermentation is extended, environmental conditions lead to depletion of microbial substrates where aeration remains important. Then new populations of bacteria grow, especially spore-forming bacteria, which produce some molecules influencing the final cocoa flavor. It is the case of *Bacillus* species. Thermophilic bacteria do not have the ability to grow in the beginning of fermentation, but they can become prevalent flora at the end of fermentation (when the temperature rises). *Bacillus cereus, Bacillus coagulans* and *Bacillus megaterium* are the species found in cocoa fermentation. However, their roles in the fermentation process are not known (Ardhana and Fleet 2003); *Bacillus licheniformis* was detected in the last phase of Ghanaian cocoa fermentation (Nielsen et al. 2007).

Non-spore-forming bacteria could also be detected in cocoa fermentation, as is the case in the family of *Enterobacteriaceae: Tatumella, Erwinia* and *Pantoea* spp. (Lefeber et al. 2011, Papalexandratou et al. 2011b).

Impact of Acetic Acid Bacteria on Cocoa Fermentation

Aeration of the fermenting mass and the temperature increase during fermentation promote the growth of AAB, which become the dominant microorganisms in cocoa fermentation (Passos et al. 1984). Their dominance could be explained by their resistance to acidity and heat during fermentation (Ndoye et al. 2006). The AAB are responsible for the oxidation of ethanol to acetic acid, which is later oxidized to carbon dioxide and water (Schwan et al. 1997) by the following reaction:

$$\underset{\text{Acetic acid}}{CH_3COOH} \longrightarrow 2\,CO_2 + 2\,H_2O$$

According to Kumiko et al. (2001), not all strains of the same species have the ability to oxidize acetic acid to CO_2 and H_2O. *Acetobacter* strains could oxidize acetic acid into CO_2 during cocoa fermentation (da Veiga Moreira et al. 2013). Some dominating AAB species in cocoa bean fermentation as seen in *A. pasteurianus* are desirable during fermentation to obtain acetic acid out of carbohydrates and/or citric acid and ethanol, respectively (De Vuyst et al. 2010).

Romero et al. (2012) compared the amounts of acetic acid produced during the traditional cocoa fermentation. When AAB strains were used in cocoa fermentation as the starter, the increase in the acetic acid level was observed after 60 hours of fermentation

(Romero et al. 2012). *A. tropicalis* could also be used in cocoa fermentation in order to accelerate the acetic acid production, which leads to increase in acetic acid production after 48 hours of fermentation. It was observed that the accumulation of acetic acid in the dicotyledon was slow in the early stages, but increased after the third day (Ardhana and Fleet 2003, Lagunes-Gálvez et al. 2007) maintaining the acetic acid content till the end of fermentation. The production of acetic acid was more influenced by the conversion of carbohydrates into ethanol by the yeast-AAB combination and subsequent conversion of ethanol into acetic acid, than by the AAB population size (De Melo Pereira et al. 2013). The combined action of LAB and AAB, in particular in the beginning of fermentation, could affect the acetic acid production (Papalexandratou et al. 2011b).

Physico-Chemical Changes during Cocoa Fermentation

The traditional cocoa fermentation duration varies from four to six days. The most important physical parameters measured during cocoa fermentation are temperature and acidity.

Physical Parameters

The initial temperature of pulp bean varies with ambient temperature between 23 to 30°C. It increases through the fermentation and reaches a maximum value of 50°C after 48 hours (Lagunes Gálvez et al. 2007, Lefeber et al. 2011, Papalexandratou et al. 2011a, b). The temperature could exceed 50°C (De Melo Pereira et al. 2013). Papalexandratou et al. (2013) showed that the temperature increased after mixing the fermenting cocoa pulp-bean masses. The mucilage acidity is dependent on the presence of citric acid and the production of acetic acid. The initial pH of the fresh pulp ranges from 3.5 to 3.9. During the fermentation period, it increases gradually, which is followed by a slight drop and then stabilizes around 4.0 to 4.4 (Lagunes Gálvez et al. 2007, Lefeber et al. 2011, Papalexandratou et al. 2011a, b, 2013).

At the beginning of fermentation, the average water content of cocoa pulp is around 76.6 per cent and decreases to 69.3 per cent after 144 hours of fermentation (Lagunes Gálvez et al. 2007).

Biochemical Parameters

Fresh cocoa pulp contains glucose, fructose and citric acid, but no sucrose (Lefeber et al. 2011, Papalexandratou et al. 2011b). In some cases, sucrose is present and hydrolyzed into glucose and fructose in the pulp by yeast invertase activity (De Melo Pereira et al. 2013). After 48 hours of fermentation, the carbohydrates are rapidly consumed and completely exhausted after 96 hours. Most of the pulp is liquefied in the first 72 hours of fermentation (Lefeber et al. 2011, Papalexandratou et al. 2011b). Citric acid naturally present in the pulp disappears during the first 72 hours of fermentation in parallel with the dominance of LAB when lactic acid reaches a maximum concentration (Papalexandratou et al. 2011b). Ethanol is produced and then converted to acetic acid by AAB. At the end of fermentation, lactic acid is also oxidized by AAB (Lefeber et al. 2011).

The main carbohydrate inside the beans is sucrose, whose hydrolysis is facilitated by the heat generated due to acetic fermentation (Lefeber et al. 2011, De Melo Pereira et al. 2013). The concentrations of glucose, fructose and citric acid stay constant during fermentation and the concentration of citric acid decreases only at the end of fermentation.

Potential Effect of Acetic Acid Bacteria on Physical Changes

The decrease in pH in the mucilage and dicotyledon is mainly due to the consumption of citric acid and production of acetic acid by AAB (Ardhana and Fleet 2003). The acetic fermentation is an exothermic reaction (Schwan and Wheals 2004). During the cocoa beans fermentation, the temperature could exceed 50°C (De Melo Pereira et al. 2013) and this allows the development of thermophilic bacteria, such as *Bacillus*, which could affect the final quality of cocoa. They may contribute to the acidity and perhaps, at times, to off-flavors of fermented cocoa beans.

Aroma Production during Fermentation

Volatile compounds that impart the special flavor of chocolate are not naturally present in cocoa. It is during the various steps of cocoa post-harvest processing that the aromas develop. Fermentation and roasting are the main steps that contribute to the production and development of cocoa aroma compounds. Volatile compounds can be developed during cocoa fermentation by the primary metabolism of microorganisms. Microorganisms can perform a multitude of reactions (using specific substrates), such as oxidations, reductions, degradation and hydrolysis reactions that participate in the production of aroma compounds.

Involvement of Microorganisms in Cocoa Aroma

The main groups of volatile compounds that are naturally present in fresh pulp-beans are aldehydes, ketones and acids (Afoakwa et al. 2009, Hamdouche 2015). Ho et al. (2014) showed that the alcohol compounds were not detectable in fresh beans as compared to aldehydes and ketones, while the esters are not detectable.

Several studies have been conducted to investigate the role of microorganisms in wine aroma production, such as the production of acetate esters by yeasts (Plata et al. 1998, Lambrechts and Pretorius 2000, Rojas et al. 2003, Molina et al. 2007, Swiegers et al. 2009, Viana et al. 2009, Saez et al. 2011). But little work has been done to understand the role of microorganisms in aroma formation during cocoa post-harvest processing, mainly during the fermentation steps. Volatile compounds from different groups were identified during cocoa fermentation, which were mainly alcohols and esters (Afoakwa et al. 2008, Rodriguez-Campos et al. 2011, 2012, Ho et al. 2014). These molecules are desirable compounds in fermented cocoa (Schwan and Wheals 2004). Alcohols and esters are mainly produced by yeast in fermented foods (Holloway and Subden 1991, Lee et al. 1995) as well as in fermented cocoa (Ho et al. 2014).

The main alcoholic compounds present during cocoa fermentation are: 3-methyl-1-butanol, 3-methyl-2-butanol, methyl-1-propanol, 2,3-butanediol, 1,3-butanediol, 2-phenylethanol and isoamyl alcohol (Rodriguez-Campos et al. 2011, 2012, Hamdouche 2015). Ho et al. (2014) showed that 3-methyl-1-butanol, 2,3-butanediol and 2-phenylethanol were produced by yeasts, such as *Kloeckera apiculata, Saccharomyces cerevisiae, Hanseniaspora guilliermondii, Pichia kudriavzevii (Issatchenkia orientalis), Klyveromyces marxianus*. These compounds are highly desired in cocoa.

The main compounds belonging to ester group detected during cocoa fermentation were acetates (Rodriguez-Campos et al. 2011, 2012, Ho et al. 2014) that were produced by acetyl alcohol transferases of yeasts (Calderbank and Hammond 1994, Rojas et al. 2002).

Compounds belonging to aldehyde and ketone groups were also detected in fermented cocoa and they did not seem to be affected by the action of yeasts (Ho et al. 2014). The main recovered aldehydes were 2-methylbutanal, 3-methylbutanal and benzaldehyde, which may be formed by LAB from amino acids precursors, such as leucine, isoleucine and phenylalanine during cocoa fermentation (Jinap et al. 1994, Groot and de Bont, 1998,

1999). In addition, aldehyde compounds, such as phenyl acetaldehyde, were reported as derivatives of amino acid catabolism performed during cocoa fermentation (Afoakwa et al. 2008). Acetoin is the most important ketone in quantity in fermented cocoa followed by other compounds: 2,3-butanedione, acetophenone and phenyl acetaldehyde. Acetoin could be produced by fermentation from pyruvate and butanediol (Pretorius 2000).

It was noticed that acetic acid and isovaleric acid were the dominant acids in the unroasted fermented cocoa (Frauendorfer and Schieberle 2008). During fermentation, other acids were also detected—propionic acid, hexanoic acid, octanoic acid, nonanoic acid, and butyric acid, which have an impact on cocoa aromatic quality (Serra-Bonvehí 2005). Organic acids content increases during cocoa fermentation because of the metabolism of sugars present in the pulp (Thompson et al. 2001, Serra-Bonvehí 2005).

In some works, pyrazines compounds were not detected during cocoa fermentation (Ho et al. 2014, Rodriguez-Campos et al. 2011, 2012). Other studies showed the presence of some pyrazines, like tetramethylpyrazine and trimethylpyrazine at the end of fermentation, could be produced by *Bacillus* (Jinap et al. 1994, Schwan and Wheals 2004). Therefore, a long period of fermentation and in particular over-fermentation can influence the aromatic quality of cocoa by producing desired molecules.

Impact of Acetic Acid Bacteria on Cocoa Aroma

When the pH drops from 6.5 to 4.5 and the temperature increases to 45-50°C, the seed cell walls become permeable (Belitz et al. 2009). Acetic acid, which is the key metabolite in cocoa bean fermentation, penetrates through the grain and causes, with the high temperature generated by the oxidation of alcohol, the death of the cocoa seed and the destruction of the internal structure of the cells. This marks the end of fermentation and initiates biochemical changes in the beans, leading to the formation of aroma precursors of cocoa (Schwan and Wheals 2004). Among the biochemical reactions taking place, polyphenols oxidation predominate. Amino acids and peptides react with this oxidation and cause brown pigments, which confer the characteristic color of cocoa. Polyphenol oxidation allows a decrease in the phenol content; this mellows the original bitter and astringent flavor of cocoa (Belitz et al. 2009).

Acetic acid bacteria have a primordial role in cocoa fermentation. In the absence of acetic fermentation, the chocolate will not possess its typical aroma. Acidification of the bean is necessary to trigger the biochemical reactions and the formation of aroma precursors. Romero et al. (2012) proposed the selection and inoculation of AAB strains that are most important during the fermentation process in order to avoid the loss of quality in fermented cocoa beans.

Acetic acid bacteria could produce desirable volatile compounds in cocoa, as acetoin is produced by the oxidation of lactic acid. *Acetobacter pasteurianus* carried out a fast oxidation of lactic acid into acetoin. Acetic acid and acetoin are the major end metabolites of cocoa bean fermentation, reflecting the final activity of AAB during the process (Moens et al. 2014). Hamdouche (2015) identified the aromatic profile of some yeast and bacteria strains in cocoa beans fermentation and observed that *A. pasteurianus* could produces esters, which were desirable compounds in cocoa. Esters usually give a fruity and flowery flavor to food (Belitz et al. 2009).

Improving Cocoa Aromatic Quality by Using Culture Starters

Cocoa bean fermentation is a natural and spontaneous operation. Therefore, the action of microorganisms is random and uncontrolled. Thus, we could think to develop starter cultures for better control of the fermentation process in order to improve the quality of fermented cocoa. Schwan et al. (1997) were the first to show that spontaneous fermentation and fermentation using mixed culture, including yeasts, LAB and AAB produced cocoa of a comparable quality.

Some recent studies investigated the use of selected bacteria and yeasts as starter cultures in order to improve the potential flavor of cocoa. For example, Lefeber et al. (2012) showed that a mixed starter culture, including strains of LAB (*L. fermentum*), AAB (*A. pasteurianus*) and yeast (*S. cerevisiae*) produced high quality cocoa and flavored chocolates compared to fermentation inoculated with a pure LAB/AAB bacterial starter culture. A bit later, Crafack et al. (2013) demonstrated the potential impact of aromatic and pectinolytic yeasts, which were *Pichia kluyveri* and *Kluyveromyces marxianus* on cocoa flavor and sensory qualities. The importance of yeast metabolism on the development of chocolate aroma has recently been elucidated by Ho et al. (2014), who showed that cocoa fermented in the absence of yeast yielded an acidic chocolate with a lack of the characteristic chocolate aroma.

Conclusion

Acetic fermentation cannot take place in cocoa fermentation without the intervention of yeast and LAB. These microorganisms are involved firstly as they degrade the pulp surrounding the beans and therefore, create an aerobic environment favorable for the development of acetic acid bacteria. Technology actions, as brewing during the fermentation, facilitate the aerobic conditions that are favorable for transformation of ethanol into acetic acid by acetic acid bacteria. Furthermore, without acetic fermentation, enzymatic reactions inside the beans cannot be triggered, which deters production of aromatic precursors that give the typical chocolate aroma. To provide a good chocolate aroma quality, it is important to select strains with aromatic potential in cocoa fermentation, like yeasts.

References

Afoakwa, E.O., Paterson, A., Fowler, M. and Ryan, A. (2008). Flavor formation and character in cocoa and chocolate: A critical review. *Critical Reviews in Food Science and Nutrition* **48:** 840-857.

Afoakwa, E.O., Paterson, A., Fowler, M. and Ryan, A. (2009). Matrix effects on flavor volatiles release in dark chocolates varying in particle size distribution and fat content using GC-mass spectrometry and GC-olfactometry. *Food Chemistry* **113:** 208-215.

Ardhana, M.M. and Fleet, G.H. (2003). The microbial ecology of cocoa bean fermentations in Indonesia. *International Journal of Food Microbiology* **86:** 87-99.

Bankoff, L., Ouattara, G.H., Karou, T.G., Guehi, S.T., Nemlin, J.G. and Diopoh, J.K. (2014). Impacts de la fermentation du cacao sur la croissance de la flore microbienne et la qualité des fèves marchandes. *Agronomie Africaine* **25:** 159-170.

Barel, M. (2013). Qualité du cacao. L'impact du traitement post-récolte. Edition Quae. Versailles, France.

Belitz, H.D., Grosch, W. and Schieberle, P. (2009). Coffee, tea, cocoa. *In:* Belitz, H.D., Grosch, W. and Schieberle, P. (Eds.). *Food Chemistry*. Springer-Verlag Berlin Heidelberg, pp. 938-970.

Calderbank, J. and Hammond, J.R.M. (1994). Influence of higher alcohol availability on ester formation by yeast. *Journal of the American Society of Brewing Chemists* **52:** 84-90.

Camu, N., De Winter, T., Verbrugghe, K., Cleenwerck, I., Vandamme, P., Takrama, J.S. and De Vuyst, L. (2007). Dynamics and biodiversity of populations of lactic acid bacteria and acetic acid bacteria involved in spontaneous heap fermentation of cocoa beans in Ghana. *Applied and Environmental Microbiology* **73:** 1809-1824.

Camu, N., Gonzalez, A., De Winter, T., Van Schoor, A., De Bruyne, K., Vandamme, P. and De Vuyst, L. (2008). Influence of turning and environmental contamination on the dynamics of populations of lactic acid and acetic acid bacteria involved in spontaneous cocoa bean heap fermentation in Ghana. *Applied and Environmental Microbiology* **74:** 86-98.

Crafack, M., Mikkelsen, M.B., Saerens, S., Knudsen, M., Blennow, A., Lowor, S. and Nielsen, D.S. (2013). Influencing cocoa flavour using *Pichia kluyveri* and *Kluyveromyces marxianus* in a defined mixed starter culture for cocoa fermentation. *International Journal of Food Microbiology* **167:** 103-116.

da Veiga Moreira, I.M., Miguel, M.G.D.C.P., Duarte, W.F., Dias, D.R. and Schwan, R.F. (2013). Microbial succession and the dynamics of metabolites and sugars during the fermentation of three different cocoa (*Theobroma cacao L.*) hybrids. *Food Research International* **54:** 9-17.

Daniel, H.M., Vrancken, G., Takrama, J.F., Camu, N., De Vos, P. and De Vuyst, L. (2009). Yeast diversity of Ghanaian cocoa bean heap fermentations. *FEMS Yeast Research* **9:** 774-783.

De Melo Pereira, G.F., Magalhaes-Guedes, K.T. and Schwan, R.S. (2013). rDNA-based DGGE analysis and electron microscopic observation of cocoa beans to monitor microbial diversity and distribution during the fermentation process. *Food Research International* **53:** 482-486.

De Vuyst, L., Lefeber, T., Papalexandratou, Z. and Camu, N. (2010). The functional role of lactic acid bacteria in cocoa bean fermentation. *In:* F. Mozzi, R.R. Raya and G.M. Vignolo (Eds.). *Biotechnology of Lactic Acid Bacteria: Novel Applications*. Wiley-Blackwell, Ames, IA, pp. 301-326.

Frauendorfer, F. and Schieberle, P. (2008). Changes in key aroma compounds of Criollo cocoa beans during roasting. *Journal of Agricultural and Food Chemistry* **56:** 10244-10251.

Groot, M.N.N. and de Bont, J.A. (1998). Conversion of phenylalanine to benzaldehyde initiated by an aminotransferase in *Lactobacillus plantarum*. *Applied and Environmental Microbiology* **64:** 3009-3013.

Groot, M.N.N. and de Bont, J.A. (1999). Involvement of manganese in conversion of phenylalanine to benzaldehyde by lactic acid bacteria. *Applied and Environmental Microbiology* **65:** 5590-5593.

Guehi, S.T., Dabonne, S., Ban-Koffi, L., Kedjebo, D.K. and Zahouli, G.I.B. (2010). Effect of turning beans and fermentation method on the acidity and physical quality of raw cocoa beans. *Advance Journal of Food Science and Technology* **2:** 163-171.

Hamdouche, Y. (2015). Discrimination des procédés de transformation post-récolte du Cacao et du Café par analyse globale de l'écologie microbienne. Thesis, Supagro, Montpellier, France.

Hamdouche, Y., Guehi, T., Durand, N., Kedjebo, K.B.D., Montet, D. and Meile, J.C. (2015). Dynamics of microbial ecology during cocoa fermentation and drying: Towards the identification of molecular markers. *Food Control* **48:** 117-122.

Ho, V., Zhao, J. and Fleet, G. (2014). Yeasts are essential for cocoa bean fermentation. *International Journal of Food Microbiology* **174:** 72-87.

Holloway, P. and Subden, R.E. (1991). Volatile metabolites produced in a Riesling must by wild yeast isolates. *Canadian Institute of Food Science and Technology Journal* **24:** 57-59.

ICCO (International Cocoa Organization) (2015). Avaible from: http://www.icco.org (Accessed: April 2016).

Jinap, S., Thien, J. and Yap, T.N. (1994). Effect of drying on acidity and volatile fatty acids content of cocoa beans. *Journal of the Science of Food and Agriculture* **65:** 67-75.

Kumiko, N., Mariko, T., Satoshi, U., Nobuhiro, I., Hirotaka, M., Hisayo, O.Y. and Yoshika, T.M. (2001). Characterization of acetic acid bacteria in traditional acetic acid fermentation of rice vinegar (Komesu) and unpolished rice vinegar (Kurosu) produced in Japan. *Applied and Environmental Microbiology* **67:** 986-1000.

Lagunes-Gálvez, S., Loiseau, G., Paredes, J.L., Barel, M., and Guiraud, J.P. (2007). Study on the microflora and biochemistry of cocoa fermentation in the Dominican Republic. *International Journal of Food Microbiology* **114:** 124-130.

Lambrechts, M.G. and Pretorius, I.S. (2000). Yeast and its importance to wine aroma. *Journal of South African Society for Enology & Viticulture* **21:** 97-129.

Lee, S., Villa, K. and Patino, H. (1995). Yeast strain development for enhanced production of desirable alcohols/esters in beer. *Journal of the American Society of Brewing Chemists* **53:** 153-156.

Lefeber, T., Gobert, W., Vrancken, G., Camu, N. and De Vuyst, L. (2011). Dynamics and species diversity of communities of lactic acid bacteria and acetic acid bacteria during spontaneous cocoa bean fermentation in vessels. *Food Microbiology* **28:** 457-464.

Lefeber, T., Papalexandratou, Z., Gobert, W., Camu, N. and De Vuyst, L. (2012). On-farm implementation of a starter culture for improved cocoa bean fermentation and its influence on the flavour of chocolates produced thereof. *Food Microbiology* **30:** 379-392.

Moens, F., Lefeber, T. and De Vuyst, L. (2014). Oxidation of metabolites highlights the microbial interactions and role of *Acetobacter pasteurianus* during cocoa bean fermentation. *Applied and Environmental Microbiology* **80:** 1848-1857.

Molina, A.M., Swiegers, J.H., Varela, C., Pretorius, I.S. and Agosin, E. (2007). Influence of wine fermentation temperature on the synthesis of yeast-derived volatile aroma compounds. *Applied Microbiology and Biotechnology* **77:** 675-687.

Ndoye, B., Cleenwerck, I., Engelbeen, K., Dubois-Dauphin, R., Guiro, A.T., Van Trappen, S. and Thonart, P. (2006). *Acetobacter senegalensis sp. nov.*, a thermotolerant acetic acid bacterium isolated in Senegal (sub-Saharan Africa) from mango fruit (*Mangifera indica L.*). *International Journal of Systematic and Evolutionary Microbiology* **57:** 1576-1581.

Nielsen, D.S., Hønholt, S., Tano-Debrah, K. and Jespersen, L. (2005). Yeast populations associated with Ghanaian cocoa fermentations analysed using denaturing gradient gel electrophoresis (DGGE). *Yeast* **22:** 271-284.

Nielsen, D.S., Teniola, O.D., Ban-Koffi, L., Owusu, M., Andersson, T.S. and Holzapfel, W.H. (2007). The microbiology of Ghanaian cocoa fermentations analyzed using culture dependent and culture-independent methods. *International Journal of Food Microbiology* **114:** 168-186.

Nielsen, D.S., Snitkjaer, P. and van den Berg, F. (2008). Investigating the fermentation of cocoa by correlating denaturing gradient gel electrophoresis profiles and near infrared spectra. *International Journal of Food Microbiology* **125:** 133-140.

Papalexandratou, Z., Camu, N., Falony, G. and De Vuyst, L. (2011a). Comparison of the bacterial species diversity of spontaneous cocoa bean fermentations carried out at selected farms in Ivory Coast and Brazil. *Food Microbiology* **28:** 964-973.

Papalexandratou, Z., Vrancken, G., De Bruyne, K., Vandamme, P. and De Vuyst, L. (2011b). Spontaneous organic cocoa bean box fermentations in Brazil are characterized by a restricted species diversity of lactic acid bacteria and acetic acid bacteria. *Food Microbiology* **28:** 1326-1338.

Papalexandratou, Z., Lefeber, T., Bahrim, B., Lee, O.S., Daniel, H.M. and De Vuyst, L. (2013). *Hanseniaspora opuntiae, Saccharomyces cerevisiae, Lactobacillus fermentum, and Acetobacter pasteurianus* predominate during well-performed Malaysian cocoa bean box fermentations, underlining the importance of these microbial species for a successful cocoa bean fermentation process. *Food Microbiology* **35:** 73-85.

Passos, M.L.F., Olzany, S.D., Lopez, A., Ferreira, L.L.F.C. and Vieira, G.W. (1984) Characterization and distribution of lactic acid bacteria from traditional cocoa bean fermentations in Bahia. *Journal of Food Science* **49:** 205-208.

Plata, M.D.C., Mauricio, J.C., Millán, C. and Ortega, J. (1998). *In vitro* specific activity of alcohol acetyltransferase and esterase in two flor yeast strains during biological aging of Sherry wines. *Journal of Fermentation and Bioengineering* **85:** 369-374.

Pretorius, I.S. (2000). Tailoring wine yeast for the new millennium: Novel approaches to the ancient art of winemaking. *Yeast* **16:** 675-729.

Rodriguez-Campos, J., Escalona-Buendía, H.B., Orozco-Avila, I., Lugo-Cervantes, E. and Jaramillo Flores, M.E. (2011). Dynamics of volatile and non-volatile compounds in cocoa (*Theobroma cacao L.*) during fermentation and drying processes using principal components analysis. *Food Research International* **44:** 250-258.

Rodriguez-Campos, J., Escalona-Buendía, H.B., Contreras-Ramos, S.M., Orozco-Avila, I., Jaramillo-Flores, E. and Lugo-Cervantes, E. (2012). Effect of fermentation time and drying temperature on volatile compounds in cocoa. *Food Chemistry* **132:** 277-288.

Rojas, V., Gil, J.V., Manzanares, P., Gavara, R., Piñaga, F. and Flors, A. (2002). Measurement of alcohol acetyltransferase and ester hydrolase activities in yeast extracts. *Enzyme and Microbial Technology* **30:** 224-230.

Rojas, V., Gil, J.V., Piñaga, F. and Manzanares, P. (2003). Acetate ester formation in wine by mixed cultures in laboratory fermentations. *International Journal of Food Microbiology* **86:** 181-188.

Romero,C.T.,Robles,O.V.,Rodríguez,J.G.andRamìrez-Lape,M.(2012).Isolationandcharacterization of acetic acid bacteria in cocoa fermentation. *African Journal of Microbiology Research* **6:** 339-347.

Saez, J.S., Lopes, C.A., Kirs, V.E. and Sangorrín, M. (2011). Production of volatile phenols by *Pichia manshurica* and *Pichia membranifaciens* isolated from spoiled wines and cellar environment in Patagonia. *Food Microbiology* **28:** 503-509.

Saltini, R., Akkerman, R. and Frosch, S. (2013). Optimizing chocolate production through traceability: A review of the influence of farming practices on cocoa bean quality. *Food Control* **29:** 167-187.

Schwan, R.F., Cooper, R.M. and Wheals, A.E. (1997). Endopolygalacturonase secretion by *Kluyveromyces marxianus* and other cocoa pulp-degrading yeasts. *Enzyme and Microbial Technology* **21:** 234-244.

Schwan, R.F. and Wheals, A.E. (2004). The microbiology of cocoa fermentation and its role in chocolate quality. *Critical Reviews in Food Science and Nutrition* **44:** 205-221.

Sengun, I.Y. (2015). Acetic acid bacteria in food fermentations. *In:* D. Montet and R.C. Ray (Eds.). *Fermented Foods: Part 1. Biochemistry and Biotechnology*. CRC Press, Boca Raton, USA, pp. 91-111.

Serra-Bonvehí Bonvehí, J. (2005). Investigation of aromatic compounds in roasted cocoa powder. *European Food Research and Technology* **221:** 19-29.

Swiegers, J.H., Kievit, R.L., Siebert, T., Lattey, K.A., Bramley, B.R., Francis, I.L. and Pretorius, I.S. (2009). The influence of yeast on the aroma of Sauvignon Blanc wine. *Food Microbiology* **26:** 204-211.

Thompson, S.S., Miller, K.B. and Lopez, A.S. (2001). Cocoa and coffee *In:* Doyle, M.P., Beuchat, L.R. and Montville, T.J. (Eds.). *Food Microbiology: Fundamentals and Frontiers*, 2nd edn. ASM Press, Washington D.C., USA, pp. 721-733.

Viana, F., Gil, J.V., Vallés, S. and Manzanares, P. (2009). Increasing the levels of 2 phenylethyl acetate in wine through the use of a mixed culture of *Hanseniaspora osmophila* and *Saccharomyces cerevisiae*. *International Journal of Food Microbiology* **135:** 68-74.

Visintin, S., Alessandria, V., Valente, A., Dolci, P. and Cocolin, L. (2016). Molecular identification and physiological characterization of yeasts, lactic acid bacteria and acetic acid bacteria isolated from heap and box cocoa bean fermentations in West Africa. *International Journal of Food Microbiology* **216:** 69-78.

12

Detrimental Effects of Acetic Acid Bacteria in Foods

Maria José Valera, Maria Jesús Torija and Albert Mas*

Oenological Biotechnology Group, Department of Biochemistry and Biotechnology, Faculty of Oenology, University Rovira i Virgili, Marcel·li Domingo, 1. 43007, Tarragona, Spain

General Remarks

Acetic Acid Bacteria (AAB) were initially described a long time ago. Pasteur defined the wine spoilers as *Mycoderma aceti* and Beijerinck used the term *Acetobacter* at the end of the 19th century and later Asai (1935) used the term *Gluconobacter*. The 1984 version of *Bergey's Manual of Systematic Bacteriology* (De Ley et al. 1984) used only these two genera—*Gluconobacter* and *Acetobacter*. However, the development of more powerful and precise techniques for molecular detection and identification of AAB allowed greater diversity due to the description of new genera and species. Many of the 'old' species were reclassified and renamed with many of them reported as spoilage microorganisms. Thus, we have tried to update the correct names of genera and species to the present nomenclature. However, some microorganisms that were identified for more general species (i.e. *Gluconobacter oxydans*) or those not completely identified at the species level have been maintained as their initial classification in the reference article.

Another relevant aspect is the cultivability of AAB. They have been considered 'fastidious' microorganisms, because they are poorly recovered on plates. The cultivability loss could even be of three orders of magnitude or not cultivable at all (Torija et al. 2010). Although this poor cultivability could be due to selective media that are not appropriate for AAB or due to the aggregation of cells, the main reason is probably related to the status of Viable But Not Cultivable (VBNC). This status refers to the extreme conditions found in wine (Millet and Lonvaud-Funel 2000), or vinegar (Torija et al. 2010). For this reason, special effort has been put into the development of culture-independent methods that provide an accurate estimation of the total AAB population in these media (González et al. 2006, Andorrà et al. 2008, Ilabaca et al. 2008, Torija et al. 2010, Valera et al. 2015). Due to the recent developments of these techniques, most of the previous references described exclusively the cultivable populations of AAB.

Concept of Spoilage and New Product

The acetic acid bacteria (AAB) are ubiquitous microorganisms that can be found in natural environments, such as fruits, vegetables, insects; but also in man-made sources, mainly drinks and foods (Raspor and Goranovič 2008). However, they are particularly described as spoilers in fruits, dairy, fermented food and beverage industries in which the technical procedures or the hygienic conditions are disregarded. One of the main concerns of the

*Corresponding author: albert.mas@urv.cat

manufacturers in food industries is the spoilage of their products. Even with modern techniques for preservation, a large amount of food is lost due to spoilage. A food item is considered spoiled when it presents organoleptic properties that make it unacceptable for the consumer. The negative characteristics can be physical damages or chemical changes. These changes can be attributed to enzymatic activities in the raw material or microbial growth and metabolism in the product and during its transformation. Among them, the most common cause of spoilage in food products is due to microbial transformation (Gram et al. 2002). In the case of AAB spoilage, the main characteristic is the acetic acid production; however, other metabolites, such as polysaccharide films or gluconic acid, can also alter the organoleptic properties of the damaged products.

It is remarkable that microbial spoilage can be carried out by bacteria that are either part of the natural microbiota present in the raw material, or appear along the elaboration, manipulation or storing processes. So, the development of spoilage is very complex and the relation between the microorganisms involved are described in some substrates. In fact, the spoilage of food is found when the population of spoiler microorganisms reaches the value of 10^7-10^9 CFU/g (Boddy and Wimpenny 1992), although lower values have also been found in spoiled wines (Bartowsky et al. 2003). Nevertheless, the control of environmental parameters of the substrate, such as temperature, atmosphere, water activity (a_w) and pH allow prediction of the microbiota that can grow on some products. For that reason, the use of mathematical modelization is a strategy used in order to prevent spoilage in industry (Battey and Shaffner 2001).

As AAB are part of the native microbiota in fruits and fermented products, when the environmental conditions are favorable, they can take over and produce metabolites that alter some of the food properties. In the case of some beverage industries and other fresh and fermented products, these changes are considered as negative. But sometimes, hurdles bring opportunities. From another point of view, spoiled products can be seen as new products which make the new properties attractive for other purposes, thus increasing its commercial value. It is well known that some AAB are able to grow in environments with a high content of ethanol and they can use it as an energy source by producing acetic acid as the final product. In fact, some AAB strains can survive in acidic environments, tolerating pH as low as 2 (Du Toit and Pretorius 2002). This is the key to wine spoilage, but also for vinegar production. Also, vinegar is generally considered inexpensive because the production of common vinegar requires low cost raw materials (such as substandard fruits or byproducts of other industries); but this is not always true. In fact, vinegar is a value-added product and some types of vinegar are more expensive than most wines, due to their special organoleptic characteristics and the process followed to prepare them (Solieri and Giudici 2009).

Wine Spoilage as the Best-Known Detrimental Effect

Wine spoilage by AAB has been known since ancient times, even though the involvement of these microorganisms was not known until the 19th century. It is likely that the first wines produced in Iran (erstwhile Persia) could be spoiled by the native bacteria from grapes when the conditions allowed it. The first winemakers added clay or resin to the vessels in which wine was made, due to their empiric knowledge, in order to prevent oxidations (and thus the growth of AAB). Babylonians discovered the properties of vinegar using it as a wine byproduct, that is, as a food preservative (Bottero 2004). In fact, for many centuries, vinegar was the most acidic solution known.

The process of winemaking generates a complex ecological niche for several microorganisms. The interaction of yeast, bacteria and fungi yields a final product with specific characteristics. However, the contribution of these microbes is not always positive for wine as a final product. During the different steps of winemaking, there are several points where microbial spoilage can occur (Fig. 1). It is noteworthy that the microbiota present in wine will spoil the product or not depends on the environmental conditions and the technological practices carried out. In fact, the main spoiler microorganisms are naturally present on grapes or in the cellar equipment and instruments used during the preparation of wine. The microbes that are usually involved in wine spoilage are yeast strains belonging to the genera *Brettanomyces, Candida, Hanseniaspora, Pichia, Zygosaccharomyces* and others, as well as lactic acid bacteria, such as *Lactobacillus, Leuconostoc, Pediococcus* and the AAB of the genera *Acetobacter* and *Gluconobacter* (Du Toit and Pretorius 2000). Today, due to the description of new genera and species, others too have to be considered as wine spoilers, such as *Gluconacetobacter, Komagataeibacter* or *Asaia*.

The wine alterations can be detected visually through the increase in turbidity, viscosity, presence of sediments and film formation. Moreover, there are changes found in both aroma and taste as bitterness and off-flavors. Mousiness, ester taint, phenolic, buttery or geranium notes are some examples of descriptors found in microbial spoiled wines. One of the best-known defects in wines is the vinegary aroma or taste. This defect is mainly due to acetic acid, which is the main volatile acid in wine (normally 0.2-0.5 g/L in sound wine), that can be produced both by yeast and by lactic and acetic acid bacteria. In yeast, production of acetic acid is part of the normal metabolism in the presence of sugar during alcoholic fermentation (Erasmus et al. 2004). Therefore, its presence in wine is unavoidable. However, the main concerns about acetic acid in wine are due to AAB, which, in contact with oxygen and temperatures between 20 and 30°C, finds the environment appropriate to grow and increase the volatile acidity. Depending on the type of wine, concentrations higher than 0.4-1.5 g/L are undesirable (Davis et al. 1985), although 0.8 g/L is generally considered as the maximum tolerable level in red wine and even lower in white wine. For this reason, it is important to control this parameter during the winemaking process. The norms in several countries limit the concentration of acetic acid in the final wines (Eglinton and Henschke 1999). Wines that present excess of acetic acid, which is under the limit of the allowance, have low commercial value and in some cases, they can be rectified by blending or after reverse-osmosis treatment to lower the acetic acid content (Bartowsky and Henschke 2008). In addition to the vinegar-like sour flavor, AAB can confer on wine glue, nutty, sherry-like, solvent or bruised apple aromas and it is also common to find a reduction in fruity characters (Bartowsky et al. 2003).

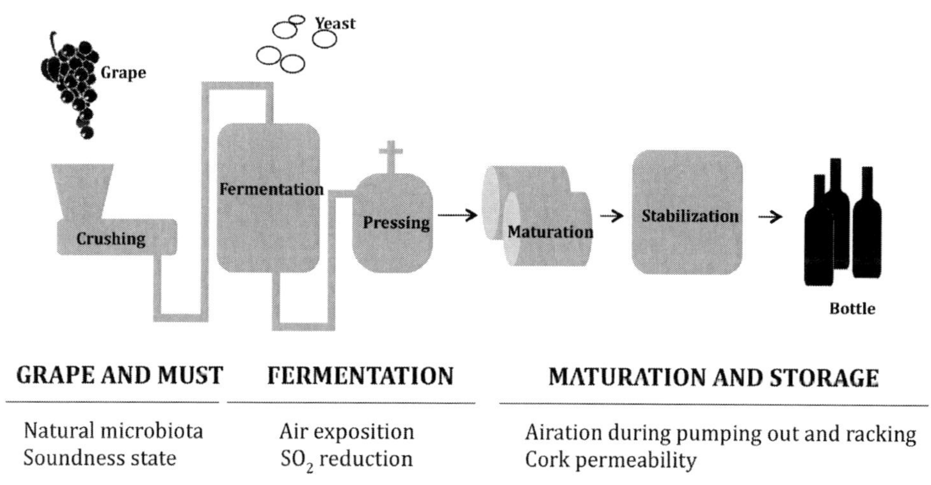

Figure 1: Schematic view of the wine making process with indication of the susceptible steps of AAB spoilage.

Spoilage of Grapes and Musts

Before the fermentation process, AAB can proliferate on grapes or must. In the cellar, a common practice in red wine making is cold soaking. The grapes, after crushing, stay with the skins for a few days at low temperature. This process increases the extraction of more color compounds from the skins to the must (Ribéreau-Gayon et al. 2000) and it can be a point of spoilage due to high oxygen contact and sugar availability.

The addition of SO_2 is not always sufficient to prevent the growth of undesirable bacteria. Another issue to control at this point is the pH of the must—the higher pH values allow higher counts of AAB because SO_2 is mostly in the form of sulfites that are not the most powerful antimicrobial form. However, the presence of sulfites, despite reducing the AAB populations, is not critical for the survival of AAB from grapes that can be detected until the final wines (González et al. 2005). The main effect of SO_2 is bacteriostatic, as the AAB population seems not to be affected by its use, as shown by the culture-independent techniques used to analyze the AAB population during alcoholic fermentation (Andorrà et al. 2008).

But undoubtedly, a key factor in the appearance of AAB prior to fermentation is the soundness of the grapes. Rotten grapes allow the recovery of higher bacterial population (Fleet 1999, Barbe et al. 2001, Barata et al. 2012) and also higher genetic diversity compared with the same grapes in a sound state (Mateo et al. 2014). Sometimes, wines are made with grapes infected with noble rot, *Botrytis cinerea*. These grapes also present higher populations of AAB than the sound ones. As glucose and fructose are the main carbon sources for bacterial growth in grapes or must, the main product in the metabolism of AAB is gluconic acid produced by the oxidation of glucose and, in minor quantity, oxofructose produced by the oxidation of fructose (Barbe et al. 2001). These compounds can change the organoleptic properties of the product and they are a characteristic of botrytized wines. Moreover, the must of grapes infected with *Botrytis cinerea* must be treated with higher amounts of SO_2 because the two compounds, gluconic acid and oxofructose, bind themselves to SO_2, thus decreasing the antimicrobial activity (Barbe et al. 2001).

Another hurdle that arises due to metabolism of glucose by AAB is the production of cellulose. The fibers of this exopolysaccharide affect the grape must and wine filterability (Drysdale and Fleet 1988).

In any case, the spoilage of grapes is controlled during the winemaking process, and if the volatile acidity of the grapes must exceed the limits predetermined, they cannot be used (Eglinton and Henschke 1999). Furthermore, high concentrations of acetic acid produced by AAB can have a toxic effect on yeast and hinder the development of alcoholic fermentation (Ribéreau-Gayon et al. 2000).

Spoilage during Wine Fermentation

Once the fermentation starts, the anaerobic environment does not allow an increase in the AAB population. Although the alcohol formed during this stage is the substrate for AAB, their proliferation is only possible when the environmental conditions are appropriate. The growth is favored in aerobic environment and a temperature higher than 10°C. In fact, AAB have an optimal temperature of growth ranging between 28-30°C. Therefore, AAB can grow during stuck fermentations if exposed to air, but this is commonly avoided using the correct oenological treatments.

Moreover, for most of the AAB strains, ethanol is toxic in concentrations of 5-10 per cent (v/v). However, some resistant strains are able to survive in concentrations up to 15 per cent (v/v) (De Ley et al. 1984). AAB present in wines, with ethanol as an energy source, can transform it to compounds that alter the organoleptic properties of the final wine. They are able to synthesize not only acetic acid but also other metabolites, such as acetaldehyde and ethyl acetate (Drysdale and Fleet 1989a, Plata et al. 2005). Acetaldehyde is the first product of ethanol oxidation and can appear when oxygen is very limited or when the oxidation is performed by yeast (for instance in sherry wines). Acetaldehyde binds SO_2 with high efficiency and reduces substantially the antimicrobial form of this product (Ribéreau-Gayon et al. 2000). Ethyl acetate can be either produced as a result of AAB metabolism or by chemical reaction between acetic acid and ethanol. Though the chemical reaction is slow, the production of ethyl acetate by AAB is seen as the mechanism to reduce the initial toxicity of acetic acid due to the intervention of either esterases or acetyl-transferases (Guillamón and Mas 2011). On the other hand, during alcoholic fermentation, yeast produces glycerol. This alcohol is also oxidized by AAB, yielding dihydroxyacetone. The positive characteristics (basically mouth feel) that glycerol confers to wine are decreased as far as it is consumed by the AAB. Also in this process, the antimicrobial activity of SO_2 is reduced due to their binding to dihydroxyacetone. Moreover, AAB can also decrease, in the final wines, the concentration of some metabolites that are either produced during alcoholic fermentation or derived from grapes and which are considered positive for their organoleptic properties. For example, AAB can metabolize organic acids, such as malic, tartaric or citric acids (Drysdale and Fleet 1989b).

Spoilage during Maturation and Storage

The most common cause for wine spoilage by AAB is the unintentional exposure to air during maturation or storage (Joyeux et al. 1984). In the barrels, the amount of oxygen available is highly reduced, even though, AAB have been commonly isolated from wine samples taken from the bottom of tanks and barrels because they can survive even under low oxygen conditions due to the presence of other electron acceptors, such as phenolic compounds and quinones, which will allow them to survive but not to proliferate (Drysdale and Fleet 1989a). Therefore, AAB should be considered opportunistic microorganisms, as they are able to proliferate when oxygen levels are suitable for their growth. In addition, they could benefit from oxygen diffusion through wooden barrels (Joyeux et al. 1984).

Common oenological practices, such as pumping over and racking, are points where the wine is in contact with small amounts of oxygen. In these conditions, AAB that were present even in a low quantity could be stimulated to grow. After racking and fining, the number of AAB can increase between 10^2 and 10^3 cell/mL (Millet and Lonvaud Funel 2000).

After bottling, the amount of oxygen is reduced and the conditions in the environment become quite anaerobic. However, during bottling, the addition of oxygen is uncontrolled and the AAB population can increase (Millet and Lonvaud-Funel 2000). In fact, although wine before bottling is subjected to several stabilization steps, such as fining or filtering, the diffusion of oxygen throughout the cork can allow AAB spoilage. The presence of a ring of biomass was visually detected in the neck of bottles packaged vertically upright. The permeability of the cork and the environment created in the biofilm by AAB can allow the growth of these bacteria and the spoilage of wines (Bartowsly et al. 2003). Furthermore, if oxygen can enter through the cork, the first product will be acetaldehyde that will combine with free SO_2 to eliminate its bacteriostatic effect, thus allowing AAB to proliferate.

Species of AAB Involved in Wine Spoilage

Originally, just three AAB species were commonly associated with grapes and wine—*Gluconobacter oxydans*, *Acetobacter aceti* and *Acetobacter pasteurianus* (Table 1); currently, other additional *Gluconobacter* species have been identified on grapes and wine as well as species of *Acetobacter* genus. On the other hand, they have also been described from

Table 1: AAB genera detected during different stages in winemaking

Stage	*Sample*	*Sample origin*	*AAB genera detected*	*Reference*
Grapes and musts	Sound grapes	Spain (Canary Islands)	*Acetobacter, Gluconobacter*	Valera et al. 2011
	Sound grapes	Spain	*Acetobacter, Gluconobacter, Kozakia*	Navarro et al. 2013
	Sound grapes	Chile	*Acetobacter, Gluconobacter*	Prieto et al. 2007
	Sound grapes	Australia	*Acetobacter, Gluconobacter, Asaia, Ameyamaea*	Mateo et al. 2014
	Rotten grapes	Australia	*Acetobacter, Gluconobacter, Asaia, Ameyamaea*	Mateo et al. 2014
	Botrytized grapes	France	*Acetobacter, Gluconobacter*	Joyeux et al. 1984
	Botrytized grapes	France	*Acetobacter, Gluconobacter*	Barbe et al. 2001
Wine fermentation	Red wine fermentation	Spain	*Acetobacter, Gluconobacter, Gluconacetobacter, Komagataeibacter*[1]	González et al. 2004
	Red wine fermentation	South Africa	*Acetobacter, Gluconobacter, Gluconacetobacter, Komagataeibacter*[1]	Du Toit and Lambrechts 2002
	Wine	Austria	*Acetobacter*	Silhavy and Mandl 2006
	Malolactic fermentation	Spain	*Asaia*	Ruiz et al. 2010
	White fermentation	Japan	*Gluconacetobacter*	Kato et al. 2011
	Spoiled red wine	Portugal	*Acetobacter*	Silva et al. 2006
Wine maturation and storage	Red wine stored in barrels	France	*Acetobacter Gluconobacter*	Joyeux et al. 1984
	Bottled red wine	Australia	*Acetobacter*	Bartowsky et al. 2003

[1]In the mentioned references described as *Acetobacter* or *Gluconacetobacter*

other genera: *Kozakia baliensis, Asaia siamensis, Asaia lannaensis, Komagataeibacter hansenii* and *Ameyamaea chiangmaiensis* (Mateo et al. 2014).

Normally, AAB species detected on grapes or in grape must are different from those in wine. The environmental conditions on grapes or musts are characterized by a high sugar concentration and by high ethanol content in wines. AAB species isolated from grapes or grape must mainly belong to the genera *Gluconobacter* and *Acetobacter*, mostly on sound and infected grapes, respectively (Joyeux et al. 1984, Barbe et al. 2001, Prieto et al. 2007). However, in recent studies, besides these two genera, the presence of *Kozakia*, *Asaia* and *Ameyamaea* has been described (Navarro et al. 2013, Mateo et al. 2014).

In the case of the wines spoiled with AAB, the genera described are *Gluconacetobacter* and *Komagataeibacter* as well as *Acetobacter* but this one in minor proportion at the beginning of fermentation. However, this genus increases gradually its population throughout the fermentation. The genus *Komagataeibacter* follows the opposite dynamics; it was recovered at the beginning of the middle fermentation, and not at the end (González et al. 2004, Du Toit and Lambrechts 2002). The study of the AAB populations present along the winemaking process, from the raw material to the stored product, can help to avoid the problem of spoilage in wine.

Prevention of Wine Spoilage

In order to prevent wine spoilage by AAB, it is necessary to adopt appropriate practices and to control the hygienic conditions during all stages of the process. Therefore, after a treatment with SO_2 and appropriate inoculation of the yeast-starter culture, it is difficult for AAB bacteria to proliferate during alcoholic fermentation. However, it will also depend on the proportion between bacteria and yeast population at the start of fermentation (Drysdale and Fleet 1988), which is low in sound grapes but high in damaged ones. When both *Acetobacter* species and *Saccharomyces cerevisiae* were inoculated with the same population, the concentration of acetic acid at the end of fermentation was nine times higher than in the wine inoculated with *S. cerevisiae* (Du Toit and Pretorius 2002).

When the initial amount of AAB in must is under control, the main aspect to take into account is aeration. AAB proliferation occurs normally after exposure to air because the bacteria are highly aerobic and require oxygen for growth. The risk of spoilage is high when wines are stored prior to bottling and during pumping over or transfer steps; however, the addition of adequate amounts of SO_2 in those points and no air exposure of wines are sufficient to avoid it (Millet and Lonvaud-Funel 2000).

Once the wines are bottled, the risk of spoilage is very low due to stabilization treatments that are regularly performed prior to this stage—the membrane filtration, addition of SO_2 and also the low oxygen content in the bottle to avoid the growth of AAB (Ough 1989, Bruer et al. 1999). However, in the last few years, winemakers have been reducing the content of SO_2 progressively. The public health issues advice to reduce the SO_2 dosage due to human sensibility and allergies to sulphites and sulphates. Moreover, some winemakers do not use the filtration treatment due to the decrease in wine quality. Both trends have increased the finished wine spoilage in the last few years (Bartowsky et al. 2003). Recently, fashionable tendencies, such as 'natural wines' or biodynamic wines are reducing both the dosage of SO_2 as well as the inoculation with selected yeast. These practices are very risky and, in fact, high volatile acidity is common among these wines.

Beer Spoilage

The physicochemical characteristics of beer make it stable, but, less than wine. Only a few microorganisms are able to grow in these conditions, not only in beer, but also in malt or wort. The ethanol content between 0.5-10 per cent v/v, a high content of CO_2 (approximately 0.5 per cent w/w), the low pH between 3.8 and 4.7, the reduced content of oxygen and the hop antimicrobial compounds altogether make beer an adverse environment for microbial growth (Sakamoto and Konings 2003). Moreover, during beer brewing there are some critical steps where the microbial content is drastically reduced, such as the malting process, hops addition and boiling, and final pasteurization. Even in beer with limited treatment, such as traditional Belgian lambic beer, although the presence of Enterobacteria in the worts and the beginning of fermentation is well described (Martens et al. 1991), bacterial pathogens like *Salmonella enterica* serovar Typhimurium and *Staphylococcus aureus* do not grow or survive in beer (Bunker 1955).

In natural conditions, the microbial population during beer formation changes with time. There is a complex community of yeast, bacteria and fungi among which some microorganisms can play a positive role during beer brewing if they are correctly controlled in the key steps of the process. In fact, some of these microorganisms produce killer toxins and other antimicrobial compounds, which together with compounds present in the raw material (especially hops) increase the microbial stability of beer. On the other hand, there are microorganisms that far from being dangerous to human health can cause spoilage in beer. They are able to increase the turbidity and produce unpleasant changes in flavor and aroma of finished beer. These changes are important as they reduce the quality of the product and therefore, they are responsible for economic losses for brewing companies (Wieme et al. 2014).

Among the microorganisms involved in beer spoilage, the main ones are: wild yeasts, that increase the turbidity and cause off-flavors in the product, even after packaging; fungi, which are able to produce mycotoxins during malting; and bacteria, such as lactic acid bacteria, and in minor proportions, AAB (Vaughan et al. 2005).

There are different stages in beer brewing that are susceptible to contamination due to beer spoiler microorganisms, from raw materials used for beer production to storage and servicing of the final beer (Fig. 2).

The role of AAB in the acetification of beer has been well described since many years ago. Pasteur pointed out in his book *Etudes sur la bière* (1876) the necessity of bacteria to

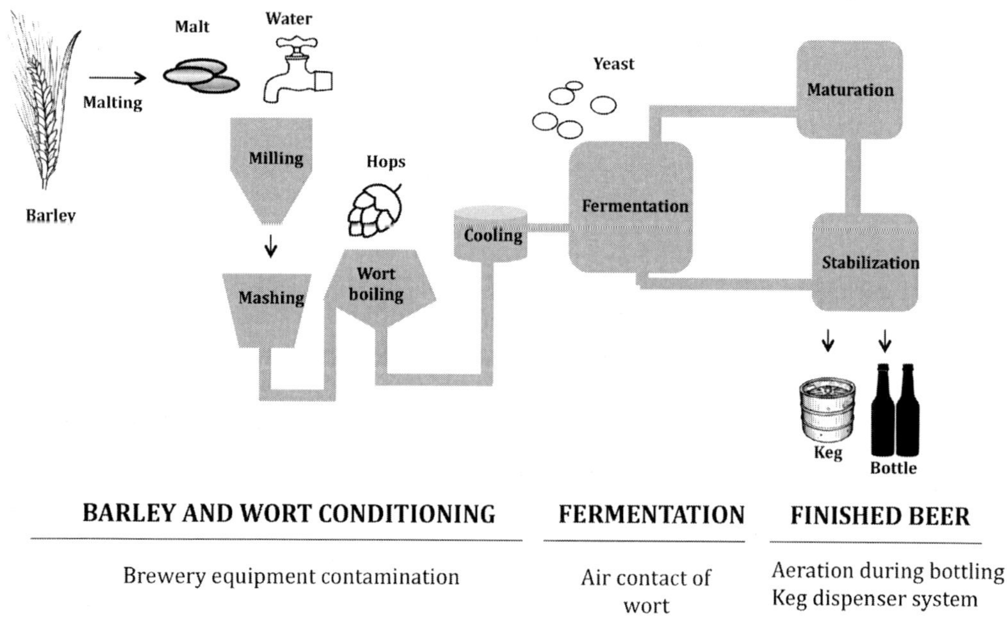

Figure 2: Schematic view of beer brewing with indication of the susceptible steps of AAB spoilage.

cause sickness in beer. He called '*Mycoderma aceti*' the bacterium involved in one of the seven forms of beer spoilage, and this fault in the beer was characterized by an acetic smell and taste. In this case, AAB are responsible for the loss of alcohol that is transformed in acetic acid, besides the increase of turbidity and the formation of films made of polysaccharides. Beer spoiled by these bacteria is accompanied by a change in their color, ropy aspect and off-flavors (Shimwell 1948). The off-flavors derived from the metabolism of AAB without acetification are probably due to oxidation of polyalcohols, such as glycerol, to dihydroxyacetone. On the other hand, the ropiness is observed as pellicle or greasy-looking covering on the surface of beer (Van Vuuren 1996).

AAB have been isolated from breweries (Ault 1965, Van Vuuren et al. 1977, Hough et al. 1982). However, the use of defined yeast starter cultures, the exhaustive cleaning practices and the improvement in beer packaging practices with the reduction of oxygen content in the product keeps AAB contamination in the beer industry under control.

Spoilage in Barley and Wort

Most often the microorganisms present in the raw material are mainly fungi, such as *Aspergillus* and *Fusarium*. On the other hand, during mashing and wort separation, the spoiler microorganisms are normally thermophilic bacteria because they have to survive the wort-boiling step.

AAB has not been described in barley and hops, and nowadays are not commonly found in worts or malt for beer. However, they were formerly detected in malt, wort and yeast contaminated samples in low populations (Ault 1965). As an example, *Gluconobacter oxydans* has been demonstrated to be able to grow in beer wort, causing ropiness (Raspor and Goranovič 2008).

Spoilage during the Fermentation

Although AAB are resistant to bacteriostatic compounds present in hop, resistant to ethanol and acid, the anaerobic conditions imposed by the yeast metabolism during fermentation do not allow their growth (Van Vuuren 1996). However, if in any step of beer production, there is the presence of oxygen, AAB can increase their population. The spoiler microorganisms can be resident in the equipment and the disinfection of all the material in contact with beer during brewing is a must. The presence of AAB in breweries is well known (Garofalo et al. 2015) and for this reason, beer exposed to air during the common treatments in brewing can be rapidly acetified. AAB have been reduced in beer spoilage due to the use of good practices of cleaning stainless steel in the brewery, define yeast-starter culture strains in an appropriate concentration and low presence of oxygen during the whole process (Kersters et al. 2006). Nevertheless, in storage vessels, pressure equilibration, maturation and filtration of samples, the recovery of AAB was quite high.

This is true in the case of industrially produced beers. However, in lambic beer, a special type of Belgian beer obtained by spontaneous fermentation, AAB have been detected in several ecological studies. The process of production includes overnight cooling of wort in shallow open vessels. During this treatment, wort is infected with different microorganisms. The next morning, the wort is pumped into large casks. In these casks, the fermentation is followed by maturation that lasts for one to three years before bottling. The initial wort is mainly infected with Enterobacteria, which are responsible for primary biotransformations. Otherwise, the AAB growth is detected after 8-16 days of fermentation, when ethanol content increases while pH decreases (Martens et al. 1991). Recent studies using both culture-dependent and independent methods obtained similar results on the succession of Enterobacteria in the beginning and AAB at the end (Bokulich et al. 2012, Spitaels et al. 2014a, b).

Moreover, AAB have been detected in acidic ales of Roeselare, which is another type of beer from Belgium. This ale beer is produced through mixed yeast-bacterial fermentation. This process starts with alcoholic fermentation followed by lactic acid fermentation. Among the different styles of acidic ales of Roeselare, those that have higher final acidity present longer fermentation. After 27 to 40 weeks, AAB can appear, depending on the permeability for oxygen in the cask used. The main concern in these cases is the control of acidity (Martens et al. 1997).

In some studies, the effect of AAB growth on the yeast performance was seen. Contamination with AAB could lead to yeast autolysis during beer fermentation, especially in former times when this step was carried out in wooden vessels (Vaughan et al. 2005). Curiously, one strain of *Acetobacter* sp. was described in bottled beers because it was able to produce toxic compounds that killed the yeast involved in brewing (Gilliland and Lacey 1966).

Spoilage in Finished Beer

In the final beer, the alcohol percentage is one of the most important factors to prevent spoilage. However, some strains of *Acetobacter* sp. and *Gluconobacter oxydans* were detected and, thus, were able to survive with 12-13 per cent (v/v) of ethanol in beer produced at high gravity (Magnus et al. 1986).

After fermentation, to avoid the growth of undesirable microorganisms, beer is subjected to stabilization treatments, such as pasteurization. Pasteurization can be carried out in two different ways—tunnel pasteurization and flash pasteurization. The first one is applied to cans or bottles, ensuring stability of the product once packaged. On the other hand, flash or plate pasteurization is normally used to treat beer before packaging in kegs or other containers that cannot be tunnel pasteurized. In order to prevent the changes in flavor, some brewers prefer package canned and bottled beers under aseptic conditions. This method is performed after treatment of stabilization by flash pasteurization or membrane filtration. However, in this case, the hygienic processing of the product must be extreme. In the case of AAB, the critical parameter is control of the oxygen contact with beer during packaging (Sakamoto and Konings 2003). Nevertheless, improvement in packaging practices that has reduced the presence of oxygen and therefore AAB spoilage, has resulted in a higher frequency of spoilage by anaerobic bacteria, such as those from the genera *Pectinatus* and *Megasphaera*. Both are responsible for undesirable flavors in the final beer. In any case, the main spoilers of beer are some species of lactic acid bacteria, though the growth of other lactic acid bacteria species is considered positive for beer brewing as they contribute to the stability of the product (Vaughan et al. 2005).

Nowadays, the beer dispensing systems are the place where AAB are commonly found as beer spoilers. The preliminary steps in brewing are currently very well controlled and the practices to prevent the spoilage are efficient. However, the dispensing system is exposed to microbes present in the environment while the tap is open and during the changing of kegs. Kegged beers are free of contamination when they are delivered, but the spoilage is detected after dispensing (Harper 1981). The presence of AAB has been detected in biofilms formed in these systems. The development of this biofilm allows survival of bacteria, even after cleaning. The AAB can cause haze and a film on the surface of the liquid (Vaughan et al. 2005). Commonly, AAB growth is inhibited when exposed to carbon dioxide, but the appearance of AAB in public houses demonstrates that this characteristic is strain dependent (Harper et al. 1980).

Species of AAB Involved in Beer Spoilage

Two genera that have been described in beer spoilage are *Acetobacter* and *Gluconobacter* (Kersters et al. 2006). *Gluconobacter oxydans* was firstly isolated in cask beer from the United

Kingdom (Shimwell 1936) as they were characterized for a ropy aspect, while *Acetobacter pasteurianus* was isolated in beer from the Nederlands (Beijerinck and Folpmers 1916) and in spoiled beer from Osaka, Japan by Kondo in 1941 (Lisdiyanti et al. 2000). Moreover, one AAB strain isolated from stored beer (ale) at Toronto, Canada (Kozulis and Parsons 1958) was initially classified as *Acetobacter pasteurianus*. However, some years later, this strain was reclassified as the strain for the species *Acetobacter cerevisiae* (Cleenwerck et al. 2002). More recently, strains of *Acetobacter cerevisiae* have also been isolated in Belgium in the traditional, spontaneously-fermented lambic beer as well as *Acetobacter lambici* (Spitaels et al. 2014a). *Acetobacter orleanensis* was isolated for the first time from beer (Lisdiyanti et al. 2001). Moreover, *Acetobacter lovaniensis* and *Acetobacter fabarum* have been recovered during spontaneous fermentation in American coolship ale—a type of beer that is produced using a process similar to traditional Belgian lambic beer preparation (Bokulich et al. 2012).

In a recent article, the use of MALDI-TOF, Mass Spectrometry profiles allowed reclassification of some AAB strains previously isolated in beer. Also, they were able to identify other species in spoiled bottled beers and worts from Belgian breweries, such as *Acetobacter cerevisiae/Acetobacter malorum* (they cannot be distinguished due to their high homology), *Acetobacter fabarum, Acetobacter indonesiensis, Acetobacter orleanensis, Acetobacter persici, Gluconobacter cerevisiae, Gluconobacter japonicus, Gluconobacter cerinus* and *Gluconobacter oxydans* (Wieme et al. 2014).

Prevention of Beer Spoilage

There are big differences in the susceptibility of beer to spoilage. Depending on the type of beer, the production process presents peculiarities that can allow or avoid the growth of undesirable microorganisms. So low acidity, low alcohol, low carbon dioxide and high sugar concentrations are characteristics that favor spoilage.

In order to control beer spoilage by AAB, the main concerns are cleaning of the equipment and the absence of oxygen during the process: AAB are resistant to ethanol, acid and hop addition and this bacterial group become resident microbiota in breweries. Furthermore, the effective disinfection of dispensing systems in public houses is also one of the most interesting points that has been taken into account (Vaughan et al. 2005).

Cider Spoilage

Cider is a fermented product derived from apple juice. Depending on the region of production, the cider display different characteristics based on the technological processes performed for its preparation and the varieties of apples used.

Wild yeast, fungi and bacteria can cause microbial spoilage of cider during fermentation, maturation and storage periods of the product. AAB are responsible for acetification of cider, but it is also possible in the production of gluconic acid in apple must. On the other hand, AAB are able to reduce the cider acidity by oxidation of citric and malic acids (Suárez and Cabranes 1989).

Apples used for cider making are allowed to fall naturally; sometimes this process is helped by using sticks. The fruits are harvested from the ground, where they are in contact with the microorganisms from the soil and grass. AAB are part of this microbiota, but also during the processing of the fruit in the press house, the microbial population increases. In some cider industries, the apple juice is stored in concentrated form to reduce the water activity and prevent the growth of spoiler microorganisms (Whiting 1973). The clarification and stabilization treatment of must can reduce the risk of spoilage at this point. However, when AAB have produced gluconic acid and their derivatives by using apple glucose, the SO_2 activity is reduced due to a combination of these compounds. The good industrial practices for cider production are to remove ropiness, acetification and enzymatic darkening. The maturation and storage are also critical points where cider can be contaminated with AAB, especially when exposed to oxygen. Filtration, pasteurization and other stabilization are treatments currently performed in industrial cider to avoid the contamination (Beech 1972). The framboisé or cider-sickness is spoilage in sweet ciders with a pH higher than 3.7, detected in English and French finished ciders in traditional or even industrial cider companies. It is detected in tanks, wood casks or in bottles. The spoiled product is characterized by the production of an off-flavor described as a 'banana-skin', 'rotten lemon' or 'raspberry' aroma, caused by high concentrations of acetaldehyde in the medium. Moreover, they present high gas formation, foam and turbidity. The studies performed on spoilage suggest that the responsible microorganisms can be *Zymomonas mobilis*, lactic acid bacteria or AAB (Maugenet 1962, Coton and Coton 2003).

In natural cider obtained from the north of Spain produced by spontaneous fermentation, this microbial group is found during all the stages of preparation. Nevertheless, they are most abundant in apple musts (10^4-10^6 CFU/mL). During alcoholic fermentation, the anaerobic conditions affect their growth and they appear again in maturation and storage stages (Suárez and Cabranes 1989). Traditionally, this type of cider is not inoculated with defined yeast starter cultures. The main concerns in order to prevent the spoilage of cider by AAB in this region are the soundness of the fruit, the hygienic condition of the equipment in house pressing, the proper clarification of must prior to fermentation and the control of oxygen during the preparation.

The genera detected in ciders are *Acetobacter* and *Gluconobacter*. *Gluconobacter* appears in the initial stages of cider making and uses the sugar to produce acids, while *Acetobacter* is able to metabolize ethanol and organic acids as well as residual sugars (Suárez and Cabranes 1989).

Other Alcoholic Fermented Products

There are several sources where AAB can be present as spoiler microorganisms. For example, sake containing up to 24 per cent ethanol was found to be spoiled by *Acetobacter pasteurianus* and presented an acetic acid smell. In spoiled palm wine, *Acetobacter indonesiensis*, *Acetobacter tropicalis* and *Komagataeibacter xylinus* were detected (Hommel and Ahnert 2004).

Spoilage of Dairy Products

The diversity of dairy products available and the economic importance of this industry make spoilage of dairy products very characteristic and the microbiota present in these products have been standardized.

Several spoilage microorganisms are able to grow in nutritive environments, such as milk or its fermented derivatives. These microorganisms include aerobic psychrotrophic Gram-negative bacteria, heterofermentative Lactobacilli, other spore-forming bacteria, yeasts and molds (Ledenbach and Marchal 2009).

There are several processes that have been traditionally developed to avoid spoilage of milk and other dairy products and most of them involve the native microbiota present in these raw materials. Sometimes, these processes result in a new product with different organoleptic properties and the industry is known to adapt them to suit their interests (Ledenbach and Marchal 2009).

In many dairy foods, the strategy to prevent spoilage is to decrease the pH, by fermenting the lactose to lactic acid or by adding acids as preservatives; on the other hand, the inoculation of defined microbial cultures can avoid the growth of undesirable microorganisms. Moreover, the control of water content in dairy products is important to decrease the growth of spoiler microorganisms and this can be done by addition of salt or sugar, removing water or freezing. Some of these techniques can be used sequentially in order to obtain different products with new emergent properties (Ledenbach and Marchal 2009).

AAB responsible for spoilage of milk whey have been used in some cultures to make whey vinegar. Once the whey is fermented by yeast to metabolize lactose, the alcohol produced can be used as a carbon source by AAB (Parrondo et al. 2003).

AAB in Kefir

AAB are present in traditional fermented milks, such as kefir. Kefir is a beverage prepared by using a complex community of microorganisms, including lactic acid bacteria, yeasts and AAB. These microorganisms are grouped in grains, and are joined by a complex exopolysaccharide and protein matrix (Farnworth and Mainville 2003). The traditional fermented milk was kept in skin bags, where it was preserved and it marked the origin of the first grains of kefir (Motaghi et al. 1997). For kefir production, milk used can be from cows, ewes, goats and buffalo (Ismail et al. 1983, Mann 1983, Wszolek et al. 2006). These grains are used as starter culture and are recovered at the end of fermentation to be reused in the next batch. During the fermentation process, the milk proteins and lipids are transformed as well as the sugars, yielding ethanol and other alcohols in low proportion, organic acids, aldehydes, volatile acids, small peptides, ester and sulfur compounds. All these molecules are responsible for the special organoleptic properties of kefir (Wszolek et al. 2006).

AAB appear in minor proportion in the grains compared with lactic acid bacteria and yeasts, which represent up to a 99 per cent of the cells. The distribution of the microbial population depends on the origin of the grains with the main group being Lactococci (10^9 cells/g). On the other hand, AAB population in kefir ranges between 10^4-10^5 cells/mL (Pogačić et al. 2013). However, while some authors consider AAB as contaminant bacteria (Rosi 1978, Rosi and Rossi 1978, Angulo et al. 1993) and attribute their presence in kefir grains to poor hygienic conditions during manipulation, other authors consider their presence as desirable during the fermentation process (Koroleva 1988), thereby showing the positive and peculiar characteristics that they confer to the final product. Besides, the production of vitamin B_{12} by *Acetobacter* spp. is supposed to stimulate the growth of other microorganisms present in the grains (Rea et al. 1996).

The unique AAB genus detected in kefir or kefir grains is *Acetobacter*. *Acetobacter pasteurianus* (Ottogalli et al. 1973) and *Acetobacter aceti* (Rosi 1978, Koroleva 1991) were the first AAB species isolated in kefir. More recently, *Acetobacter fabarum* was recovered in kefir from Tibet (Gao et al. 2012) and *Acetobacter syzygii* and *Acetobacter lovaniensis* were detected in kefir by metagenomic approaches (Unsal 2008). This result agrees with that for Brazilian kefir (Magalhães et al. 2011) in which *Acetobacter lovaniensis* was isolated in grains with a quite high population of AAB (8.9 per cent).

However, by using culture-dependent and independent techniques for the detection of bacteria, it is possible to describe the presence of *Gluconobacter japonicus* (Miguel et al. 2010).

AAB in Frozen Desserts

Other products that can be colonized and spoiled by AAB are fruit ice creams. Here, the bacteria come from the fruits that are previously infected. Most of the microorganisms that are present on fruits are yeasts, some lactic acid bacteria and also AAB from the genera *Gluconobacter* and *Acetobacter*.

The treatment of freezing is not enough to control all the microorganisms present in fruits. These fruits must be processed before being added to the ice cream during preparation. Peeling, washing and bleaching are the first steps to decrease the content of microorganisms. After that, the fruits can be conserved, either frozen or at room temperature if subjected to a heat treatment and aseptically packed until the moment of being used. Moreover, it is possible to candy them by adding progressively higher amounts of sugar in order to reduce the water activity before mixing them in the ice cream preparation (Marshall 2001).

Spoilage of Other Products

Fruits and Fruit Juices

The microbiota present in fruits is influenced by the environment where they are harvested. Normally, these microorganisms come from the soil, airborne spores and irrigation water. In general, they survive on the surface of the fruit when they are sound and not crushed. In fact, the complex microbiota of the fruit protects it as a natural biological barrier against microorganisms that can cause crop damage (Andrews and Harris 2000, Janisiewicz and Korsten 2002). However, there are microorganisms which induce spoilage in fruits under certain conditions. These spoiler microorganisms can be part of the natural microbiota from the seed of the plant introduced during growth, or at some point in harvesting, or after harvesting as a result of handling, storage or distribution (Barth et al. 2009). For this reason, all the steps must be controlled and carried out in order to avoid any contamination along the process.

The internal environment and chemical characteristics of fruits encourage the growth of microorganisms. However, polymers and tissues protecting the surface of the fruit make it especially difficult for bacteria to penetrate inside. For this reason, their internalization is easier in damaged products.

Fresh fruits are characterized by high water activity and acidity where the latter is a hurdle in the growth of most microorganisms, especially molds and yeasts. This is the case with apples, pineapples, strawberries, grapes and tomatoes. The acidity does not allow the primary growth of fungi (Splittstoesser 1982) and they are normally spoiled by resistant bacteria, such as lactic acid bacteria, AAB and Enterococci species.

Spoilage caused by bacteria is observed visually as formation of colonies, enzymatic pectolyzation and discoloration, as also by the production of off-flavors, mainly due to fermentation of sugars. AAB are able to produce acids derived from the metabolization of sugars, which are reported as being responsible for fruit discoloration (Barth et al. 2009). It is remarkable that the species of *A. pasteurianus*, frequently isolated from fermented products, are not usually found in fruits or fruit juices. However, other *Acetobacter* species are commonly recovered. In fact, several AAB species have been described in tropical fruits from Southeast Asia; probably due to their sweetness, they represent a good source for their growth (Table 2).

Moreover, different AAB species have been recovered from other fruits and fruit juices besides those described in grapes and grape musts.

Observance of hygienic conditions during the steps ranging from growth to harvesting and post-harvesting operations, such as storage conditions and packaging, is important to ensure the quality of the fruits. On the other hand, the selection of sound fruit in order to avoid the damaged ones is a key step, especially when they are used for juice preparation. In damaged fruits, the microbial population increases rapidly, compromising the safety

Table 2: AAB species recovered from different fruits and fruit juices

Source	*Species*	*Reference*
Ash berries	*Komagataeibacter xylinus*	Yamada et al. 2012
Coconut	*Acetobacter lovaniensis, Acetobacter indonesiensis, Acetobacter tropicalis, Acetobacter cibinongensis, Acetobacter orientalis*	Lisdiyanti et al. 2003
Guava	*Acetobacter orleanensis, Acetobacter tropicalis*	Lisdiyanti et al. 2003
Mango	*Acetobacter lovaniensis, Acetobacter indonesiensis*	Lisdiyanti et al. 2003
Persimmon dried fruit	*Gluconacetobacter liquefaciens*	Yamada et al. 2012
Pineapple	*Gluconacetobacter diazotrophicus*	Tapia-Hernández et al. 2000, Lisdiyanti et al. 2003 Gosselé and Swings 1986
Pinneapple affected by pink-disease	*Komagataeibacter hansenii, Gluconacetobacter liquefaciens, Gluconobacter oxydans*	
Starfruit	*Acetobacter orleanensis, Acetobacter lovaniensis, Acetobacter indonesiensis, Acetobacter syzygii, Acetobacter orientalis*	Lisdiyanti et al. 2003
Strawberries	*Gluconobacter frateuri*	Mason and Claus 1989
Coconut juice	*Acetobacter lovaniensis, Acetobacter tropicalis, Gluconobacter oxydans*	Lisdiyanti et al. 2003
Organic apple juice	*Komagataeibacter rhaeticus, Komagataeibacter swingsii*	Dellaglio et al. 2005

and organoleptic properties of juices. The storage of concentrated fruit juices decreases the water activity and act as a good alternative for prevention of spoilage of these products.

Soft Drinks

Soft drinks are non-alcoholic beverages that can be carbonated or not, and usually they are sweetened, present edible acids and natural or artificial flavors. They are served cold. Some of these soft drinks contain fruit pulp or fruit juice. The high water activity provides a good environment for microbial growth, but the low pH of these beverages decreases the risk of spoilage by microorganisms. However, some yeasts, molds and bacteria are able to survive in these conditions (Jayalakshmi et al. 2011). The effects of spoilage can be observed in the form of turbidity, presence of particulates, taints and excessive gas. Contamination in soft drinks is due to microbiota arising from the raw material, the use of returned bottles or aerial vectors, such as insects (Byrne 1994). *Acetobacter* and *Gluconobacter* are able to colonize these environments, causing spoilage of soft drinks (Vasavada and Herpekan 2002).

The use of preservatives in still beverages, such as potassium sorbate and sodium benzoate, is not sufficient to prevent the growth of *Gluconobacter oxydans,* which is able to survive in these conditions (Sand and Van Grinsven 1976, Eyles and Warth 1989). The required concentrations of these preservatives is determined using mathematical models to control microbial spoilage without altering the organoleptic properties of the product (Battey and Shaffner 2001). In this model, the features taken into account include pH, titratable acidity, sugar content (°Brix), sodium benzoate and potassium sorbate concentrations.

Due to the cultivability hurdle that present the AAB, other strategies are used in order to detect their presence and avoid spoilage. The quantitative PCR was successfully used to report and quantify the presence of *Gluconobacter* and *Gluconacetobacter* in an electrolyte replacement drink (Gammon et al. 2007).

Fermented Vegetables

Vegetables are characterized by high water activity and a not very acidic pH—conditions that allow the growth of yeast and molds. However, after fermentation, used as a conservative practice, there are changes in the organoleptic and chemical properties of the product. Lowering pH allows the growth of acid tolerant microorganisms, such as AAB. Also, if the product is conserved in a pickling solution or vinegar, this common environment for AAB, it can be a source of spoilage by this microbial group (Sakamoto et al. 2011).

In a recent article (Medina et al. 2016) the microbial spoilage of fermented cucumber was analyzed. Sodium chloride-fermented cucumbers can present spoilage characterized by rising pH and manure- and cheese-like aromas, which is problematic for the pickling industry. It is speculated that oxygen availability may be the determining factor in the initiation of spoilage and leading microbiota. *Acetobacter pasteurianus* and *Acetobacter peroxydans* were recovered from fermented cucumber and *Acetobacter* sp. and *Gluconobacter* sp. were also detected by culture-independent techniques.

Future Perspectives

The microbiological control of processes that lead to spoilage is a constant since the first production of transformed foods. Nowadays the food industry has many tools to prevent spoilage and has invested in high technology and plenty of resources to be most effective. However, food products still suffer spoilage and high economical losses are reported due to food spoilage. Thus, not all the mechanisms of food spoilage are well known at this moment and are impossible to eliminate completely. Nevertheless, still some improvements and precautions can be adopted to reduce the level of spoilage in the food industry.

The most common genera of spoiler AAB are *Acetobacter* and *Gluconobacter* that are present in almost all the sources damaged by this bacterial group (Table 3). Although AAB are ubiquitous microorganisms, the correct selection of raw material quality, performance of hygienic practices and good technological processing allow prevention of spoilage in foods and beverages. Thus, there is no need to mention here that good hygienic practices are a must in the food industry. However, two main aspects are critical for the development of AAB in foods: oxygen and appropriate substrates (in terms of sugars, ethanol, etc.). The prevention of oxygen is probably the easiest way to avoid the development of AAB and thus, reduction or elimination of exposure to oxygen should be a goal for the industry where AAB is one of the main spoilers.

The substrates could be fundamental to the product itself (ethanol in wines or sugar in fruit juices, for instance). In this case, the use of antimicrobial compounds, such as SO_2, is highly recommended, despite its limited effect on AAB. The actual trend in winemaking is to get a product less processed with minimum intervention (from vineyard to bottling) and maintain high quality. For this reason, it is necessary to find new preservative agents or stabilization techniques to avoid microbial spoilage. Other products with antimicrobialproperties isolated from plants, animals or other microorganisms can substitute the use of chemical products. For that reason, other compounds, such as chitosan, could open new perspectives, although its use is still under consideration.

Furthermore, other strategies, such as reducing pH, could be appropriate, provided the legislation allows this practice.

Table 3: Genera of spoiler AAB in different sources

Genera of AAB	*Wine*	*Beer*	*Dairy*	*Fruits*	*Cider*	*Soft drinks*	*Fermented vegetables*
Acetobacter	x	x	x	x	x	x	x
Gluconobacter	x	x	x	x	x	x	x
Gluconacetobacter	x			x			
Komagataeibacter	x			x			
Asaia				x			
Ameyamaea				x			
Kozakia				x			

Another line of prevention could be the use of appropriate starters to control the AAB population. The available starters are selected to perform the basic transformation (i.e. transformation of sugars into ethanol in the alcoholic fermentation). However, the starter producing industry can select more microorganisms that possess other attributes, such as the production of antimicrobial compounds. This is a new line of research that is producing quick results. Nonetheless, no research on compounds that limit AAB growth has been reported so far.

Finally, one of the key points in microbiological control is the monitoring of microbial population. It has already been mentioned the problem that AAB presents by way of the status of VBNC and, thus the poor information that plating can provide. So far the culture-independent methods are expensive, time consuming and require specialized personnel. All these specifications limit the availability of these techniques in the food industry. However, they are the only reliable techniques to confront the AAB population. Thus, further research is needed to make them available to the food industry through some specific biosensor that permits the estimation of AAB population above the critical level. Thus, these techniques could match the culture-independent techniques to meet the everyday needs of the food industry.

References

Andorrà, I., Landi, S., Mas, A., Guillamón, J.M. and Esteve-Zarzoso, B. (2008). Effect of oenological practices on microbial populations using culture-independent techniques. *Food Microbiology* **25:** 849-856.

Andrews, J. H. and Harris, R.F. (2000). The ecology and biogeography of microorganisms on plants. *Annual Review of Phytopathology* **38:** 145-180.

Angulo, L., López, E. and Lema, C. (1993). Microflora present in kefir grains of the Galician region (north-west of Spain). *Journal of Dairy Research* **60:** 263-267.

Asai, T. (1935). Taxonomic studies on acetic acid bacteria and allied oxidative bacteria isolated from fruits: A new classification of the oxidative bacteria. *Agricultural Chemical Society of Japan* **11:** 674-708.

Ault, R.G. (1965). Spoilage bacteria in brewing—A review. *Journal of the Institute of Brewing* **71:** 376-391.

Barata, A., Malfeito-Ferreira, M. and Loureiro, V. (2012). Changes in sour rotten grape berry microbiota during ripening and wine fermentation. *International Journal of Food Microbiology* **154:** 152-161.

Barbe, J.-C., de Revel, G., Joyeux, A., Bertrand, A. and Lonvaud-Funel, A. (2001). Role of botrytized grape micro-organisms in SO_2 binding phenomena. *Journal of Applied Microbiology* **90:** 34-42.

Barth, M., Hankinson, T.R., Zhuang, H. and Breidt, F. (2009). Microbiological spoilage of fruits and vegetables. *In:* W.H. Sperber and M.P. Doyle (Eds.). *Compendium of the Microbiological Spoilage of Foods and Beverages*. Springer, New York, pp. 135-183.

Bartowsky, E.J., Xia, D., Gibson, R.L., Fleet, G.H. and Henschke, P.A. (2003). Spoilage of bottled red wine by acetic acid bacteria. *Letters in Applied Microbiology* **36:** 307-314.

Bartowsky, E.J. and Henschke, P.A. (2008). Acetic acid bacteria spoilage of bottled red wine—A review. *International Journal of Food Microbiology* **125:** 60-70.

Battey, A.S. and Schaffner, D.W. (2001). Modelling bacterial spoilage in cold-filled ready-to-drink beverages by *Acinetobacter calcoaceticus* and *Gluconobacter oxydans*. *Journal of Applied Microbiology* **91:** 237-247.

Beech, F.W. (1972). Cider making and cider research: A review. *Journal of the Institute of Brewing* **78:** 477-491.

Beijerinck, M.W. and Folpmers, T. (1916). Formation of pyruvic acid from malic acid by microbes. Verslag van de Gewone Vergaderingen der Afdeeling Natuurkunde/*Koninklijke Akademie van Wetenschappen Amsterdam* **18:** 1198-2000.

Boddy, L. and Wimpenny, J.W.T. (1992). Ecological concepts in food microbiology. *Journal of Applied Bacteriology Symposium Series* **73:** 23S-38S.

Bokulich, N., Bamforth, C.W. and Mills, D. (2012). Brewhouse-resident microbiota are responsible for multi-stage fermentation of American coolship ale. *PloS One*, **7:** e35507.

Bottero, J. (2004). *The Oldest Cuisine in the World: Cooking in Mesopotamia*. University of Chicago Press, Chicago.

Bruer, N.G.C., Coulter, A.D. and Graves, P.J. (1999). Microbiological spoilage of bottled red wine. *In:* R.J. Blair, A.N. Sas, P.F. Hayes and P.B. Høj (Eds.). *Proceedings of the Tenth Australian Wine Industry Technical Conference*, Sydney, pp. 253-254.

Bunker, H.J. (1955). The survival of pathogenic bacteria in beer. *Proceedings of the European Brewery Convention* **5:** 330-341.

Byrne, M. (1994). The beverage boom continues. *Food Engineering International* **19:** 54-57.

Cleenwerck, I., Vandemeulebroecke, K., Janssens, D. and Swings, J. (2002). Re-examination of the genus *Acetobacter*, with descriptions of *Acetobacter cerevisiae* sp. nov. and *Acetobacter malorum* sp. nov. *International Journal of Systematic and Evolutionary Microbiology* **52:** 1551-1558.

Coton, E. and Coton, M. (2003). Microbiological origin of 'Framboisé' in French ciders. *Journal of the Institute of Brewing* **109:** 299-304.

Davis, C.R., Wibowo, D.J., Eschenbruch, R.E., Lee, T.H. and Fleet, G.H. (1985). Practical implications of malolactic fermentation: A review. *American Journal of Enology and Viticulture* **36:** 290-301.

De Ley, J., Gillis, M. and Swings, J. (1984). Family VI. *Acetobacteraceae. Bergey's Manual of Systematic Bacteriology* **1:** 267-278.

Dellaglio, F., Cleenwerck, I., Felis, G.E., Engelbeen, K., Janssens, D. and Marzotto, M. (2005). Description of *Gluconacetobacter swingsii* sp. nov. and *Gluconacetobacter rhaeticus* sp. nov., isolated from Italian apple fruit. *International Journal of Systematic and Evolutionary Microbiology* **55:** 2365-2370.

Drysdale, G.S. and Fleet, G.H. (1988). Acetic acid bacteria in winemaking: A Review. *American Journal of Enology Viticulture* **39:** 143-154.

Drysdale G.S. and Fleet, G.H. (1989a). The growth and survival of acetic acid bacteria in wines at different concentrations of oxygen. *American Journal of Enology Viticulture* **40:** 99-105.

Drysdale, G.S. and Fleet, G.H. (1989b). The effect of acetic acid bacteria upon the growth and metabolism of yeast during the fermentation of grape juice. *Journal of Applied Bacteriology* **67:** 471-481.

Du Toit, M. and Pretorius, I. S. (2000). Microbial spoilage and preservation of wine: Using weapons from Nature's own arsenal—A Review. *South African Society for Enology & Viticulture* **21:** 74-96.

Du Toit, W.J. and Lambrechts, M.G. (2002). The enumeration and identification of acetic acid bacteria from South African red wine fermentations. *International Journal of Food Microbiology* **74:** 57-64.

Du Toit, W.J. and Pretorius, I.S. (2002). The occurrence, control and esoteric effect of acetic acid bacteria in winemaking. *Annals of Microbiology* **52:** 155-179.

Eglinton, J.M. and Henschke, P.A. (1999). The occurrence of volatile acidity in Australian wines. *The Australian Grapegrower and Winemaker Annual Technical Issue* **426a:** 7-12.

Erasmus, D.J., Cliff, M. and Van Vuuren, H.J. (2004). Impact of yeast strain on the production of acetic acid, glycerol and the sensory attributes of icewine. *American Journal of Enology and Viticulture* **55:** 371-378.

Eyles, M.J. and Warth, A.D. (1989). The response of *Gluconobacter oxydans* to sorbic and benzoic acids. *International Journal of Food Microbiology* **8:** 335-342.

Farnworth, E.R. and Mainville, I. (2003). Kefir: A fermented milk product. *In:* E.R. Farnworth (Ed.). *Handbook of Fermented Functional Foods*. CRC Press, Florida, pp. 77-112.

Fleet, G.H. (1999). Microorgansims in food ecosystems. *International Journal of Food Microbiology* **50:** 101-117.

Gammon, K.S., Livens, S., Pawlowsky, K., Rawling, S.J., Chandra, S. and Middleton, A.M. (2007). Development of real-time PCR methods for the rapid detection of low concentrations of *Gluconobacter* and *Gluconacetobacter* species in an electrolyte replacement drink. *Letters in Applied Microbiology* **44:** 262-267.

Gao, J., Gu, F., Abdella, N.H., Ruan, H. and He, G. (2012). Investigation on culturable microflora in Tibetan kefir grains from different areas of China. *Journal of Food Science* **77:** M425-M433.

Garofalo, C., Osimani, A., Milanović, V., Taccari, M., Aquilanti, L. and Clementi, F. (2015). The occurrence of beer spoilage lactic acid bacteria in craft beer production. *Journal of Food Science* **80:** M2845-2852.

Gilliland, R.B. and Lacey, J.P. (1966). An *Acetobacter* lethal to yeasts in bottled beer. *Journal of the Institute of Brewing* **72:** 291-303.

González, A., Hierro, N., Poblet, M., Rozès, N., Mas, A. and Guillamón, J.M. (2004). Application of molecular methods for the differentiation of acetic acid bacteria in a red wine fermentation. *Journal of Applied Microbiology* **96:** 853-860.

González, A., Hierro, N., Poblet, M., Mas, A. and Guillamón, J.M. (2005). Application of molecular methods to demonstrate species and strain evolution of acetic acid bacteria population during wine production. *International Journal of Food Microbiology* **102:** 295-304.

González, A., Hierro, N., Poblet, M., Mas, A. and Guillamón, J. M. (2006). Enumeration and detection of acetic acid bacteria by real-time PCR and nested PCR. *FEMS Microbiology Letters* **254(1):** 123–128.

Gosselé, F. and Swings, J. (1986). Identification of *Acetobacter liquefaciens* as causal agent of pink-disease of pineapple fruit. *Journal of Phytopathology* **116:** 167 175.

Gram, L., Ravn, L., Rasch, M., Bartholin, J., Christensen, A.B. and Givskov, M. (2002). Food spoilage—interactions between food spoilage bacteria. *International Journal of Food Microbiology* **78:** 79-97.

Guillamón, J.M. and Mas, A. (2011). Acetic acid bacteria. *In:* A.V. Carrascosa, R. Muñoz and R. González (Eds.). *Molecular Wine Microbiology*. Elsevier, New York, pp. 227-255.

Harper, D.R., Hough, J.S. and Young, T.W. (1980). Microbiology of beer dispensing systems. *Brewers Guardian* **109:** 24-28.

Harper, D.R. (1981). Microbial contamination of draught beer in public houses. *Proceedings in Biochemistry* **16:** 2-7.

Hommel, R.K. and Ahnert, P. (2004). *Acetobacter*. *In:* C. Batt, P. Patel and R. Robinson (Eds.). *Encyclopedia of Food Microbiology*, Vol. 1. Elsevier, New York, pp. 1-7.

Hough, J.S., Briggs, D.E., Stevens, R. and Young, T.W. (1982). *Malting and Brewing Science*. Chapman and Hall, London, pp. 741-775.

Ilabaca, C., Navarrete, P., Mardones, P., Romero, J. and Mas, A. (2008). Application of culture independent molecular methods to evaluate acetic acid bacteria diversity in vinegar-making process. *International Journal of Food Microbiology* **126:** 245-249.

Ismail, A.A., El-Nockrashy, S.A. and Khorshid, M.A. (1983). A beverage from separated buffalo milk fermented with kefir grains. *International Journal of Dairy Technology* **36:** 117-118.

Janisiewicz, W.J. and Korsten, L. (2002). Biological control of postharvest diseases of fruits. *Annual Review of Phytopathology* **40:** 411-441.

Jayalakshmi, T., Krishnamoorthy, P., Kumar, G.R. and Sivamani, P. (2011). The microbiological quality of fruit-containing soft drinks from Chennai. *Journal of Chemical and Pharmaceutical Research* **3:** 626-630.

Joyeux, A., Lafon-Lafourcade, S. and Ribéreau-Gayon, P. (1984). Evolution of acetic acid bacteria during fermentation and storage of wine. *Applied and Environmental Microbiology* **48:** 153-156.

Kato, S., Ishihara, T., Hemmi, H., Kobayashi, H. and Yoshimura, T. (2011). Alterations in D-amino acid concentrations and microbial community structures during the fermentation of red and white wines. *Journal of Bioscience and Bioengineering* **111:** 104-108.

Kersters, K., Lisdiyanti, P., Komagata, K. and Swings, J. (2006). The Family Acetobacteraceae: The Genera *Acetobacter, Acidomonas, Asaia, Gluconacetobacter, Gluconobacter,* and *Kozakia. In:* M. Dworkin, S. Falkow, E. Rosenberg, K.-H. Schleifer and E. Stackebrandt (Eds.). *The Prokaryotes.* Springer, New York, pp. 163-200.

Koroleva, N.S. (1988). Technology of kefir and kumys. Fermented Milks—Sciences and technology. Document No. 227. International Diary Federation, Brussels, pp. 35-40.

Koroleva, N.S. (1991). Products prepared with lactic acid bacteria and yeasts. *In:* R.K. Robinson (Ed.). *Therapeutic Properties of Fermented Milks.* Elsevier Applied Sciences Publishers, London, UK, pp. 159-179.

Kozulis, J.A. and Parsons, R.H. (1958). *Acetobacter alcoholophilus* n. sp.—a new species isolated from storage beer. *Journal of the Institute of Brewing* **64:** 47-50.

Ledenbach, L.H. and Marshall, R.T. (2009). *Compendium of the Microbiological Spoilage of Foods and Beverages.* Springer, New York, NY, USA.

Lisdiyanti, P., Kawasaki, H., Seki, T., Yamada, Y., Uchimura, T. and Komagata, K. (2000). Systematic study of the genus *Acetobacter* with descriptions of *Acetobacter indonesiensis* sp. nov., *Acetobacter tropicalis* sp. nov., *Acetobacter orleanensis* (Henneberg 1906) comb. nov., *Acetobacter lovaniensis* (Frateur, 1950) comb. nov., and *Acetobacter estunensis* (Carr, 1958) comb. nov. *The Journal of General and Applied Microbiology* **46:** 147-165.

Lisdiyanti, P., Kawasaki, H., Seki, T., Yamada, Y., Uchimura, T. and Komagata, K. (2001). Identification of *Acetobacter* strains isolated from Indonesian sources and proposals of *Acetobacter syzygii* sp. nov., *Acetobacter cibinongensis* sp. nov. and *Acetobacter orientalis* sp. nov. *The Journal of General and Applied Microbiology* **47:** 119-131.

Lisdiyanti, P., Katsura, K., Potacharoen, W., Navarro, R.R., Yamada, Y., Uchimura, T. and Komagata, K. (2003). Diversity of acetic acid bacteria in Indonesia, Thailand, and the Philippines. *Microbiology and Culture Collections* (Japan) **19:** 91-99.

Magalhães, K.T., Pereira, G.V.D.M., Campos, C.R., Dragone, G. and Schwan, R.F. (2011). Brazilian kefir: Structure, microbial communities and chemical composition. *Brazilian Journal of Microbiology* **42:** 693-702.

Magnus, C.A., Ingledew, W.M. and Casey, G.P. (1986). High-gravity brewing: Influence of high-ethanol beer on the viability of contaminating brewing bacteria. *Journal of the American Society of Brewing Chemists* (USA) **44:** 158-161.

Mann, E.J. (1983). Kefir and koumiss [literature review]. *Dairy Industries International* **48:** 9-10.

Martens, H., Dawoud, E. and Verachtert, H. (1991). Wort enterobacteria and other microbial populations involved during the first month of lambic fermentation. *Journal of the Institute of Brewing* **97:** 435-439.

Martens, B.H., Iserentant, D., Verachtert, H. and Methods, E. (1997). Microbiological aspects of a mixed yeast—bacterial fermentation in the production of a special Belgian acidic ale. *Journal of the Institute of Brewing* **103:** 85-91.

Marshall, R.T. (2001). Frozen Desserts. *In:* E.H. Marth and J. Steel (Eds.). *Applied Dairy Microbiology.* Marcel Dekker, New York, pp. 93-126.

Mason, L.M. and Claus, G.W. (1989). Phenotypic characteristics correlated with deoxyribonucleic acid sequence similarities for three species of *Gluconobacter: G. oxydans* (Henneberg, 1897) De Ley, 1961, *G. frateurii* sp. nov. and *G. asaii* sp. nov. *International Journal of Systematic and Evolutionary Microbiology* **39:** 174-184.

Mateo, E., Torija, M.J., Mas, A. and Bartowsky, E.J. (2014). Acetic acid bacteria isolated from grapes of South Australian vineyards. *International Journal of Food Microbiology* **178:** 98-106.

Maugenet, J. (1962). Métabolisme de l'acide lactique par *Acetobacter* dans les cidres 'framboisés'. *Compte-rendua l'Academie d'Agriculture* **48:** 214-217.

Medina, E., Pérez-Díaz, I.M., Breidt, F., Hayes, J., Franco, W., Butz, N. and Azcarate-Peril, M.A. (2016). Bacterial ecology of fermented cucumber rising pH spoilage as determined by nonculture-based methods. *Journal of Food Science* **81:** M121-129.

Miguel, M.G.D.C.P., Cardoso, P.G., de Assis Lago, L. and Schwan, R.F. (2010). Diversity of bacteria present in milk kefir grains using culture-dependent and culture-independent methods. *Food Research International* **43:** 1523-1528.

Millet, V. and Lonvaud-Funel, A. (2000). The viable but non-culturable state of wine micro-organisms during storage. *Letters in Applied Microbiology* **30:** 136-141.

Motaghi, M., Mazaheri, M., Moazami, N., Farkhondeh, A., Fooladi, M.H. and Goltapeh, E.M. (1997). Kefir production in Iran. *World Journal of Microbiology and Biotechnology* **13:** 579-581.

Navarro, D., Mateo, E., Torija, Maj. and Mas, A. (2013). Acetic acid bacteria in grape must. *Acetic Acid Bacteria* **2:** 19-23.

Ottogalli, G., Galli, A., Resmini, P. and Volonterio, G. (1973). Composizione microbiologica, chimica ed ultrastruttura dei granuli di Kefir. *Annali di Microbiologia ed Enzimologia* **23:** 109-121.

Ough, C.S. (1989). The changing California wine industry. *Journal of the Science of Food and Agriculture* **47:** 257-268.

Parrondo, J., Herrero, M., García, L.A. and Díaz, M. (2003). A Note—Production of Vinegar from Whey. *Journal of the Institute of Brewing* **109:** 356-358.

Pasteur, L. (1876). *Études sur la bière: Ses maladies, causes qui les provoquent, procédé pour la rendre inaltérable*. Gauthier-Villars, France.

Plata, C., Mauriacio, J.C., Millan, C. and Ortega, J.M. (2005). Influence of glucose and oxygen on the production of ethyl acetate and isoamyl acetate by a *Saccharomyces cerevisiae* strain during alcoholic fermentation. *World Journal of Microbiology and Biotechnology* **21:** 115-121.

Pogačić, T., Šinko, S., Zamberlin, Š. and Samaržija, D. (2013). Microbiota of kefir grains. *Mljekarstvo* **63:** 3-14.

Prieto, C., Jara, C., Mas, A. and Romero, J. (2007). Application of molecular methods for analyzing the distribution and diversity of acetic acid bacteria in Chilean vineyards. *International Journal of Food Microbiology* **115:** 348-355.

Raspor, P. and Goranovič, D. (2008). Biotechnological applications of acetic acid bacteria. *Critical Reviews in Biotechnology* **28:** 101-124.

Rea, M.C., Lennartsson, T., Dillon, P., Drinan, F.D., Reville, W.J., Heapes, M. and Cogan, T. M. (1996). Irish kefir-like grains: Their structure, microbial composition and fermentation kinetics. *Journal of Applied Bacteriology* **81:** 83-94.

Ribéreau-Gayon P., Dubourdieu, D., Doneche, B. and Lovaud, A. (2000). Handbook of Enology. Vol. 1. *The Microbiology of Wine and Vinifications*. John Wiley & Sons Ltd., Chichester, England.

Rosi, J. (1978). Microrganismi del kefir: Gli acetobatteri. *Sciezae Tecnica Lattiero-Casearia* **29:** 221-227.

Rosi, J. and Rossi, J. (1978). The kefir microorganisms: The lactic acid bacteria. *Sciezae Tecnica Lattiero-Casearia* **29:** 291-305.

Ruiz, P., Seseña, S., Izquierdo, P.M. and Palop, M.L. (2010). Bacterial biodiversity and dynamics during malolactic fermentation of Tempranillo wines as determined by a culture-independent method (PCR-DGGE). *Applied Microbiology and Biotechnology* **86:** 1555-1562.

Sakamoto, K. and Konings, W.N. (2003). Beer spoilage bacteria and hop resistance. *International Journal of Food Microbiology* **89:** 105-124.

Sakamoto, N., Tanaka, S., Sonomoto, K. and Nakayama, J. (2011). 16S rRNA pyrosequencing-based investigation of the bacterial community in nukadoko, a pickling bed of fermented rice bran. *International Journal of Food Microbiology* **144:** 352-359.

Sand, F.E.M.J. and Van Grinsven, A.M. (1976). Comparison between the yeast flora of Middle Eastern and Western European soft drinks. *Antonie van Leeuwenhoek* **42:** 523-532.

Shimwell, J.L. (1936). Study of a new species of *Acetobacter* (*A. capsulatum*) producing ropiness in beer and beer-wort. *Journal of the Institute of Brewing* **42:** 585-595.

Shimwell, J.L. (1948). A study of ropiness in beer. *Journal of the Institute of Brewing* **54:** 237-244.

Silhavy, K. and Mandl, K. (2006). *Acetobacter tropicalis* in spontaneously fermented wines with vinegar fermentation in Austria. *Mitteilungen Klosterneuburg* **56:** 102-107.

Silva, L.R., Cleenwerck, I., Rivas, R., Swings, J., Trujillo, M.E., Willems, A., Velázquez, E. (2006). *Acetobacter oeni* sp. nov., isolated from spoiled red wine. *International Journal of Systematic and Evolutionary Microbiology* **56:** 21-24.

Solieri, L. and Giudici, P. (2009). *Vinegars of the World*. Springer-Verlag, Italy.

Spitaels, F., Li, L., Wieme, A., Balzarini, T., Cleenwerck, I., Van Landschoot, A., De Vuyst, L. and Vandamme, P. (2014a). *Acetobacter lambici* sp. nov., isolated from fermenting lambic beer. *International Journal of Systematic and Evolutionary Microbiology* **64:** 1083-1089.

Spitaels, F., Wieme, A.D., Janssens, M., Aerts, M., Daniel, H.-M., Van Landschoot, A., De Vuyst, L. and Vandamme, P. (2014b). The microbial diversity of traditional spontaneously fermented lambic beer. *PloS One* **9:** e95384.

Splittstoesser, D.F. (1982). Microorganisms involved in the spoilage of fermented fruit juices. *Journal of Food Protection* **45:** 874-877.

Suárez, A. and Cabranes, C. (1989). Los microorganismos de la sidra natural asturiana. Avances de un estudio microbiológico. *Información Técnica del Centro de Experimentación Agraria de Villaviciosa* **3/89:** 1-21.

Tapia-Hernández, A., Bustillos-Cristales, M.R., Jimenez-Salgado, T., Caballero-Mellado, J. and Fuentes-Ramirez, L.E. (2000). Natural endophytic occurrence of *Acetobacter diazotrophicus* in pineapple plants. *Microbial Ecology* **39:** 49-55.

Torija, M.J., Mateo, E., Guillamón, J.M. and Mas, A. (2010). Identification and quantification of acetic acid bacteria in wine and vinegar by TaqMan-MGB probes. *Food Microbiology* **27:** 257-265.

Unsal, B. (2008). *Phylogenetic Analysis of Bacterial Communities in Kefir by Metagenomics.* M.S. Thesis. Graduate School of Engineering and Sciences. Izmir Institute of Technology, Izmir, Turkey.

Valera, M.J., Laich, F., González, S.S., Torija, M.J., Mateo, E. and Mas, A. (2011). Diversity of acetic acid bacteria present in healthy grapes from the Canary Islands. *International Journal of Food Microbiology* **151:** 105-112.

Valera, M.J., Torija, M.J., Mas, A. and Mateo, E. (2015). Acetic acid bacteria from biofilm of strawberry vinegar visualized by microscopy and detected by complementing culture-dependent and culture-independent techniques. *Food Microbiology* **46:** 452-462.

Van Vuuren, H.J., Louw, H.A., Loos, M.A. and Meisel, R. (1977). Procedures involving lipid media for detection of bacterial contamination in breweries. *Applied and Environmental Microbiology* **33:** 246-248.

Van Vuuren, H.J.J. (1996). Gram-negative spoilage bacteria. *In:* F.G. Priest and I. Campbell (Eds.). *Brewing Microbiology*, 2nd edn. Elsevier, London, pp. 163-191.

Vasavada, P.C. and Herpekan, D. (2002). Non-thermal alternative processing technologies for the control of spoilage bacteria in fruit juices and fruit-based drinks. *Food Safety Magazine* **8:** 46-47.

Vaughan, A., Sullivan, T.O., Sinderen, D. Van and Brew, J.I. (2005). Enhancing the Microbiological Stability of Malt and Beer—A Review. *Journal of the Institute of Brewing* **111:** 355-371.

Whiting, G.C. (1973). Acetification in ciders and perries. *Journal of the Institute of Brewing* **79:** 218-226.

Wieme, A.D., Spitaels, F., Aerts, M., De Bruyne, K., Van Landschoot, A. and Vandamme, P. (2014). Identification of beer-spoilage bacteria using matrix-assisted laser desorption/ionization time-of-flight mass spectrometry. *International Journal of Food Microbiology* **185:** 41-50.

Wszolek, M., Kupiec-Teahan, B., Guldager, H.S. and Tamine, A.Y. (2006). Production of kefir, koumiss and other related products. *In:* A.Y. Tamine (Ed.). *Fermented Milks.* Blackwell Publishing, Oxford, pp. 174-216.

Yamada, Y., Yukphan, P., Vu, H.T.L., Muramatsu, Y., Ochaikul, D. and Nakagawa, Y. (2012). Subdivision of the genus *Gluconacetobacter* Yamada, Hoshino and Ishikawa 1998: The proposal of *Komagatabacter* gen. nov., for strains accommodated to the *Gluconacetobacter xylinus* group in the α-Proteobacteria. *Annals of Microbiology* **62:** 849-859.

13

Beneficial Effects of Acetic Acid Bacteria and Their Food Products

Seniz Karabiyikli[1] and Ilkin Yucel Sengun[2,*]

[1] Gaziosmanpaşa University, Faculty of Engineering and Natural Science, Department of Food Engineering, 60250, Tokat, Turkey

[2] Ege University, Faculty of Engineering, Department of Food Engineering, 35100, Bornova, Izmir, Turkey

Introduction

Acetic acid bacteria are widely spread in nature and in a variety of food products. The main characteristic of acetic acid bacteria is the oxidation of ethanol, sugars and sugar alcohols to different kinds of organic acids, in general and in particular, to acetic acid. Therefore, the end product of the fermentation is mainly acetic acid; this oxidation reaction is also called oxidative fermentation, which is the main production mechanism of several foods and beverages, such as vinegar, cocoa and Kombucha tea. This group is also important to produce specific products, such as bacterial cellulose and ascorbic acid. The large interest in food and beverage production is mainly due to the ability of acetic acid bacteria to oxidize ethanol to acetic acid. However, this reaction could also induce the spoilage of some food products, such as alcoholic beverages, especially wine and beer, soft drinks, fruits and fruit-based products (Raspor and Goranovic 2008, Guillamón and Mas 2009, Sengun and Karabiyikli 2011).

Industrial production of acetic acid is performed either synthetically or biologically (Cheung et al. 2000). The biological production of acetic acid is only about 10 per cent of world production. However, it is important for vinegar production, which should be of biological origin according to the many food legislations in the world (Bala 2003). Although vinegar is used as a flavoring agent for foods, it also has medicinal uses. The most commonly known benefits of vinegar is antimicrobial effect, which mainly originates from the organic acids and phenolic content of the product. Hence, vinegar has a special importance in terms of food safety (Sengun and Karapinar 2004, 2005a, b). Other food products in which acetic acid bacteria play a role during their fermentation processes, have also some beneficial effects.

In this chapter, the importance of acetic acid bacteria and their products will be described in terms of food safety. Health benefits of acetic acid-derived products are also outlined, especially by focusing on vinegar.

Importance of Acetic Acid and Vinegar in Food Safety

Organic acids (e.g. acetic, citric, lactic and malic acid) have been used extensively as antimicrobial agents and/or surface disinfectants for a long time due to their inhibitory

*Corresponding author: ilkin.sengun@ege.edu.tr

activity on survival and growth of microorganisms. They control microbial growth as a consequence of the change in the concentration of hydrogen ions (pH). Most of the bacteria need neutral pH environment for optimal growth, but they could also tolerate the pH values between 4 and 9. The antimicrobial activity of organic acids depends on various factors, such as type and concentration of the acid, environmental conditions, target microorganism and the pH value, which is most important among the others (Raybaudi-Massilia et al. 2009). The mechanism of the efficiency changes depending on whether the acid is strong or weak. Strong acids (e.g. hydrochloric, sulfuric acid) are fully dissociated and do not penetrate into the cell membrane. Thus, they basically cause denaturation of the enzymes on the cell surface. Weak acids have lipophilic characteristics and penetrate through the cell membrane; thus they decrease the pH value of the cytoplasm (Rahman 2007). However, the weak acids exist in pH-dependent equilibrium between the two states as fully dissociated or un-dissociated. The organic acids show their antimicrobial effect optimally at low pH values where they are in an un-dissociated state. In this way, they can penetrate through the cell membrane and reach to the cytoplasm easily. Because of the neutral interior cell environment, the acid is forced to dissociate into anions and protons. The bacteria pump $H+$ ions outside the cell to protect the intracellular pH. Hence the membrane-transport proteins and transport enzymes are denaturized in turn, the membrane permeability is increased. Antimicrobial activity of organic acids mainly depends on disruption of membrane and cell signaling, accumulation of toxic anions, inhibition of glycolysis, inhibition of active transport and inhibition of essential metabolic reactions (Brul and Coote 1999, Davidson 2001, Raybaudi-Massilia et al. 2009).

Acetic acid as one of the most common organic acids is a carboxylic acid, which means that its acidity is associated with the carboxyl group in its structure (Dibner and Butin 2002). The main principle of the antimicrobial effect of acetic acid is related to its influence on the surface tension or that the effect was due to the whole molecule rather than to hydrogen ions alone (Ewadh et al. 2013). Antibacterial activity was connected with pH reduction of the substrate, depression of the intracellular pH by ionization of the un-dissociated acid molecule or disruption of substrate transport by alteration of cell membrane permeability; therefore, it is pH dependent (Ewadh et al. 2013).

Vinegar is a special kind of condiment produced from a variety of raw materials containing fermentable carbohydrates through the activity of yeasts and acetic acid bacteria. Acetic acid bacteria have the ability to produce organic acids, such as acetic, tartaric, lactic, malic, formic, citric and succinic acids. Hence, vinegar is an important source of organic acids produced, depending on the type of vinegar and inherent microflora. Acetic acid, which is responsible for the basic sensorial characteristic of vinegar, is the predominant acid found in all types of vinegar (Sengun and Karabiyikli 2011). Although vinegar is generally used as a flavoring agent in foods, it is also widely used as a natural preservative (Sengun and Karapinar 2004). The antimicrobial effect of vinegar on saprophytic and/or pathogenic microorganisms has been investigated by several researchers and it is concluded that vinegar is an effective solution to inhibit a variety of foodborne pathogens present on the surface of manufacturing area or food (Table 1).

Vinegar could be used for disinfection of a variety of equipment, foods and food preparation surfaces. It is applicable for complete upper denture wearers as an antifungal agent and more commonly used as disinfectant in fresh fruits and vegetables (Table 1). Fresh fruits and vegetables are good vehicles for transmission of most of the pathogens and play an important role in a variety of foodborne diseases (Ramos et al. 2014). Hence, sanitization of fresh produces as raw consumed products is a very important application in ensuring food safety. There are a number of solutions used to enhance the safety of fresh produce, mainly chlorine-based solutions and acids such as acetic, ascorbic, citric

Table 1: Antimicrobial effect of vinegar

Microorganisms (inoculated product)	*Type of vinegar*	*Influences and reduction rates*	*References*
E. coli, *L. monocytogenes*, *Lo. elongisporus*, *P. aeruginosa*, *St. aureus*, *S.* Typhimurium, *Y. enterocolitica*	Black vinegar (4.22-4.95 per cent a.a*)	The effect of vinegar was shown to be higher than that of the commercial antibiotics carbenicillin and tetracycline against test microorganisms (inhibition zones: 12-22 mm)	Choi et al. 2015
Total bacteria, lactic acid bacteria, coliform bacteria, *Salmonella*	Bamboo vinegar supplemented feed (0.3 per cent) (1.78 per cent a.a)	Treatment decreased the number of *Salmonella* (4.5 log reduction) and coliform bacteria (3.4 log reduction) in fecal microflora of fattening pigs, while it increased the total bacteria (2.4 log increase) and lactic acid bacteria (2.4 log increase) counts	Chu et al. 2013
Candida spp.	Alcohol vinegar (4 per cent a.a)	Minimum inhibitory concentration: 75 per cent, Minimum fungicidal concentration: 62.5 per cent	de Castro et al. 2015
Kl. pneumoniae	Liquid phase vinegar (2.4 per cent), vapor phase vinegar (8 per cent)	2.56 log reduction was achieved after 6 h exposure	Krusong et al. 2015
B. cereus, *E. coli*, *E. coli* O157:H7, *Kl. pneumoniae*, *L. monocytogenes*, *P. aeruginosa*, *Pr. vulgaris*, *S.* Typhimurium, *St. aureus*, *Y. enterocolitica*	Industrial vinegar (3.38-5.68 per cent a.a) Traditional vinegar (0.32-7.20 per cent a.a)	All tested bacteria had sensitivity to all vinegars at different levels. Two traditional vinegar samples did not show any inhibitory effect on the bacterial strains. Traditional vinegars showed lower antimicrobial activity according to the industrial vinegars (inhibition zones: 6.18-23.56 mm)	Ozturk et al. 2015
As. fumigatus, *Pe. chrysogenum*	Vinegar 4.0–4.2 per cent a.a	Effective only to inhibit the growth of *Pe. chrysogenum* (inhibition zone: 15 mm)	Rogawansamy et al. 2015
L. monocytogenes (lettuce)	Balsamic vinegar of Modena (50 per cent)	2.15 log reduction in 15 min	Ramos et al. 2014
	White wine vinegar (50 per cent)	1.16 log reduction in 15 min	

(*Contd.*)

Table 1: (*Contd.*)

Microorganisms (inoculated product)	*Type of vinegar*	*Influences and reduction rates*	*References*
S. Typhimurium (Ground beef)	Liquid buffered vinegar (2 per cent), powdered buffered vinegar (2.5 per cent)	Reduction ranged between 0.36-0.70 log	Stelzleni et al. 2013
Natural microflora of domestic catfish fillets	Vinegar (0.5 per cent)	Approximately 4 log reduction	Lingham et al. 2012
E. coli O157:H7, *L. monocytogenes*, *S.* Typhimurium	Vinegar (2.5 per cent and 5 per cent a.a)	>5 log reduction of *S.* Typhimurium. The sensitivity of pathogens to vinegar: *S.* Typhimurium > *E. coli* O157:H7 > *L. monocytogenes*	Yang et al. 2009a
Candida spp.	Vinegar	Approximately 1 log reduction	Pinto et al. 2008
E. coli O157:H7 (lettuce)	Rice vinegar (5 per cent a.a)	3 log reduction in 5 min.	Chang and Fang 2007
E. coli O157:H7, *L. monocytogenes*, *S.* Enteritidis	Vinegar + virgin olive oil extract	3 log reduction was observed	Medina et al. 2007
S. Typhimurium (rocket leaves and spring onion)	Grape vinegar (3.95 per cent a.a)	Reduction changed depending on vegetable used: 2.49 log for rocket leaves and 2.10 log for spring onion	Sengun and Karapinar 2005a
	Grape vinegar-lemon juice mix (1:1) (8.57 per cent a.a)	Reductions ranged between 6.55 log (for rocket leaves) and 2.31 log (for spring onion)	
Y. enterocolitica (carrot)	Grape vinegar-lemon juice mix (1:1) (6.6 per cent a.a)	7.14 log reduction in 30 min.	Sengun and Karapinar 2005b
S. Typhimurium (carrot)	Grape vinegar (4.03 per cent a.a)	Reduction ranged between 1.57-3.58 log depending on exposure time	Sengun and Karapinar 2004
E. coli	White vinegar (1.9 per cent a.a)	5.4 log reduction in 10 min.	Vijayakumar and Wolf-Hall 2002
Sh. sonnei	Vinegar (7.6 per cent a.a)	7.07 log reduction in 5 min.	Wu et al. 2000
Y. enterocolitica	Grape vinegar (4.9 per cent a.a)	7 log reduction in 15 min.	Karapinar and Gonul 1992

*a.a.: acetic acid, *As.*: *Aspergillus*, *B.*: *Bacillus*, *E.*: *Escherichia*, *Kl.*: *Klebsiella*, *L.*: *Listeria*, *Lo.*: *Lodderomyces*, *P.*: *Pseudomonas*, *Pe.*: *Penicillium*, *Pr.*: *Proteus*, *S.*: *Salmonella*, *Sh.*: *Shigella*, *St.*: *Staphylococcus*, *Y.*: *Yersinia* and lactic acids. On the other hand, there are a limited number of natural antimicrobial agents that could be used as a sanitizer for foods and vinegar, especially for fresh fruits and vegetables (Sengun and Karapinar 2004, 2005a, b).

Antimicrobial effect of vinegars depends on the pH value, acid concentration and the content of vinegar. Vinegars contain mainly acetic acid, but are also rich in other bioactive components (Table 2). Types and quantities of microorganisms, attachment of the cells on surfaces, the surface properties of the produce used, treatment time and temperature are also important factors that influence the antimicrobial activity of vinegar.

Vinegars made from various raw materials could have different phenolic compounds (Table 2). The production method also affects the phenolic composition of vinegar (Budak and Guzel-Seydim 2010). Although antimicrobial activity of vinegar mainly originates from the acetic acid content of the product, other components found in vinegar have also important contribution. The strong antimicrobial activity of balsamic vinegar has been connected to the presence of phenolic compounds that possess antimicrobial properties, such as vanillic acid, gallic acid and caffeic acid (Ramos et al. 2014). On the other hand, high amounts of phenolic compounds of vinegar indicate not only the antimicrobial potential, but also the antioxidant potential of the product.

Table 2: Phenolic compounds in different types of vinegar

Vinegar type	*Phenolic compounds*	*References*
Acacia vinegar	Gallic acid, protocatechuic acid, tyrosol, caftaric acid, vanillic acid, caffeic acid, syringic acid, gallic ethyl ester, resveratrol glucoside, ellagic acid	Cerezo et al. 2008
Apple vinegar	Gallic acid, chlorogenic acid, catechin, caffeic acid	Aykın et al. 2015
Black vinegar	Gallic acid, catechin, chlorogenic acid, p-hydroxybezoic acid, p-cumeric acid, ferulic acid, sinapic acid	Chou et al. 2015
Cherry vinegar	Gallic acid, protocatechuic acid, tyrosol, caftaric acid, vanillic acid, caffeic acid, syringic acid, gallic ethyl ester, resveratrol glucoside, ellagic acid, (+)-catechin	Cerezo et al. 2008
Chestnut vinegar	Gallic acid, protocatechuic acid, tyrosol, caftaric acid, vanillic acid, caffeic acid, syringic acid, gallic ethyl ester, resveratrol glucoside, ellagic acid, (+)-catechin, (–)-epicatechin	Cerezo et al. 2008
Grape vinegar	Gallic acid, catechin, epicatechin, chlorogenic acid, caffeic acid, syringic acid, ferulic acid	Budak and Guzel-Seydim 2010
Kiwi vinegar	Gallic acid, chlorogenic acid, vanillic acid, catechin, phlorizin, P-coumaric acid, caffeic acid, ferulic acid	Li et al. 2013
Oak vinegar	Gallic acid, protocatechuic acid, tyrosol, caftaric acid, vanillic acid, caffeic acid, syringic acid, gallic ethyl ester, resveratrol glucoside, ellagic acid, (+)-catechin	Cerezo et al. 2008
Pomegranate vinegar	Gallic acid, catechin, caffeic acid	
Shanxi aged vinegar	Protocatechuic acid, dihydroferulic acid, dihydrosinapic acid, P-hydroxybenzoic acid, salicylic acid, P-coumaric acid, ferulic acid, sinapic acid	Chen et al. 2015
Traditional Balsamic vinegar	Furan-2-carboxylic acid, 5-hydroxyfuran-2-carboxylic acid, 4-hydroxybenzoic acid, vanillic acid, protocatechuic acid, syringic acid, isoferulic acid, p-coumaric acid, gallic acid, ferulic acid, caffeic acid	Plessi et al. 2006

Health Benefits of Food Products of Acetic Acid Bacteria

In today's world, the consumers prefer to consume natural, safe, nutritive and functional foods. The increasing demand for functional foods results in a huge sector worldwide and in their extended utility. Consumption of functional foods for improvement of targets in human body are gastrointestinal function, including the balancing of colonic microflora and control of nutrient bioavailability, food transit time, immune activity, endocrine activity, mucosal motility and epithelial cell proliferation (Ranadheera et al. 2010, Rivera-Espinoza and Gallardo-Navarro 2010). Antioxidant and redox systems need vitamins and non-vitamin components like polyphenols in certain amounts and their functions are important for all cells and tissues. Xenobiotic metabolic activities and their control by some food components, such as glucosinolates, which are mostly non-nutritive, are also important. These functions may be important in controlling toxicity and carcinogenicity caused by chemical contaminants of the food and/or the environment. Moods and behavior as well as cognitive and physical performances may be influenced by food components, though there is a fine line between nutrition and pharmacology in this category (Saarela et al. 2000, Mattila-Sandholm et al. 2002).

Probiotics can be defined as 'live microorganisms that, when administered in adequate amounts, confer a health benefit on the consumer' (FAO/WHO 2002). The definition can include many types of probiotic cultures (mono- and mixed-strain cultures, multiple probiotic species), applications (gastrointestinal topical) and probiotic activity (live cells, dead cells and cellular components). Consuming viable probiotic cells provides protection against enteric pathogens, modulates the host immune system, supplies enzymes to help metabolize some food nutrients, influences the microbial composition of intestine and detoxifies some harmful food components and metabolites in the intestine (Ceapa et al. 2013). Generally, it is known that probiotics have been added to dairy products, especially fermented ones, such as yogurt or fermented milk. However, non-dairy-based probiotic products in the form of liquids, tablets, granules and capsules have been gaining popularity (Rivera-Espinoza and Gallardo-Navarro 2010, Doyle et al. 2013).

The greater part of probiotics consists of lactic acid bacteria, but acetic acid bacteria, which can be introduced as probiotics because of their strong fermentation and acidification activities and vitamin C production, were also isolated from fermented dairy products (Gonzalez and Mas 2011, Haghshenas et al. 2015). *Acetobacter aceti, A. indonesiensis* and *A. syzygii* are the species that are introduced as probiotics (Lefeber et al. 2011, Haghshenas et al. 2015). In the study carried out by Haghshenas et al. (2015), *Acetobacter* strains isolated from Iranian traditional dairy products were evaluated for their probiotic properties and potential in anticancer activity. They reported that *A. indonesiensis* and *A. syzygii* have desirable tolerance for low pH and high bile salt concentration, appropriate antipathogenic activity against certain pathogenic bacteria and acceptable antibiotic susceptibility. The cytotoxic findings also showed that *A. syzygii* secretions have high anticancer activity on treated cancerous cell lines (Haghshenas et al. 2015). Moreover, acetic acid bacteria have the ability to produce levan, which has noticeable prebiotic effects and also biomedical properties, such as anti-inflammatory, antioxidant, anticarcinogenic and hypocholesterolaemic effects (Srikanth et al. 2015). However, more detailed studies are needed to display the comprehensive probiotic and therapeutic properties of acetic acid bacteria and their derived products.

Acetic acid bacteria involved in the manufacture of fermented foods produce mainly acetic acid that possesses antimicrobial properties. One of the most important food applications of acetic acid bacteria is vinegar. Acetic acid production is common for most of the acetic acid bacteria, whereas the production of other bioactive compounds is strain dependent. For example, 3-dehydroshikimate, which is the direct precursor of protocatechuic acid and has an important antioxidant and anti-inflammatory effect, is produced by some *Gluconobacter* strains (Adachi et al. 2008).

Vinegar, cocoa and Kombucha tea, the well-known products of acetic acid bacteria, are important sources for their valuable contents that have various health benefits.

Therapeutic Effects of Vinegar

The use of vinegar in traditional and natural folk medicine for treating a variety of diseases, such as nail fungus, head lice, warts, wounds and ear infections can be traced far back to ancient times (Rund 1996, Rutala et al. 2000, Shay 2000, Dohar 2003, Budak et al. 2014). Antimicrobial, antioxidant, antidiabetic, anticarcinogenic effects of vinegar and its positive effects on weight loss, heart health, heartburn and acid reflux, injuries, brain health and increasing nutrient adsorption are well-known (Mercola 2014, Chen et al. 2016). In several *in vivo/vitro* studies on vinegar, its antihypertensive properties (Kondo et al. 2001, Nakamura et al. 2010, Nandasiri 2012), anticarcinogenic effect (Shizuma et al. 2011), contribution to enhancement of glycogen repletion (Fushimi and Sato 2005) and stimulating effect for reduction of serum cholesterol (Setorki et al. 2010, Setorki et al. 2011) were demonstrated. Some of these studies are given in Table 3.

Antiglycemic Effects

Vinegar has beneficial effects on blood sugar levels; hence, the medicinal use of vinegar in diabetic treatment has increased in recent years. Some researchers evaluated the hypoglycemic property of vinegar on insulin-resistant individuals or risk groups for type 2 diabetes (Table 3). The findings showed that vinegar consumption reduced postprandial glycemia 64 per cent as compared to placebo values in the insulin-resistant incidences (Johnston et al. 2004) and caused 60 min. reduction in glucose response of a group of healthy adult women and weakly affected later energy consumption (Johnston and Buller 2005). The blood glucose response of rats was significantly decreased by 0.3 per cent and 2 per cent acetic acid solution-supplemented feeds (Xibib et al. 2003, Sakakibara et al. 2006). White rice vinegar also improved fasting hyperglycemia and body weight loss through attenuating insulin deficiency, pancreatic beta-cell deficit and hepatic glycogen depletion and fatty changes in streptozotocin-induced diabetic rats (Gu et al. 2012). Caffeoylsophorose, a new natural α-glucosidase inhibitor isolated from the purple sweet potato vinegar showed hypoglycemic effect in rats (Matsui et al. 2014). It was confirmed that the antiglycemic effect of vinegar is comparable to the standard antidiabetic drugs (Yusoff et al. 2015).

The antiglycemic effect could be linked to carbohydrate maldigestion (Ogawa et al. 2000, Johnston et al. 2013), which is accomplished either by accelerating gastric emptying or promoting the glucose uptake by tissues (Budak et al. 2014). The hypoglycemic effect of acetic acid might be due to activation of 5′-AMP-activated protein kinase (AMPK) in the liver (Sakakibara et al. 2006). However, more research is needed to clarify the exact mechanism of vinegar action.

Cardiovascular Effects

There are many studies concluding the protective effect of vinegar from cardiovascular diseases. Apple cider vinegar shows cholesterol-lowering effect in rats fed high-cholesterol diets (Budak et al. 2011), inhibit low density lipoprotein (LDL) oxidation and prevent cardiovascular diseases (Laranjinha et al. 1994). High doses of vinegar significantly reduced total serum cholesterol (TC) and LDL concentrations in experimental rabbits (Setorki et al.

Table 3: Beneficial effects of vinegar consumption

Beneficial effects	*Vinegar type*	*References*
Antiglycemic effects	Apple cider vinegar	Johnston et al. 2004, 2013
	Nipa palm vinegar	Yusoff et al. 2015
	White vinegar	Brighenti et al. 1995
	White rice vinegar	Gu et al. 2012
	Wine vinegar	Liatis et al. 2010
Cardiovascular effects	Apple cider vinegar	Budak et al. 2011
	Black malt vinegar	Odahara et al. 2008
	Komesu and kurosu	Murooka and Yamshita 2008, Chou et al. 2015
	Rice vinegar	Ohnami et al. 1985, Kondo et al. 2001, Nishikawa et al. 2001
	Sorghum vinegar	Li et al. 2014
	Tomato vinegar	Lee et al. 2013
	Wine vinegar	Sugiyama et al. 2003, Honsho et al. 2005
Antioxidant effects	Kurosu	Nishidai et al. 2000, Shimoji et al. 2002, Chou et al. 2015
	Oat vinegar	Qiu et al. 2010
	Persimmon vinegar	Ubeda et al. 2011
	Rice vinegar	Seki et al. 2008
	Wine vinegar	Dávalos et al. 2005, Budak and Guzel-Seydim 2010
	Zhenjiang aromatic vinegar	Xu et al. 2007
Anticarcinogenic effects	Apple vinegar	Abe et al. 2014
	Kurosu	Nanda et al. 2004, Shimoji et al. 2004, Tong et al. 2010
	Rice vinegar	Seki et al. 2004
	Sugarcane vinegar	Mimura et al. 2004
	Sweet potato vinegar	Morimura et al. 2004
	Traditional balsamic vinegars	Daglia et al. 2013
Lipid lowering effects	Apple cider vinegar	Budak et al. 2011
	Kurosu	Chou et al. 2015
	Permisson vinegar	Moon et al. 2010
	Pomegranate vinegar	Park et al. 2014
	Tomato vinegar	Lee et al. 2013

2010). Li et al. (2014) reported that alditol and monosaccharide from sorghum vinegar significantly induced antiplatelet aggregation activity, which could be beneficial for the treatment of cardiovascular diseases.

Cardiovascular effects of wine vinegar beverage and grape juice on pentobarbital-anesthetized rats were researched and the recommended dose of the beverage (3 ml/kg, p.o.) decreased the heart rate and mean blood pressure. According to the results, it was suggested that this beverage may be useful for people who suffer from palpitation and/or hypertension (Sugiyama et al. 2003).

Antihypertensive effects of acetic acid and vinegar on hypertensive rats were evaluated by some researchers. Acetic acid itself significantly reduced both blood pressure and renin activity and it was concluded that the antihypertensive effect of vinegar is mainly due to the acetic acid in it (Kondo et al. 2001). The effect of black malt vinegar on the blood pressure of spontaneously hypertensive rats was investigated. It was found that single and continuous administration of black malt vinegar showed a significant hypotensive effect (Odahara et al. 2008). Some extracts of rice vinegar prevent angiotensin-converting enzyme activity in spontaneously hypertensive rats (Ohnami et al. 1985) and rice vinegar prevents angiotensin-converting enzyme activity in the blood pressure regulatory system (Nishikawa et al. 2001).

Komesu (Japanese polished amber rice vinegar) and Kurosu (unpolished black rice vinegar) are both known for their antihypertensive and antiinflammatory effects (Murooka and Yamshita 2008), and organic acids in vinegar play an important role in the prevention and cure of cardiovascular diseases (Jing et al. 2015). Ligustrazine, which improves blood circulation and used in the treatment of cerebrovascular and cardiovascular diseases, was identified from Zhenjiang aromatic vinegar (He et al. 2004).

Antioxidant Effects

Significant antioxidant capacity was reported in various types of vinegars by researchers. The antioxidant activity is closely related to the bioactive components of vinegar. It was declared that Kurosu, which is rich in phenolic compounds, has higher antioxidant capacity than wine and apple vinegars (Nishidai et al. 2000). Moreover, oat vinegar contains several phytochemicals, such as tocotrienols, phenolic compounds, flavonoids, phytic acids, sterols and avenanthramides, showed stronger free-radical scavenging activity, reducing power and inhibition of lipid peroxidation than rice vinegar (Qiu et al. 2010). Mother vinegar had also high antioxidant capacity and total phenolic substance (Aykın et al. 2015). However, antioxidant activity of Zhenjiang aromatic vinegar extracts was not correlated with phenolic content, whereas melanoidins, that are brown polymers formed through Maillard reaction during vinegar production process, may have health-promotion activity (Xu et al. 2007). Similarly, the antioxidant activities of balsamic vinegars were found related with melanoidins present in the product (Verzelloni et al. 2010). Two major antioxidant compounds (tyrosol and ferulic acid) were identified from the rice shochu distilled residue-derived vinegar and the results of *in vitro* and *in vivo* studies showed that this type of vinegar is a healthy food with an antioxidant effect in the prevention of oxidative injury and cancer (Seki et al. 2008).

It was reported that traditional grape wine vinegar had higher chlorogenic and syringic acid content than industrial grape wine vinegar and antioxidant capacity of traditional vinegar was higher than that of industrial vinegar. Antioxidant activity of persimmon vinegar was higher than wine vinegars because of the wild yeast strain used in persimmon vinegar production (Ubeda et al. 2011). Hence, different types of vinegar and production methods used affect their functional constituents.

Anticarcinogenic Effects

Anticarcinogenic properties of various types of vinegars have been demonstrated *in vivo* and *in vitro* studies. It was reported that daily consumption of vinegar decreased the risk of oesophageal cancer (Xibib et al. 2003). *Kurosu*, which is rich in phenolic compounds and has significant antioxidant activity, inhibited the proliferation of human cancer cells in a dose-dependent manner (Nanda et al. 2004) and decreased the adipocyte size in rat models

(Tong et al. 2010). In another *in vivo* study, drinking water supplement with *Kurosu* extract (0.05 per cent to 0.1 per cent w/v) significantly inhibited the incidence (–60 per cent) and multiplicity (–50 per cent) of azoxymethane-induced colon cancer in rats (Shimoji et al. 2004).

The α-dicarbonyl compounds present in balsamic and traditional balsamic vinegars were analyzed and the cytotoxic activity of glyoxal, methylglyoxal, 2,3-butanedione was evaluated against an intestinal human cell line. It was concluded that dietary intake of the tested α-dicarbonyl compounds is a significant source of toxicity (Daglia et al. 2013). Tryptophol, known as anticancer compound, was isolated from Japanese black soybean vinegar (Inagaki et al. 2007, Baba et al. 2013)

Sugar cane vinegar, called Kibizu, induced apoptosis in human leukemia cells (Mimura et al. 2004). Vinegar produced from sweet potato shochu distillery wastewater (post-distillation slurry) by *A. aceti* showed antitumor activity in a mouse model (Morimura et al. 2004). Neutral alpha-glycan content formed during apple vinegar fermentation acts against experimental mouse tumors (Abe et al. 2014). Sarcoma tumor cell volumes decreased when mice were fed vinegar-fortified feed (Seki et al. 2004). All these findings demonstrate that vinegar consumption has an important potential to inhibit the proliferation of cancer cells and show antitumor activity.

Lipid-lowering Effects

Lipid-lowering effects of different kinds of vinegars have been reported by researchers. The effects of persimmon-vinegar supplementation on blood lipid profiles, carnitine concentrations and hepatic mRNA levels of enzymes involved in fatty acid metabolism were investigated by Moon et al. (2010). The results of the study showed that persimmon-vinegar increased serum and hepatic acid-insoluble acylcarnitines (AIAC) concentrations, while decreasing the epididymis fat weight, serum and hepatic total cholesterol (TC) concentrations. Therefore, the consumption of persimmon-vinegar can be a useful intervention in obesity and hyperlipidemia caused by ingestion of a high fat diet.

The addition of vinegar in a high-cholesterol diet increases the expression of the acyl-coenzyme A oxidase gene. In this way, fatty acid oxidation is increased and the elevation of plasma triglycerides is suppressed (Yamashita et al. 2007). The lipid-lowering and antioxidant effects of black vinegar were investigated on hamsters fed with black vinegar supplements. It decreased weight gain, relative size of visceral fat, serum/liver TC levels, serum cardiac index and hepatic thiobarbituric acid reactive substances (TBARS) values and damage indices but increased faecal lipid contents and hepatic antioxidant capacities (Chou et al. 2015).

Tomato vinegar containing phytochemicals was tested on rats fed a high-fat diet. Tomato vinegar decreased hepatic triglyceride and cholesterol levels and lowered plasma LDL-cholesterol level, thereby suggesting that it is an anti-obesity therapeutic agent or functional food (Lee et al. 2013). Beneficial effects of pomegranate vinegar beverage on adiposity were also reported by Park et al. (2014). These results prove the potential of lipid-lowering effects of a variety of vinegars.

Other Health Benefits

It was demonstrated that the mother of vinegar is therapeutical for burns because of its antibacterial properties (Krystynowicz et al. 2000). Some extracellular metabolites of acetic acid bacteria are effective in repairing tissue scars in rats (Bielecki et al. 2000). For example, cellulose produced by acetic acid bacteria can be used to manufacture wound dressings (Raspor and Goranovič 2008). Moreover, hiccups occurred during chemotherapy stopped or decreased in intensity after drinking vinegar (Gonella and Gonella 2015). Vinegar consumption restores ovulatory function in polycystic over-syndrome patients and could reduce pharmacological treatment. It could also reduce the treatment time for insulin resistance and ovulation in infertility patients (Wu et al. 2013). Addition of probiotics and acetic acid to feed poultry as an alternative to antibiotics was investigated and slow increase in weight by using organic acid in the diet was observed (Král et al. 2011).

In spite of all these positive and contributive effects of vinegar consumption, Petsiou et al. (2014) reviewed the studies on the effect of vinegar on hyperglycemia, hyperinsulinemia, hyperlipidemia and obesity. They concluded that although some case reports support the addition of vinegar to the daily diet of patients with glucose and lipid abnormalities, large-scale and long-term trials with impeccable methodology are needed before definitive health claims can be made.

Therapeutic Effects of Cocoa

The biochemical reactions on fermentation of cocoa beans are mostly related with lactic and acetic acids generally formed as a result of microbial activity (Evina et al. 2016). Cocoa can be defined as a functional food with a high content of monomeric (epicatechin and catechin) and oligomeric (procyanidins) flavanols (Adamson et al. 1999, Hammerstone et al. 1999, Natsume et al. 2000).

Cocoa and chocolate may have beneficial health effects against chronic inflammation, oxidative stress, carcinogenic factors, cardiovascular disease and other chronic diseases (Table 4). Cocoa products and chocolate have recently been recognized as a potent antioxidant and antiinflammatory agents with their rich flavonoid contents, and have benefits on cardiovascular health, but largely unproven effects on neurocognition and behavior. A variety of studies show the effects of chocolate intake on cholesterol concentration, cardiovascular system, skin and the release of neurotransmitters anandamide and serotonin and on the health-related properties of dark chocolate (Sokolov et al. 2013).

Recently, studies on psychology have demonstrated cocoa-based beverages contain flavanols which can improve the performance at cognitive function tests (Smith 2013). However, treatment of rats with cocoa extract containing high levels of flavanols influenced the antidepressant action (Messaoudi et al. 2008). The benefits of flavanol-rich cocoa-derived products in the prevention and treatment of mood disorders were reviewed and it was defended that flavanol-rich cocoa-derived products, such as dark chocolate may have beneficial effects as add-on items together with traditional antidepressant regimes (Smith 2013).

Table 4: Beneficial effects of cocoa

Beneficial effect	*References*
Anticarcinogenic effects	Balkwill and Mantovani 2001, Taubert et al. 2007, Cooper et al. 2008, Maskarinec 2009
Antiinflammatory effect	Selmi et al. 2008, Ramos-Romero et al. 2012
Antioxidant effect	Lee et al. 2003, Maleyki and Ismail 2010, McCarty et al. 2010
Cardiovascular effects	Selmi et al. 2008, Janszky et al. 2009, Grassi et al. 2010, Djoussé et al. 2011, Pucciarelli 2013
Lipid-lowering effect	Alspach 2007, Grassi et al. 2008, Mellor et al. 2010
Neurobiological effects	Valente et al. 2009, Sokolov et al. 2013
Psychological effects	Michener and Rozin 1994, Macht and Dettmer 2006, Parker et al. 2006, Parker and Crawford 2007, Messaoudi et al. 2008, Smith 2013, Sokolov et al. 2013
Risk of stroke	Larsson et al. 2011

Patients of hypertension and poor glucose tolerance consumed 100 g flavanol-rich dark chocolate bars or 100 g flavanol-free white chocolate bars in two daily portions and it was detected that flavanol-rich dark chocolate bars decreased systolic and diastolic blood pressure and improved insulin sensitivity (Grassi et al. 2008). Grassi et al. (2010) concluded that cocoa flavonoids could reduce cardiovascular risk.

The anti-inflammatory impact of cocoa defined as a functional food due to its high content of flavanols was studied by Selmi et al. (2008) and it was declared that there are lots of experimental data to support the beneficial effects of cocoa flavanols on cardiovascular health in the general population. Additionally, Pucciarelli (2013) focused on the bioactive mechanistic linkages between cocoa and heart health. Furthermore, cocoa is effective on the immune system and cocoa intake modifies the functionality of gut-associated lymphoid tissue (Pérez-Cano 2013).

The epidemiologic evidence on protective effects of cocoa and chocolate products against cancer was investigated and it was mentioned that there are limited studies and only a few have emerged on the effects of cocoa products on endpoints that may be related to cancer risk (Maskarinec 2009).

Therapeutic Effects of Kombucha

Kombucha is a traditional fermented beverage produced by fermentation of sweetened black tea by a symbiosis of acidophilic yeast and acetic acid bacteria. This microbiological population called 'tea fungus' is situated in a microbial cellulose layer (Greenwalt et al. 2000, Battikh et al. 2012). Microbiological profile of tea fungus was identified by several researchers as *A. aceti, A. pasteurianus, K. xylinus, Brettanomyces bruxellensis, Brettanomyces custersii, Brettanomyces lambicus, Candida stellate, Hanseniaspora uvarum, Kloeckera apiculata, Kluyveromyces marxianus, Lachancea thermotolerans, Lachancea fermentati, Saccharomyces cerevisiae, Saccharomycopsis fibuligera, Saccharomycodes ludwigii, Schizosaccharomyces pombe, Torulaspora delbrueckii, Zygosaccharomyces bailii,* and some *Candida* and *Pichia* species (Reiss 1994, Liu et al. 1996, Balentine 1997, Zhang et al. 2011, Chakravorty et al. 2016). During fermentation of Kombucha, ethanol is produced by yeasts and then it is oxidized to acetic acid by acetic acid bacteria. Acetic acid bacteria commonly found in tea fungus belong to the genera of *Acetobacter, Gluconacetobacter, Gluconobacter* and *Komagataeibacter. K. xylinus* was the most important species present during Kombucha fermentation because of its ability to synthesize a floating cellulose (Yamada et al. 2012). Glucose is also metabolized by acetic acid bacteria in synthesis of cellulose and gluconic acid (Dufresne and Farnworth 2000). Microbial community and dynamics, which may vary between different Kombucha fermentations, affect the biochemical properties of Kombucha. Hence, the microbial community may also affect the health benefits of Kombucha, such as antimicrobial, antioxidant and antidiabetic properties (Malbaša et al. 2011, Chakravorty et al. 2016).

Kombucha is quite a popular drink because of its beneficial effects on human health. It can be regarded as an innovative functional product for human health. The positive contributions by Kombucha consumption in weight loss, asthma, cataracts, diabetes, diarrhea, gout, herpes, prostate, insomnia, rheumatism, hypertension, hemorrhoids, general well-being, nervous system, cancer and AIDS were claimed, but only a few of them were proved scientifically (Table 5).

This drink includes acetic acid, lactic acid, glucuronic acid and gluconic acid as the major compounds (Teoh et al. 2004, Jayabalan et al. 2007). High acidity of Kombucha (approximately 33 g/L total acidity) causes inhibition of the growth of harmful bacteria in the gastrointestinal tract (Greenwalt et al. 1997, 1998). It has an antimicrobial effect on several microorganisms, such as *Aeromonas hydrophila, Agrobacterium tumefaciens, Bacillus*

Table 5: Beneficial effects of Kombucha tea

Beneficial effects	*References*
Antimicrobial effect	Steinkraus et al. 1996, Greenwalt et al. 1997, 1998, Sai Ram et al. 2000, Pauline et al. 2001, Sreeramulu 2000, 2001, Cetojevic-Simin et al. 2008, Battikh et al. 2012
Antioxidant effect	Sai Ram et al. 2000, Chu and Chen 2006, Chan et al. 2011, Jayabalan et al. 2008, Malbaša et al. 2011, Chakravorty et al. 2016
Anticarcinogenic effect	Jayabalan et al. 2011
Antidiabetic effect	Aloulou et al. 2012, Bhattacharya et al. 2013
Antiglycation effect	Chakravorty et al. 2016
Immunopotentiating effect	Sai Ram et al. 2000
Gastric ulcer healing effect	Banerjee et al. 2010
Treatment for high cholesterol	Yang et al. 2009b
Detoxification effect	Petrović et al. 2000, Vina et al. 2013a, b, Jayabalan et al. 2014, Nguyen et al. 2015

cereus, Campylobacter jejuni, Escherichia coli, Helicobacter pylori, Listeria monocytogenes, Salmonella Typhimurium, *Salmonella* Enteritidis, *Shigella sonnei* and *Staphylococcus aureus* (Steinkraus et al. 1996, Greenwalt et al. 1998, Sai Ram et al. 2000, Pauline et al. 2001, Sreeramulu 2001). Although acetic acid is mainly responsible for the antimicrobial activity, neutralized or thermally processed Kombucha can also exert antimicrobial effect, which suggests the presence of antimicrobial compounds other than acetic acid or large proteins in Kombucha (Sreeramulu et al. 2000, 2001).

One of the most significant organic compounds found in Kombucha is glucuronic acid, which is a precursor for the biosynthesis of vitamin C and has detoxifying properties (eliminates pollutants, excess steroid hormones, exogenous chemicals and bilirubin from the human body via the urinary system) (Merchie et al. 1997, Vina et al. 2013a, b, Jayabalan et al. 2014, Nguyen et al. 2015). Glucuronic acid can be converted into glucosamine, a beneficial substance associated with cartilage, collagen and fluids related to the treatment of osteoarthritis (Yavari et al. 2011). The other organic compound commonly found in Kombucha is usnic acid. It has an important antimicrobial effect, especially on *Staphylococcus, Streptococcus, Pneumococcus, Mycobacterium* and some pathogenic fungi (Cocchietto et al. 2002).

Antioxidant activity of Kombucha is another subject that must be taken into consideration. Theaflavin and thearubigin, which are the main polyphenols found in Kombucha, have been reported to be responsible for the antioxidant capacity of tea. The total phenol content increases with fermentation time, which might be due to biodegradation of polyphenols during fermentation. Hence, antioxidant activities of tea differ from each other, depending on the source variety and fermentation time (Chu and Chen 2006, Chan et al. 2011). Antioxidant activity of Kombucha can also be linked to the microbial community of tea. Increase in the microbial diversity leads to an increase in antioxidant properties of the beverage (Chakravorty et al. 2016). This activity can be improved by modifying the formulation of Kombucha by using wheatgrass juice and starter, comprising yeast (*Dekkera bruxellensis*) and acetic acid bacteria (*Gluconacetobacter rhaeticus* and *Gluconobacter roseus*) (Sun et al. 2015). Kombucha can also be used as an innovative starter culture for various fermented milks to improve antioxidant capacity, vitamin C content and sensory properties of the products (Hrnjez et al. 2014).

Conclusion

Acetic acid bacteria are a group of microorganisms widespread in Nature, which can produce various organic acids from different sources. Most of the species have the capability to transform ethanol into acetic acid and thus, have a significant role to play in vinegar production. Wide variety of vinegars have been used not only to flavor and preserve the foods, but also to cure some diseases. Therapeutic properties of vinegar described by numerous researchers include antimicrobial, antioxidative, antiglycemic, antiobesity, anticarcinogenic, antihypertensive, cardiovascular and lipid-lowering effects. These health benefits can also be derived from consumption of cocoa and Kombucha, which are products in which acetic acid bacteria play a role during fermentation. However, it is hard to expect the same profit from all these products unless produced under controlled process conditions. As a conclusion, daily intake of vinegar, cocoa and Kombucha may affect human health in a positive way, depending on the product properties mainly because of the inherent components found in the product, such as organic acids and phenolic contents. However, further studies are needed for detailed information on the functional properties of acetic acid bacteria and their products.

References

Abe, K., Kushibiki, T., Matsue, H., Furukawa, K.I. and Motomura, S. (2014). Generation of antitumor active neutral medium-sized α-glycan in apple vinegar fermentation. *Bioscience, Biotechnology and Biochemistry* **71:** 2124-2129.

Adachi, O., Ano, Y., Toyama, H. and Matsushita, K. (2008). A novel 3-dehydroquinate dehydratase catalyzing extracellular formation of 3-dehydroshikimate by oxidative fermentation of *Gluconobacter oxydans* IFO 3244. *Bioscience, Biotechnology and Biochemistry* **72:** 1475-1482.

Adamson, G.E., Lazarus, S.A., Mitchell, A.E., Prior, R.L., Cao, G., Jacobs, P.H. and Schmitz, H.H. (1999). HPLC method for the quantification of procyanidins in cocoa and chocolate samples and correlation to total antioxidant capacity. *Journal of Agricultural and Food Chemistry* **47:** 4184-4188.

Aloulou, A., Hamden, K., Elloumi, D., Ali, M.B., Hargafi, K., Jaouadi, B., Ayadi, F., Elfeki, A. and Ammar, E. (2012). Hypoglycemic and antilipidemic properties of Kombucha tea in alloxan-induced diabetic rats. *BMC Complementary and Alternative Medicine* **12:** 63.

Alspach, G. (2007). The truth is often bittersweet...chocolate does a heart good. *Critical Care Nurse* **27:** 11-15.

Aykın, E., Budak, N.H. and Güzel-Seydim, Z.B. (2015). Bioactive components of mother vinegar. *Journal of the American College of Nutrition* **34:** 80-89.

Baba, N., Higashi, Y., Kanekura T. (2013). Japanese black vinegar "Izumi" inhibits the proliferation of human squamous cell carcinoma cells via necroptosis. *Nutrition and Cancer* **65:** 1093-1097.

Bala, S. (2003). Acetic acid. *In: Chemicals Economic Handbook*. SRI International.

Balentine, D.A. (1997). Special issue: Tea and health. *Critical Reviews in Food Science and Nutrition* **8:** 691-692.

Balkwill, F. and Mantovani, A. (2001). Inflammation and cancer: Back to Virchow. *The Lancet* **357:** 539-545.

Banerjee, D., Hassarajani, S., Maity, B., Narayan, G., Bandyopadhyay, S.K. and Chattopadhyay, S. (2010). Comparative healing property of Kombucha tea and black tea against indomethacin-induced gastric ulceration in mice: Possible mechanism of action. *Food and Function* **1:** 284-293.

Battikh, H., Bakhrouf, A. and Ammar, E. (2012). Antimicrobial effect of Kombucha analogues. *LWT-Food Science and Technology* **47:** 71-77.

Bhattacharya, S., Gachhui, R. and Sil, P.C. (2013). Effect of Kombucha, a fermented black tea in attenuating oxidative stress mediated tissue damage in alloxan induced diabetic rats. *Food and Chemical Toxicology* **60:** 328-340.

Bielecki, S., Krystynowicz, A., Turkiewicz, M. and Kalinowska, H. (2000). Bacterial cellulose. *In:* A. Steinbuchel (Ed.). *Biopolymers: Polysaccharides I*. Wiley-VCH Verlag GmbH, Munster, Germany, pp. 37-90.

Brighenti, F., Castellani, G., Benini, L., Casiraghi, M.C., Leopardi, E., Crovetti, R. and Testolin, G. (1995). Effect of neutralized and native vinegar on blood glucose and acetate responses to a mixed meal in healthy subjects. *European Journal of Clinical Nutrition* **49:** 242-247.

Brul, S. and Coote, P. (1999). Preservative agents in foods: Mode of action and microbial resistance mechanisms. *International Journal of Food Microbiology* **50:** 1-17.

Budak, H.N. and Guzel-Seydim, Z.B. (2010). Antioxidant activity and phenolic content of wine vinegars produced by two different techniques. *Journal of the Science of Food and Agriculture* **90:** 2021-2026.

Budak, N.H., Kumbul Doguc, D., Savas, C.M., Seydim, A.C., Kok Tas, T., Ciris, M.I. and Guzel-Seydim, Z.B. (2011). Effects of apple cider vinegars produced with different techniques on blood lipids in high-cholesterol-fed rats. *Journal of Agricultural and Food Chemistry* **59:** 6638-6644.

Budak, N.H., Aykin, E., Seydim, A.C., Greene, A.K. and Guzel-Seydim, Z.B. (2014). Functional properties of vinegar. *Journal of Food Science* **79:** 757-764.

Ceapa, C., Wopereis, H., Rezaïki, L., Kleerebezem, M., Knol, J. and Oozeer, R. (2013). Influence of fermented milk products, prebiotics and probiotics on microbiota composition and health. *Best Practice & Research. Clinical Gastroenterology* **27:** 139-155.

Cerezo, A.B., Tesfaye, W., Torija, M.J., Mateo, E., García-Parrilla, M.C. and Troncoso, A.M. (2008). The phenolic composition of red wine vinegar produced in barrels made from different woods. *Food Chemistry* **109:** 606-615.

Cetojevic-Simin, D.D., Bogdanovic, G.M., Cvetkovic, D.D. and Velicanski, A.S. (2008). Antiproliferative and antimicrobial activity of traditional Kombucha and *Satureja montana* L. Kombucha. *Journal of Balkan Union of Oncology* **13:** 395-401.

Chakravorty, S., Bhattacharya, S., Chatzinotas, A., Chakraborty, W., Bhattacharya, D. and Gachhu, R. (2016). Kombucha tea fermentation: Microbial and biochemical dynamics. *International Journal of Food Microbiology* **220:** 63-72.

Chan, E.W.C., Soh, E.Y., Tie, P.P. and Law, Y.P. (2011). Antioxidant and antibacterial properties of green, black, and herbal teas of *Camellia sinensis*. *Pharmacognosy Research* **3:** 266-272.

Chang, J. and Fang, T.J. (2007). Survival of *Escherichia coli* O157:H7 and *Salmonella enterica* serovars Typhimurium in iceberg lettuce and the antimicrobial effect of rice vinegar against *E. coli* O157:H7. *Food Microbiology* **24:** 745-751.

Chen, H., Zhou, Y., Shao, Y. and Chen, F. (2015). Free phenolic acids in Shanxi aged vinegar: changes during aging and synergistic antioxidant activities. *International Journal of Food Properties* **19:** 1183-1193.

Chen, H., Chen, T., Giudici, P. and Chen, F. (2016). Vinegar functions on health: Constituents, sources, and formation mechanisms. *Comprehensive Reviews in Food Science and Food Safety* **15:** 1124-1138.

Cheung, H., Tanke, R.S. and Torrence, G.P. (2000). Acetic acid. *Ullmann's Encyclopedia of Industrial Chemistry*. Weinheim: Wiley-VCH, doi:10.1002/14356007.a01_045.pub2.

Choi, H., Gwak, G., Choi, D., Park, J. and Cheong, H. (2015). Antimicrobial efficacy of fermented dark vinegar from unpolished rice. *Microbiology and Biotechnology Letters* **43:** 97-104.

Chou, C.H., Liu, C.W., Yang, D.J., Wu, Y.H.S. and Chen, Y.C. (2015). Amino acid, mineral and polyphenolic profiles of black vinegar, and its lipid-lowering and antioxidant effects *in vivo*. *Food Chemistry* **168:** 63-69.

Chu, G.M., Jung, C.K., Kim, H.Y., Ha, J.H., Kim, J.H., Jung, M.S., Lee, S.J., Song, Y., Ibrahim, R.I.H., Cho, J.H., Lee, S.S. and Song, Y.M. (2013). Effects of bamboo charcoal and bamboo vinegar as antibiotic alternatives on growth performance, immune responses and fecal microflora population in fattening pigs. *Animal Science Journal* **84:** 113-120.

Chu, S.C. and Chen, C. (2006). Effects of origins and fermentation time on the antioxidant activities of Kombucha. *Food Chemistry* **98:** 502-507.

Cocchietto, M., Skert, N., Nimis, P. and Sava, G. (2002). A review on usnic acid, an interesting natural compound. *Naturwissenschaften* **89:** 137-146.

Cooper, K.A., Donovan, J.L., Waterhouse, A.L. and Williamson, G. (2008). Cocoa and health: A decade of research. *British Journal of Nutrition* **99:** 1-11.

Daglia, M., Amoroso, A., Rossi, D., Mascherpa, D. and Maga, G. (2013). Identification and quantification of α-dicarbonyl compounds in balsamic and traditional balsamic vinegars and their cytotoxicity against human cells. *Journal of Food Composition and Analysis* **31:** 67-74.

Dávalos, A., Bartolomé, B. and Gómez-Cordovés, C. (2005). Antioxidant properties of commercial grape juices and vinegars. *Food Chemistry* **93:** 325-330.

Davidson, P.M. (2001). Chemical preservatives and naturally antimicrobial compounds. *In:* M.P. Doyle, L.R. Beuchat and T.J. Montville (Eds.). *Food Microbiology Fundamentals and Frontiers.* ASM Press, Washington DC, pp. 593-627.

de Castro, R.D., Mota, A.C.L.G., de Oliveira Lima, E., Batista, A.U.D., de Araújo Oliveira, J. and Cavalcanti, A.L. (2015). Use of alcohol vinegar in the inhibition of *Candida* spp. and its effect on the physical properties of acrylic resins. *BMC Oral Health,* **15:** 52.

Dibner, J.J. and Butin, P. (2002). Use of organic acids as model to study the impact of gut microflora on nutrition and metabolism. *The Journal of Applied Poultry Research* **11:** 453-463.

Djoussé, L., Hopkins, P.N., North, K.E., Pankow, J.S., Arnett, D.K. and Ellison, R.C. (2011). Chocolate consumption is inversely associated with prevalent coronary heart disease: The National Heart, Lung, and Blood Institute Family Heart Study. *Clinical Nutrition* **30:** 182-187.

Dohar, J.E. (2003). Evolution of management approaches for otitis externa. *Pediatric Infectious Disease Journal* **22:** 299-308.

Doyle, M.P., Steenson, L.R. and Meng, J. (2013). Bacteria in food and beverage production. *In:* E. Rosenberg (Ed.). *The Prokaryotes, Bacteriology and Biotedchnology* 4th edn. Springer-Verlag, New York, pp. 241-256.

Dufresne, C. and Farnworth, E. (2000). Tea, Kombucha and health: A review. *Food Research International* **33:** 409-421.

Evina, V.J.E., De Taeye, C., Niemenak, N., Youmbi, E. and Collin, S. (2016). Influence of acetic and lactic acids on cocoa flavan-3-ol degradation through fermentation-like incubations. *LWT—Food Science and Technology* **68:** 514-522.

Ewadh, M., Hasan, H., Bnyan, I., Mousa, F., Sultan, J. and Ewadh, M. (2013). Antibacterial activity of 2-(2-hydroxy phenylimino) acetic acid. *Advances in Life Science and Technology* **7:** 15-19.

FAO/WHO (Food and Agriculture Organization/World Health Organization) (2002). *Guidelines for the Evaluation of Probiotics in Food.* London, Ontario.

Fushimi, T. and Sato, Y. (2005). Effect of acetic acid feeding on the circadian changes in glycogen and metabolites of glucose and lipid in liver and skeletal muscle of rats. *British Journal of Nutrition* **94:** 714-719.

Gonella, S. and Gonella, F. (2015). Use of vinegar to relieve persistent hiccups in an advanced cancer patient. *Journal of Palliative Medicine* **18:** 467-470.

Gonzalez, A. and Mas, A. (2011). Differentiation of acetic acid bacteria based on sequence analysis of 16S–23S rRNA gene internal transcribed spacer sequences. *International Journal of Food Microbiology* **147:** 217-222.

Grassi, D., Desideri, G., Necozione, S., Lippi, C., Casale, R., Properzi, G., Blumberg, J.B. and Ferri, C. (2008). Blood pressure is reduced and insulin sensitivity increased in glucose-intolerant, hypertensive subjects after 15 days of consuming high-polyphenol dark chocolate. *The Journal of Nutrition* **138:** 1671-1676.

Grassi, D., Desideri, G. and Ferri, C. (2010). Blood pressure and cardiovascular risk: What about cocoa and chocolate? *Archives of Biochemistry and Biophysics* **501:** 112-115.

Greenwalt, C.J. (1997). *Antibiotic Activity of the Fermented Tea Kombucha.* Cornell University, January.

Greenwalt, C.J., Ledford, R.A. and Steinkraus, K.H. (1998). Determination and characterization of the antimicrobial activity of the fermented tea Kombucha. *LWT—Food Science and Technology* **31:** 291-296.

Greenwalt, C.J., Steinkraus, K.H. and Ledford, R.A. (2000). Kombucha, the fermented tea: Microbiology, composition, and claimed health effects. *Journal of Food Protection* **63:** 976-981.

Gu, X., Zhao, H.L., Sui, Y., Guan, J., Chan, J.C. and Tong, P.C. (2012). White rice vinegar improves pancreatic beta-cell function and fatty liver in streptozotocin-induced diabetic rats. *Acta Diabetologica* **49:** 185-191.

Guillamón, J.M. and Mas, A. (2009). *Acetic Acid Bacteria—Biology of Microorganisms on Grapes, in Must and in Wine.* Springer, Berlin Heidelberg, pp. 31-46.

Haghshenas, B., Nami, Y., Abdullah, N., Radiah, D., Rosli, R., Barzegari, A. and Yari Khosroushahi, A. (2015). Potentially probiotic acetic acid bacteria isolation and identification from traditional dairies microbiota. *International Journal of Food Science & Technology* **50:** 1056-1064.

Hammerstone, J.F., Lazarus, S.A., Mitchell, A.E., Rucker, R. and Schmitz, H.H. (1999). Identification of procyanidins in cocoa (*Theobroma cacao*) and chocolate using high-performance liquid chromatography/mass spectrometry. *Journal of Agricultural and Food Chemistry* **47:** 490-496.

He, Z.Y., Ao, Z.H., Wu, J. (2004). Study on the mensuration of tetramethylpyrazine in Zhenjiang vinegar and its generant mechanism. *China Condiment* **2:** 36-39.

Honsho, S., Sugiyama, A., Takahara, A., Satoh, Y., Nakamura, Y. and Hashimoto, K. (2005). A red wine vinegar beverage can inhibit the renin-angiotensin system: Experimental evidence *in vivo*. *Biological and Pharmaceutical Bulletin* **28:** 1208-1210.

Hrnjez, D., Vaštag, Ž., Milanović, S., Vukić, V., Iličić, M., Popović, L. and Kanurić, K. (2014). The biological activity of fermented dairy products obtained by Kombucha and conventional starter cultures during storage. *Journal of Functional Foods* **10:** 336-345.

Inagaki, S., Morimura, S., Gondo, K., Tang, Y., Akutagawa, H., Kida, K. (2007). Isolation of tryptophol as an apoptosis-inducing component of vinegar produced from boiled extract of black soybean in human monoblastic leukemia U937 cells. *Bioscience, Biotechnology, and Biochemistry* **71:** 371-379.

Janszky, I., Mukamal, K.J., Ljung, R., Ahnve, S., Ahlbom, A. and Hallqvist, J. (2009). Chocolate consumption and mortality following a first acute myocardial infarction: The Stockholm Heart Epidemiology Program. *Journal of Internal Medicine* **266:** 248-257.

Jayabalan, R., Marimuthu, S. and Swaminathan, K. (2007). Changes in content of organic acids and tea polyphenols during Kombucha tea fermentation. *Food Chemistry* **102:** 392-398.

Jayabalan, R., Subathradevi, P., Marimuthu, S., Satishkumar, M. and Swaminathan, K. (2008). Changes in free-radical scavenging ability of Kombucha tea during fermentation. *Food Chemistry* **109:** 227-234.

Jayabalan, R., Chen, P.N., Hsieh, Y.S., Prabhakaran, K., Pitchai, P., Marimuthu, S., Thangaraj, P., Swaminathan, K. and Yun, S.E. (2011). Effect of solvent fractions of Kombucha tea on viability and invasiveness of cancer cells-characterization of dimethyl 2-(2-hydroxy-2-methoxypropylidine) malonate and vitexin. *Indian Journal of Biotechnology* **10:** 75-82.

Jayabalan, R., Malbasa, R.V., Loncar, E.S., Vitas, J.S. and Sathishkumar, M. (2014). A review on Kombucha tea—Microbiology, composition, fermentation, beneficial effects, toxicity and tea fungus. *Comprehensive Reviews in Food Science and Food Safety* **13:** 538-550.

Jing, L., Yanyan, Z. and Junfeng, F. (2015). Acetic acid in aged vinegar affects molecular targets for thrombus disease management. *Food & Function* **6:** 2845-2853.

Johnston, C.S., Kim, C.M. and Buller, A.J. (2004). Vinegar improves insulin sensitivity to a high-carbohydrate meal in subjects with insulin resistance or type 2 diabetes. *Diabetes Care* **27:** 281-282.

Johnston, C.S. and Buller, A.J. (2005). Vinegar and peanut products as complementary foods to reduce postprandial glycemia. *Journal of the American Dietetic Association* **105:** 1939-1942.

Johnston, C.S., Quagliano, S. and White, S. (2013). Vinegar ingestion at mealtime reduced fasting blood glucose concentrations in healthy adults at risk for type 2 diabetes. *Journal of Functional Foods* **5:** 2007-2011.

Karapinar, M. and Gonul, S.A. (1992). Removal of *Yersinia enterocolitica* from fresh parsley by washing with acetic acid or vinegar. *International Journal of Food Microbiology* **16:** 261-264.

Kondo, S., Tayama, K., Tsukamoto, Y., Ikeda, K. and Yamori, Y. (2001). Antihypertensive effects of acetic acid and vinegar on spontaneously hypertensive rats. *Bioscience, Biotechnology and Biochemistry* **65:** 2690-2694.

Král, M., Angelovičová, M., Mrázová, Ľ., Tkáčová, J. and Kliment, M. (2011). Probiotic and acetic acid effect on broiler chickens performance. *Scientific Papers Animal Science and Biotechnologies* **44:** 62-64.

Krusong, W., Teerarak, M. and Laosinwattana, C. (2015). Liquid and vapor-phase vinegar reduces *Klebsiella pneumoniae* on fresh coriander. *Food Control* **50:** 502-508.

Krystynowicz, A., Czaja, W., Pomorski, L., Kolodziejczyk, M. and Bielecki, S. (2000). *The Evaluation of Usefulness of Microbial Cellulose as a Wound Dressing Material*. 14th Forum for Applied Biotechnology, Proceedings Part 1. Meded Fac Land-bouwwet-Rijksuniv Gent, Belgium, pp. 213-220.

Laranjinha, J.A.N., Almeida, L.M. and Madeira, V.M.C. (1994). Reactivity of dietary phenolic acids with peroxyl radicals: Antioxidant activity upon low density lipoprotein peroxidation. *Biochemical Pharmacology* **48:** 487-494.

Larsson, S.C., Virtamo, J. and Wolk, A. (2011). Chocolate consumption and risk of stroke in women. *Journal of the American College of Cardiology* **58:** 1828-1829.

Lee, J.H., Cho, H.D., Jeong, J.H., Lee, M.K., Jeong, Y.K., Shim, K.H. and Seo, K.I. (2013). New vinegar produced by tomato suppresses adipocyte differentiation and fat accumulation in 3T3-L1 cells and obese rat model. *Food Chemistry* **141:** 3241-3249.

Lee, K.W., Kim, Y.J., Lee, H.J. and Lee, C.Y. (2003). Cocoa has more phenolic phytochemicals and a higher antioxidant capacity than teas and red wine. *Journal of Agricultural and Food Chemistry* **51:** 7292-7295.

Lefeber, T., Gobert, W., Vrancken, G., Camu, N. and De Vuyst, L. (2011). Dynamics and species diversity of communities of lactic acid bacteria and acetic acid bacteria during spontaneous cocoa bean fermentation in vessels. *Food Microbiology* **28:** 457-464.

Li, X.J., Wang, X.Y., Yuan, J., Lu, M., Wang, R.H., Quan, Z.J. (2013). The determination and comparison of phenolics in apple vinegar, persimmon vinegar and kiwifruit vinegar. *Food and Ferment Industries* **39:** 186-190.

Li, J., Yu, G. and Fan, J. (2014). Alditols and monosaccharides from sorghum vinegar can attenuate platelet aggregation by inhibiting cyclooxygenase-1 and thromboxane-A2 synthase. *Journal of Ethnopharmacology* **155:** 285-292.

Liatis, S., Grammatikou, S., Poulia, K.A., Perrea, D., Makrilakis, K., Diakoumopoulou, E. and Katsilambros, N. (2010). Vinegar reduces postprandial hyperglycaemia in patients with type II diabetes when added to a high, but not to a low, glycaemic index meal. *European Journal of Clinical Nutrition* **64:** 727-732.

Lingham, T., Besong, S., Ozbay, G. and Lee, J-L. (2012). Antimicrobial activity of vinegar on bacterial species isolated from retail and local channel catfish (*Ictalurus punctatus*). *Journal of Food Processing and Technology* S11-001.

Liu, C.H., Hsu, W.H., Lee, F.L. and Liao, C.C. (1996). The isolation and identification of microbes from a fermented tea beverage, Haipao, and their interactions during Haipao fermentation. *Food Microbiology* **13:** 407-415.

Macht, M. and Dettmer, D. (2006). Everyday mood and emotions after eating a chocolate bar or an apple. *Appetite* **46:** 332-336.

Malbaša, R.V., Lončar, E.S., Vitas, J.S. and Čanadanović-Brunet, J.M. (2011). Influence of starter cultures on the anti-oxidant activity of Kombucha beverage. *Food Chemistry* **127:** 1727-1731.

Maleyki, M.A. and Ismail, A. (2010). Antioxidant properties of cocoa powder. *Journal of Food Biochemistry* **34:** 111-128.

Maskarinec, G. (2009). Cancer protective properties of cocoa: A review of the epidemiologic evidence. *Nutrition and Cancer* **61:** 573-579.

Matsui, T., Ebuchi, S., Fukui, K., Matsugano, K., Terahara, N., Matsumoto, K. (2014). Caffeoylsophorose, a new natural α-glucosidase inhibitor, from red vinegar by fermented purple-fleshed sweet potato. *Bioscience, Biotechnology, and Biochemistry* **68:** 2239-2246.

Mattila-Sandholm, T., Myllärinen, P., Crittenden, R., Mogensen, G., Fondén, R. and Saarela, M. (2002). Technological challenges for future probiotic foods. *International Dairy Journal* **12:** 173-182.

McCarty, M.F., Barroso-Aranda, J. and Contreras, F. (2010). Potential complementarity of high-flavanol cocoa powder and spirulina for health protection. *Medical Hypotheses* **74:** 370-373.

Medina, E., Romero, C., Brenes, M. and de Castro, A. (2007). Antimicrobial activity of olive oil, vinegar, and various beverages against foodborne pathogens. *Journal of Food Protection* **70:** 1194-1199.

Mellor, D.D., Sathyapalan, T., Kilpatrick, E.S., Beckett, S. and Atkin, S.L. (2010). High-cocoa polyphenol-rich chocolate improves HDL cholesterol in Type 2 diabetes patients. *Diabetic Medicine* **27:** 1318-1321.

Merchie, G., Lavens, P. and Sorgeloos, P. (1997). Optimization of dietary vitamin C in fish and crustacean larvae: A review. *Aquaculture* **155:** 165-181.

Mercola, J. (2014). *Functional Health Properties of Vinegar*. Available from: http://articles.mercola.com/sites/articles/archive/2014/06/14/vinegar-health-properties.aspx (Accessed: June 2016).

Messaoudi, M., Bisson, J.F., Nejdi, A., Rozan, P. and Javelot, H. (2008). Antidepressant-like effects of a cocoa polyphenolic extract in Wistar-Unilever rats. *Nutritional Neuroscience* **11:** 269-276.

Michener, W. and Rozin, P. (1994). Pharmacological versus sensory factors in the satiation of chocolate craving. *Physiology & Behavior* **56:** 419-422.

Mimura, A., Suzuki, Y., Toshima, Y., Yazaki, S., Ohtsuki, T., Ui, S. and Hyodoh, F. (2004). Induction of apoptosis in human leukemia cells by naturally fermented sugar cane vinegar (*kibizu*) of Amami Ohshima Island. *Biofactors* **22:** 93-97.

Moon, Y.J., Choi, D.S., Oh, S.H., Song, Y.S. and Cha, Y.S. (2010). Effects of persimmon-vinegar on lipid and carnitine profiles in mice. *Food Science and Biotechnology* **19:** 343-348.

Morimura, S., Kawano, K., Han, L.S., Seki, T. and Shige, T. (2004). Production of vinegar from sweet potato-shochu, post-distillation slurry and evaluation of its antitumor activity via oral administration in a mouse model. *Seibutsu-Kogaku Kaishi* **82:** 573-578.

Murooka, Y. and Yamshita, M. (2008). Traditional healthful fermented products of Japan. *Journal of Industrial Microbiology & Biotechnology* **35:** 791.

Nakamura, K., Ogasawara, Y., Endou, K., Fujimori, S., Koyama, M. and Akano, H. (2010). Phenolic compounds responsible for the superoxide dismutase-like activity in high-brix apple vinegar. *Journal of Agricultural Food Chemistry* **58:** 10124-10132.

Nanda, K., Miyoshi, N., Nakamura, Y., Shimoji, Y., Tamura, Y., Nishikawa, Y., Uenakai, K., Kohno, H. and Tanaka, T. (2004). Extract of vinegar 'Kurosu' from unpolished rice inhibits the proliferation of human cancer cells. *Journal of Experimental and Clinical Cancer Research* **23:** 69-76.

Nandasiri, H.M.A.R. (2012). *Antioxidant, Antihypertensive and Lipid Lowering Properties of Fruit Vinegar Beverages*. M.S. Thesis, Dalhousie University Halifax, Nova Scotia.

Natsume, M., Osakabe, N., Yamagishi, M., Takizawa, T., Nakamura, T., Miyatake, H. and Yoshida, T. (2000). Analyses of polyphenols in cacao liquor, cocoa and chocolate by normal-phase and reversed-phase HPLC. *Bioscience, Biotechnology and Biochemistry* **64:** 2581-2587.

Nguyen, N.K., Nguyen, P.B., Nguyen, H.T. and Le, P.H. (2015). Screening the optimal ratio of symbiosis between isolated yeast and acetic acid bacteria strain from traditional Kombucha for high-level production of glucuronic acid. *LWT—Food Science and Technology* **64:** 1149-1155.

Nishidai, S., Nakamura, Y. and Torikai, K. (2000). Kurosu, a traditional vinegar produced from unpolished rice, suppresses lipid peroxidation *in vitro* and in mouse skin. *Bioscience, Biotechnology and Biochemistry* **64:** 1909-1914.

Nishikawa, Y., Takata, Y., Nagai, Y., Mori, T., Kawada, T. and Ishihara, N. (2001). Antihypertensive effect of Kurosu extract, a traditional vinegar produced from unpolished rice, in the SHR rats. *Journal—Japanese Society of Food Science and Technology* **48:** 73-75.

Odahara, M., Ogino, Y., Takizawa, K., Kimura, M., Nakamura, N. and Kimoto, K. (2008). Hypotensive effect of black malt vinegar on spontaneously hypertensive rats. *Journal of the Japanese Society for Food Science and Technology* **55:** 81-86.

Ogawa, N., Satsu, H., Watanabe, H., Fukaya, M., Tsukamoto, Y., Miyamoto, Y. and Shimizu, M. (2000). Acetic acid suppresses the increase in disaccharidase activity that occurs during culture of caco-2 cells. *Journal of Nutrition* **130:** 507-513.

Ohnami, K., Matsuoka, E. and Okuda, T. (1985). Effects of Kurosu on the blood pressure of the spontaneously hypertension rats. *Kiso to Rinsho* **19:** 237-241.

Ozturk, I., Calıskan, O., Tornuk, F. and Sagdıc, O. (2015). Antioxidant, antimicrobial, mineral, volatile, physicochemical and microbiological characteristics of traditional home-made Turkish vinegars. *Lebensmittel-Wissenschaft und-Technologie* **63:** 144-151.

Park, J.E., Kim, J.Y., Kim, J., Kim, Y.J., Kim, M.J., Kwon, S.W. and Kwon, O. (2014). Pomegranate vinegar beverage reduces visceral fat accumulation in association with AMPK activation in overweight women: A double-blind, randomized and placebo-controlled trial. *Journal of Functional Foods* **8:** 274-281.

Parker, G. and Crawford, J. (2007). Chocolate craving when depressed: A personality marker. *The British Journal of Psychiatry* **191:** 351-352.

Parker, G., Parker, I. and Brotchie, H. (2006). Mood state effects of chocolate. *Journal of Affective Disorders* **92:** 149-159.

Pauline, T., Dipti, P., Anju, B., Kavimani, S., Sharma, S.K., Kain, A.K. and Selvamurthy, W. (2001). Studies on toxicity, anti-stress and hepato-protective properties of Kombucha tea. *Biomedical and Environmental Sciences* **14:** 207-213.

Pérez-Cano, F.J., Massot-Cladera, M., Franch, À., Castellote, C. and Castell, M. (2013). The effects of cocoa on the immune system. *Frontiers in Pharmacology* **4:** 71.

Petrović, S.E., Malbasa, R.V. and Verac, R.M. (2000). Biosynthesis of glucuronic acid by means of tea fungus. *Nahrung* **44:** 138-139.

Petsiou, E.I., Mitrou, P.I., Raptis, S.A. and Dimitriadis, G.D. (2014). Effect and mechanisms of action of vinegar on glucose metabolism, lipid profile and body weight. *Nutrition Reviews* **72:** 651-661.

Pinto, T.M.S., Neves, A.C.C., Leão, M.V.P. and Jorge, A.O.C. (2008). Vinegar as an antimicrobial agent for control of *Candida* spp. in complete denture wearers. *Journal of Applied Oral Science* **16:** 385-390.

Plessi, M., Bertelli, D. and Miglietta, F. (2006). Extraction and identification by GC-MS of phenolic acids in traditional Balsamic vinegar from Modena. *Journal of Food Composition and Analysis* **19:** 49-54.

Pucciarelli, D. (2013). Cocoa and heart health: A historical review of the science. *Nutrients* **5:** 3854-3870.

Qiu, J., Ren, C., Fan, J. and Li, Z. (2010). Antioxidant activities of aged oat vinegar *in vitro* and in mouse serum and liver. *Journal of the Science of Food and Agriculture* **90:** 1951-1958.

Rahman, M.S. (2007). pH in food preservation. *In:* M.S. Rahman (Ed.). *Handbook of Food Preservation.* CRC Press, New York, pp. 287-296.

Ramos-Romero, S., Pérez-Cano, F.J., Castellote, C., Castell, M. and Franch, A. (2012). Effect of cocoa-enriched diets on lymphocytes involved in adjuvant arthritis in rats. *British Journal of Nutrition* **107:** 378-387.

Ramos, B., Brandão, T.R.S., Teixeira, P. and Silva, C.L.M. (2014). Balsamic vinegar from Modena: An easy and effective approach to reduce *Listeria monocytogenes* from lettuce. *Food Control* **42:** 38-42.

Ranadheera, R.D.C.S., Baines, S.K. and Adams, M.C. (2010). Importance of food in probiotic efficacy. *Food Research International* **43:** 1-7.

Raspor, P. and Goranovič, D. (2008). Biotechnological applications of acetic acid bacteria. *Critical Reviews in Biotechnology* **28:** 101-124.

Raybaudi-Massilia, R.M., Mosqueda-Melgar, J., Soliva-Fortuny, R. and Martin-Belloso, O. (2009). Control of pathogenic microorganisms in fresh-cut fruits and fruit juices by traditional and alternative natural antimicrobials. *Comprehensive Reviews in Food Science and Food Safety* **8:** 157-178.

Reiss, J. (1994). Influence of different sugars on the metabolism of the tea fungus. *Zeitschrift fuer Lebensmittel-Untersuchung und Forschung* **198:** 258-261.

Rivera-Espinoza, Y. and Gallardo-Navarro, Y. (2010). Non-dairy probiotic products. *Food Microbiology* **27:** 1-11.

Rogawansamy, S., Gaskin, S., Taylor, M. and Pisaniello, D. (2015). An evaluation of antifungal agents for the treatment of fungal contamination in indoor air environments. *International Journal of Environmental Research and Public Health* **12:** 6319-6332.

Rund, C.R. (1996). Non-conventional topical therapies for wound care. *Ostomy/Wound Management* **42:** 18-20.

Rutala, W.A., Barbee, S.L., Agular, N.C., Sobsey, M.D. and Weber, D.J. (2000). Antimicrobial activity of home disinfectants and natural products against potential human pathogens. *Infection Control & Hospital Epidemiology* **21:** 33-38.

Saarela, M., Mogensen, G., Fondén, R., Mättö, J. and Mattila-Sandholm, T. (2000). Probiotic bacteria: Safety, functional and technological properties. *Journal of Biotechnology* **84:** 197-215.

Sai Ram, M., Anju, B., Pauline, T., Dipti, P., Kain, A.K., Mongia, S.S., Sharma, S.K., Singh, B., Singh, R., Ilavazhagan, G., Kumar, D. and Selvamurthy, W. (2000). Effect of Kombucha tea on chromate (VI)-induced oxidative stress in albino rats. *Journal of Ethnopharmacology* **71:** 235-240.

Sakakibara, S., Yamauchi, T., Oshima, Y., Tsukamoto, Y. and Kadowaki, T. (2006). Acetic acid activates hepatic AMPK and reduces hyperglycemia in diabetic KK-A (y) mice. *Biochemical and Biophysical Research Communications* **344:** 597-604.

Seki, T., Morimura, S., Shigematsu, T., Maeda, H. and Kida, K. (2004). Antitumor activity of rice-shochu post-distillation slurry and vinegar produced from the post-distillation slurry via oral administration in a mouse model. *Biofactors* **22:** 103-105.

Seki, T., Morimura, S., Tabata, S., Tang, Y., Shigematsu, T. and Kida, K. (2008). Antioxidant activity of vinegar produced from distilled residues of the Japanese liquor shochu. *Journal of Agricultural and Food Chemistry* **56:** 3785-3790.

Selmi, C., Cocchi, C.A., Lanfredini, M., Keen, C.L. and Gershwin, M.E. (2008). Chocolate at heart: The anti-inflammatory impact of cocoa flavanols. *Molecular Nutrition & Food Research* **52:** 1340-1348.

Sengun, I.Y. and Karapinar, M. (2004). Effectiveness of lemon juice, vinegar and their mixture in the elimination of *Salmonella* Typhimurium on carrots (*Daucus carota* L.). *International Journal of Food Microbiology* **96:** 301-305.

Sengun, I.Y. and Karapinar, M. (2005a). Effectiveness of household natural sanitizers in the elimination of *Salmonella* Typhimurium on rocket (*Eruca sativa* Miller) and spring onion (*Allium cepa* L.). *International Journal of Food Microbiology* **98:** 319-323.

Sengun, I.Y. and Karapinar, M. (2005b). Elimination of *Yersinia enterocolitica* on carrots (*Daucus carota* L.) by using household sanitisers. *Food Control* **16:** 845-850.

Sengun, I.Y. and Karabiyikli, S. (2011). Importance of acetic acid bacteria in food industry. *Food Control* **22:** 647-656.

Setorki, M., Asgary, S., Eidi, A., Haerirohani, A. and Majid, K. (2010). Acute effects of vinegar intake on some biochemical risk factors of atherosclerosis in hypercholesterolemic rabbits. *Lipids in Health and Disease* **9:** 1-8.

Setorki, M., Asgary, S., Haghjooyjavanmard, S. and Nazari, B. (2011). Reduces cholesterol induced atherosclerotic lesions in aorta artery in hypercholesterolemic rabbits. *Journal of Medicinal Plants Research* **5:** 1518-1525.

Shay, K. (2000). Denture hygiene: A review and update. *The Journal of Contemporary Dental Practice* **1:** 1-8.

Shimoji, Y., Tamura, Y., Nakamura, Y., Nanda, K., Nishidai, S., Nishikawa, Y., Ishihara, N., Uenakai, K. and Ohigashi, H. (2002). Isolation and identification of DPPH radical scavenging compounds in Kurosu (Japanese unpolished rice vinegar). *Journal of Agricultural and Food Chemistry* **50:** 6501-6503.

Shimoji, Y., Kohno, H., Nanda, K., Nishikawa, Y., Ohigashi, H., Uenakai, K. and Tanaka, T. (2004). Extract of *Kurosu*, a vinegar from unpolished rice, inhibits azoxymethane-induced colon carcinogenesis in male F344 rats. *Nutrition and Cancer* **49:** 170-173.

Shizuma, T., Ishiwata, K., Nagano, M., Mori, H. and Fukuyama, N. (2011). Protective effects of fermented rice vinegar sediment (*Kurozu moromimatsu*) in a diethylnitrosamine-induced hepatocellular carcinoma animal model. *Journal of Clinical Biochemistry and Nutrition* **49:** 31-35.

Smith, D.F. (2013). Benefits of flavanol-rich cocoa-derived products for mental well-being: A review. *Journal of Functional Foods* **5:** 10-15.

Sokolov, A.N., Pavlova, M.A., Klosterhalfen, S. and Enck, P. (2013). Chocolate and the brain: Neurobiological impact of cocoa flavanols on cognition and behavior. *Neuroscience & Biobehavioral Reviews* **37:** 2445-2453.

Sreeramulu, G., Zhu, Y. and Knol, W. (2000). Kombucha fermentation and its antimicrobial activity. *Journal of Agricultural and Food Chemistry* **48:** 2589-2594.

Sreeramulu, G., Zhu, Y. and Knol, W. (2001). Characterization of antimicrobial activity in Kombucha fermentation. *Acta Biotechnologica* **21:** 49-56.

Srikanth, R., Siddartha, G., Reddy, C.H.S.S.S., Harish, B.S., Ramaiah, M.J. and Uppuluri, K.B. (2015). Antioxidant and anti-inflammatory levan produced from *Acetobacter xylinum* NCIM2526 and its statistical optimization. *Carbohydrate Polymers* **123:** 8-16.

Steinkraus, K.H., Shapiro, K.B., Hotchkiss, J.H. and Mortlock, R.P. (1996). Investigations into the antibiotic activity of tea fungus/Kombucha beverage. *Acta Biotechnologica* **16:** 199-205.

Stelzleni, A.M., Ponrajan, A. and Harrison, M.A. (2013). Effects of buffered vinegar and sodium dodecyl sulfate plus levulinic acid on *Salmonella* Typhimurium survival, shelf-life and sensory characteristics of ground beef patties. *Meat Science* **95:** 1-7.

Sugiyama, A., Saitoh, M., Takahara, A., Satoh, Y. and Hashimoto, K. (2003). Acute cardiovascular effects of a new beverage made of wine vinegar and grape juice, assessed, using an *in vivo* rat. *Nutrition Research* **23:** 1291-1296.

Sun, T.-Y., Li, J.-S. and Chen, C. (2015). Effects of blending wheatgrass juice on enhancing phenolic compounds and antioxidant activities of traditional Kombucha beverage. *Journal of Food and Drug Analysis* **23:** 709-718.

Taubert, D., Roesen, R. and Schömig, E. (2007). Effect of cocoa and tea intake on blood pressure: A meta-analysis. *Archives of Internal Medicine* **167:** 626-634.

Teoh, A.L., Heard, G. and Cox, J. (2004). Yeast ecology of Kombucha fermentation. *International Journal of Food Microbiolology* **95:** 119-126.

Tong, L.T., Katakura, Y., Kawamura, S., Baba, S., Tanaka, Y., Udono, M., Kondo, Y., Nakamura, K., Imaizumi, K. and Sato, M. (2010). Effects of Kurozu concentrated liquid on adipocyte size in rats. *Lipids in Health and Disease* **9:** 134-142.

Ubeda, C., Hidalgo, C., Torija, M.J., Mas, A., Troncoso, A.M. and Morales, M.L. (2011). Evaluation of antioxidant activity and total phenols index in persimmon vinegars produced by different processes. *LWT—Food Science and Technology* **44:** 1591-1596.

Valente, T., Hidalgo, J., Bolea, I., Ramirez, B., Anglés, N., Reguant, J. and Unzeta, M. (2009). A diet enriched in polyphenols and polyunsaturated fatty acids, LMN diet, induces neurogenesis in the subventricular zone and hippocampus of adult mouse brain. *Journal of Alzheimer's Disease* **18:** 849-865.

Verzelloni, E., Tagliazucchi, D. and Conte, A. (2010). From balsamic to healthy: Traditional balsamic vinegar melanoidins inhibit lipid peroxidation during simulated gastric digestion of meat. *Food and Chemical Toxicology* **48:** 2097-2102.

Vijayakumar, C. and Wolf-Hall, C. (2002). Evaluation of household sanitizers for reducing levels of *E. coli* on iceberg lettuce. *Journal of Food Protection* **65:** 1646-1650.

Vina, I., Linde, R., Patetko, A. and Semjonovs, P. (2013a). Glucuronic acid from fermented beverages: Biochemical functions in humans and its role in health protection. *International Journal of Research and Reviews in Applied Science* **14:** 217-230.

Vīna, I., Semjonovs, P., Linde, R. and Patetko, A. (2013b). Glucuronic acid containing fermented functional beverages produced by natural yeasts and bacteria associations. *International Journal of Research and Reviews in Applied Science* **14:** 17-25.

Wu, F.M., Doyle, M.P., Beuchat, L.R., Wells, J.G., Mintz, E.D. and Swaminathan, B. (2000). Fate of *Shigella sonnei* on parsley and methods of disinfection. *Journal of Food Protection* **63:** 468-572.

Wu, D., Kimura, F., Takashima, A., Shimizu, Y., Takebayashi, A., Kita, N., Zhang, G.M. and Murakami, T. (2013). Intake of vinegar beverage is associated with restoration of ovulatory function in women with polycystic ovary syndrome. *The Tohoku Journal of Experimental Medicine* **230:** 17-23.

Xibib, S., Meilan, H., Moller, H., Evans, H.S., Dixin, D., Wenjie, D. and Jianbang, L. (2003). Risk factors for oesophageal cancer in Linzhou, China: A case-control study. *Asian-Pacific Journal of Cancer Prevention* **4:** 119-124.

Xu, Q., Tao, W. and Ao, Z. (2007). Antioxidant activity of vinegar melanoidins. *Food Chemistry* **102:** 841-849.

Yamada, Y., Yukphan, P., Lan Vu, H.T., Muramatsu, Y., Ochaikul, D., Tanasupawat, S. and Nakagawa, Y. (2012). Description of *Komagataeibacter* gen. nov., with proposals of new combinations (*Acetobacteraceae*). *The Journal of General and Applied Microbiology* **58:** 397-404.

Yamashita, H., Fujisawa, K., Ito, E., Idei, S., Kawaguchi, N., Kimoto, M., Hiemori, M. and Tsuji, H. (2007). Improvement of obesity and glucose tolerance by acetate in Type 2 diabetic Otsuka Long-Evans Tokushima Fatty (OLETF) rats. *Bioscience, Biotechnology and Biochemistry* **71:** 1236-1243.

Yang, H., Kendall, P.A., Medeiros, L. and Sofos, J.N. (2009a). Inactivation of *Listeria monocytogenes, Escherichia coli* O157:H7, and *Salmonella typhimurium* with compounds available in households. *Journal of Food Protection* **72:** 1201-1208.

Yang, Z.W., Ji, B.-P., Zhou, F., Li, B., Luo, Y., Yang, L. and Li, T. (2009b). Hypocholesterolaemic and antioxidant effects of Kombucha tea in high-cholesterol fed mice. *Journal of the Science of Food and Agriculture* **89:** 150-156.

Yavari, N., Assadi, M.M., Moghadam, M.B. and Larijani, K. (2011). Optimizing glucuronic acid production using tea fungus on grape juice by response surface methodology. *Australian Journal of Basic and Applied Sciences* **5:** 1788-1794.

Yusoff, N.A., Yam, M.F., Beh, H.K., Razak, K.N.A., Widyawati, T., Mahmud, R., Ahmad, M. and Asmawi, M.Z. (2015). Antidiabetic and antioxidant activities of Nypa fruticans Wurmb. vinegar sample from Malaysia. *Asian-Pacific Journal of Tropical Medicine* **8:** 595-605.

Zhang, H., Zhang, Z. and Xin, X. (2011). Isolation and identification of microorganisms from Kombucha fungus culture. *Journal of Beijing Union University* **2:** 11.

Index